ADVANCES IN
ORGAN BIOLOGY

Volume 5B • 1998

MOLECULAR AND CELLULAR BIOLOGY OF BONE

MOLECULAR AND CELLULAR BIOLOGY OF BONE

Guest Editor

MONE ZAIDI
Medical College of Pennsylvania
School of Medicine and
Veterans Affairs Medical Center
Philadelphia, Pennsylvania

Associate Guest Editors

OLUGBENGA A. ADEBANJO
Medical College of Pennsylvania
School of Medicine and
Veterans Affairs Medical Center
Philadelphia, Pennsylvania

CHRISTOPHER L. -H. HUANG
Department of Physiology
University of Cambridge
Cambridge, England

ADVANCES IN ORGAN BIOLOGY

MOLECULAR AND CELLULAR BIOLOGY OF BONE

Guest Editor: MONE ZAIDI
Veterans Affairs Medical Center

Series Editor: E. EDWARD BITTAR
Department of Physiology
University of Wisconsin-Madison

Associate Guest Editors: OLUGBENGA A. ADEBANJO
Veterans Affairs Medical Center

CHRISTOPHER L. -H. HUANG
Department of Physiology
University of Cambridge

VOLUME 5B • 1998

Stamford, Connecticut *London, England*

100 Prospect Street
Stamford, Connecticut 06901

JAI PRESS LTD.
38 Tavistock Street
Covent Garden
London WC2E 7PB
England

ISBN: 0-7623-0390-5

Printed and bound by CPI Antony Rowe, Eastbourne

CONTENTS (Volume 5B)

CONTENTS (Volume 5A)

CONTENTS (Volume 5C)

LIST OF CONTRIBUTORS

E.M. Aarden — Research Scientist
Department of Cell Biology
Faculty of Medicine, Leiden University
Leiden, The Netherlands

Etsuko Abe, PhD — Research Professor of Medicine
Department of Medicine
University of Arkansas for Medical Sciences
Little Rock, Arkansas

A.B. Abou-Samra, MD — Associate Professor of Medicine
Endocrine Unit, Department of Medicine
Massachusetts General Hospital
Harvard Medical School
Boston, Massachusetts

Olugbenga A. Adebanjo, MD — Assistant Professor of Medicine
Department of Medicine
Medical College of Pennsylvania
School of Medicine and Veterans Affairs Medical Center
Philadelphia, Pennsylvania

N.E. Ajubi — Research Scientist
Department of Cell Biology
Faculty of Medicine
Leiden University
Leiden, The Netherlands

David J. Baylink, MD
Distinguished Professor of Medicine
Loma Linda University
and Associate Vice President
for Medical Affairs for Research
J.L. Pettis Veterans Affairs Medical Center
Loma Linda, California

Paolo Bianco, MD
Dipartmento di Biopatologia Umana
Universita La Sapienza
Rome, Italy

L.F. Bonewald, PhD
Associate Professor of Medicine
Department of Medicine
University of Texas Health Science Center
San Antonio, Texas

Brendan F. Boyce
Professor of Pathology
Department of Medicine
Division of Endocrinology and Metabolism
University of Texas Health Science Center
San Antonio, Texas

Alan Boyde, PhD
Professor of Mineralized Tissue Biology
Department of Anatomy and
Developmental Biology
University College London
London, England

Edward M. Brown, MD
Professor of Medicine
Endocrine-Hypertension and Renal
Divisions
Brigham and Women's Hospital
Boston, Massachusetts

Elisabeth H. Burger, PhD
Professor
Department of Oral Cell Biology
ACTA-Vrije Universiteit
Amsterdam, The Netherlands

T.J Chambers, PhD, MBBS, MRCPath
Professor and Chairman
Department of Histopathology
St. George's Hospital Medical School
London, England

Chantal Chenu, PhD
Staff Research Fellow
INSERM
Hôpital Edouard Herriot
Lyon, France

Roberto Civitelli, MD
Associate Professor of Medicine and Orthopedic Surgery and Assistant Professor of Cell Biology and Physiology
Division of Bone and Mineral Diseases
Washington University School of Medicine
St. Louis, Missouri

Thomas L. Clemens, PhD
Professor of Medicine
Department of Molecular and Cellular Physiology and Orthopedic Surgery
University of Cincinnati Medical Center
Cincinnati, Ohio

Silvia Colucci, PhD
Assistant Professor of Histology
Institute of Human Anatomy
University of Bari
Bari, Italy

Stephen C. Cowin
Department of Mechanical Engineering
City University of New York
New York, New York

C.G. Dacke, B.Tech, PhD, FIBiol
Reader and Head, Pharmacology Division
School of Pharmacy and Biomedical Science
University of Portsmouth
Portsmouth, England

Sarah L. Dallas, PhD
Assistant Professor of Medicine
Department of Medicine
University of Texas Health Science Center
San Antonio, Texas

Pietro De Togni, MD
Assistant Professor of Pathology
Immunogenetics and Transplantation Laboratory
University of Arkansas for Medical Sciences
Little Rock, Arkansas

P.D. Delmas, MD, PhD
Professor of Medicine
INSERM
Hôpital Edouard Herriot
Lyon, France

S.J. Dixon, DDS, PhD
Associate Professor of Physiology and Oral Biology
Department of Physiology
Faculty of Dentistry
The University of Western Ontario
London, Ontario, Canada

S. Epstein, MD, FRCP
Professor of Medicine and Chief Division of Endocrinology
Medical College of Pennsylvania
Hahnemann School of Medicine
Philadelphia, Pennsylvania

R. J. Fitzsimmons, PhD
Assistant Research Professor of Medicine and Director Mineral Metabolism
Jerry L. Pettis Veterans Affairs Medical Center
Loma Linda University
Loma Linda, California

Herbert Fleisch, MD
Professor and Chairman
Department of Pathophysiology
University of Berne
Berne, Switzerland

Steven R. Goldring, MD
Associate Professor of Medicine and Chief of Rhematology
Beth Israel-Deaconess Hospital
Harvard Medical School
Boston, Massachusetts

David Goltzman, MD
Professor and Chairman
Department of Medicine
McGill University,Royal Victoria Hospital
Montréal, Québec, Canada

Grant R. Goodman, MD
Research Associate
Department of Medicine
Albert Einstein Medical Center
Philadelphia, Pennsylvania

Maria Grano, PhD	Assistant Professor of Histology Institute of Human Anatomy University of Bari Bari, Italy
Ted S. Gross, PhD	Assistant Professor Departments of Medicine and Molecular and Cellular Physiology and Orthopedic Surgery University of Cincinnati Medical Center Cincinnati, Ohio
Theresa A. Guise, MD	Assistant Professor of Medicine Division of Endocrinology and Metabolism University of Texas Health Science Center San Antonio, Texas
Steven C. Hebert, MD	Professor of Medicine and Chief, Division of Nephrology Vanderbelt University Nashville, Tennessee
Janet E. Henderson, PhD	Assistant Professor of Medicine Department of Medicine McGill University Montréal, Québec, Canada
M. Horton, MD, FRCP, FRCPath	Professor Rayne Institute Bone and Mineral Center University of London London, England
Osamu Ishibashi, MS	Scientist Ciba-Geigy Japan Limited International Research Laboratories Takarazuka, Japan
Sheila Jones, PhD	Professor of Anatomy Department of Anatomy and Developmental Biology University College London London, England

J. Klein-Nulend, PhD
Assistant Professor
Department of Oral Cell Biology
ACTA-Vrije Universiteit
Amsterdam, The Netherlands

Toshio Kokubo, PhD
Group Leader
International Research Laboratories
Ciba Geigy Japan Limited
Takarazuka, Japan

Masayoshi Kumegawa, DDS
Professor
Department of Oral Anatomy
Meikai University School of Dentistry
Saitama, Japan

Pierre J. Marie, PhD
Professor
Cell and Molecular Biology of Bone and Cartilage
Lariboisière Hospital
Paris, France

T.J. Martin, MD, DSC, FRCPA, FRACP
Professor of Medicine
St. Vincent's Institute of Medical Research
University of Melbourne
Fitzroy, Victoria, Australia

Toshio Matsumoto, MD
Professor and Chairman
First Department of Medicine
Tokushima University School of Medicine
Tokushima, Japan

Cedric Minkin, PhD
Professor
Department of Basic Sciences
University of Southern California School of Dentistry
Los Angeles, California

Ambrish Mithal, MD, DM
Professor
Department of Medical Endocrinology
Sanjay Gandhi Post Graduate Institute of Medical Sciences
Lucknow, India

Hanna Mocharla, PhD
Research Instructor
Department of Medicine
University of Arkansas for Medical Sciences
Little Rock, Arkansas

S. Mohan, PhD
Research Professor of Medicine, Biochemistry, and Physiology
J.L. Pettis Veterans Affairs Medical Center
Loma Linda University
Loma Linda, California

Baljit Moonga, PhD
Assistant Professor of Medicine
Medical College of Pennsylvania
School of Medicine and
Veterans Affairs Medical Center
Philadelphia, Pennsylvania

K.W. Ng, MBBS, MD, FRACP
Associate Professor
Department of Medicine
The University of Melbourne
St. Vincent's Hospital
Fitzroy, Victoria, Australia

Peter J. Nijweide
Professor
Department of Cell Biology
Faculty of Medicine
Leiden University
Leiden, The Netherlands

Richard O.C. Oreffo, D. Phil.
MRC Research Fellows
MRC Bone Research Laboratory
Nuffield Orthopedic Center
University of Oxford
Headington
Oxford, England

Roberto Pacifici, MD
Associate Professor of Medicine
Division of Bone and Mineral Diseases
Washington University Medical Center
St. Louis, Missouri

Michael Pazianas, MD
Associate Professor of Medicine
Division of Geriatric Medicine and Institute on Aging
University of Pennsylvania
Philadelphia, Pennsylvania

J. Wesley Pike, PhD
Professor of Medicine
Department of Molecular and Cellular Physiology
University of Cincinnati Medical Center
Cincinnati, Ohio

James T. Ryaby, PhD
Director of Research
Orthologic Corporation
Phoenix, Arizona

Ian R. Reid, MD
Associate Professor of Medicine
Department of Medicine
University of Auckland
Auckland, New Zealand

Barry Rifkin, DDS, PhD
Professor and Dean
State University of New York Dental School
Stony Brook, New York

Pamela Gehron Robey, PhD
Chief
Craniofacial and Skeletal Diseases
National Institute of Dental Research
National Institutes of Health
Bethesda, Maryland

G. David Roodman, MD
Professor of Medicine and Chief of Hematology
Audie Murphy Veterans Affairs Medical Center
University of Texas Health Science Center
San Antonio, Texas

F. Patrick Ross, PhD
Associate Professor of Pathology
Department of Pathology
Barnes-Jewish Hospital
St. Louis, Missouri

Dennis Sakai, PhD
Research Professor
Department of Basic Sciences
University of Southern California School of Dentistry
Los Angeles, California

Edna Schwab, MD
Assistant Professor of Medicine
Division of Geriatric Medicine and Institution Aging
University of Pennsylvania
Philadelphia, Pennsylvania

Geetha Shankar, PhD
Scientist
NPS Pharmaceuticals Inc.
Salt Lake City, Utah

Jay Shapiro, MD
Professor
Department of Medicine
Walter Reed Army Medical Center
Bethesda, Maryland

Stephen M. Sims, PhD
Associate Professor
Department of Physiology
Faculty of Medicine and Dentistry
The University of Western Ontario
London, Ontario, Canada

Li Sun, MD, PhD
Research Fellow
Medical College of Pennsylvania
School of Medicine and Veterans Affairs Medical Center
Philadelphia, Pennsylvania

Yasuto Taguchi, MD
Research Fellow
Department of Medicine
University of Arkansas for Medical Sciences
Little Rock, Arkansas

Yasuhiro Takeuchi, MD
Assistant Professor
Fourth Department of Internal Medicine
University of Tokyo School of Medicine
Tokyo, Japan

James T. Triffitt, PhD
Head of Department
MRC Bone Research Laboratory
Nuffield Orthopedic Center
University of Oxford
Headington
Oxford, England

A. Van der Plas
Head of Technical Staff
Department of Cell Biology
Faculty of Medicine
Leiden University
Leiden, The Netherlands

Anthony Vernillo, PhD DDS
Associate Professor
Department of Oral Medicine and Pathology
New York University College of Dentistry
New York, New York

A. Frederik Weidema, PhD
Research Associate
Laboratorium voor Fysiologie
Katholieke Universiteit Leuven
Herestraat, Leuven, Belgium

Matsuo Yamamoto, PhD
Research Fellow
Department of Medicine
University of Arkansas for Medical Sciences
Little Rock, Arkansas

Tomoo Yamate, MD, PhD
Instructor
Department of Medicine
University of Arkansas for Medical Sciences
Little Rock Arkansas

Toshiyuki Yoneda, DDS, PhD
Professor of Medicine
Department of Medicine
Division of Endocrinology and Metabolism
University of Texas Health Science Center
San Antonio, Texas

Alberta Zambonin Zallone, PhD
Professor of Histology
Institute of Human Anatomy
University of Bari
Bari, Italy

M. Zaidi, MD, PhD, FRCP, FRCPath
Professor of Medicine and Associate Dean
Medical College of Pennsylvania
School of Medicine
Associate Chief of Staff and Chief, Geriatrics and Extended Care
Veterans Affairs Medical Center
Philadelphia, Pennsylvania

FOREWORD

These volumes differ from the current conventional texts on bone cell biology. Biology itself is advancing at breakneck speed and many presentations completely fail to present the field n a truly modern context. This text does not attempt to present detailed clinical descriptions. Rather, after discussion of basic concepts, there is a concentration on recently developed findings equally relevant to basic research and a modern understanding of metabolic bone disease. The book will afford productive new insights into the intimate inter-relation of experimental findings and clinical understanding. Modern medicine is founded in the laboratory and demands of its practitioners a broad scientific understanding: these volumes are written to exemplify this approach. This book is likely to become essential reading equally for laboratory and clinical scientists.

Ian MacIntyre, FRS
Research Director
William Harvey Research Institute
London, England

DEDICATION

To Professor Iain MacIntyre,
MBChB, PhD, Hon MD, FRCP, FRCPath, DSc, FRS

In admiration of his seminal contributions to bone and mineral research that have spanned over more than four decades, and

In gratitude for introducing us into the field of bone metabolism and for his continued encouragement, assistance, and friendship over many years

PREFACE

The intention of putting this book together has been not to develop a full reference text for bone biology and bone disease, but to allow for an effective dissemination of recent knowledge within critical areas in the field. We have therefore invited experts from all over the world to contribute in a way that could result in a complete, but easily readable text. We believe that the volume should not only aid our understanding of basic concepts, but should also guide the more provocative reader toward searching recent developments in metabolic bone disease.

For easy reading and reference, we have divided the text into three subvolumes. Volume 5A contains chapters outlining basic concepts stretching from structural anatomy to molecular physiology. Section I in Volume 5B is devoted to understanding concepts of bone resorption, particularly in reference to the biology of the resorptive cell, the osteoclast. Section II in Volume 5B contains chapters relating to the formation of bone with particular emphasis on regulation. Volume 5C introduces some key concepts relating to metabolic bone disease. These latter chapters are not meant to augment clinical knowledge; nevertheless, these do emphasize the molecular and cellular pathophysiology of clinical correlates. We do hope that the three subvolumes, when read in conjunction, will provide interesting reading for those dedicated to the fast emerging field of bone biology.

We are indebted to the authors for their significant and timely contributions to the field of bone metabolism. We are also grateful to Christian Costeines (JAI Press) and Michael Pazianas (University of Pennsylvania) for their efforts in ensuring the creation of quality publication. The editors also acknowledge the support and perseverance of their families during the long hours of editing.

Mone Zaidi
Guest Editor
Olugbenga A. Adebanjo
Christopher L.-H. Huang
Associate Guest Editors

SECTION II

BONE RESORPTION

OSTEOCLASTOGENESIS, ITS CONTROL, AND ITS DEFECTS

Etsuko Abe, Tomoo Yamate, Hanna Mocharla,
Yasuto Taguchi, and Matsuo Yamamoto

Advances in Organ Biology
Volume 5B, pages 289-313.

ISBN: 0-7623-0390-5

I. INTRODUCTION

Bone is remodeled continuously throughout life; this involves the resorption of old bone by osteoclasts and the subsequent formation of new bone by osteoblasts (for review see Parfitt et al., 1996). These two tightly regulated and coupled events maintain the anatomical and structural integrity of skeletal tissues. Osteotropic hormones, such as vitamin D [1,25$(OH)_2D_3$], parathyroid hormone (PTH) and calcitonin, preferentially modulate the process of bone remodeling by promoting osteoclast production, a process known as osteoclastogenesis. Ontogenically, osteoblasts are believed to be derived from undifferentiated mesenchymal cells, which further differentiate into osteocytes that are embedded in calcified tissues. Osteoclasts are multinucleated cells present only in bone and it is believed that the progenitors are of hematopoietic origin. Osteoclast progenitors proliferate and differentiate into mononuclear preosteoclasts that fuse to form multinucleated osteoclasts. Macrophage polykaryons are also formed from hematopoietic macrophage-like cells; however, these cells have different biochemical and histological characteristics and have no bone-resorbing activity (Chambers, 1985; Tong et al., 1994). Osteoclasts have unique morphological and biological features that are suited for their resorptive function (Baron et al., 1993). Recent studies on osteoclast development indicate that the early stage of osteoclastogenesis is regulated by the same cytokines and colony-stimulating factors that are involved in hematopoiesis, and that stromal cells/osteoblasts are also required for osteoclast formation. The successful development of *in vitro* culture systems for culturing osteoclasts has aided studies of the origin and differentiation of osteoclasts from hematopoietic precursors (Suda et al., 1992, 1995). The life-span of osteoclasts is also regulated by hormones and cytokines (Hughes et al., 1996; Suda et al., 1997). In this chapter, we review recent findings regarding the origin and differentiation of osteoclasts and the role of hormones and cytokines in regulating this process, and the cloning of osteoclast differentiation factor (ODF). In addition, we introduce and discuss osteopetrotic bone disease caused by a defect in osteoclast development or function.

II. ORIGIN OF OSTEOCLASTS

It is now well established that osteoclasts are of hematopoietic origin. Osteoclasts develop from multipotential hematopoietic stem cells, including colony forming unit-granulocyte/erythrocyte/ megakaryocyte/macrophage (CFU-GEMM) and CFU-granulocyte/macrophage (CFU/GM) mixed colonies (Hagenaars et al., 1989, 1990; Kurihara et al., 1990; Hattersley et al., 1991). However, it is debatable whether osteoclasts can be derived from the unipotential CFU-macrophages and mature macrophages. Burger et al. (1982) reported that no osteoclasts were formed in cocultures of fetal mouse bone rudiments and mature macrophages. Also, Kerby et al. (1992) identified different colonies by cell characteristics after pre-culturing mouse spleen cells in semisolid medium in the presence of interleukin-3 (IL-3) and erythropoietin. When cocultured with ST2 (marrow derived stromal) cells in the presence of $1,25(OH)_2D_3$, nonmacrophage colonies (multilineage precursor) formed numerous osteoclasts, but macrophage colonies did not produce osteoclasts. These findings were supported by the results of Hagenaars et al. (1989). In addition, none of the monocytic cell lines (M1, p388D1, J744A1, IC-21) could differentiate into osteoclasts in coculture systems. In contrast to the above findings, Udagawa and colleagues (1989, 1990) and Takahashi and colleagues (1991) reported that either alveolar macrophages or bone marrow–derived macrophage colonies that were precultured in methylcellulose and colony-stimulating factor (M-CSF), IL-3 or GMCSF could form osteoclasts in co-culture systems in the presence of osteoblasts, respectively. It is therefore inconclusive what stages of macrophages are capable of differentiating into osteoclasts and whether cells that are committed by IL-3 or GMCSF are able to transform to osteoclast progenitors in co-cultures with osteoblasts.

III. THE ROLE OF OSTEOBLASTS IN OSTEOCLASTOGENESIS

Bone marrow contains progenitors for both osteoclasts and osteoblasts. Testa et al. (1981) first successfully generated osteoclast-like cells from feline marrow cultures. Ibbotson et al. (1984) later showed that osteoclast formation in feline bone marrow culture was strongly stimulated by the osteotropic hormones, $1,25(OH)_2D_3$, PTH, and prostaglandin E (PGE). Subsequently, osteoclasts were derived from bone marrows of baboon (Roodman et al., 1985), humans (Takahashi et al., 1995a, b; MacDonald et al., 1987), rabbits (Fuller and Chambers, 1987), mice (Takahashi et al., 1988a,b; Hattersley and Chambers, 1989; Shinar et al., 1990) and rats (Kukita et al., 1993). Roodman et al. (1985) and Kurihara et al. (1989) showed that, in the presence of cytokines and osteotropic hormones, but without osteoblastic support, hematopoietic blast cells derived from human bone marrow or mouse spleen could form osteoclast-like cells expressing vitronectin receptors identified by the antibody 23C6 (Horton et al., 1985; Horton, 1988). However, osteoclasts derived under these conditions did not show sufficient pit formation, although they displayed bone-resorbing activity on being cultured with vital bone (Kurihara et al., 1989). These studies may suggest that in the absence of stromal/osteoblast support or that factors required for osteoclast formation in vitro may be different among spices and cell sources and preparation (Kanatani et al., 1995; Lader et al., 1998).

In 1988a,b, Takahashi et al. reported that osteoclasts derived from mouse bone marrow cells in the presence of osteotropic hormones were in contact with stromal/osteoblasts, suggesting that osteoclast formation might be supported by osteoblasts. To examine the importance of osteoblasts in osteoclast development, the investigators established coculture systems of primary calvarial osteoblasts and spleen cells. Using these coculture systems, osteoclasts characterized by tartrate resistant acid phosphatase (TRAP) expression, positive calcitonin binding, and pit-forming ability were obtained. It has now been established that, in the presence of supporting stromal/osteoblast cell lines (i.e., bone marrow-derived stromal cells and calvarial osteoblasts), certain hematopoietic cell lines can differentiate into osteoclast-like cells (Table 1) in the presence of various osteotropic hormones and cytokines. In the case of stromal cell lines characterized as preadipocytes, the presence of glucocorticoids with osteotropic hormones is required to support osteoclastogenesis.

Studies on the mechanism of osteoclast formation was extensively progressed by cloning of novel factors in the past two years. Simonet and colleagues (1997) and Yasuda and colleagues (1998a) are the first to clone an inhibitor of osteoclast formation, osteoprotegerin. This molecule is a member of the TNF receptor superfamily and is produced by lung, liver and brain. Overexpression of osteoprotegerin in vivo causes osteopetrosis, coinciding with a decreased number of osteoclasts in bone tissue. Osteoprotegerin is a receptor for the cytotoxic ligand, TRAIL, which induces apoptosis in T cells (Emery et al., 1998). Subsequently, osteoprotegerin ligand (OPGL), which is expressed on stromal/osteoblasts by exposure to osteotropic factors and hormones, has been cloned by the same groups (Lacey et al., 1998; Yasuda et al. 1998b). This molecule was identified as RANKL (receptor activator of NF-κB ligand) and TRANCE. Because hematopoietic cells

Table 1. Osteoblastic and Hemopoetic Cell Lines That Can Support Osteoclastogenesis

Cell Line	*Cell Phenotypes*	*References*
Supporting osteoblasts		
ST2	Mouse bone marrow stromal cells	Udagawa, N. (1989)
MC3T3G2/PA6	Mouse bone marrow stromal cells	Udagawa, N. (1989)
KS-4	Mouse calvarial cells	Yamashita, T. (1990)
MB1.8	Mouse calvarial cells	Wesolowski, G. (1995)
TMS-14	Mouse bone marrow derived cells	Kurachi, T. (1994)
Saka	SV 40 infected human stromal cells	Takahashi, S. (1995)
MS1, MS2	mouse bone marrow stromal cells (temperature sensitive T antigen)	Liu, B.Y. (1998)
Nonsupporting cells		
MCC3T3 E1, ST13, BALB3T3, NIH3T3, +/+LDA11		
Hematopoietic cell lines		
FDCP-mix	IL-3 dependent stem cells	Hagenaars, C.E. (1989)
ts series	MCSF dependent cells, temperature sensitive SV40 T antigen	Chambers, T.J. (1993)
BDM-1	MCSF dependent macrophages	Shin, J.H. (1995)
C7	MCSF dependent macrophages	Yasuda, H. (1998)

are differentiated into osteoclasts in the presence of two factors, MCSF and soluble form of OPGL/RANKL/TRANCE/ODF (RANKL), the role of osteoblasts in osteoclast development is confirmed. Interestingly, hematopoietic cells cultured in the presence of GMCSF, IL-3, IL-6, or SCF could not form osteoclasts. Similarly to the case of T cells–dendric cell interaction (Anderson et al., 1997), the signal from RANKL on osteoblasts is transduced through RANK expressing in hematopoietic cells to induce osteoclast formation. So far how the signal of RANK/RANKL is transduced into osteoclast-specific genes, such as TRAP, c-src, vitronectin receptor, catepsin K, and carbonic anhydrase II, is not known.

IV. OSTEOCLAST DIFFERENTIATION: ROLE OF HORMONES AND FACTORS

Table 2 summarizes some commonly known and recently described bone-resorbing stimulators and inhibitors. Bone-resorbing hormones and factors induce osteoclast formation not only *in vivo* but also *in vitro* either in bone marrow cultures or coculture systems of osteoblasts and hematopoietic cells. The osteoclast-inducing agents, $1,25(OH)_2D_3$ (Takahashi et al., 1988a,b), PTH (Akatsu et al., 1989a), PGE (Akatsu et al., 1989b), and IL-1 (Akatsu et al., 1991) are well known.

The ability of $1,25(OH)_2D_3$ to induce osteoclastogenesis was first demonstrated *in vitro* systems of bone marrow cultures and cocultures of osteoblast and hematopoietic cells. However, it is not clear whether the target cells for osteoclast-inducing factors are the stromal/osteoblasts or the osteoclast progenitors. It is known that $1,25(OH)_2D_3$ can induce several matrix proteins, including osteocalcin (Pike et al., 1993), osteopontin (Noda et al., 1988), and the third component of complement (C3) (Hong et al., 1991; Jin et al., 1992) in bone marrow cells or osteoblasts. Osteocalcin mRNA is expressed during osteoblast differentiation, whereas both osteopontin and C3 mRNAs are expressed in the presence of $1,25(OH)_2D_3$ in three cell types, namely osteoblasts, macrophages, and osteoclast (Ikeda et al., 1992; Sato et al., 1993; Yamate et al., 1995). Since antibodies against either osteopontin or C3 could inhibit osteoclast formation in bone marrow cultures, both proteins appear to be required for osteoclastogenesis.

Table 2. Bone Resorbing and Antiresorptive Agents

Bone resorbing agents	
Systemic hormones	$1,25(OH)_2D_3$, PTH
Local factors and cytokines	PTHrP, prostaglandin, Il-1 TNF-α, IL-11, IL-6/sIL-6R, OSM, LIF
Inhibitors for bone resorption	
Hormone factors:	Calcitonin, bisphosphonates
	IL-4, IL-10, IL-13, IL-18

Osteoclast precursors possess the C3 receptor, CR1, which is recognized by the monoclonal antibody Mac-1 (Sato et al., 1993), and osteoclasts express vitronectin receptor ($\alpha_v\beta_3$) that also binds osteopontin (Helfrich et al., 1992; Horton et al., 1993). Another mechanism involves CD44-dependent osteoclast adhesion to cellular matrix (Nakamura et al., 1995; Weber et al., 1996). In contrast to in vitro studies, knock-out of osteopontin or C3 showed normal osteoclast formation (Rottling et al., 1998) and a deficiency of vitamin D receptor showed normal osteoclast formation in vivo and in vitro (Kato et al., 1998). Therefore, these molecules are not critical for osteoclast formation in vivo, indicating that other molecules may be replaced in vivo.

The findings that cell adhesion molecules in addition to cytokines are crucial for osteoclastogenesis are important advances in studying osteoclast differentiation. A recent report indicated that leukocyte function-associated antigen-1 (LFA-1) and intracellular adhesion molecule-1 (ICAM-1) are also involved in the interaction of hematopoietic cells and stromal/osteoblasts during osteoclast formation (Kurachi et al., 1993). When antibodies to LFA-1 and ICAM-1 were added together with $1,25(OH)_2D_3$ in cocultures of spleen cells and osteoblasts, osteoclast formation was suppressed. The inhibitory effect of the antibodies on osteoclast formation could be observed during all stages of culture, but were more notable at the later stages. The expression of ICAM-1 was observed on both spleen cells and stromal/osteoblasts. Notably, both ICAM-1 and ICAM-2 are ligands for LFA-1, but ICAM-1 also interacts with other molecules, including Mac-1 and CD34. Mac 1 and LFA-1 are known to bind to discrete domains on ICAM-1. Because Mac-1 is also involved in osteoclast differentiation in vitro, it appears that the interaction of stromal cells and osteoclast progenitors may be a more complex event. Specifically, RANKL expression on the surface of COS cells could transduce signals to hematopoietic cells to form osteoclasts in the presence of MCSF (Yasuda et al., 1998b), indicating other molecules than MCSF and RANKL are not essential for osteoclast supporting activity.

Osteoclast formation involves a mechanism that promotes cAMP production. The ability of PTH and PTH related protein (PTHrP) (Akatsu et al., 1989a) to induce osteoclast formation in bone marrow cultures was shown to be mediated through cAMP dependent mechanism. The addition of dibutyryl cAMP to mouse marrow cultures could induce osteoclast-like cell formation, and isobutylmethylxanthine, or IBMX, a potent phosphodiesterase inhibitor, enhanced osteoclast-like cell formation induced by PTH. The action of cAMP may be mainly on stromal/osteoblasts, since PTH could not stimulate adenylate cyclase activity during the first four days of culture. However, the action of PTH was observed after four days when stromal cells appeared in the culture. Moreover, stromal/osteoblasts but not osteoclasts possess PTH receptors (Fermor and Skerry, 1995) and stromal/osteoblasts expressed various cytokines and proteins in response to PTH (Horowitz et al., 1989; Lowik et al., 1989; Rouleau et al., 1990; Greenfield et al., 1993). PGE (Akatsu et al., 1989b; Kaji et al., 1996) also induces osteoclast development by a mechanism that involves cAMP. Thus, the potency for osteoclast induction by PGE was highly correlated with the ability to increase cAMP production in bone marrow cells and IBMX potentiated PGE-induced osteoclast-like cell formation. Furthermore, it has been shown that IL-1 stimulated osteoclast formation through PGE_2 production in bone marrow cultures and cocultures. Therefore, osteoclast formation induced by IL-1, PTH, and PGE is mediated via cAMP production, whereas $1,25(OH)_2D_3$ induced osteoclast formation by a mechanism that appears to be independent of cAMP production.

Recent reports have indicated that stimulators of bone formation can also increase osteoclast formation. Thus, when unfractionated bone cells from between 10 and 15 day-old mice that contained bone marrow cells, osteoblasts, and TRAP-positive osteoclasts, were cultured for five days without any treatment, the TRAP-positive osteoclasts disappeared. When the precultured bone cells were treated with bone-resorbing or bone-forming agents, it was found that IGF-1 (Mochizuki et al., 1992) and bone morphogenic protein (BMP-2) (Kanatani et al., 1995), both of which are potent bone-forming agents, as well as $1,25(OH)_2D_3$, PTH, and PGE, that are powerful bone-resorbing agents, induced TRAP-positive osteoclast-like cells after seven days. The effects of BMP-2 and $1,25(OH)_2D_3$ on osteoclast formation were found to be additive. Furthermore, receptors for BMP-2 were demonstrated on hematopoietic blast cells. Since BMP-2 or IGF-1 did not induce osteoclast formation in bone marrow cultures or cocultures of hematopoietic cells and osteoblasts, bone forming and resorbing agents may act on different type of precursors, with $1,25(OH)_2D_3$ acting on bone marrow derived cells, while BMP-2 acts on other cell types besides bone marrow derived cells.

IL-4 and IL-13 are immunoregulatory cytokines recently found to influence skeletal metabolism. Both cytokines are secreted by activated T lymphocytes and are well recognized growth and differentiation factors for a wide variety of hematopoietic cells. Since the γ chain of the IL-2 receptor is shared by receptors for IL-2, IL-4, IL-7, and IL-13 (Kondo et al., 1993; Russel et al., 1993; Zurawski et al., 1993) in the signal transduction pathway, it is thought that these cytokines may function similarly in bone metabolism. Both IL-4 (Shioi et al., 1991; Nakano et al., 1994; Kawaguchi et al., 1996) and IL-13 (Onoue et al., 1996) were found to inhibit the *in vitro* bone resorption and osteoclast formation that is induced in cocultures by bone resorbing agents, such as IL-1. The inhibitory effect of IL-4 or IL-13 on IL-1 and tumor necrosis factor-α (TNF-α) induced bone resorption could be explained by prostaglandin-mediated mechanism. IL-1 was found to stimulate dramatically cyclooxygenase 2 (COX2) mRNA, but not the constitutively expressed COX1 mRNA in osteoblasts (Sato et al., 1996). Both IL-4 and IL-13 suppressed the IL 1 induced stimulation of COX2. Xu et al. (1995) reported that IL-10 also suppressed the formation of osteoclasts in rat bone marrow cultures by inhibiting GM-CSF colony formation from hematopoietic progenitor cells. Similarly, Horwood and colleagues (1998) have reported that IL-18 has inhibitory activity on osteoclast formation through GMCSF production in T cells.

V. OSTEOCLAST FORMATION THROUGH gp130 SIGNALS

A. Cytokine Signaling: An Overview

The role of cytokine induced signal transduction via gp130 on bone metabolism has been examined. IL-6, IL-11, oncostatin M (OSM) and leukemia inhibitory factor (LIF) produced by osteoblasts have been found to transduce signals through gp130 to stimulate osteoclast development (Ishimi et al., 1990, 1992; Yang and Yang, 1994; Mata et al., 1995, Bellido et al., 1996). Members of the subfamily of cytokines that include IL-6 and IL-11 bind to their receptors (the α subunit) and this induces the dimerization of gp130 to form a homodimer responsible for signal transduction. Other cytokines, such as OSM and LIF, bind to the heterodimeric form of LIF receptor (LIFR) and gp130. It appears that OSM may also bind to the heterodimeric form of OSM receptor (OSMR) and gp130 (Figure 1; Kishimoto et al., 1995). Ligand-receptor complex-induced dimerization of gp130 initiates intracellular signaling by activating members of a family of receptor-associated tyrosine kinases, known as the Janus kinases (JAKs), through a process that involves phosphorylation. The activated JAKs induce tyrosine phosphorylation of several proteins including gp130, the kinases themselves ,and a series of cytoplasmic proteins termed signal transducers and activators of transcription (STATs) (Stahl et al., 1994, 1995).

B. gp130 and Osteoclastogenesis

IL-6 alone does not induce osteoclast-like cells, but a complex of IL-6 and its soluble receptor (sIL-6R) is an effective signal for osteoclastogenesis (Tamura et al., 1993; Udagawa et al., 1995). The action of sIL-6R could be replaced by the

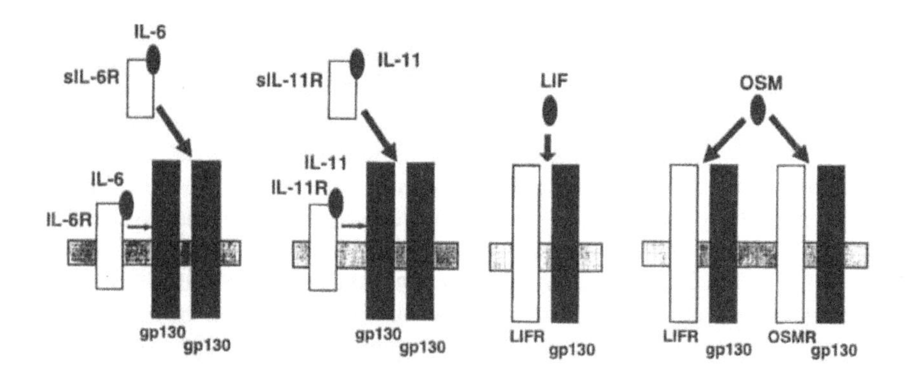

Figure 1. Complexes of cytokines and their receptors. Cytokines such as IL-6, IL-11, LIF, and OSM sharing gp130 transduce signals in osteoblasts leading to osteoclast differentiation.

treatment with dexamethasone since a combination of IL-6 and dexamethasone was also found to promote osteoclastogenesis, an effect that was comparable in magnitude to a combination of IL-6 and sIL-6R. Therefore, it is thought that dexamethasone may induce increased IL-6R expression on osteoblasts. In fact, the expression of IL-6Rs was not abundant on normal stromal/osteoblasts as compared to bone marrow hematopoietic cells, but dexamethasone was found to increase the expression level of sIL-6R on osteoblasts. The importance in osteoclastogenesis, of the level of IL-6 receptor expression in stromal/osteoblasts has been confirmed in transgenic mice for the IL-6 receptor (Udagawa et al., 1995; Tsujinaka et al., 1996). Osteoblastic cells from transgenic mice that constitutively expressed abundant human IL-6 receptors could support osteoclast development in the presence of IL-6 alone in cocultures with normal spleen cells. These results clearly indicate that the ability of IL-6 to induce osteoclast differentiation depends on IL-6 receptor expression on osteoblastic cells, not on hematopoietic osteoclast progenitors. The results also suggest that IL-6 may play less of a role in osteoclastogenesis during normal physiologic conditions, but could become relevant when there is increased expression of IL-6 receptor under certain conditions such as during dexamethasone treatment or when the IL-6 gene is transfected into cells in vivo. By in vitro studies using osteoblast cell lines and primary cells of embryonic fibroblasts, we have reported that IL-6 type cytokines with their soluble receptors promote osteoblast differentiation expressing typical osteoblast markers such as osteocalcin, type-I collagen, and mineral deposition (Taguchi et al., 1998). Further studies of the mechanism of IL-6-mediated signal transduction by Nishimura et al. (1998) indicated that IL-6 and sIL-6R transduce signals through JAK1, JAK2, STAT1 and STAT3 in the osteosarcoma cell line, MG63. These data suggest that temporal and spacial expression of IL-6-type cytokines may stimulate both osteoclast formation and osteoblast differentiation in an appropriate site.

Estrogen deficiency results in a marked bone loss due to increased stimulation of osteoclastic bone resorption. The mechanism involved in osteoclast induction during estrogen deficiency can be explained by the increase in two types of cytokines, IL-6 and/or IL-1 (Pacifici et al., 1991; Girasole et al., 1992; Jilka et al., 1992; Kimble et al., 1994; Miyaura et al., 1995). The addition of anti-IL-6 antibody to bone marrow cultures from ovariectomized mice inhibited osteoclast formation to the level of sham control, but the antibody did not influence osteoclastogenesis in the estrogen-repleted mice. This would suggest that IL-6 is not responsible for osteoclast formation during normal physiological circumstances, but may become involved in osteoclastogenesis under pathologic conditions characterized by abnormally high levels of bone resorption, such as estrogen deficiency, multiple myeloma, Paget's disease, rheumatoid arthritis, and Gorham-Stout disease. Notably, there are a normal number of osteoclasts in IL-6 deficient mice, and ovariectomy does not cause an increase in the number of osteoclasts in bone in the IL-6 deficient mice as is seen in IL-6 sufficient mice (Bellido et al., 1995).

IL-11 (Paul et al., 1990), a member of the cytokine family that mediate signaling via gp130, is also produced by stromal/osteoblasts and induces osteoclast formation (Girasole et al., 1994). An anti-IL-11 antibody suppresses the osteoclast development that is induced by $1,25(OH)_2D_3$, PTH, IL-1, or TNF-α. The production of IL-11 by primary osteoblasts is also upregulated by IL-1, TNF-α, and by the osteotropic hormones, $1,25(OH)_2D_3$ and PTH (Yang and Yang, 1994; Romas et al., 1995). Receptors for IL-11 (Hilton et al., 1994) have been shown to be expressed on osteoblastic cells (Romas et al., 1995) and their levels are decreased in senescence accelerated mice (SAM) (Kodama et al., 1995). Osteoclast formation in bone marrow cultures from these mice is decreased when compared with control mice. These collective studies suggest that IL-11 may also be involved in osteoclastogenesis under pathological conditions.

VI. CELL FUSION DURING OSTEOCLASTOGENESIS

Osteoclasts are formed by the fusion of mononuclear preosteoclasts that are identified as TRAP-positive cells. Wesolowski et al. (1995) have recently established a high yield purification technique for obtaining preosteoclasts and have studied the mechanisms involved in the induction of cell fusion and bone resorption by osteotropic hormones. In these experiments, bone marrow cells and an osteoblastic cell line were cocultured in the presence of $1,25(OH)_2D_3$ for six days. Osteoblastic cells were first removed by the treatment with collagenase/dispase and the preosteoclasts that remained attached to the dish were then harvested by treatment with echistatin, a known inhibitor of $\alpha_v\beta_3$ integrin (vitronectin receptor) function. The harvested preosteoclasts were subsequently studied to dissect the process of fusion and bone resorption. When preosteoclasts were cultured with osteoblastic cells in the presence of $1,25(OH)_2D_3$, preosteoclasts fused with each other within one day and the multinucleated osteoclasts so formed were able to resorb bone. Optimal fusion and bone resorption activity required the presence of both osteoblasts and $1,25(OH)_2D_3$ in culture. Therefore, cell commitment, differentiation, and fusion processes are sequentially programed events and these may not be dissociated under normal conditions. It is not known what kinds of factors are produced from osteoblasts to promote the total process in osteoclastogenesis. IL-6 may be a candidate for the cytokines to stimulate the early stage of osteoclastogenesis and also stimulate the cell fusion process under pathological conditions because it promoted the formation of macrophage polykaryons from murine alveolar macrophages (Abe et al., 1991).

The involvement of cell fusion processes during differentiation is found not only during osteoclast formation, but also during muscle cell differentiation, macrophage polykaryon, and egg-sperm fertilization. Several cellular proteins with extracellular domains, including cadherins, integrins, and other cell adhesion molecules have been implicated in these fusion events, but the detailed mechanism is not known. Fusion of virus and mammalian cells is also a critical initial step in human immunodeficiency virus (HIV) infection. Recent studies have shown that the expression of high mannose-oligosaccharides in mammalian cells is important in such HIV target cell interaction (Ezekowitz et al., 1989). Pradimicin derivatives, which are antiviral and antifungal agents, recognize and bind to specific sugars such as mannose residues on target cells.

To evaluate the role of mannose residues in osteoclastogenesis induced by cocultures of mouse spleen cells and stromal cells, the effect of pradimicin on osteoclast formation and its binding site on osteoclasts was examined (Kurachi et al., 1994). Osteoclast formation was inhibited in a dose-dependent manner by pradimicin at the later stages of osteoclast differentiation and this inhibition was abrogated specifically by the mannose-rich yeast, mannan. Furthermore, pradimicin was bound specifically to osteoclast progenitors at the fusion stage, but not to progenitors at their early stage or to already differentiated osteoclasts. Detailed analysis showed that mannose residues were expressed on the outer membranes of monocyte or osteoclast progenitors. These results suggest that cell surface mannose-oligosaccharides are involved in cell-cell and virus-cell fusion. The high mannose-oligosaccharides are also important for myotube formation by L6 myoblasts that involve cell-cell fusion (Jamieson et al., 1992).

Yagami-Hirosawa et al. (1995) recently cloned a new class of proteins, the meltrins, in muscle cells; these show a sequence homology to the fertilin proteins (Blobel et al., 1992; Miles et al., 1994) that are implicated as fusion proteins in the binding and fusion of sperm with egg. The expression of α-meltrin was regulated in parallel with the fusion process involved in myotube formation, but the β- and γ-meltrins were expressed constitutively in muscle cells. Meltrins contain a disintegrin and a metalloprotease domain. Transfection of the shortened α-meltrin (after cleavage of the metalloprotease domain) was found to enhance the frequency of the fusion induced by low concentrations of fetal bovine serum (FBS), but the transfection of a full length α-meltrin inhibited that process. Since α-meltrin was found not to induce fusion of fibroblasts and myoblast (C2C12) cells at normal concentration of FBS, α-meltrin might be a modulator in the fusion process. In fact, we (Mocharla et al., 1996) and Inoue and colleagues (1998) have found that α-meltrin is expressed not only in fused cells (myotubes or osteoclasts) but also in nonfused cells (osteoblasts) in bone tissues. Therefore, further investigations are needed to clarify the role of α-meltrin.

VII. THE CAUSE OF OSTEOPETROSIS

Osteopetrosis is an inherited disease characterized by an increase in bone mass due to reduced bone resorption, a marked increase in skeletal density, and a decrease in bone marrow cavity. Four types of congenital osteopetrotic mice (gl/gl, mi/mi, oc/oc, and op/op) are known at present (Table 3) (Marks, 1989). The chromosomal, genetic and biochemical characteristics of the defects are distinct for each osteopetrotic mouse.

A. Op/op Mice

A complete defect of macrophages and osteoclasts in op/op mice (Marks et al., 1984) as caused by a failure of M-CSF secretion from stromal cells was described by Wiktor-Jedrzejczak et al. (1990). The defects in the differentiation of macrophages and osteoclasts were not cured by transplantation of normal bone marrows in op/op mice. However, administration of the recombinant human M-CSF restored osteoclast formation with bone resorbing activity in these mice (Felix et al., 1990; Kodama et al., 1991). Further analysis of the M-CSF gene in op/op mice by Yoshida et al. (1990) revealed an insertion mutation in the middle of a coding region that defined a stop codon. This is believed to have caused the production of a truncated inactive translation product.

Interestingly, the macrophage and osteoclast deficiencies in op/op mice have been shown to be cured spontaneously without treatment at the age of 40 weeks, even in mice carrying a double mutational deficiency of GM-CSF and M-CSF (Nilsson et al., 1995), suggesting that the age-dependent spontaneous correction was not due to the compensatory effect of GM-CSF. Although both GM-CSF and M-CSF have been shown to be crucial for committing primitive hematopoietic stem cells to differentiate along the macrophage lineage or the osteoclastic pathway, it would appear that alternative regulatory factor(s) may also be involved in this process.

B. Mi/mi Mice

Mi/mi mice are also congenitally osteopetrotic. Transgenic mice harboring a promoter region of vasopressin gene fused with β-galactosidase (Tachibana et al., 1992) or a human globin gene have a complete loss of skin pigmentation, mi-

Table 3. Murine Models of Congenital Osteopetrosis

Mice	*Chromosome Location*	*Primary Cause*	*Osteoclast in Bone*	*Hematopoietic Cell Transplantation*	*Tissue Abnormalities*
op (osteopetrotic)	3	M-CSF mutation	–	No effect	Bone, teeth, macrophages
oc (osteoscelerotic)	19	?	+	No effect (?)	Bone, teeth, kidney
mi (microphthalmic)	6	mi protein mutation	+/–	Effective	Bone, teeth, skin, ear, eyes
gl (gley-lethal)	10	?	+	Effective	Bone, teeth, eyes

crophthalmia, and cochlear abnormalities similar to mi/mi mice. This suggests that the vasopressin or the globin genes may be integrated at the mi locus of the transgenic mice; these genes serve as a tag for the direct cloning of genomic sequences flanking the inserting sites. Using these transgenic mice, Hodgkinson et al. (1993) and Hughes et al. (1993) cloned the gene at mi locus that causes osteopetrosis. The gene was found to encode a novel member of the basic-helix-loop-helix-leucine zipper (bHLH-ZIP) family of transcription factors (Vinson and Garcia, 1992). Mi mutant mice are characterized by small eyes and loss of melanin pigments. Also note that tyrosinase is a rate limiting enzyme in melanin biosynthesis and is responsible for the pigment cell-specific transcription that is regulated by a large family of transcription factors with bHLH-ZP structure. It has been established that the mi protein bind to the CATGTG motif in the upstream region of the tyrosinase gene to transactivate the gene (Yasumoto et al., 1994). However, the target genes for mi protein in the regulation of osteoclastogenesis are yet to be described.

Since osteopetrotic disorders in mi/mi mice are curable by hematopoietic transplantation, their defect appears to be in the hematopoietic precursors of osteoclasts. Reports of the presence of osteoclasts in bone tissues of mi/mi mice have been inconclusive. While, earlier studies by Marks and Walker (1976) could not detect multinucleated osteoclasts, mononuclear preosteoclasts have been detected. These observations led to the suggestion that the defect in mi/mi mice was in fusion disability of preosteoclasts (Thesingh and Scherft, 1985). However, later studies by Marks et al. (1984) suggested that the pathogenesis of osteopetrosis could be associated with a functional defect in osteoclasts. This is because normal or slightly elevated numbers of multinucleated osteoclasts expressing the appropriate biochemical markers were detected in bone tissues of mi/mi mice. In partial support of the latter finding, Graves and Jilka (1990) reported that TRAP+ multinucleated osteoclast formation from calvarial cells cultured with PTH was significantly decreased in mi/mi mice as compared to their normal littermates. However, multinucleated osteoclasts could be derived from calvarial cells of mi/mi when cultured with bone particles. Therefore, it was concluded that the lack of osteoclast differentiation in mi/mi mice was due to a lack of responsiveness of preosteoclasts to bone resorbing agents.

Contrary to the above reports, our studies (Abe et al., 1995) indicated that bone tissues in mi/mi mice are apparently deficient of cells committed to the osteoclastic lineage due to the histological evidence of the absence of TRAP+ mononuclear and multinucleated osteoclasts. Also, osteoclasts could not be derived from *in vitro* cocultures of osteoblasts and spleen cells from mi/mi mice in the presence of $1,25(OH)_2D_3$, or even with osteoblasts from the normal littermates. On the other hand, osteoblasts from mi/mi mice supported osteoclastogenesis from spleen cells from normal littermates. The difference between the results of our studies and those of others previously discussed may be associated with possible allelic variation of the mi mutation in the mice. The mi^{wh} is indistinguishable from the wild-type littermate by Southern blot, but mi^{tg} or mi^{ws} has a discrete intragenic deletion in the coding region of mi gene which is detectable by Southern blot analysis. The mi gene in the original mi mouse has two mutations in coding region that could not be identified by Southern blot. One is an A to G transition which is silent, and the other is a three nucleotide deletion encoding an arginine residue. The arginine residue is located at the basic region of bHLH in mi gene and is critical for the helix conformation that binds to a particular DNA sequence. Thus, the different observations reported for the mi/mi mouse may be explained by different types of mutation with different phenotypic outcomes.

C. Gene Deletions

Targeted disruptions of certain genes have unexpectedly produced osteopetrotic mice (see also Chapter 22, part B). Examples of these include mice deficient of c-*src* protooncogene, c-*fos*, and PU.1. The manifestation of osteopetrotic disease in these mice was either due to autonomous defects of osteoclast function or formation. In *src*-deficient mice, the osteoclasts formed do not develop ruffled borders on the membrane to resorb bone (Soriano et al., 1991), indicating a functional defect. However, transplantation of fetal liver cells into *src*-deficient mice could cure the osteopetrotic disorders. The ability of *src*-deficient osteoblasts to support osteoclasts formation in cocultures with normal spleen cells indicates that *src*-deficient osteopetrosis is due to a defect of osteoclast progenitors, not osteoblasts. The $p60^{c\text{-}src}$ is expressed on ruffled border membranes and membranes of intracellular organelles in osteoclasts (Boyce et al., 1992; Horne et al., 1992). Osteopontin stimulates bone resorption of osteoclasts through c-*src* kinase activity associated with phosphatidylinositol 3-hydroxyl ($PtdIns_3$-OH) kinase (Hruska et al, 1995). Since herbimycin A, a tyrosine kinase inhibitor, suppresses bone resorbing activity by osteoclasts, tyrosine phosphorylation by $p60^{c\text{-}src}$ may be involved in osteoclastic bone resorption. The major substrates for tyrosine phosphorylation in osteoclasts are p125 and p120, which were identified as focal adhesion kinase ($p125^{FAK}$) (Tanaka et al., 1995a) and c-Cbl (Tanaka et al., 1995b), respectively. Indeed, p120 is not phosphorylated in *src*-deficient mice. Additionally, the p120 protein is associated with receptors for epidermal growth factor and M-CSF (c-*fms*) in macrophages and osteoclasts to regulate bone resorption activity through tyrosine phosphorylation.

Deletion of the c-*fos* gene has also produced severe osteopetrotic disorders in mice (Johnson et al., 1992; Grigoriadis et al., 1994). Although c-*fos* is not required for replication of most types of cells in prenatal and postnatal development, it plays important roles in bone metabolism. Osteopetrotic disorders in c-*fos* deficient mice are due to a defect in osteoclast progenitors. The injection of hematopoietic progenitor cells transfected with mutant c-*fos* gene into irradiated newborn wild mice leads to a defect in bone resorption and remodeling characterized by a complete absence of the secondary ossification centers in metaphysis. Transplantation of wild-type bone marrow into mutant mice or the infection of c-*fos* deficient spleen cells with retrovirus plasmid encoding c-*fos* in *in vitro* coculture systems overcomes the deficiency in osteoclast formation. The lack of osteoclasts due to c-*fos* deficiency is associated with a lineage shift between osteoclasts and macrophages. Normal numbers of putative macrophage progenitors have been found to be present in the bone of c-*fos* mutant mice; however, there is an increased number of macrophages. It appears that the deficiency of osteoclast formation in c-*fos* mutant mice results in the increased development of cells of macrophage lineage that cause osteopetrosis.

The transcription factor, PU.1, a member of the *ets* family, is a hematopoietic cell-specific factor, which is expressed abundantly in macrophages and B lymphocytes (Scott et al., 1994; Voso et al., 1994). In these cells, it regulates tissue-specific gene expression, for example, the myeloid differentiation-associated genes encoding CD11b and the M-CSF receptor (c-*fms*), by binding to the promoter regions of the genes. The disruption of PU.1 causes death of embryos at a later stage, which is associated with a defect of lymphoid and myeloid cell lineages. Mutant mice exhibit osteopetrotic features due to a deficiency of multipotential myeloid progenitors (Tondravi et al., 1995).

Thus, several factors and proteins are involved in the process of osteoclast formation. In osteopetrosis, the factors affected may be indispensable for osteoclastogenesis (and hence bone resorption.) Lack of resorption would result in the pathological outcome. Animal models of osteopetrosis may therefore shed some light on the biochemical and molecular defect(s) associated with clinical disorders in osteoclastogenesis. However, at present there are no reports of mutational defects in special genes in human osteopetrosis.

VIII. CONCLUSION

Osteoclast formation from hematopoietic precursor cells is controlled by osteoblasts and osteotropic hormones (Figure 2). The influence of osteoblasts in osteoclast differentiation appears to be exerted at all stages of the osteoclastogenesis process. In the absence of osteoblasts, precursor cells in the bone differentiate preferentially along the macrophage lineage. A deficiency of c-*fos* gene function may also lead to defective osteoclast formation and increased macrophage and macrophage polykaryons. Doses of MCSF are also critical for determining differentiation pathway of hematopoietic cells; high doses of MCSF can induce macrophages, whereas low doses of MCSF can induce osteoclast progenitors. To date, a precise role of osteoblasts in osteoclast differentiation is understood at molecular levels. Because a combination of MCSF and soluble form of RANKL can induce osteoclast development from osteoclast progenitors, both factors are essential for osteoclastogenesis. MCSF is constitutively produced by osteoblasts, whereas intact form of RANKL is induced and expressed on the membrane of osteoblasts after stimulation with osteotropic hormones and factors such as 1,25(OH)2D3,

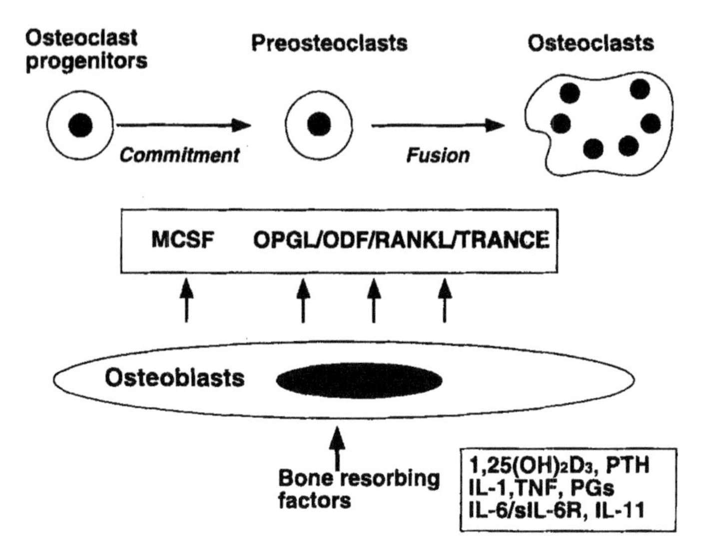

Figure 2. Osteoclast differentiation is supported by osteoblasts. Osteoblasts support differentiation and fusion of osteoclast lineage cells through cell-cell contact. MCSF and OPGL/ODF/RANKL/TRANCE expression by osteoblast are essential for osteoclastogenesis.

IL-1, IL-6-type cytokines, and TNF-α. The necessity of direct contact between hematopoietic cells and osteoblasts for osteoclast development may be explained by the in vitro observation that these molecules are bound on osteoblast membrane surface or cell matrix and therefore cell–cell contact is needed for effective transfer of the signals from osteoblasts to hematopoietic cells. However, inhibition of cell–cell contact using specific antibodies for integrin family (ICAM1 and LFA-1), adhesion molecules, osteopontin, or C3 (McNally et al., 1994) is not enough to inhibit osteoclast formation in vivo, maybe due to overexpression of other similar types of molecules.

The roles of cytokines whose functions are mediated by gp130 signals (e.g., IL-6 and sIL-6R, IL-11, OSM, and LIF) in osteoclast formation have recently been highlighted. The target cells for these cytokines in osteoclastogenesis appear to be osteoblasts, and high expression of their receptors are crucial for their function. Upon receptor engagement, these cytokines transduce their signals through gp130, JAKs, and STAT kinases; subsequently, STAT kinases are translocated into the nucleus to bind specific genes that are supposedly involved in osteoclastogenesis. Thus, to understand osteoclastogenesis induction by different hormones and cytokines, analysis of RANKL gene regulation is required. In addition, it is important to know the downstream events of RANKL and RANK signaling pathways in hematopoietic cells that induce specific gene expression in osteoclasts.

In conclusion, studies in osteoclast biology will undoubtedly offer new insights for investigating metabolic bone diseases caused by abnormal osteoclast recruitment and function such as osteopetrosis, osteoporosis, Pagets disease, rheumatoid arthritis, and periodontal disease.

ACKNOWLEDGMENTS

The authors would like to thank Drs. S.C. Manolagas and M. Zaidi for their excellent suggestions and discussions. We also acknowledge Dr. Igietseme, J.U. and C.J. Simmons in the Office of Grants and Scientific Publications at UAMS for helping to write this review.

REFERENCES

Abe, E., Ishimi, Y., Jin, C.H., Hong, M.H., Sato, H., and Suda, T. (1991). Granulocyte-macrophage colony-stimulating factor is a major macrophage fusion factor present in conditioned medium of concanavalin A-stimulated spleen cell cultures. J. Immunol. 147, 1810-1815.

Abe, E., Yamate, T., Mocharla, H., Lin, S.C., Ding, L.,Munshi, N., Jilka., and Manolagas, S.C. (1995). Correction of the defective osteoclastogenesis of the osteopetrotic mi/mi mice by transfection with the wild-type *mi* gene. J. Bone Miner. Res. (Abstract 79).

Akatsu, T., Takahashi, N., Udagawa, N., Sato, K., Nagata, N., Moseley, J.M., Martin, T.J., and Suda, T. (1989a). Parathyroid hormone (PTH)-related protein is a potent stimulator of osteoclastlike, multinucleated cell formation to the same extent as PTH in mouse bone marrow cultures. Endocrinology 125, 20-27.

Akatsu, T., Takahashi, N., Debari, K., Morita, I., Murota, S., Nagata, N., Takatani, O., and Suda, T. (1989b). Prostaglandins promote osteoclastlike cell formation by a mechanism involving cyclic adenosine 3,5-monophosphate in mouse bone marrow cell cultures, J. Bone Miner. Res. 4, 29-35.

Akatsu, T., Takahashi, N., Udagawa, N., Imamura, K., Yamaguchi, A., Sato, K., Nagata, N., and Suda, T. (1991). Role of prostaglandins in interleukin-1-induced bone resorption in mice in vitro. J. Bone Miner. Res. 6, 183-190.

Anderson, D.M., Maraskovsky, E., Billingsley, W.L., Dougall, W.C., Tometsko, M.E., Roux, E.R., Teepe, M.C., DuBose, R.F., Cosman, D., and Galibert, L. (1997). A homologue of the TNF receptor and its ligand enhance T-cell growth and dendric–cell function. Nature 390, 175-179.

Baron, R., Ravesloot, J-H., Neef, L., Chakrabotry, C., Chatteerjee, D., Lomri, A., and Horne, W. (1993). Cellular and Molecular Biology of the osteoclast. In: Cellular and Molecular Biology of Bone. (Noda, M., Ed.), pp. 445-495. Academic Press, San Diego.

Bellido, T., Jilka, R.L., Boyce, B.F., Girasole, G., Broxmeyer, H., Dalrymple, S.A., Murry, R., and Manolagas, S.C. (1995). Regulation of interleukin-6, osteoclastogenesis, and bone mass by androgens: The role of the androgen receptor. J. Clin. Invest. 95, 2886-2895.

Bellido, T., Stahl, N., Farruggella, T.F., Borba, V., Yancopoulos, G.D., and Manolagas, S.C. (1996). Detection of receptors for interleukin-6, interleukin-11, leukemia inhibitory factor, oncostatin M, and ciliary neurotropic factor in bone marrow stromal/osteoblastic cells. J. Clin. Invest. 97, 431-437.

Blobel, C.P., Wolfsberg, T.G., Turck, C.W., Myles, D.G., Primakoff, P., and White, J.M. (1992). A potential fusion peptide and an integrin ligand domain in a protein active in sperm-egg fusion. Nature 356, 248-252.

Boyce, B.F., Yoneda, T., Lowe, C., Soriano, P., and Mundy, G.R. (1992). Requirement of pp60$^{c\text{-}src}$ expression for osteoclasts to form ruffled borders and resorb bone in mice. J. Clin. Invest. 90, 1622-1627.

Burger, E.H., Van der Meer, J.W.M., Van de Gevel, J.S., Gribnau, J.C., Thesingh, C.W., and Van Furth, R. (1982). In vitro formation of osteoclasts from long-term cultures of bone marrow mononuclear phagocytes. J. Exp. Med. 156, 1604-1614.

Chambers, T.J. (1985). The pathology of the osteoclast. J. Clin. Pathol. 38, 241-252.

Chambers, T.J., Owens, J.M., Hattersley, G., Jat, P.S., and Nobel M.D. (1993). Generation of osteoclast-inactive and osteoclastogenetic cell lines from the *H-2K^b tsA58* transgenic mice. Proc. Natl. Acad. Sci. USA 90, 5578-5582.

Emery, J.D., McDonnell, P, Burker, M.B., Deen, K.C., Lyn, S., Silverman, C., Dul, E., Appelbaum, R., Eichma, C., DiPrinzio, R., Dodds, R.A., James, I.E., Rosenberg, M., Lee, J.C., and Young, P.R. (1998). Osteoprotegerin is a receptor for the cytotoxic ligand TRAIL. J. Biol. Chem. 273, 14363-14367.

Ezekowitz, R.A.B, Kuhlman, M., Groopman, J.E., and Byrn, R.A. (1989). A human serum mannose-binding protein inhibits in vitro infection by the human immunodeficiency virus. J. Exp. Med. 169, 185-196.

Fermor, B. and Skerry, T.M. (1995). PTH/PTHrp receptor expression on osteoblasts and osteocytes but not resorbing bone surfaces in growing rats. J. Bone Miner. Res. 10, 1935-1943.

Felix, R., Cecchini, M.G., and Fleish, H. (1990). Macrophage colony stimulating factor restores in vivo bone resorption in the op/op osteotropic mouse. Endocrinology 127, 2592-2594.

Fuller, K. and Chambers, T.J. (1987). Generation of osteoclasts in cultures of rabbit bone marrow and spleen cells. J. Cell. Physiol. 132, 441-452.

Gilpin, B.J., Loeche, F., Mattei, M.-G., Engvall, E., Albrechtsen, R., and Wewer U.M. (1998). A novel. secreted form of human ADAM 12 (Meltrin α) provokes myogenesis in vivo. J. Biol. Chem. 273, 157-166.

Girasole, G., Jilka, R.L., Passeri, G., Loswell, S., Boder, G., Williams, D.C., and Manolagas, S.C. (1992). 17β-Estradiol inhibits interleukin-6 production by bone marrow-derived stromal cells and osteoblasts in vitro: A potential mechanism for the antiosteotropic hormone effect of estrogens. J. Clin. Invest. 89, 883-891.

Girasole, G., Passeri, G., Jilka, R.L., and Manolagas, S.C. (1994). Interleukin 11: A new cytokine critical for osteoclast development. J. Clin. Invest. 93, 1516-1524.

Graves, III, L. and Jilka, R.L. (1990). Comparison of bone and parathyroid hormone as stimulators of osteoclast development and activity in calvarial cell cultures from normal and osteotropic (mi/mi) mice. J. Cell. Physiol. 145, 102-109.

Greenfield, E.M., Gornik, S.A., Horowitz, M.C., Donahue, H.J., and Shaw, S.M. (1993). Regulation of cytokine expression in osteoblasts by parathyroid hormone: Rapid stimulation of interleukin-6 and leukemia inhibitory factor mRNA. J. Bone Miner. Res. 8, 1163-1171.

Grigoriadis, A.E., Wang, Z-Q., Gecchini, M.G., Hofstetter, W., Felix, R., Fleish, H.A., and Wagner, E.F. (1994). *c-fos*: A key regulator of osteoclast-macrophage lineage determination and bone remodeling. Science 266, 443-448.

Hagenaars, C.E., Van der Kraan, A.M.M., Kawilarang-de Haas, E.W.M., Visser, J.W.M., and Nijweide, P.J. (1989). Osteoclast formation from cloned pluripotent haemopoietic stem cells. Bone Miner. 6, 187-190.

Hagenaars, C.E., Van der Kraan, A.M.M., Kawilarang-de Haas, E.W.M., Spooncer, E., Dexter, T.M., and Nijweide, P.J. (1990). Interleukin 3-dependent hemopoietic stem cell lines are capable of osteoclast formation in vitro: A model system for the study on osteoclast formation. In: Calcium regulation and bone metabolism. (Cohn, D.V., Glorieux, E.H., and Martin, T.J., Eds.), pp. 280-290. Elsevier Science Publishers, New York.

Hattersley, G. and Chambers, T.J. (1989). Calcitonin receptors as markers for osteoclastic differentiation; correlation between generation of bone-resorptive cells and cells that express calcitonin receptors in mouse bone marrow cultures. Endocrinology 126, 1606-1612.

Hattersley, G., Kerby, J.A., and Chambers, T.J. (1991). Identification of osteoclast precursors in multilineage hematopoietic colonies. Endocrinology 128, 259-262.

Helfrich M.H., Nesbitt, S.A., Dorey, E.L., and Horton, M.A. (1992). Rat osteoclasts adhere to a wide range of RGD (Arg-Gly-Asp) peptide-containing proteins, including the bone sialoproteins and fibronectin, via a β3 integrin. J. Bone Miner. Res. 7, 335-343.

Hilton, D.J., Hilton, A.A., Raicevic, A., Rakar, S., Harrison-Smith, M., Gough, N.M., Nicola, N.A., and Willson, T.A. (1994). Cloning of a murine IL-11 receptor α-chain; requirement for gp130 for high affinity binding and signal transduction. EMBO J. 13, 4765-4775.

Hodgkinson, C.A., Moore, K.J., Nakayama, A., Steingrimsson, E., Copeland, N.G., Jenkins, N.A., and Arnheiter, H. (1993). Mutations at the mouse microphthalmia locus are associated with defects in a gene encoding a novel basic-helix-loop-helix-zipper protein. Cell 74, 395-404.

Hong, M.H., Jin, C.H., Sato, T., Ishimi, Y., Abe, E., and Suda, T. (1991). Transcriptional regulation of the production of the third component of complement (C3) by 1α, 25-dihydroxyvitamin D3 in bone marrow-derived stromal cells (ST2) and primary osteoblastic cells. Endocrinology 129, 2774-2779.

Horne, W.C, Neff, L., Chatterjee, D., Lomri, A., Levy, J.B., and Baron, R. (1992). Osteoclasts express high levels of $pp60^{c\text{-}src}$ in association with intracellular membranes. J. Cell Biol. 119, 1003-1013.

Horowitz, M.C., Coleman, D.L. Flood, P.M., Kupper, T.S., and Jilka, R.L. (1989). Parathyroid hormone and lipopolysaccharide induce murine osteoblastlike cells to secrete a cytokine indistinguishable from granulocyte-macrophage colony-stimulating factor. J. Clin. Invest. 83, 149-157.

Horton, M.A., Lewis, D., Mcnulty, K., Pringle, J.A.S., and Chambers, T.S. (1985). Monoclonal antibodies to osteoclastomas (giant cell bone tumor), definition of osteoclast specific antigens. Cancer Res. 45, 5663-5669.

Horton, M.A. (1988). Osteoclast-specific antigens. ISI Atlas of Science. Immunology 35-43.

Horton, M.H., Dorey, E.L., Nesbitt, S.A., Samanen, J., Ali, F.E., Stadel, J.M., Nichols, A., Greig, R., and Helfrich, M.H. (1993). Modulation of vitronectin receptor-mediated osteoclast adhesion by Arg-Gly-Asp peptide analogs: A structure-function analysis. J. Bone Miner. Res. 8, 239-247.

Horwood, N.J., Udagawa, N., Elliott, J., Grail, D., Okamura, H., Kurimoto, M., Dunn, A.R., Martin, T., and Gillespie, M.T. (1998). Interleukin 18 inhibits osteoclast formation via T cell production of granurocyte macrophage colony-stimulating factor. J. Clin. Invest. 101, 595-603.

Hughes D.E., Dai, A., Tiffee, J.C., Li, H.H., Mundy, G.R. and Boyce, B.F. (1996). Estrogen promote apoptosis of murine osteoclasts mediated by TGF-β. Nat. Med. 2, 1132-1136.

Hughes, M.J., Lingrel, J.B., Krakowsky, J.M., and Anderson, K.P. (1993). A helix-loop-helix transcription factor-like gene is located at the *mi* locus. J. Biol. Chem. 268, 20687-20690.

Hruska, K.A., Rolnick F., Huskey, M., Alvarez, U., and Cheresh, D. (1995). Engagement of the osteoclast integrin αvβ3 by osteopontin stimulates phosphatidylinositol 3-hydroxy kinase activity. Endocrinology 136, 2984-2992.

Ibbotson, K.J., Roodman, G.D., McManus, L.M., and Mundy, G.R. (1984). Identification and characterization of osteoclastlike cells and their progenitors in cultures of feline marrow mononuclear cells. J. Cell Biol. 99, 471-480.

Ikeda, T., Nomura, S., Yamaguchi, A., Suda, T., and Yoshiki, S. (1992). In situ hybridization of bone matrix proteins in undecalcified adult rat bone section. J. Histochem. Cytochem. 40, 1079-1088.

Ishimi, Y., Miyaura, C., Jin, C.H., Akatsu, T., Abe, E., Nakamura, Y., Yamaguchi, A., Yoshiki, S., Matsuda, T., Hirano, T., Kishimoto, T., and Suda, T. (1990). IL-6 is produced by osteoblasts and induce bone resorption. J. Immunol. 145, 3297-3303.

Ishimi, Y., Abe, E., Jin, C.H., Miyaura, C., Hong, M.H., Oshida, M., Kurosawa, H., Yamaguchi, Y., Yamaguchi, Y., Tomida, M., Hozumi, M., and Suda, T. (1992). Leukemia inhibitory factor/differentiation-stimulating factor (LIF/D-factor): Regulation of its production and possible roles in bone metabolism. J. Cell. Physiol. 152, 71-78.

Jamieson, J.C., Wayne, S., Belo, R.S., Wright, J.A., and Spearman, M. (1992). The importance of N-linked glycoproteins and dolichyl phosphate synthesis for fusion of L6 myoblasts. Biochem. Cell Biol. 70, 408-412.

Jilka, R.L., Hangoc, G., Girasole, G., Passeri, G., Williams, D.C., Adams, J.S., Boyce, B., Brocmeyer, H., and Manolagas, S.C. (1992) Increased osteoclast development after estrogen loss: Mediation by interleukin-6. Science 257, 88-91.

Jin, C. H., Shinki, T., Hong, M. H., Sato, T., Yamaguchi, A., Ikeda, T., Yoshiki, S., Abe, E., and Suda, T. (1992). 1α, 25-Dihydroxyvitamin D3 regulates in vivo production of the third component of complement (C3) in bone. Endocrinology 131, 2468-2475.

Johnson R.S, Spiegelman, B.M., and Papaioannoou, V. (1992). Pleiotropic effects of a null mutation in the *c-fos* proto-oncogene. Cell 71, 577-586.

Kaji, H., Sugimoto, T., Kanatani, M., Fukase, M., Kumegawa, M., and Chihara, K. (1996). Prostaglandin E_2 stimulates osteoclastlike cell formation and bone-resorbing activity via osteoblast: Role of cAMP-dependent protein kinase. J. Bone Miner. Res. 11, 62-71.

Kanatani, M., Sugimoto, T., Kaji, Kobayashi, T., Nishiyama, K., Fukase, M., Kumegawa, M., and Chihara, K. (1995). Stimulatory effect of bone morphogenetic protein-2 on osteoclastlike cell formation and bone-resorbing activity. J. Bone Miner. Res. 10, 1681-1690.

Kato, S., Sekine, K., Matsumoto, T., and Yoshizawa, T. (1998). Molecular genetics of vitamin D receptor acting in bone. J Bone Miner. Metab. 16, 65-71.

Kawaguchi, H., Nemoto, K., Raisz, L.G., Harrison, J.R., Voznesensky, O.S., Alander, C.B., and Pilbeam, C.C. (1996). Interleukin-4 inhibits prostaglandin G/H synthase-2 and cyclic phosphodiesterase A_2 induction in neonatal mouse parietal bone cultures. J. Bone Miner. Res. 11, 358-366.

Kerby, J.A., Hattersley, G., Collins, D.A., and Chambers, T.J. (1992). Derivation of osteoclasts from hematopoietic colony-forming cells in culture. J. Bone Min. Res. 7, 353-362.

Kimble, R.B., Vannice, J.L., Bloedow, D.C., Thompson R.C., Hopfer, W., Kung, V.T., Brownfield, C., and Pacifici, R. (1994) Interleukin-1 receptor antagonist decreases bone loss and bone resorption in ovariectomized rats. J. Clin. Invest. 93, 1959-1967.

Kishimoto, T., Akira, S., Narazaki, M., and Taga, T. (1995). Interleukin-6 family of cytokines and gp130. Blood 86, 1243-1254.

Kodama, H., Yamaaki, A Nose, M., Niidam S., Ohgame, Y., Abe, M., Kumegawa, M., and Suda, T. (1991). Congenital osteoclast deficiency in osteopetrotic (op/op) mice is cured by injections of macrophage colony-stimulating factor. J. Exp. Med. 173, 269-272.

Kodama, Y., Takeuchi, Y., Murayama, H., Yamato, H., and Matsumoto, T. (1995). Deficiency in interleukin-11 action is involved in impaired osteoblastic differentiation of marrow cells from senescence-accelerated mouse. J. Bone Miner. Res. (Abstract 16).

Kondo, M., Takeshita, T., Ishii, N., Nakamura, M., Watanabe, S., Arai, K., and Sugamura, K. (1993). Sharing of the interleukin 2 (IL-2) receptor γ chain between receptors for IL-2 and IL-4. Science 262, 1874-1880.

Kukita, A., Kukita, T., Shin., J.H., and Kohashi, O. (1993). Induction of mononuclear precursor cells with osteoclastic phenotype in a rat bone marrow culture system depleted stromal cells. Biochem. Biophys. Res. Commun. 196, 1383-1389.

Kurachi, T., Morita, I., and Murota, S. (1993). Involvement of adhesion molecules LFA-1 and ICAM-1 in osteoclast development. Biochem. Biophys. Acta. 13, 259-266

Kurachi, T., Morita, I., Oki, T., Ueki, T., Sakaguchi, K., Enomoto, S., and Murota, S. (1994). Expression on outer membranes of mannose residues, which are involved in osteoclast formation via cellular fusion events. J. Biol. Chem. 269, 17572-17576.

Kurihara, N., Suda, T., Miura, Y., Nakauchi, H., Kodama, H., Hiura, K., Hakeda, Y., and Kumegawa, M. (1989). Generation of osteoclasts from isolated hematopoietic progenitor cells. Blood 74, 1295-1302.

Kurihara, N., Chunu, C., Miller, M., Civin, C., and Roodman, G.D. (1990). Identification of committed mononuclear precursors for osteoclastlike cells formed in long-term human bone marrow cultures. Endocrinology 126, 2733-2741.

Lacey, D.L., Timms, E., Tan, H.-L., Kelley, M.J., Dunstan, C.R., Burgess, T., Elliott, R., Colombero, A., Elliott, G., Scully, S., Hsu, H., Sullivan, J., Hawkins, N., Davy, E., Capparelli, C., Eli, A.,

Qian, Y.-X., Kaufman, S., Sarosi, I., Shalhoub, V., Senaldi, G., Guo, J., Delaney, J., and Boyle, W.J. (1998) Osteoprotegerin ligand is a cytokine that regulates osteoclast differentiation and activation. Cell 93, 165-176.

Lader, C.S. and Flanagan, A.M. (1998). Prostaglandin E2, interleukin 1α, and tumor necrosis factor-α increase human osteoclast formation and bone reosorption in vitro. Endocrinology 139, 3157-3164.

Lowik, C., Pluijum G.M., Bloys, H., Hoekman, K., Bijvoet, O., Aarden, L., and Papapoulos, S. (1989). Parathyroid hormone and PTH-like protein stimulate interleukin-6 in osteoclastogenesis. Biochem. Biophys. Res. Commun. 162, 1546-1552.

MacDonald, B.R., Takahashi, N., McManus, L.M., Holahan, J., Mundy, G.R., and Roodman, G.D. (1987). Formation of multinucleated cells that respond to osteotropic hormones in long-term human bone marrow cultures. Endocrinology 120, 2326-2333.

Marks, C.R., Seifert, M.F., and Marks III, S.C. (1984). Osteoclast populations in congenital osteopetrosis: Additional evidence of heterogeneity. Metab. Bone Dis. Rel. Res. 5, 259-264.

Marks, S.C.Jr. and Walker, D.G. (1976). Mammalian osteopetrosis—a model for studying cellular and humoral factors in bone resorption. In: The Biochemistry and Physiology of Bone. (Bourne, G.H., Ed.), pp.227-301. Academic Press, New York.

Marks, Jr., S.C., Seifert, M.F., and McGuire, J.L. (1984). Congenitally osteopetrotic (op/op) mice are not cured by transplants of spleen cells or bone marrow cells from normal littermates. Metab. Bone Dis. Relat. Res. 5, 183-186.

Marks, Jr., S.C.(1989) Osteoclast biology: Lessons from mammalian mutations. Am. J. Med. Genet. 34, 43-54.

Mata, J.D.L., Uy, H.L., Guise, T.A., Story, B., Boyce, B., and Mundy, G. (1995). Interleukin-6 enhances hypercalcemia and bone resorption mediated by parathyroid hormone-related protein in vivo. J. Clin. Invest. 95, 2846-2852.

McNally, A.K. and Anderson, J.M. (1994). Complement C3 participation in monocyte adhesion to different surfaces. Proc. Natl. Acad. Sci. USA 91, 10119-10123.

Miles, D.G., Kimmel, L.H., Blobel, C.P., White, J.M., and Primakoff, P. (1994). Identification of a binding site in the disintegrin domain of fertilin required for sperm-egg fusion. Proc. Natl. Acad. Sci. USA 91, 4195-4198.

Miyaura, C., Kusano, K., Masuzawa, T, Chaki, O., Onoue, Y., Aoyagi, M., Sasaki, T., Tamura, T., Koishihara, Y., Ohsugi, Y., and Suda, T. (1995). Endogenous bone-resorbing factors in estrogen deficiency: Cooperative effects of IL-1 and IL-6. J. Bone Miner. Res. 10, 1365-11373.

Miyamoto, A., Kunisada, T., Hemmi, H., Yamana, T., Yasauda, H., Miyake, K., Yamazaki, H., and Hayashi, S.-I. (1998). Establishment and characterization of an immortal macrophagelike cell line inducible to differentiate to osteoclasts. Biochem. Biophys. Res Commun. 242, 703-709.

Mocharla, H., Yamate, T., Taguchi, Y., O'Brian, C.A., Manolagas, S.C., and Abe, E. (1998). α-Meltrin, a new protein involved in multinucleated giant cell and osteoclast formation. J. Bone Miner. Res. 11, S140.

Mochizuki, H., Hakeda, Y., Wakazuki, N., Usui, N., Akashi, S., Sato, T., Tanaka, K., and Kumegawa, M. (1992). Insulin-like growth factor-I supports formation and activation of osteoclasts. Endocrinology 131, 1075-1080.

Nakamura, H., Kenmotsu, S., Sakai, H., and Ozawa, H. (1995). Localization of CD44, the hyaluronate receptor, on the plasma membrane of osteocytes and osteoclasts in rat tibiae. Cell Tissue Res. 280, 225-233.

Nakano, Y., Watanabe, K., Morimoto, I., Okada, Y., Ura, K., Sato, K., Kasono, K., Nakamura, T., and Eto, S. (1994). Interleukin-4 inhibits spontaneous and parathyroid hormone related protein-stimulated osteoclast formation in mice. J. Bone Miner. Res. 9, 1533-1539.

Nilsson, S.K., Lieschke, G.L., Garcia-Wijnen, C.C., Williams B., Tzelepis, D., Hodgson, G., Grail, D., Dunn, A.R., and Bertoncello, I. (1995). Granulocyte-macrophage colony-stimulating factor is not responsible for the correction of hematopoietic deficiencies in the maturing op/op mouse. Blood 86, 66-72.

Nishimura, R., Yasukawa, K., Mundy, G.R., and Yoneda, T. (1998). Combination of interleukin-6 and soluble interleukin-6 receptors induces differentiation and activation of JAK-STAT and MAP kinase pathways in MG-63 human osteoblastic cells. J. Bone Miner. Res. 13, 777-785.

Noda, M., Yoon, K., Prince, C.W., Butler, W.T., and Rodan, G.A. (1988). Transcriptional regulation of osteopontin production in rat osteosarcoma cells by TGF β transforming growth factor.J. Biol. Chem. 263, 13916-13921.

Onoue, Y., Miyaura, C., Kaminakayashiki, T., Nagai, Y. Noguchi, K., Chen, Q-R., Seo, H., Ohta, H., Nozawa, S., Kudo, I., and Suda, T. (1996). IL-13 and IL-4 inhibit bone resorption by suppressing cycloxygenase-2-dependent prostaglandin synthesis in osteoblasts. J. Immunol. 156, 758-764.

Pacifici, R., Brown, C., Puscheck, E., Friederick, E., McCracken, R., Maggio, D., Slatopolsky, E., and Avioli, L.V. (1991) Effect of surgical menopause and estrogen replacement on cytokine release from human blood mononuclear cells. Proc. Natl. Acad. Sci. USA 88, 5134-5138.

Parfitt, A.M. (1996). A new model for the regulation of bone resorption, with particular reference to the effects of bisphosphonates. J. Bone Miner. Res. 11, 150-159.

Paul, S.R., Bennett, F., Calvetti, J., Kelleher, K., Wood, C.R., Ohara, Jr., R.M., Leary, A.C., Sibley, B., Clark, S.C., Williams, D.A., and Yang, Y-C. (1990). Molecular cloning of a cDNA encoding interleukin 11, a stromal cell-derived lymphopoietic and hematopoietic cytokine. Proc. Natl. Acad. Sci. USA 87, 7512-7516.

Pike, J.W., Sone, T., Ozono, K., Kesterson, R.A., and Kerner, S.A. (1993). The osteocalcin gene as a molecular model for tissue-specific expression and 1, 25-dihydroxyvitamin D3 regulation. In: Cellular and Molecular Biology of Bone. (Noda, M., Ed.), pp.235-256. Academic Press, San Diego.

Romas, E., Udagawa, N., Zhou, H., Tamura, T., Saito, M., Taga, Suda, T., Hilton, D.J., Ng, K.W., and Martin, T.J. (1995). The role of gp130-mediated signals in osteoclast development: Regulation of interleukin 11 production by osteoblasts and distribution of its receptor in bone marrow cultures. J. Exp. Med. 183, 2581-2591.

Roodman, G.D., Ibbotson, K.J., MacDonald, B.R., Kuehl, T.J., and Mundy, G.R. (1985). 1, 25-Dihydroxyvitamin D_3 causes formation of multinucleated cells with several osteoclast characteristics in cultures of primate marrow. Proc. Natl. Acad. Sci. USA 82, 8213-8217.

Rottling, S.R., Matsumoto, H.N., McKee, M.D., Nanci, A., An, X., Novick, K.E., Kowalski, A.J., Noda, M., and Denhardt, D.T. (1998). Mice lacking osteopontin show normal development and bone structure but display altered osteoclast formation in vitro. J. Bone Miner. Res. 13, 1101-1111.

Rouleau, M.F., Mitchell, J., and Golzman, D. (1990). Characterization of the major parathyroid hormone target cell in endometaphysis of rat long bones. J. Bone Miner. Res. 5, 1043-1053.

Russell, S.M., Keegan, A.D., Harada, N., Nakamura, Y., Noguchi, M., Leland, P., Friemann, M.C., Miyazima, A., Puri, R.K., Paul, W.E., and Leonald, W.L. (1993). Interleukin-2 receptor g chain: a functional component of the interleukin-4 receptor. Science 262, 1880-1883.

Sato, T., Abe, E., Jin, C. H., Hong, M. H., Katagiri, T., Kinoshita, T., Amizuka, N., Ozawa, H., and Suda, T. (1993). The biological roles of the third component of complement (C3) in osteoclast formation. Endocrinology 133, 397-404.

Sato, T., Morita, I., Şakaguchi, K., Nakahama, K-I., Smith, W., Dewitt, D.L., and Murota, S-I. (1996). Involvement of prostaglandin endoperoxide H syntase-2 in osteoclastlike cell formation by interleukin-1β. J. Bone Miner. Res. 11, 392-400.

Scott, E.W., Simon, C., Anastasi, J., and Harinder, S. (1994). Requirement of transcription factor PU.1 in the development of multiple hematopoietic lineages. Science 265, 1573-1577.

Shin, J.H., Kukita, A., Ohki, K., Katsuki, T., and Kohashi, O. (1995). In vitro differentiation of the murine macrophage cell line BDM-1 into osteoclastlike cells. Endocrinology 136, 4285-4292.

Shinar, D.M., Sato, M., and Rodan, G.A. (1990). The effect of hemopoietic growth factors on the generation of osteoclastlike cells in mouse bone marrow cultures. Endocrinology 126, 1728-1735.

Shioi, A., Teitelbaum, S.L., Ross, F.P., Welgus, H.G., Suzuki, H., Ohara, J., and Lacey, D.L. (1991). Interleukin 4 inhibits murine osteoclast formation in vitro. J. Cell. Biochem. 47, 272-277.

Simonet, W.S., Lacey, D.L., Dunstan, C.R., Kelley, M., Chang, M.-S., Luthy, R., Nguyen, H.Q., Wooden, S., Bennett, L., Boone, T., Shimamoto, G., DeRose, M., Elliott, R., Colombero, A., Tan, H.-L., Trail, G., Sullivan, J., Davy, E., Bucay, N., Renshaw-Gegg, L., Hughes, T.M., Hill, D., Patteson, W., Cambell, P., Sander, S., Van, G., Tarpley, J., Derby, P., Lee, R., Amgen EST program, and Boyle, W.J. (1997). Osteoprotegerin: a novel secreted protein involved in the regulation of bone density. Cell 89, 309-319.

Soriano, P., Montgomery, C., Geske, R., and Bradley, A. (1991). Targeted disruption of the *c-src* proto-oncogene leads to osteopetrosis in mice. Cell 64, 693-702.

Stahl, N., Boulton, T.G., Farruggella, T., Davis, S., Witthuhn, B.A., Quelle, F.W., Silvennonen, G., Barbieri, S., Pellegrini, J., Ihle, J.N., and Yancopoulos, G.D. (1994). Association and activation of Jak-Tyk kinases by CNTF-LIF-OSM-IL6β receptor component. Science 263, 92-95.

Stahl, N., Farruggella, T.J., Boulton, T.G., Zhong, Z., Darnell, Jr., J.E., and Yancopoulos, G.D. (1995). Choice of STATs and other substrates specified by molecular tyrosine-based motifs in cytokine receptors. Science 267, 1349-1353.

Suda, T., Takahashi, N., and Martin, T.J. (1992). Modulation of osteoclast differentiation. Endocr. Rev. 13, 66-80.

Suda, T., Takahashi, N., and Martin T.J. (1995). Modulation of osteoclast differentiation: Update 1995. Endocrine Rev. 4, 266-270.

Suda, T., Nakamura, I., Jimi, E., and Takahashi, N. (1997). Regulation of osteoclast function. J. Bone Miner. Res., 12, 869-879.

Tachibana, M., Hara, Y., Vyas, Y., Hodgikinson, C., Fex, J., Grundfast, K., and Arnheiter, M. (1992). Cochelear disorder associated with melanocyte anatomy in mice with a transgenic insertion mutation. Mol. Cell. Neurosci. 3, 433-445.

Taguchi, Y., Yamamoto, M., Yamate, T., Lin, S.-C., Mocharla, H., Detogni, P., Nakayama N., Boyce, B.F., Abe, E., and Manolagas, S.C. (1998). IL-6–type cytokines stimulate mesenchymal progenitor differentiation toward the osteoblastic lineage. Proc. Assoc. Amer. Phys. (In press).

Takahashi, N., Akatsu, T., Sasaki, T., Nicholson, G.C., Moseley, J.M., Martin, T.J., and Suda, T. (1988a). Induction of calcitonin receptors by 1α, 25-dihydroxyvitamin D_3 in osteoclastlike multinucleated cells formed from mouse bone marrow cells. Endocrinology 123, 1504-1510.

Takahashi, N., Akatsu, T., Udagawa, N., Sasaki, T., Yamaguchi, A., Moseley, J.M., Martin, T.J., and Suda, T. (1988b). Osteoblastic cells are involved in osteoclast formation. Endocrinology 123, 2600-2602.

Takahashi, N., Udagawa, N., Akatsu, T., Yanaka, H., Shionome, M., and Suda, T. (1991). Role of colony-stimulating factors in osteoclast development. J. Bone Min. Res. 6, 977-985.

Takahashi, S., Reddy, S.V., Dallas, M., Devlin, R., Chou, J.Y., and Roodman, G.D. (1995a). Development and characterization of a human marrow stromal cell line that enhances osteoclastlike cell formation. Endocrinology 136, 1441-1449.

Takahashi, S., Goldring, S., Katz, M., Hilsenbeck, S., Williams, R., and Roodman G.D. (1995b). Downregulation of calcitonin receptor mRNA expression by calcitonin during human osteoclastlike cell differentiation. J. Clin. Invest. 95, 167-171.

Tamura, T., Udagawa, N., Takahashi, N., Miyaura, C., Tanaka, S., Yamada, Y, Koishihara, Y., Ohsugi, Y., Kumaki, K., Taga, T., Kishimoto, T., and Suda, T. (1993). Soluble interleukin-6 receptor triggers osteoclast formation by interleukin 6. Proc. Natl. Acad. Sci. USA 90, 11924-11928.

Tanaka, S., Takahashi, N., Udagawa, N., Murakami, H., Nakamura, I., Kurokawa, T., and Suda, T. (1995a). Possible involvement of focal adhesion kinase, $p125^{FAK}$, in osteoclastic bone resorption. J. Cell. Biochem. 58, 424-435.

Tanaka, S., Amling, M., Neff, L., Peyman A., Uhlmann, E., Levy, J.B., and Baron, R. (1995b) c-Cbl is downstream of c-Src in a signalling pathway necessary for bone resorption. Nature 383, 528-531.

Testa, N.G., Allen, T.D., Lajtha, L.G., Onions, D, and Jarret, O. (1981) Generation of osteoclasts in vitro. J. Cell Sci. 47, 127-137.

Thesingh, C.W. and Scherft, J.P. (1985). Fusion disability of embryonic osteoclast precursor cells and macrophages in the microphthalmic osteopetrotic mouse. Bone 6, 43-52

Tondravi, M.M., Erdmann, J.M., Mckercher, S., Anderson, K., Maki, R., and Teitelbaum, S.L. (1995). Novel osteopetrosis mutation caused by the knock-out of the hematopoietic transcription factor PU.1 in transgenic mice. J. Bone Miner. Res. (Abstract 148).

Tondravi, M.M., Mckercher, S.R., Anderson, K., Erdmann, J.M., Quiroz, M., Maki, R., and Teitelbaum, S.L. (1997). Osteopetrosis in mice lacking haematopoietic transcription factor PU.1. Nature 386, 81-84.

Tong, H-S., Sakai D.D., Sims, S.M., Dixon, J., Yamin, M., Goldring, S.R., Snead, M.L., and Minkin, C. (1994). Murine osteoclasts and spleen cell polykaryons are distinguished by mRNA phenotyping. J. Bone Min. Res. 9, 577-583.

Tsujinaka, T., Fujita, J., Ebisu, C., Yano, M., Kominami, E., Suzuki, K., Tanaka, K., Katsume, A., Ohsugi, Y., Shiozaki, H., and Monden, M. (1996). Interleukin 6 receptor antibody inhibits muscle atrophy and modulates proteolytic systems in interleukin 6 transgenic mice. J. Clin. Invest. 97, 244-249.

Udagawa, N., Takahashi, N., Akatsu, T., Sasaki, T., Yamaguchi, A., Kodama, H., Martin, T.J., and Suda, T. (1989) The bone marrow-derived stromal cell lines MC3T3-G2/PA6 and ST2 support osteoclastlike cell differentiation in co-cultures with mouse spleen cells. Endocrinology 125, 1805-1813.

Udagawa, N., Takahashi, N., Akatsu, T., Tanaka, H., Sasaki, T., Nishihara, T., Koga, T., Martin, T.J., and Suda, T. (1990). Origin of osteoclast: Mature monocyte and macrophages are capable of differentiating into osteoclasts under a suitable microenvironment prepared by bone marrow-derived stromal cells. Proc. Natl. Acad. Sci. USA 87, 7260-7264.

Udagawa, N., Takahashi, N., Katagiri, T., Tamura, T., Wada, S., Findlay, D.M., Martin, T.J., Hirota, H., Taga, T., Kishimoto, T., and Suda, T. (1995). Interleukin (IL)-6 induction of osteoclast differentiation depends on IL-6 receptors expressed on osteoblastic cells but not on osteoclast progenitors. J. Exp. Med. 182, 1461-1468.

Vinson, C.R. and Garcia, K.C. (1992). Molecular model for DNA recognition by the family of basic helix-loop-helix-zipper proteins. The New Biologist 4, 396-403.

Voso, M.T., Burn, T.C., Wulf, G., Lim, B., Leone, G., and Tenen, D.G. (1994) Inhibition of hematopoiesis by competitive binding of transcription factor PU.1. Proc. Natl. Acad. Sci. USA 91, 7932-7936.

Weber, G.F., Ashkar, S., Glimcher, M.J., and Cantor, H. (1996). Receptor-ligand interaction between CD44 and osteopontin (eta-1). Science 271, 509-512.

Wesolowski, G., Duong, L.T., Lakkakirpi, P.T., Magy, R.M., Tezuka, K-I., Tanaka, H., Rodan G.A., and Rodan, S.B. (1995). Isolation and characterization of highly enriched, perfusion mouse osteoclastic cells. Exp. Cell Res. 219, 679-686.

Wiktor-Jedrzejczak, W., Barticci, A., Ferrante, Jr., A.W., Ahmed-Ansari, A., Sell, K.W., Pollard, J.W., and Stanley, E.R. (1990). Total absence of colony-stimulating factor I in the macrophage-deficient osteopetrotic (op/op) mouse. Proc. Natl. Acad. Sci. USA 87, 4828-4832.

Xu, L-X., Kukita, T., Kukita, A., Otsuka, T., Niho, Y., and Iijima, T. (1995). Interleukin-10 selectively inhibits osteoclastogenesis by inhibiting differentiation of osteoclast progenitors into preosteoclastlike cells in rat bone marrow culture system. J. Cell. Physiol. 165, 624-629.

Yamashita, T., Asano, K., Takahashi, N., Akatsu, T., Udagawa, N., Sasaki, T., Martin, T.J., and Suda, T. (1990) Cloning of an osteoblastic cell line involved in the formation of osteoclastlike cells. J. Cell. Physiol. 145, 587-595.

Yamate, T., Mocharla, H., Taguchi, Y., Igietseme, J.U., Manolagas, S.C., and Abe, E., (1997). Osteopontin expression by osteoclast and osteoblast progenitors in the murine bone marrow: Demonstration of its requirement for osteoclastogenesis and its increase after ovariectomy. Endocrinology 138, 3047-3055.

Yagami-Hirosawa, T., Sata, T., Kurisaki, T., Kamijo, K., Nabeshima, Y-I., and Fujisawa-Sehara, A. (1995). A metalloprotease-disintegrin participating in myoblast fusion. Nature, 377, 652-656.

Yang, L. and Yang, Y-C. (1994). Regulation of interleukin (IL)-11 gene expression in IL-1-induced primate bone marrow stromal cells. J. Biol. Chem. 269, 32732-32739.

Yasuda, H., Shima, N., Nakagawa, N., Yamaguchi, K., Kinosaki, M., Mochizuki, A.-I., Tomoyasu, A., Yano, K., Goto, M., Murakami, A., Tsuda, E., Morinaga, T., Higashio, K., Udagawa, N., Takahashi, N., and Suda, T. (1998b). Osteoclast differentiation factor is a ligand for osteoprotegerin/osteoclastogenesis-inhibitory factor is identical to TRANCE/RANKL. Proc. Natl. Acad. Sci. USA 95, 3597-3602.

Yasumoto, K-I., Yokoyama, K., Shibata, K., Tomita, Y., and Shibayama, S. (1994). Microphthalmia-associated transcription factor as a regulator for melanocyte-specific transcription of the human tyrosinase gene. Mol. Cell. Biol. 14, 8058-8070.

Yoshida, H., Hayashi, S.,Kunisada, T., Pgawa, M., Nishikawa, S., Okamura, H., Sudo, T., Shultz, L.D., and Nishikawa, S. (1990). The murine mutation osteopetrosis is in the colony-stimulating factor gene. Nature 345, 442-444.

Zurawski, S.M., Vega, Jr., F., Huyghe, B., and Zurawski, G. (1993). Receptors for interleukin-13 and interleukin-4 are complex and share a noble component that functions in signal transduction. EMBO J. 12, 2663-2670.

OSTEOCLAST INTEGRINS: ADHESION AND SIGNALING

Geetha Shankar and Michael Horton

Advances in Organ Biology
Volume 5B, pages 315-329.

ISBN: 0-7623-0390-5

I. INTRODUCTION: ADHESION EVENTS IN BONE

Osteoclasts are responsible for the normal and pathological breakdown of the extracellular matrix of bone (Rifkin and Gay, 1992). This process involves a series of stages which include the proliferation and homing to bone of hemopoietic precursors, their differentiation into cells which express features of mature osteoclasts, multinucleation by cell fusion, and migration of osteoclasts to the area of bone to be remodeled. Osteoclasts attach to the bone surface and polarize to create three areas of plasma membrane: the basolateral membrane, which faces the marrow space and is not in contact with the bone; the tight sealing zone (or clear zone), which is closely applied to the bone surface; and the ruffled border, the highly convoluted plasma membrane which faces the bone matrix and is surrounded by the sealing zone. The sealing zone forms a diffusion barrier and allows the localized accumulation of high concentrations of acid and proteases; these are secreted via the ruffled border into the space underneath the cell. Many of these steps can be postulated to involve adhesion between developing or mature osteoclasts, other cell types in bone and the extracellular matrix of bone.

The best defined of these adhesive interactions are mediated by members of a particular class of cell adhesion molecule, the integrins. Integrin receptors are now known to be major functional proteins of osteoclasts. The rest of the chapter reflects this bias and two main topics are covered by this review. The first concerns the role of integrins in osteoclast adhesive processes during bone resorption; this is of particular interest as considerable effort is being made in the pharmaceutical industry to develop drugs which may reduce osteoclastic resorption (in, for example, osteoporosis) by modifying integrin-ligand interaction. Second, the process of signal transduction via integrins in osteoclasts is reviewed and set in the context of its possible relevance to the regulation of osteoclast function.

II. ADHESION PROTEINS, INTEGRINS, AND THEIR LIGANDS: AN OVERVIEW

Molecular and immunological approaches have led to considerable advances in our understanding of the range of cell membrane molecules which are capable of mediating cell adhesion. Detailed sequence and structural analysis (reviewed in Barclay et al., 1993) has enabled them to be grouped into families, with related structure based upon their content of highly homologous domains. An example of such a family are the integrins.

Integrins (Hynes, 1992) are heterodimeric proteins whose constituent polypeptide chains, α and β, are noncovalently linked. So far, 16 different mammalian α subunits and eight β subunits have been identified, forming 22 distinct heterodimers. Both integrin subunits are transmembrane, N-glycosylated glycoproteins with a large extracellular domain, a single hydrophobic transmembrane region and, generally, a short cytoplasmic domain. Electron microscopy of several purified integrin dimers shows an extended structure with dimensions of approximately 10 by 20 nm, and formed by an N-terminal globular head, composed by the association of the two subunits, connected to the membrane by two stalks.

Analysis of the cDNA sequences of the α subunits reveals several common features. All contain seven homologous, tandem repeat sequences, the last three or four containing putative divalent cation-binding sites showing similarity to the EF-hand loop structure seen in calmodulin. These sites are of critical importance to both ligand binding and subunit association. Some integrins (e.g., the β_2-associated leukocyte function-associated antigen α chains, and α_2) contain an inserted, or I, domain between the second and third repeats; this shows homology to procollagen and is involved in ligand binding. β subunits have a high cysteine content largely concentrated in four segments which are internally disulphide bonded. Other conserved regions include the PEGG domain of unknown function, which is absolutely conserved from mammals to invertebrates. Cross-linking studies (e.g., using radioactively-labeled RGD peptide probes for $\alpha_v\beta_3$ integrin, the vitronectin receptor (see Mould and Humphries, 1995), and analysis of experimental gene mutations and patients with Glanzmann's Syndrome (an inherited deficiency of the platelet integrin α IIbβ3), have shown that the ligand binding site is composed of distinct, relatively short elements in the N-termini of both the α and β subunits. When taken with the requirement for an I domain for ligand binding in some integrins, this suggests that the ligand interaction site depends upon the composite structure formed by both of the two chains of the receptor, with specificity reflecting subunit usage.

A range of structural domains are also identifiable within the ligands recognized by integrin receptors. Some integrins, including those in bone (*vide infra*) interact with the well characterized Arg-Gly-Asp (RGD) peptide motif, originally described in the protein fibronectin and now known to be present widely in many extracellular matrix proteins (Mould and Humphries, 1995). These include the some of the major noncollagenous proteins synthesized by bone cells, in addition to plasma proteins passively adsorbed by the mineral phase of bone.

A large number of functions have been ascribed to the cell adhesive activity of integrin molecules in a wide range of cells and tissues. Signal transduction has been increasingly shown to be mediated by many classes of adhesion receptors, including integrins. The theme that these activities also occur in bone is developed further in this chapter.

III. CELL ADHESION MOLECULES IN OSTEOCLASTS

There is recent and fairly extensive literature on the expression of cell adhesion molecules by the stromal and matrix-forming components of the skeleton: osteoblasts, osteocytes, and chondrocytes. For each cell type, a number of receptors, including integrins, have been detected but there is no clear consensus as to the molecular phenotype, let alone their functional attributes. In contrast, there is a clearer picture for osteoclasts. Here only three integrins have been described and there is little evidence for expression of other adhesion proteins by mature osteoclasts, though they may well be in developing osteoclast precursors (see below). Moreover, there is a strong functional correlate by which means antagonism of osteoclast integrins leads to a downregulation of osteoclastic bone resorption, an effect with clinical implications (*vide infra*).

A. Integrins and Osteoclastic Bone Resorption

The first suggestions that adhesion receptors played a functional role in osteoclastic bone resorption was obtained from two lines of evidence. First, monoclonal antibody 13C2 (Horton et al., 1985) was found to inhibit bone resorption *in vitro* by human osteoclasts from giant cell tumour of bone (osteoclastoma) (Chambers et al., 1986). Second, Beckstead et al. (1986) and Horton (1986) demonstrated that osteoclasts express the platelet glycoprotein, gpIIIa (now identified as the integrin β3 chain). It was established later that the antibody 13C2 recognized the vitronectin receptor ($\alpha_v\beta_3$), a member of the integrin family of cell adhesion molecules (Davies et al., 1989). Subsequent detailed phenotypic (Horton and Davies, 1989; Hughes et al., 1993; Clover et al., 1992; reviewed in Horton and Rodan, 1996) and biochemical analysis (Nesbitt et al., 1993) has demonstrated that mammalian osteoclasts express three integrin dimers: $\alpha_v\beta_3$, the classical vitronectin receptor; $\alpha_2\beta_1$, a collagen/laminin receptor; and $\alpha_v\beta_1$, a further vitronectin receptor (data summarized in Horton and Rodan, 1995). There have been occasional reports showing the expression of some other integrins (α3, β5) but this has not been a general finding (see Horton and Rodan, 1996 for details). Some differences have been noted with avian osteoclasts which express $\alpha_5\beta_1$ and $\alpha_v\beta_5$ in addition to $\alpha_v\beta_3$ (see Ross et al, 1993; Horton and Rodan, 1996), and possibly β_2 integrins (Athanasou et al., 1992).

Adhesion of osteoclasts to the bone surface involves the interaction of integrins with extracellular matrix proteins within the bone matrix. This has been studied in *in vitro* adhesion assays using cells from several species (Flores et al., 1992; Helfrich et al., 1992; Ross et al., 1993; reviewed in Horton and Rodan, 1996). The vitronectin receptor mediates RGD-peptide dependent adhesion to a wide variety of proteins containing the RGD sequence, including the bone sialoproteins and several plasma proteins. In addition, mammalian, but not avian, osteoclasts (Ross et al., 1993) adhere to type I collagen but utilize a different integrin, $\alpha_2\beta_1$ (Helfrich et al., 1996). Interestingly, osteoclast integrin-mediated adhesion to collagen is sensitive to RGD peptides, unlike collagen-binding by integrins of other cells (reviewed in Helfrich et al., 1996).

The demonstration that antibodies recognizing the vitronectin receptor block osteoclast adhesion suggested that it may be possible to influence bone resorption *in vitro*, either by RGD-containing peptides or function-blocking antibodies to osteoclast integrins (reviewed in Horton and Rodan, 1996). The observation that the

RGD-sequence containing snake venom protein, echistatin, blocked bone resorption confirmed this hypothesis (Sato et al., 1990). Subsequently, these findings have been extended using RGD peptides, peptidomimetic agents, other snake venom proteins and antibodies to α_v and β_3 components of the vitronectin receptor in a variety of *in vitro* systems and using osteoclasts from several species (reviewed in Horton and Rodan, 1996). Osteoclasts also express the β_1 integrins, $\alpha_2\beta_3$ and $\alpha_v\beta_1$; antibodies to β_1 and α_2 inhibit bone resorption in isolated osteoclast assays, presumably by modifying interactions with collagen (see Helfrich et al, 1996).

Recently the snake venom proteins, echistatin and kistrin, have both been shown to induce hypocalcemia in rats *in vivo* (Fisher et al., 1993; King et al., 1994), the former in the parathyroid hormone-infused thyroparathyroidectomy model and the latter in parathyroid hormone-related protein-induced hypercalcemia. Small cyclic RGD-containing peptides and peptidomimetics and antibodies to the rat β_3 integrin (Crippes et al., 1996) have also recently been shown to induce hypocalcaemia in the former model. The results with anti-integrin antibodies, taken with the RGD-sequence specificity observed with non-RGD sequence-containing variants of echistatin (Fisher et al., 1993; Sato et al., 1994) suggest that integrins are mediating the hypocalcemic effect by a direct action on osteoclasts in bone and not by modifying intestinal absorption or renal excretion of calcium. The finding that echistatin (and peptidomimetics) block the acute loss of trabecular bone seen following ovariectomy in the mouse (Yamamoto et al., 1998; Engleman et al., 1997) suggests that the inhibitory effect of RGD occurs via a direct action on bone, most likely via the $\alpha_v\beta_3$ integrin on osteoclasts.

B. Nonintegrin Receptors in Osteoclasts

Several studies have been carried out to assess the expression of nonintegrin adhesion receptors in osteoclasts. We failed to identify a range of other receptor types in our earlier studies (Horton and Davies, 1989). More recently, though, data indicating that osteoclasts express E-cadherin (Mbalaviele et al., 1995), the 67 kDa laminin receptor Mac-2 (Takahashi et al., 1993), and CD44 (Nakamura et al., 1995) have been published. Some of these proteins are not dominant in mature osteoclasts and it is possible that they are mainly involved in osteoclast development, fusion, and functional maturation from hemopoietic stem cells (see below).

C. Adhesion Molecules and Osteoclast Development

The involvement of adhesion receptors during the development of osteoclasts from hemopoietic stem cells to mature functional osteoclasts has begun to be analyzed in short-term rodent bone marrow cultures in which osteoclast development is stimulated with 1α,25-dihydroxyvitamin D_3. One caveat concerning such an approach is that interpretation of the results can be difficult: inhibitory effects of test agents (e.g., peptides or anti-integrin antibodies) may be indirect, occurring via other cell types critical for osteoclast differentiation, in addition to direct effects on osteoclasts and their precursors. Early studies showed that rodent osteoclast development can be inhibited by the RGD-containing snake venom proteins, implying a role for vitronectin receptor. Studies with antibodies to β_1 (Helfrich et al, 1996) suggest that a role for this class of integrin is a distinct possibility. E- (but not P- or N-) cadherin has recently been reported to be expressed by human and rodent osteoclasts (Mbalaviele et al., 1995); function-blocking antibodies to E-cadherin and adhesion blocking HAV peptide inhibit osteoclast formation and fusion *in vitro* implying that this class of receptor may be active *in vivo* (Mbalaviele et al., 1995). There is also evidence for the involvement of β_2 and α_4 integrins, and their respective counter-receptors intracellular adhesion molecule-1 and vascular cell adhesion molecule-1 (VCAM-1), in osteoclast development in marrow culture systems (reviewed in Horton and Rodan, 1996). The question of whether the CD44 hyaluronidate receptor, expressed by osteoclasts and osteocytes, is involved in osteoclast development has not yet been addressed.

D. The Osteoclast Clear Zone

The clear zone, or tight sealing zone, is the organelle-free, actin-rich part of the osteoclast that forms a tight attachment to mineralized bone matrix. The finding that osteoclast attachment to matrix proteins and bone is interrupted by integrin inhibitors led to the suggestion that the osteoclast tight seal may be mediated by integrins. Some published data has supported the view that vitronectin receptor is enriched in clear zones of resorbing osteoclasts (Reinholt et al., 1990; Neff et al., 1995), as well as podosomes of osteoclasts cultured on glass (Zambonin-Zallone et al., 1989; reviewed in Aubin, 1992). Others, however, have been unable to confirm the former observation, reporting that vitronectin receptor is undetectable in the sealing zone (Lakkakorpi et al., 1991, 1993; Masarachia et al., 1995). We (Väänänen and Horton, 1995) have argued that the dimensions of the integrin molecule (see above), when compared to a membrane-to-bone gap of less than 5 nm, preclude an involvement of integrins in the maintenance of the tight seal during resorption. This contrasts with the clear evidence for a role for integrins in the initial attachment and, possibly, movement of osteoclasts. The molecular mechanism of the attachment process in the established clear zone of a resorbing, non-migratory osteoclast thus remains to be established (Väänänen and Horton, 1995).

IV. SIGNAL TRANSDUCTION VIA OSTEOCLAST INTEGRINS

While there is much information on the extracellular interactions between integrins and their ligands, little is known about intracellular pathways and regulation of cellular function and behavior by these receptors. However, as more data is forthcoming in this area of research, it is clear that, despite the short cytoplasmic tails and lack of intrinsic enzymatic activity, these receptors are capable of transducing signals within a variety of cells. Pertinent data in osteoclasts, being more difficult to study, is even more limited. Emerging data from more accessible cell types suggests a commonality and a convergence in downstream effects, and these data may well provide answers to some key questions regarding signaling via osteoclast integrins. The following sections briefly discuss signal transduction pathways by integrins in general (for reviews see Richardson and Parsons, 1995; Clark and Brugge, 1995), followed by recent data relating to integrin-ligand interactions in osteoclasts.

A. Integrins and Signaling

Integrin-ligand interactions have been shown to induce biochemical changes within cells ("outside-in" signaling), or regulate receptor activation states ("inside-out" signaling) (Hynes, 1992). Early evidence for a signaling role for integrins came from studies in endothelial cells and neutrophils where integrin receptor engagement induced changes in intracellular pH and calcium (Ca^{2+}) (Ingber et al., 1990; Richter et al., 1990). Kornberg et al. (1991) provided the first evidence for tyrosine phosphorylation events when they demonstrated that cell adhesion and subsequent clustering of β_1 integrins resulted in enhanced phosphorylation of certain intracellular proteins. Since then it has become increasingly clear that phosphorylation may be a downstream effect in many cell types. Many of the target intracellular proteins that regulate these events have also been identified. Thus, in many cases, intracellular signal mediation following ligand occupancy is accompanied by integrin receptor clustering. Integrin receptor clustering leads to the formation of focal adhesions, areas within the cell where integrins link to intracellular cytoskeletal elements. Chimeric and mutational analyses of integrin receptors have shown that the cytoplasmic domains of β chains are sufficient for targetting integrins to focal adhesions, while the α subunit of the receptor confers ligand specificity to the interaction (Sastry and Horwitz, 1993; LaFlamme et al., 1994). Insight into integrin signaling pathways has come from biochemical analysis of proteins associated with focal adhesions upon receptor engagement. As mentioned above, one of the earliest events detected in integrin receptor signaling is protein phosphorylation. Tyrosine phosphorylation following integrin receptor occupancy occurs in a variety of cell types, including fibroblasts, carcinoma cells, and leukocytes (Rosales et al., 1995; Arroyo et al., 1994). Other intracellular events that have been demonstrated with integrin ligand receptor interactions include activation of serine-threonine kinases such as protein kinase C (PKC) and mitogen-activated protein (MAP) kinases, intracellular calcium elevations, elevation of intracellular pH, changes in gene expression, and regulation of programmed cell death (apoptosis), the latter also being a consequence of tyrosine phosphorylation.

One of the key players in tyrosine kinase activation is focal adhesion kinase (FAK), which is itself tyrosine phosphorylated, while also having enhanced tyrosine kinase activity when the integrin receptor is occupied by an appropriate ligand (Schaller and Parsons, 1994). Signals generated by integrins can be regulated by FAK in one of many ways: FAK can be localized by a focal adhesion targeting (FAT) sequence (Hildebrand et al., 1993); the cytoskeletal protein paxillin also helps mediates FAK localization (reviewed in Clark and Brugge, 1995); and cellular proteins that contain *src* homology 2 (SH2) domains bind to FAK via tyrosine phosphoryation sites on FAK (Pawson, 1995). The *src* family of tyrosine kinases have also been implicated in integrin-mediated signaling, as have protein tyrosine phosphatases (PTPs) (Arroyo et al., 1994). FAK has also been shown to be associated with growth factor receptor-bound protein (Grb2) and mSOS1, key players in the *ras*-MAP kinase pathway (van der Geer et al., 1994). This association suggests that integrins may interact with signals generated by other receptors and may synergize or inhibit downstream effects of growth factors which commonly utilize this signaling pathway. Examples of regulation of integrin signaling by growth factors include association of the vitronectin receptor $\alpha_v\beta_3$ with an intracellular protein insulin receptor substrate-1 (IRS-1), an intracellular protein that mediates signaling by insulin and binds other signaling molecules including phosphatidylinositol(PI)-3-kinase (PI-3-K) and Grb2 (Vuori and Ruoslahti, 1994).

Phospholipid kinases such as PI-3K and phosphatidylinositol(4)phosphate-kinase (PIP-5K) have also been directly implicated in the integrin signaling processes. The former enzyme has been shown to coprecipitate with FAK and may play an important role in mediating integrin-mediated cytoskeletal rearrangements (Chen and Guan, 1994), further conferring a central role for FAK in signaling via integrins. Integrin signaling pathways can interact with other receptor mediated pathways. Integration of this type of cross talk appears to be controlled by a small molecular weight guanosine triphosphatase (GTPase), *rho* (Ridley and Hall, 1992). Clearly, although signaling via integrins is proving to be increasingly complex, the convergence to certain key molecules in all signaling pathways may well simplify the picture.

B. Signaling via Osteoclast Integrins

In osteoclasts, integrin-ligand interactions can trigger intracellular signaling in a manner similar to that observed in other cell types. Thus, studies have demonstrated that integrin engagement can cause changes in intracellular calcium, pH, and tyrosine phosphorylation of intracellular proteins. In avian osteoclasts, Miyauchi et al. (1991) demonstrated that osteopontin (OPN) and synthetic RGD peptides from the OPN and bone sialoprotein sequences caused immediate reductions in intracellular free calcium ($[Ca^{2+}]_i$) levels, an effect that was blocked by the $\alpha_v\beta_3$ antibody, LM609. This effect was attributed to the activation of a plasma membrane Ca^{2+}-ATPase. Subsequent studies in rat, mouse, and human osteoclasts and osteoclast-like cells have shown the contrary. Addition of RGD peptides and proteins caused transient elevations in $[Ca^{2+}]_i$ (Paniccia et al., 1993; Shankar et al.,

1993, 1995; Zimolo et al., 1994), a signal that was localized to a novel intranuclear calcium pool (Shankar et al., 1993, 1995). The reasons for the discrepancy between the avian and mammalian osteoclasts are not clear. While the obvious explanation would be on the basis of a species difference, it is likely that the chick study was indicative of an "attachment" signal, and the transient Ca^{2+} spikes in the mammalian cells were "detachment" signals. This would be in keeping with the general observation that transient elevations in $[Ca^{2+}]_i$ are often inhibitory to osteoclastic bone resorption, as are RGD peptides and proteins. The exact role of the highly expressed vitronectin receptor ($\alpha_v\beta_3$) in osteoclast function is of particular interest in light of recent data showing that physiological levels of Ca^{2+} significantly reduce the affinity of OPN for $\alpha_v\beta_3$ and block cell adhesion (Hu et al., 1995). OPN, a major noncollagenous protein in bone, is believed to play a key role in osteoclast-matrix adhesion events. It has long been speculated that sensing of extracellular Ca^{2+} by osteoclasts is regulated by specific Ca^{2+} receptors, which provide the necessary negative feedback loop to turn off osteoclastic bone resorption (reviewed in Zaidi et al., 1993). The finding that the OPN-$\alpha_v\beta_3$ interaction is sensitive to elevated Ca^{2+} suggests that the vitronectin receptor in osteoclasts could also function as a calcium sensor on these cells. OPN binding to $\alpha_v\beta_3$ was found to stimulate the production of phosphatidyl inositol(4)phosphate (PIP) and phosphatidyl inositol(3,4,5)trisphosphate (PI(3,4,5)P3) in avian osteoclasts. Further, antibodies to $\alpha_v\beta_3$ were found to immunoprecipitate PI-3-K, suggesting that the latter enzyme may be mediating this effect (Hruska et al., 1995).

The role of PKC in osteoclasts is unclear. Teti et al. (1995) report that, in rabbit and chick osteoclasts, activation of PKC by the phorbol ester enhanced calcium signaling by elevated extracellular calcium ($[Ca^{2+}]_o$), but attenuated the response in human osteoclasts. It is likely that PKC-α is involved in the Ca^{2+} sensing mechanism in osteoclasts, and may therefore also mediate vitronectin receptor stimulated signaling.

RGD protein matrices can alter intracellular pH in osteoclasts. For example, adhesive substrates were found to influence the acid secreting abilities of osteoclasts (Hashizume et al., 1995; Zimolo et al, 1995). Osteoclasts on dentine were found to be more vacuolated and have more acidic organelles than osteoclasts settled on glass. Further, osteoclasts on collagen type I or vitronectin substrates also had more acidic organelles than osteoclasts on glass alone. These data suggest that osteoclast-matrix interactions may influence the bone resorbing capacity or activity of osteoclasts.

Perhaps one of the most significant findings in osteoclast biology over the past several years has been the role of the protooncogene, c-*src*. Mutant mice in which the c-*src* encoding gene was disrupted were found to exhibit an osteopetrotic phenotype (Soriano et al., 1991). The fact that these mice were otherwise normal is indicative of the high level of redundancy in the c-*src* family of tyrosine kinases. Thus, mice that lacked both c-*src* and c-*yes* did not survive beyond birth (Stein et al., 1994). Unlike in c-*fos* knockout mice where osteopetrosis resulted from a lack of osteoclasts, the c-*src* deficient mice appeared to have dysfunctional osteoclasts. *In vitro* studies have demonstrated that osteoclasts express high levels of c-*src*, c-*fyn*, c-*yes*, and c-*lyn* (Horne et al., 1992). However, these other *src*-related proteins do not compensate for the functional deficiency in c-*src*⁻ osteoclasts, suggesting that this protooncogene has a specific function in osteoclasts. It was also found that the c-*src* protein is mostly associated with intracellular organelles, suggesting that it might play a role in fusion of membrane vesicles. In avian osteoclasts, $\alpha_v\beta_3$ was found to associate with c-*src* which might link the integrin receptor to PI-3-K. Thus, the overall effects of c-*src* disruption appear to be lack of ruffled border formation (resulting in dysfunctional osteoclasts) and inhibition of vesicular exocytosis (Yoneda et al., 1993; Hall et al., 1994). Human and avian osteoclasts have also been shown to express FAK abundantly (Berry et al., 1994), a finding which suggests that a similar scenario to growth factor signaling may also feature in osteoclasts.

In the absence of a representative osteoclast cell line, biochemical analyses of signaling pathways in osteoclasts has been slow and often indirect. However, with the use of methods that result in significantly higher yields of these fragile and inaccessible cells, such as the rabbit osteoclast preparation (Tezuka et al., 1992) or by the use of the newly discovered growth factor, TRANCE (Yasuda et al., 1998), it is likely that many of these signaling pathways will be well defined over the next few years. Based on studies carried out in other cell types, it is also likely that integrin signaling in osteoclasts may follow a sequence of events involving c-*src*, FAK, PI-3-K, and the *ras*-MAPK pathway. It remains to be seen whether integrin receptor engagement results in (i) changes in immediate early gene expression via elevations in intranuclear calcium and MAPK activation, (ii) specific signals for phosphorylation, resulting in apoptotic pathways, or (iii) involvement of c-*src* via VNR activation in an osteoclast specific manner.

V. INTEGRINS AS THERAPEUTIC TARGETS IN BONE DISEASE

This chapter has focused on the role of integrin cell adhesion receptors in the regulation of bone resorption and in signal transduction. The finding that inhibition of $\alpha_v\beta_3$ integrin function in animal models results in reduced bone resorption has suggested that antagonists of this receptor could be developed for use in bone disease. Indeed, it is likely that agents will be generated by the pharmaceutical industry with sufficient activity and specificity for the $\alpha_v\beta_3$ integrin dimer of osteoclasts (Engleman et al., 1997). This expectation is based upon the successful development of analogous compounds for the inhibition of platelet fibrinogen receptor, gpIIbIIIa (i.e., the αIIbβ3 integrin), in the prevention of thrombosis (Cox et al., 1994; Gadek and Blackburn, 1995). The next few years will tell if this novel strategy will have clinical value in the treatment of bone diseases such as osteoporosis.

VI. SUMMARY

A limited repertoire of adhesion receptors of the integrin class, including the $\alpha_v\beta_3$, or the vitronectin receptor, have been shown to be expressed by mature osteoclasts. These mediate cell attachment to bone matrix proteins and are involved actively in the process of bone resorption. Interference with integrin-ligand interaction leads to the inhibition of bone resorption; this finding is being exploited to develop drugs for the treatment of osteoporosis. Osteoclast integrins are also involved in signal transduction on interaction with matrix proteins or RGD-containing peptides. The functional significance of this remains to be clarified.

ACKNOWLEDGMENTS

The authors wish to thank the Wellcome Trust for financial support of some of the work outlined in this review.

REFERENCES

Athanasou, N.A., Alvarez, J.I., Ross, F.P., Quinn, J.M. and Teitelbaum, S.L. (1992). Species differences in the immunophenotype of osteoclasts and mononuclear phagocytes. Calcif. Tissue Int. 50, 427-432.

Arroyo, A.G., Campanero, M.R., Sánchez-Mateos, P., Zapata, J.M., Ursa, M.A., del. Poza, M.A., and Sánchez-Madrid, F. (1994). Induction of tyrosine phosphorylation during ICAM-3 and LFA-1–mediated intercellular adhesion, and its regulation by the CD45 tyrosine phosphatase. J. Cell Biol. 126, 1277-1286.

Aubin, J.E. (1992). Perspectives—Osteoclast adhesion and resorption: The role of podosomes. J. Bone Min Res. 7, 365-368.

Barclay, A.N., Birkeland, M.L., Brown, M.H., Beyers, A.D., Davis, S.J., Somoza, C., and Williams, A.F. (Eds.) (1993). The Leucocyte Antigen Factsbook. Academic Press, London.

Beckstead, J.H., Stenberg, P.E., McEver, R.P., Shuman, M.A., and Bainton, D.F. (1986). Immunohistochemical localization of membrane and α-granule proteins in human megakaryocytes: Application to plastic embedded bone marrow biopsy specimens. Blood 67, 285-293.

Berry, V., Rathod, H., Pulman, L.B., and Datta, H.K. (1994). J. Endocrinol.

Chambers, T.J., Fuller, K., Darby, J.A., Pringle, J.A.S., and Horton, M.A. (1986). Monoclonal antibodies against osteoclasts inhibit bone resorption in vitro. Bone Min. 1, 127-135.

Chen, H.-C. and Guan, J.-L. (1994). Association of focal adhesion kinase with its potential substrate phosphatidylinositol 3-kinase. Proc. Natl. Acad. Sci. 91, 10148-10152.

Clark, E.A. and Brugge, J.S. (1995). Integrins and signal transduction pathways: The road taken. Science 268, 233-239.

Clover, J., Dodds, R.A., and Gowen, M. (1992). Integrin subunit expression by human osteoblasts and osteoclasts in situ and in culture. J. Cell Sci. 103, 267-271.

Cox, D., Aoki, T., Scki, J., Motoyama, Y., and Yoshida, K. (1994). The pharmacology of integrins. Medicinal Res. Revs. 14, 195-228.

Crippes, B.A., Engleman, V.W., Settle, S.L., Ornberg, R.L., Helfrich, M.H., Horton, M.A., and Nickols, G.A. (1996). Antibody to β_3 integrin inhibits the calcemic response to PTHrP in the thyroparathyroidectomized rat. Endocrinol. 137, 918-924.

Davies, J., Warwick, J., Totty, N., Philp, R., Helfrich, M., and Horton, M. (1989). The osteoclast functional antigen implicated in the regulation of bone resorption is biochemically related to the vitronectin receptor. J. Cell Biol. 109, 1817-1826.

Engleman, V.W., Nickols, G.A., Ross, F.P., Horton, M.A., Griggs, D.W., Settle, S.L., Ruminski, P.G., and Teitelbaum, S.L. (1997). A peptidomimetic antagonist of the $\alpha(v)\beta3$ integrin inhibits bone resorption in vitro and prevents osteoporosis in vivo. J. Clin. Invest. 99, 2284-2292.

Fisher, J.E., Caulfield, M.P., Sato, M., Quartuccio, H.A., Gould, R.J. Garsky, V.M., Rodan, G.A., and Rosenblatt, M. (1993). Inhibition of osteoclastic bone resorption in vivo by echistatin, an "arginyl-glycyl-aspartyl" (RGD) –containing protein. Endocrinol. 132, 1411-1413.

Flores, M.E., Norgard, M., Heinegard, D., Reinholt, F.P., and Andersson, G. (1992). RGD-directed attachment of isolated rat osteoclasts to osteopontin, bone sialoprotein and fibronectin. Exp. Cell Res. 201, 526-530.

Gadek, T. and Blackburn, B.K. (1996). Identification and development of integrin/ligand antagonists for the treatment of human disease. In: *Adhesion Receptors as Therapeutic Targets.* (Horton, M.A., Ed.), 247-272. CRC Press Inc., Boca Raton, Florida.

van der Geer, P., Hunter, T., and Lindberg, R.A. (1994). Receptor protein-tyrosine kinases and their signal transduction pathways. Ann. Rev. Cell Biol. 10, 251-337.

Hall, T.J., Schaueblin, M., and Missbach, M. (1994). Evidence that c-src is involved in the process of osteoclastic bone resorption. Biochem. Biophys. Res. Comm. 199, 1237-1244.

Hashizume, Y., Araki, S., Sawada, K., Yamada, K., and Katayama, K. (1995). Adhesive substrates influence acid-producing activities of cultured rabbit osteoclasts: Cultured osteoclasts with large vacuoles have enhanced acid-productive activities. Exp. Cell Res. 218, 452-459.

Helfrich, M.H., Nesbitt, S.A., Dorey, E.L., and Horton, M.A. (1992). Rat osteoclasts adhere to a wide range of RGD (Arg-Gly-Asp) peptide–containing proteins, including the bone sialoproteins and fibronectin, via a β_3 integrin. J. Bone Min. Res. 7, 335-343.

Helfrich, M.H., Nesbitt, S.A., Lakkakorpi, P., Barnes, M.J., Bodary, S.C., Shankar, G., Mason, W.T., Mendrick, D.L., Väänänen, H.K., and Horton, M.A. (1996). β_1 integrin and osteoclast function: involvement in collagen recognition and bone resorption. Bone, 4, 317-328.

Hildebrand, J.D., Schaller, M.D., and Parsons, J.T. (1993). Identification of sequences required for the efficient localization of the focal adhesion kinase pp125^{FAK} to cellular focal adhesions. J. Cell Biol. 123, 993-1005.

Horne, W.C., Neff, L., Chatterjee, D., Lomri, A., Levy, J.B., and Baron, R. (1992). Osteoclasts express high levels of pp60$^{c\text{-}src}$ in association with intracellular membranes. J. Cell Biol. 119, 1003-1013.

Horton, M.A. and Davies, J. (1989). Adhesion receptors in bone. J. Bone Min. Res. 4, 803-807.

Horton, M.A., Lewis, D., McNulty, K., Pringle, J.A.S., and Chambers, T.J. (1985). Monoclonal antibodies to osteoclastomas (giant cell bone tumours): Definition of osteoclast-specific antigens. Cancer Res. 45, 5663-5669.

Horton, M.A. (1986). Expression of platelet glycoprotein IIIa by human osteoclasts. Blood 68, 595.

Horton, M.A. and Rodan, G.A. Integrins as therapeutic targets in bone disease. (1996). In: *Adhesion Receptors as Therapeutic Targets*. (Horton, M.A., Ed.), pp. 223-245. CRC Press Inc., Boca Raton, Florida.

Hruska, K.A., Rolnick, F., Huskey, M., Alvarez, U., and Cheresh, D. (1995). Engagement of the osteoclast integrin $\alpha_v\beta_3$ by osteopontin stimulates phosphatidylinositol 3-hydroxyl kinase activity. Ann. N.Y. Acad. Sci. 760, 151-165.

Hu, D.D., Hoyer, J.R., and Smith, J.W. (1995). Ca^{2+} suppresses cell adhesion to osteopontin by attenuating binding affinity for integrin $\alpha_v\beta_3$. J. Biol. Chem. 270, 9917-9925.

Hughes, D.E., Salter, D.M., Dedhar, S., and Simpson, R. (1993). Integrin expression in human bone. J. Bone Min. Res. 8, 527-533.

Hynes, R.O. (1992). Integrins: Versatility, modulation, and signaling in cell adhesion. Cell 69, 11-25.

Ingber, D.E., Prusty, D., Frangioni, J.V., Cragoe, Jr., E.J., Lechene, C., and Schwartz, M.A. (1990). Control of intracellular pH and growth by fibronectin in capillary endothelial cells. J. Cell Biol. 110, 1803-1811.

King, K.L., D'Anza, J.J., Bodary, S., Pitti, R., Siegel, M., Lazarus, R.A., Dennis M.S., Hammonds Jr., R.G., and Kukreja S.C. (1994). Effects of kistrin on bone resorption in vitro and serum calcium in vivo. J. Bone Min. Res. 9, 381-387.

Kornberg, L.J., Earp, H.S., Turner, C.E., Prockop, C., and Juliano, R.L. (1991) Signal transduction by integrins: Increased protein tyrosine phosphorylation caused by clustering of β_1 integrins. Proc. Natl. Acad. Sci. 88, 8392-8396.

LaFlamme, S.E., Thomas, L.A., Yamada, S.S., and Yamada, K.M. (1994). Single subunit chimeric integrins as mimics and inhibitors of endogenous integrin functions in receptor localization, cell spreading and migration, and matrix assembly. J. Cell Biol. 126, 1287-1298.

Lakkakorpi, P.T., Helfrich, M.H., Horton, M.A., and Väänänen, H.K. (1993). Spatial organization of microfilaments and vitronectin receptor, $\alpha_v\beta_3$, in osteoclasts. A study using confocal laser scanning microscopy. J. Cell Sci. 104, 663-670.

Lakkakorpi, P.T., Horton, M.A., Helfrich, M.H., Karhukorpi, E.-K., and Väänänen, H.K. (1991). Vitronectin receptor has a role in bone resorption but does not mediate tight sealing zone attachment of osteoclasts to the bone surface. J. Cell Biol. 115, 1179-1186.

Masarachia, P., Yamamoto, M., Rodan, G.A., and Duong, L.T. (1995). Co-localization of the vitronectin receptor $\alpha_v\beta_3$ and echistatin in osteoclasts during bone resorption in vivo. J.Bone Min. Res. 10 (Suppl. 1), S164.

Mbalaviele, G., Chen, H., Boyce, B.F., Mundy, G.R., and Yoneda, T. (1995). The role of cadherin in the generation of multinucleated osteoclasts from mononuclear precursors in murine marrow. J. Clin. Invest. 95, 2757-2765.

Miyauchi, A., Alvarez, J., Greenfield, E.M., Teti, A., Grano, M., Colucci, S., Zambonin-Zallone, A., Ross, F.P., Teitelbaum, S.L., Cheresh, D., and Hruska, K.A. (1991). Recognition of osteopontin and related peptides by an $\alpha_v\beta_3$ integrin stimulates immediate cell signals in osteoclasts. J. Biol. Chem. 266, 20369-20374.

Mould, P.A. and Humphries, M.J. (1995). Functional domains of adhesion molecules. In: Adhesion Receptors as Therapeutic Targets. (Horton, M.A., Ed.), pp. 75-105. CRC Press, Boca Raton, Florida.

Nakamura, H., Kenmotsu, S., Sakai, H., and Ozawa, H. (1995). Localization of CD44, the hyaluronidate receptor, on the plasma membrane of osteocytes and osteoclasts in rat tibiae. Cell Tiss. Res., 280, 225-233.

Neff, L., Gaiht, J., and Baron, R. (1995). Ultrastructural demonstration of the α_v subunit of the vitronectin receptor in the sealing zone of resorbing osteoclasts. J.Bone Min. Res. 10, (Suppl. 1), S329.

Nesbitt, S., Nesbit, A., Helfrich, M., and Horton, M. (1993). Biochemical characterization of human osteoclast integrins. J. Biol. Chem. 268, 16737-16745.

Paniccia, R., Colucci, S., Grano, M., Serra, M., Zallone, A.Z., and Teti, A. (1993). Immediate cell signal by bone-related peptides in human osteoclastlike cells. Am. J. Physiol. 265, C1289-C1297.

Pawson, T. (1995). Protein-tyrosine kinases. Nature 373, 477-478.

Reinholt, F.P., Hultenby, K., Oldberg, A., and Heinegard, D. (1990). Osteopontin—a possible anchor of osteoclasts to bone. Proc. Natl. Acad. Sci. USA 87, 4473-4475.

Richardson, A. and Parsons, J.T. (1995). Signal transduction through integrins: A central role for focal adhesion kinase? BioEssays 17, 229-236.

Richter, J., Ng-Sikorski, J., Olsson, I., and Andersson, T. (1990). Tumor necrosis factor-induced degranulation in adherent human neutrophils is dependent on CD11b/CD18-integrin-triggered oscillations of cytosolic free Ca^{2+}. Proc. Natl. Acad. Sci. 87, 9472-9476.

Ridley, A.J. and Hall, A. (1992). The small GTP-binding protein rho regulates the assembly of focal adhesions and actin stress fibers in response to growth factors. Cell 70, 401-410.

Rifkin, B.R. and Gay, C.V. (1992). Biology and Physiology of the Osteoclast. CRC Press, Boca Raton, Florida.

Rosales, C., O'Brien, V., Kornberg, L., and Juliano, R. (1995). Signal tranduction by cell adhesion receptors. Biochim. Biophys. Acta. 1242, 77-98.

Ross, F.P., Chappel, J., Alvarez, J.I., Sander, D., Butler, W.T., Farach-Carson, M.C., Mintz, K. A., Gehron Robey, P., Teitelbaum, S.L., and Cheresh, D.A. (1993). Interactions between the bone matrix proteins osteopontin and bone sialoprotein and the osteoclast integrin $\alpha_v\beta_3$ potentiate bone resorption. J. Biol. Chem. 268, 9901-9907.

Sastry, S.K. and Horwitz, A.F. (1993). Integrin cytoplasmic domains: Mediators of cytoskeletal linkages and extra- and intracellular–initiated transmembrane signaling. Curr. Opin. Cell Biol. 5, 819-831.

Sato, M., Garsky, V., Majeska, R.J., Einhorn, T.A., Murray, J., Tashjian, A.H., and Gould, R.J. (1994). Structure-activity studies of the s-echistatin inhibition of bone resorption. J.Bone Min. Res. 9, 1441-1449.

Sato, M., Sardana, M.K., Grasser, W.A., Garsky, V.M., Murray, J.M., and Gould, R.J. (1990). Echistatin is a potent inhibitor of bone resorption in culture. J. Cell Biol. 111, 1713-1723.

Schaller, M.D. and Parsons, J.T. (1994). Focal adhesion kinase protein and associated proteins. Curr. Opin. Cell. Biol. 6, 705-710.

Shankar, G., Davison, I., Helfrich, M.H., Mason, W.T., and Horton, M.A. (1993). Integrin receptor–mediated mobilization of intranuclear calcium in rat osteoclasts. J. Cell Sci. 105, 61-68.

Shankar, G., Gadek, T.R., Burdick, D.J., Davison, I, Mason, W.T., and Horton, M.A. (1995). Structural determinants of calcium signaling by RGD peptides in rat osteoclasts: Integrin-dependent and -independent actions. Exptl. Cell Res. 219, 364-371.

Soriano, P., Montgomery, C., Geske, R., and Bradley, A. (1991). Targeted disruption of the c-*src* proto-oncogene leads to osteopetrosis in mice. Cell 64, 693-702.

Stein, P.L., Vogel, H., and Soriano, P. (1994). Combined deficiencies of src, fyn, and yes tyrosine kinases in mutant mice. Genes Devel. 8, 1999-2007.

Takahashi, N., Udagawa, N., Tanaka, S., Murakami, H., Owan, I., Tamura, T., and Suda, T. (1993). Postmitotic osteoclast precursors express macrophage-associated phenotypes. J. Bone Min. Res. 8, S396.

Tanaka, H., Sato, M., Shinar, D., and Rodan, G.A. (1991). Echistatin inhibits osteoclast-like cell generation in the co-culture system of mouse osteoblastic cells and bone marrow cells. J. Bone Min. Res. 6, (Suppl. 1), S148.

Teti, A., Huwiler, A., Paniccia, R., Sciortini, G., and Pfeilschifter, J. (1995). Translocation of protein kinase-C isoenzymes by elevated extracellular Ca^{2+} concentration in cells from a human giant cell tumor of bone. Bone 17, 175-183.

Tezuka, K., Sato, T., Kamioka, H., Nijweide, P.J., Tanaka, K., Matsuo, T., Ohta, M., Kurihara, N., Hakeda, Y., and Kumegawa, M. (1992). Identification of osteopontin in isolated rabbit osteoclasts. Biochem. Biophys. Res. Comm. 186, 911-917.

Väänänen, H.K. and Horton, M.A. (1995). The osteoclast clear zone is a specialized cell-matrix adhesion structure. J.Cell Sci. 108, 2729-2732.

Vuori, K. and Ruoslahti, E. (1994). Association of insulin receptor substrate-1 with integrins. Science 266, 1576-1578.

Yamamoto, M., Fisher, J.E., Gentile, M., Seedor, J.G., Leu, C.T., Rodan, S.B., and Rodan, G.A. (1998). The integrin ligand echistatin prevents bone loss in ovariectomized mice and rats. Endocrinology. 139, 1411-1419.

Yasuda, H., Shima, N., Nakagawa, N., Yamaguchi, K., Kinosaki, M., Mochizuki, S., Tomoyasu, A., Yano, K., Goto, M., Murakami, A., Tsuda, E., Morinaga, T., Higashio, K., Udagawa, N., Takahashi, N., and Suda, T. (1998). Osteoclast differentiation factor is a ligand for osteoprotegerin/osteoclastogenesis-inhibitory factor and is identical to TRANCE/RANKL. Proc. Natl. Acad. Sci. USA 95, 3597-3602.

Yoneda, T., Lowe, C., Lee, C.H., Guiterrez, G., Niewolna, M., Williams, P.J., Izbicka, E., Uehara, Y., and Mundy, G.R. (1993). Herbimycin A, a $pp60^{c\text{-}src}$ tyrosine kinase inhibitor, inhibits osteoclastic bone resorption in vitro and hypercalcemia in vivo. J. Clin. Invest. 91, 2791-2795.

Zambonin-Zallone, A., Teti, A., Grano, M., Rubinacci, A., Abbadini, M., Gaboli, M., and Marchisio, P.C. (1989). Immunocytochemical distribution of extracellular matrix receptors in human osteoclasts: A β_3 antigen is colocalized with vinculin and talin in the podosomes of osteoclastoma giant cells. Exp. Cell Res. 182, 645-652.

Zaidi, M., Alam, A.S.M.T., Shankar, V.S., Bax, B.E., Bax, C.M.R., Moonga, B.S., Bevis, P.J.R., Stevens, C., Blake, D.R., Pazianas, M., and Huang, C.L.-H. (1993). Cellular biology of bone resorption. Biol. Rev. 68, 197-264.

Zimolo, Z., Wesolowski, G., Tanaka, H., Hyman, J.L., Hoyer, J.R., and Rodan, G.A. (1994). Soluble $\alpha_v\beta_3$-integrin ligands raise $[Ca^{2+}]i$ in rat osteoclasts and mouse-derived, osteoclastlike cells. Am. J. Physiol. 266, C376-C381.

Zimolo, Z., Wesolowski, G., and Rodan, G.A. (1995). Acid extrusion is induced by osteoclast attachment to bone: Inhibition by alendronate and calcitonin. J. Clin. Invest. 96, 2277-2283.

HORMONAL REGULATION OF FUNCTIONAL OSTEOCLAST PROTEINS

F. Patrick Ross

I. MOLECULAR MECHANISMS OF BONE RESORPTION

The initial step in bone resorption is attachment of the osteoclast to matrix, followed by creation of an isolated extracellular resorptive microenvironment con-

Advances in Organ Biology
Volume 5B, pages 331-346.

ISBN: 0-7623-0390-5

taining a highly convoluted structure apposed to bone, known as the ruffled membrane (Kallio et al., 1971). Attachment involves interactions between integrins and matrix proteins in bone (see also Chapter 12). A number of groups provided evidence that the integrin $\alpha_v\beta_3$ on osteoclasts binds one or more bone matrix proteins containing the motif Arg-Gly-Asp (RGD). The majority of these experiments, using antibodies to $\alpha_v\beta_3$ or RGD-containing peptides which block the function of the integrin, were performed *in vitro* (Chambers, 1988; Horton et al., 1991; Ross et al., 1993; Gronowicz and Derome, 1994). On the other hand, blunting of osteoclast-mediated hypercalcemia by *in vivo* administration of the disintegrins kistrin or echistatin (Fisher et al., 1993; King et al., 1994) lends support to the hypothesis that $\alpha_v\beta_3$ plays a critical role in osteoclast function. Further proof of the role of the integrin $\alpha_v\beta_3$ in bone resorption has been provided using the model of the oophorectomized rat, which loses 55% bone mass within six weeks of ovariectomy. This loss is prevented by *in vivo* administration of a small molecule RGD mimetic, shown to block, *in vitro*, both $\alpha_v\beta_3$-mediated attachment to osteopontin as well as bone resorption by murine osteoclasts bearing $\alpha_v\beta_3$ as their almost exclusive α_v integrin (Nickols et al., 1995). Attachment of the osteoclast to bone gives rise to intracellular signals. Thus, ligation of $\alpha_v\beta_3$ on freshly isolated osteoclasts results in changes in intracellular calcium (Miyauchi et al., 1991; Zimolo et al., 1994), the net result being cytoskeletal reorganization, an important, but largely unexplored, aspect of osteoclast biology.

Studies in which RGD peptides have been used to inhibit bone resorption *in vitro* suggest that the ligand for $\alpha_v\beta_3$ contains this sequence. However, since a number of bone matrix proteins, including osteopontin, bone sialoprotein, thrombospondin, type 1 collagen, fibronectin and vitronectin, all containing the RGD motif, the identity of the bone matrix protein(s) which bind to the integrin $\alpha_v\beta_3$ is not clear. Immunoelectron microscopy suggests that $\alpha_v\beta_3$ and osteopontin are colocalized in bone (Reinholt et al., 1990), but the only functional data rest on the ability of an immunopurified osteopontin antibody to inhibit osteoclast-bone interaction (Ross et al., 1993).

Collagen is primarily a ligand for β_1 integrins, but an important recent observation is that, once the protein collagen is denatured, the previously cryptic RGD sequence becomes available for $\alpha_v\beta_3$ ligation, leading to the generation of a yet-unidentified cell survival signal (Montgomery et al., 1994). While these findings were obtained using a melanoma cell line, the possible implications are clear for the osteoclast, a cell which both degrades collagen and undergoes apoptosis (Boyce et al., 1995).

The next step in osteoclastic bone resorption is the generation of a polarized bone-cell interface, whose unique morphological feature is a highly ruffled membrane (Kallio et al., 1971). Two groups (Baron et al., 1988, 1990; Blair et al., 1988) showed that this membrane arises by insertion of protein-bearing intracellular vesicles into the intially-formed bone-osteoclast interface (Figure 1). While it is known that the overall process involves migration and subsequent fusion of lysosomal vesicles, detailed information about the process in the osteoclast is limited. Studies in epithelial cells, which are also polarized, have revealed a role for the other major

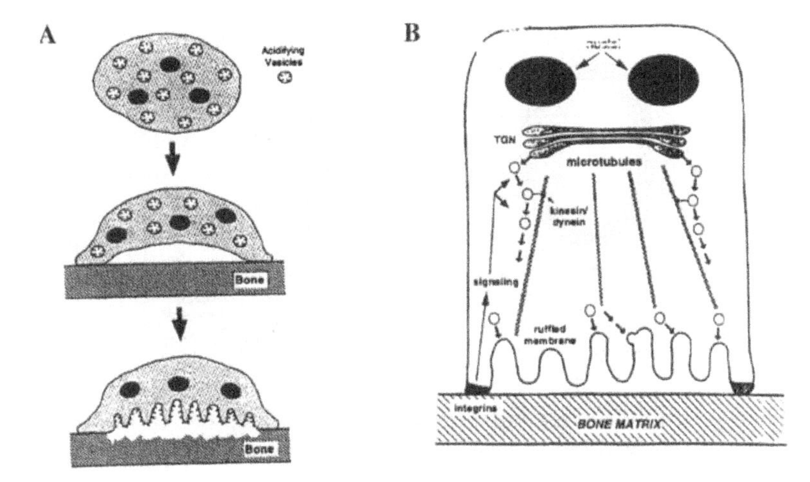

Figure 1. Model of osteoclast polarization. (**A**). The osteoclast, when not on bone, is nonpolarized, with numerous acidifying vesicles in the cytoplasm. Attachment to bone generates a signal, probably via the integrin $\alpha_v\beta_3$, resulting in movement of the vesicles to the bone-apposed plasma membrane. Vesicle insertion leads to generation of the characteristic ruffled membrane, containing high levels of the proton pump, shown as spikes. (**B**). Vesicle movement is postulated to occur by trafficking along microtubules (Raff, 1994), oriented towards the bone-cell interface following attachment. Vesicle movement is probably driven by molecular motors, such as dynein or kinesin (Collins, 1994). It is also possible vesicles may arise by direct budding from the trans-Golgi system, thereby incorporating newly synthesized proteins into the ruffled membrane (Fath et al., 1993).

element of the intracellular architecture, the microtubular network, in vesicular movement (Elferink and Scheller, 1993; Fath et al., 1993), but it is not known if the same applies in the osteoclast.

Fusion of lysosomal vesicles with an existing plasma membrane is analogous to the process of regulated exocytosis in other systems, including neuronal depolarization, release of granules by neuroendocrine cells, and the movement of a specific glucose transporter to the surface of adipocytes (reviewed in Sudhof, 1995). In all instances, targeting and insertion of the appropriate vesicle involves members of the rab protein subfamily, a member of the *ras* superfamily (Figure 2). *Rabs* are involved in all aspects of vesicle targeting, including movements between intracellular membranes and both endo- and exocytosis (Ferro-Novick and Novick, 1993; Zerial and Stenmark, 1993; Novick and Garrett, 1994). However, a small *rab* subset and most notably the proteins *rab*3A–D appear to be important in regulated exocytosis (Lledo et al., 1993; Geppert et al., 1994; Weber et al., 1994).

The set of interwoven biochemical reactions by which osteoclasts degrade bone (summarized in Figure 3) are now well understood. Studies with isolated osteoclasts

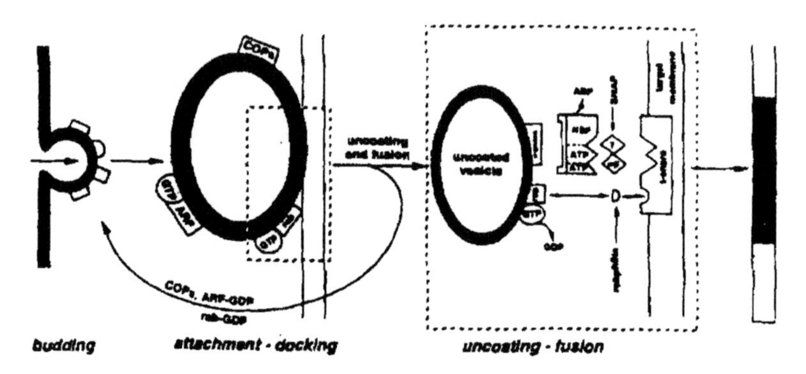

Figure 2. Role or small GTPases in vesicle targeting. The first step in vesicle movement is generation of a vesicle, an event which involves several families of proteins, including coatamers [COPs (cytosolic coat proteins); Schekman and Orci, 1996] and ARFs (ADP-ribosylation factors), one subfamily of the small GTPases (Nuoffer and Balch, 1995). Once generated the vesicle is transported to its acceptor membrane, where it docks and fuses. These latter events are regulated by a range of cytosolic adaptor proteins and their corresponding receptors [called SNAPs (soluble NSF attachment proteins) and SNAREs (SNAP receptors)] respectively (Rothman and Warren, 1994)) and *rabs*, other members of the same family of low molecular weight GTPases, (Fisher von Mollard et al., 1994; Pfeffer, 1994; Nuoffer and Balch, 1995). As a nascent vesicle forms, the appropriate *rab*, in its GTP-bound form, binds. Recruitment of SNAPs and SNAREs is directed by association of an ATP-NSF (N-ethylmaleimide sensitive factor) complex with the vesicle. A poorly understood sequence of ATP and GTP hydrolysis leads to release of coatamers, SNAPs, and *rabs*, with the net consequence being vesicle docking and fusion. Additional proteins on both vesicles and membranes are likely involved in the overall process. Panel **B** is an enlargement of a portion of panel **A**.

reveal that dissolution of the inorganic phase precedes that of protein (Blair et al., 1986). Demineralization involves acidification of the extracellular space, a process which is mediated by a vacuolar H^+-ATPase in the ruffled membrane of the polarized cell (Blair et al., 1989). The structure and functional activity of this multi-enzyme complex is very similar, if not identical to the analogous proton pump in the intercalated cell of the kidney. The pump is a multimer containing eight subunits, some of which are intrinsic membrane proteins. Others, including the 70 kDa protein containing the ATPase activity, are attached noncovalently to the subunits buried in the membrane. In support of the identity of this complex as the critical moiety in osteoclast acidification, the fungal metabolite bafilomycin A, which is a potent and specific inhibitor of all vacuolar proton pumps, has been shown to inhibit bone resorption (Mattsson et al., 1991). The intact proton pump complex has recently been isolated from avian osteoclasts and the identity of several subunits to those present in other vacuolar pumps has been established by Western analysis. Importantly, the activity of the isolated complex was restored by incorporation into lipid vesicles (Mattsson et

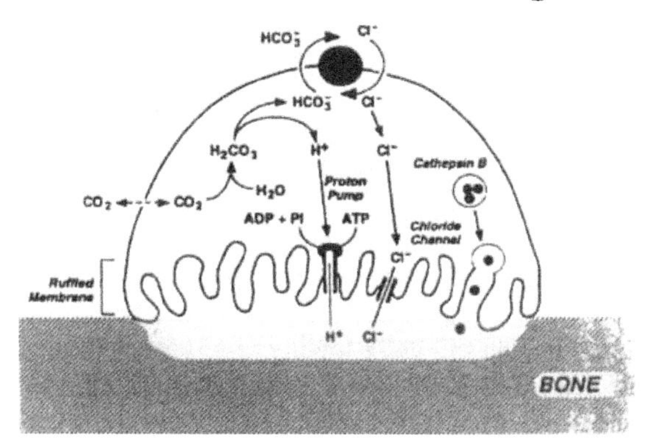

Figure 3. Model of the major individual steps in osteoclastic bone resorption. See text for details.

al., 1994). As with all members of the vacuolar pump family, extrusion of the proton through the plasma membrane is accompanied by hydrolysis of one molecule of ATP. Since the osteoclast transports protons extracellularly by an electrogenic mechanism the question arises as to how intracellular pH is maintained. Teti et al. (1989) found that osteoclasts express on their antiresorptive border an energy-independent Cl^-/HCO_3^- exchanger similar to band 3 of the erythrocyte. Finally, electroneutrality is preserved by a plasma membrane Cl^- channel, charge coupled to the H^+-ATPase, resulting in secretion of HCl into the resorptive microenvironment (Blair et al., 1991).

Acidification of the isolated resorptive environment is critical, permitting not only mineral mobilization, but subsequent solubilization of the organic phase of bone (Blair et al., 1986). Two families of proteases with very different pH optima have been proposed to play a role (see Chapter 15). Both the 92 kDa neutral collagenase and a number of cathepsins have been detected in osteoclasts by a combination of *in situ* hybridization, immunochemical or biochemical methods (Kakegawa et al., 1993; Ohsawa et al., 1993; Sasaki and Ueno-Matsuda, 1993; Goto et al., 1994; Reponen et al., 1994; Wucherpfennig et al., 1994; Shi et al., 1995). For several reasons cathepsins are strong candidates as the major bone-degrading enzyme. First, collagen-degrading members of this family have been purified from both freshly isolated avian osteoclasts (Blair et al., 1993) and human osteoclastoma tissue (Page et al., 1992). Second, since the pH in the isolated microenvironment where bone resorption occurs is between 4 and 5 (Silver et al., 1988), within the optimal range for cathepsin activity, it seems likely that this process represents the major mechanism by which the organic component of bone is removed. On the other hand, it is possible that neutral collagenases play a role either in removal of osteoid prior to the cycle of acid- and cathepsin-mediated degradation of mineralized bone, or to degrade collagen fragments following osteoclast detachment, at which point the pH will have risen to a value compatible with greater collagenase activity (Vaes, 1988).

II. MODELS OF OSTEOCLAST FUNCTION

Insights into the role of the osteoclast in bone biology require that suitable model systems be available for experimentation. Given the numbers of osteoclasts in any tissue are small, achieving this goal has been difficult. Studies of osteoclasts *in vivo*, while clearly examining the cell in the totality of its environment, suffer from the disadvantage that there is no direct access to the living cell. Several strategies, discussed in more detail elsewhere (Teitelbaum et al., 1996), have been developed to overcome these problems. In brief, these approaches involve whole animal or organ culture studies, both beset with problems of not knowing whether the result observed arises from direct effects on the osteoclast, or is mediated via accessory cells; gene targeting experiments, which have yielded important insights into osteoclast ontogeny; and isolation of viable osteoclasts, either by physical methods based on their large size, or by immunopanning, using antibodies to markers found at high levels on osteoclasts.

Of greater relevance to the present discussion is the approach of several groups, aimed at developing methods for *in vitro* generation of osteoclasts. The first successful attempts utilized mononuclear cells derived from the same calcium-deficient hens known to have many mature osteoclasts (Alvarez et al., 1991). When these tartrate-resistant acid phosphatase-negative monocytic cells were cultured at high density they fused within 5–6 days, yielding an almost homogeneous population of polykaryons having an osteoclastic phenotype. The large numbers of both precursors and fused cells allowed for the performance of a range of biochemical and cell biological studies.

Early attempts to generate mammalian osteoclasts *in vitro* met with limited success. Unfractionated marrow cultures, treated with 1,25 dihydroxyvitamin D_3 ($1,25(OH)_2D_3$), led to the production of small numbers of feline, canine, primate, and human osteoclast-like cells (Roodman et al., 1985; MacDonald et al., 1986). The cells, most notably those of human origin, exhibit limited ability to produce characteristic resorption pits. While early experiments with mouse marrow cells yielded similar results, recent progress has allowed the generation of sufficient cells to perform biochemical experiments. When either primary cultures of murine osteoblasts (Takahashi et al. 1988), or several clonal stromal cells lines (Udagawa et al. 1989, 1990), are cocultured with purified murine monocytic precursors, many multinucleated cells are generated, with the capacity to avidly resorb bone. This system has been used to demonstrate that the hormone, prostagandin E_2, and cytokine, interleukin (IL)-4, have opposite effects on osteoclastogenesis (Lacey et al., 1994). Using the same assay and a function-blocking antibody to macrophage colony stimulating factor (M-CSF), it was possible to reproduce *in vitro* the *in vivo* data concerning the essential nature of this cytokine for osteoclast production (Wiktor-Jedrzejczak et al., 1990; Kodama et al., 1991).

Recent advances in the isolation and culture of human osteoprogenitor and hematopoietic stem cells (Haylock et al., 1992; Gronthos et al., 1994; Huang and Terstappen, 1994), plus the growing knowledge of, and access to, recombinant human cytokines, suggests that the same approach used in the murine system may soon lead to an analogous human model on *in vitro* osteoclastogenesis.

III. STEROID HORMONE AND CYTOKINE REGULATION OF FUNCTIONAL OSTEOCLAST PROTEINS

A. Steroids, Osteoclastogenesis, and Osteoclast Function

Several steroids influence osteoclast generation and/or function. Thus, levels of the sex steroid estrogen determine production, by accessory cells, of the hematopoietic cytokines IL-1, IL-6, and tumor necrosis factor, all of which are capable of stimulating proliferation and/or differentiation of osteoclast precursors (Horowitz, 1993; Manolagas and Jilka, 1995; Pacifici, 1996). Similarly, glucocorticoids act on osteoblasts, increasing expression of the IL-6 receptor, a protein which complexes with two molecules of gp130 to mediate IL-6 signaling (Udagawa et al., 1995; Suda et al., 1996).

Retinoic acid and $1{,}25(OH)_2D_3$, two members of the steroid hormone superfamily (Wahli and Martinez, 1991), both stimulate bone resorption. For the retinoid this is true both *in vivo* (Hough et al., 1988) and *in vitro* (Scheven and Hamilton, 1990; Togari et al., 1991), but little is known about the mechanism of action of this effect (see below, however, for the effect of retinoic acid on integrin expression). In contrast, several genes critical to osteoclast function are known to be regulated by $1{,}25(OH)_2D_3$. The secosteroid, as part of its maturational capacity for osteoclast precursors, modulates surface levels of the M-CSF receptor (Perkins et al., 1991; Perkins and Teitelbaum, 1991). This result has important consequences for growth factor-dependent proliferation and differentiation, a key role of M-CSF. Additionally, this cytokine increases the rate of synthesis and surface expression of the integrins $\alpha_4\beta_1$ and $\alpha_5\beta_1$, both receptors for fibronectin, a major hematopoietic matrix protein. As a result of increased integrin expression the cells have increased capacity to interact with components of the marrow microenvironment, a first and necessary step for cell maturation (Shima et al., 1995).

A second protein regulated by $1{,}25(OH)_2D_3$ is carbonic anhydrase (Lomri and Baron, 1992), whose role in osteoclasts, discussed earlier, is to generate high levels of carbonic acid, the source of protons excreted by the cell. Analysis of the carbonic anhydrase promoter reveals a canonical vitamin D response element, capable of binding a complex containing the vitamin D receptor/retinoid X receptor (Quelo et al., 1994). The significance of this finding is that all steroids mediate their genomic actions through such heterodimers, acting as transcription factors (Glass, 1994).

B. Regulation of Integrin Expression on Avian Osteoclast Precursors

While isolated studies on steroid activation of osteoclast-related genes are of interest, the most detailed examination of steroid-induced osteoclast proteins comes from examination of integrin expression on osteoclast precursors. Given that $\alpha_v\beta_3$ expression is critical to osteoclast function, we turned to regulation of the heterodimer on precursor cells. Our initial studies, using the avian model, focused on modulation by steroid hormones. We demonstrated that $1,25(OH)_2D_3$, a hormone which enhances osteoclastogenesis *in vivo*, stimulates transcription of the α_v gene (Medhora et al., 1993). Furthermore, the same secosteroid also increases β_3 transcription, $\alpha_v\beta_3$ expression, and matrix attachment by early precursors. To study β_3 regulation, we cloned, by homology polymerase chain reaction, a full length cDNA for avian β_3, whose sequence demonstrated high homology with β_3 from other species (Mimura et al., 1994). Since retinoic acid also accelerates osteoclast formation and function, we asked if the retinoid influences $\alpha_v\beta_3$ expression. Our data demonstrate that, in contrast to $1,25(OH)_2D_3$, which increases transcription of both α_v and β_3, the retinoid augments surface appearance of $\alpha_v\beta_3$ by increasing β_3 and not α_v transcription. Furthermore, while treatment with retinoic acid enhances formation of tartrate resistant acid phophastase-positive polykaryons, these cells do not exhibit increased bone resorption. Thus, treatment with the retinoid alone fails to produce fully differentiated osteoclasts (Chiba et al., 1996). Since oophorectomy increases bone resorption *in vivo*, we explored the capacity of the sex steroid to regulate $\alpha_v\beta_3$ expression. Whereas nanomolar levels of hormone, typical of those seen in cycling women, have no effect on heterodimer expression, picomolar concentrations lead to increased $\alpha_v\beta_3$ appearance. In contrast to both $1,25(OH)_2D_3$ and retinoic acid there is no change in gene transcription at any level of estrogen. Rather picomolar, and not nanomolar, steroid increases stability of β_3 mRNA without altering that of α_v (Li et al., 1995). These data, when combined with differential regulation of β_3 and α_v and the fact that the α_v subunit can combine with multiple β partners, suggested to us that β_3 represents the rate-limiting subunit in regulating $\alpha_v\beta_3$ expression. For this reason we cloned the promoter region of the avian β_3 gene and demonstrated the presence of a canonical vitamin D response element (Cao et al., 1993). Examination of the proximal region of the promoter revealed a nearly perfect AP-1 site, suggesting the β_3 gene may be responsive to phorbol esters such as phorbol myristyl acetate. This hypothesis, and the expected increase in surface $\alpha_v\beta_3$ was confirmed by a combination of immunoprecipitation and transient transfection studies in an avian monocytic cell line. Furthermore, $1,25(OH)_2D_3$ and phorbol esters synergize in both gene transcription and integrin expression (Zhu et al., 1996). Finally, we defined a novel composite response element in a different region of the avian β_3 promoter. This DNA fragment, comprising three half sites separated by three and nine nucleotides binds both the vitamin D receptor/retinoid X receptor and retinoic acid receptor/retinoid X receptor complexes and, in transient transfection experiments, responds to both $1{,}25(OH)_2D_3$ and retinoic acid (Cao et al., 1996).

C. Cytokine Regulation of Mammalian Osteoclast Formation and Integrin Expression

In addition to steroids, osteoclastogenesis in mammals is also regulated by complex, hierarchical interactions of a number of cytokines (Suda et al., 1996). While avian osteoclast precursors are useful for examining steroid hormone regulation of integrin expression, this system cannot be used to study the role of cytokines, since avian counterparts of mammalian osteoclastogenic proteins are not available. Our initial studies were directed by the finding that a mouse overexpressing IL-4 exhibits low turnover osteoporosis accompanied by decreased osteoclast number (Lewis et al., 1993). To examine the possible mechanisms involved we turned to a murine model of osteoclast formation involving coculture of a stromal cell line, ST2, with nonadherent bone marrow-derived osteoclast precursors (Shioi et al., 1994) and showed that IL-4 inhibits osteoclastogenesis in a dose-dependent manner (Lacey et al., 1995). By addition of IL-4 to either ST2 or bone marrow cells prior to mixing, we established that inhibition of osteoclast formation arises as a result of the action of the cytokine on osteoclast precursors. A second cytokine, interferon γ, mimics the action of IL-4, while either prostaglandin E or cyclic AMP analogues stimulate osteoclastogenesis (Lacey et al., 1993). To determine whether IL-4 has a direct effect on mature osteoclasts, cocultures were continued for 8–10 days, at which time IL-4 was added and multinucleated cells were examined for alterations in both intracellular ionized calcium (Ca^{2+}) and the ability to resorb bone. IL-4 increased intracellular Ca^{2+} by both blocking influx of Ca^{2+} through a voltage operated channel and inhibiting release of Ca^{2+} from intracellular stores. The cytokine also decreased bone resorption in a dose-dependent manner (Bizzarri et al., 1994).

We next screened a panel of hematopoietic cytokines to identify those which regulate α_v and/or β_3 mRNA levels in osteoclast precursors and found that, of those tested, IL-4 was most potent, increasing β_3 with no change in α_v. The change in mRNA levels was accompanied by enhanced surface $\alpha_v\beta_3$, leading in turn to greater adhesion to RGD-containing matrix (Kitazawa et al., 1995). Subsequent studies revealed a second osteoclastogenic cytokine, granulocyte-macrophage colony stimulating factor, also increases β_3 mRNA levels and $\alpha_v\beta_3$ appearance on the surface of murine osteoclast precursors, while at the same time decreasing expression of the structurally and functionally related integrin $\alpha_v\beta_5$ (Inoue et al., 1995). Our data on β_3 regulation led us to initiate a strategy for cloning the murine β_3 promoter. We defined conditions whereby the stromal cells which contaminate the coculture system can be removed by proteolytic digestion, leading to the production of substantially pure, viable osteoclasts (Shioi et al., 1994) and used mRNA isolated from these cells to generate a cDNA library, from which we obtained by standard molecular techniques a short, partial 5' untranslated region. A combination of primer extension and S1-nuclease protection assays led to identification of the transcriptional start site and the complete 5' untranslated region was used to screen a genomic library. Several clones containing potential regions were isolated and a 3.5 kb DNA fragment, containing a number of potential binding sites for transcription factors was isolated and shown to act as an IL-4 inducible promoter (McHugh et al., 1994, 1995).

D. Regulation of *Rab3* Expression in Murine Osteoclast Precursors

We used the same murine coculture system described previously (Shioi et al., 1994) to determine if *rab* proteins are expressed during osteoclastogenesis. Given members of the *rab3* family play a role in exocytosis in a variety of other cell systems (reviewed in Ferro-Novick and Novick, 1993; Fisher von Mollard et al., 1994; Sudhof, 1995), we focused our attention on *rab3* isoforms. Our preliminary data indicate immature precursors contain little or no *rab3* proteins, but levels of two members, migrating with sizes of 25 and 27 kilodaltons (likely, but not proven to be *rab3A* and *rab3C*, respectively), increase as osteoclasts form. Treatment of precursors with IL-1, IL-3, IL-6, or tumor necrosis factor increases expression of the same *rabs* (Abu-Amer et al., 1995), suggesting yet another role for cytokines during osteoclastogenesis.

IV. CONCLUSIONS

The current model of osteoclast formation and function proposes that precursors, arising by differentiation of marrow-derived stem cells, respond in an hierarchical manner to a variety of stimuli which include both steroid hormones and hematopoietic cytokines. The net outcome is the proliferation and maturation of cells committed to the osteoclast lineage. Our findings, which have begun to define the molecular mechanisms by which functionally-important proteins are regulated in osteoclasts and their precursors, identify specific areas of osteoclast biology representing potential targets for the development of compounds capable of inhibiting bone resorption.

V. SUMMARY

The osteoclast is a physiological polykaryon which arises by fusion of precursors of the monocyte/macrophage lineage (Suda et al., 1992). The function of this cell is to resorb bone, a composite matrix consisting of both inorganic and organic elements. The inorganic component is largely substituted hydroxyapatite, while the organic phase contains 20 or more proteins, with type 1 collagen the single major species (>90% of total protein by weight (Boskey, 1989)). The molecular mechanisms underlying the activity of the osteoclast have become apparent largely as a result of studies conducted during the past decade, a subject briefly summarized. Thereafter we focused on the role of steroid hormones and cytokines in modulating macromolecules involved in osteoclast formation and/or function. Specifically, we reviewed the role and regulation of two families of proteins. The main emphasis was on integrins, a group of cell surface molecules, orginally shown to mediate cell-matrix and cell-cell interactions (Hynes, 1992), but now known to regulate many other aspects of cell behavior. Thus, integrin ligation leads to altered protein secretion, cell cycling, intracellular signaling pathways, and of particular importance in the context of bone, the capacity to sense mechanical stress (Werb et al., 1989; Ingber et al., 1994; Parsons, 1996). Second, *rabs*, whose number exceeds 20, are a subgroup of the cytosolic GTPase family (Nuoffer and Balch, 1995). *Rab* proteins are involved in targeted vesicular transport to cell membranes (Fisher von Mollard et al., 1994; Pfeffer, 1994). Given osteoclast polarization involves movement of intracellular vesicles towards the bone-osteoclast interface, followed by vesicle insertion thereby generating the characteristic ruffled membrane, *rab* proteins may well play a role in bone resorption.

ACKNOWLEDGMENTS

The studies performed and reported by the author are the result of a long-standing, close collaboration with Steven L. Teitelbaum, M.D.

REFERENCES

Abu-Amer, Y., Teitelbaum, S.L., Chappel, J., Zong, Q., and Ross, F.P. (1995). Regulation of a Rab protein, which mediates regulates exocytosis and co-localizes with ruffled membrane-associated proteins. J. Bone Miner. Res. 10, S150.

Alvarez, J.I., Teitelbaum, S.L., Blair, H.C., Greenfield, E.M., Athanasou, N.A., and Ross, F.P. (1991). Generation of avian cells resembling osteoclasts from mononuclear phagocytes. Endocrinol. 128, 2324-2335.

Baron, R., Neff, L., Brown, W., Courtoy, P.J., Louvard, D., and Farquhar, M. (1988). Polarized secretion of lysosomal enzymes: Co-distribution of cation-independent mannose-6-phosphate receptors and lysosomal enzymes along the osteoclast exocytic pathway. J. Cell Biol. 106, 1863-1872.

Baron, R., Neff, L., Brown, W., Louvard, D., and Courtoy, P.J (1990). Selective internalization of the apical plasma membrane and rapid redistribution of lysosomal enzymes and mannose-6-phosphate receptors during osteoclast inactivation by calcitonin. J. Cell Sci. 97, 439-447.

Bizzarri, C., Shioi, A., Teitelbaum, S.L., Ohara, J., Harwalker, V.A., Erdmann, J.M., Lacey, D.L., and Civitelli, R. (1994). Interleukin-4 inhibits bone resorption and acutely increase cytosolic Ca^{2+} in murine osteoclasts. J. Biol. Chem. 269, 13817-13824.

Blair, H.C., Kahn, A.J., Crouch, E.C., Jeffrey, J.J., and Teitelbaum, S.L. (1986). Isolated osteoclasts resorb the organic and inorganic components of bone. J. Cell Biol. 102, 1164-1172.

Blair, H.C., Teitelbaum, S.L., Schimke, P.A., Konsek, J.D., Koziol, C.M., and Schlesinger, P.H. (1988). Receptor-mediated uptake of mannose-6-phosphate bearing glycoprotein by isolated chicken osteoclasts. J. Cell. Physiol. 137, 476-482.

Blair, H.C., Teitelbaum, S.L., Ghiselli, R., and Gluck, S. (1989). Osteoclastic bone resorption by a polarized vacuolar proton pump. Science 245, 855-857.

Blair, H.C., Teitelbaum, S.L., Tan, H., Koziol, C.M., and Schlesinger, P.H. (1991). Passive chloride permeability charge coupled to H^+-ATPase of avian osteoclast ruffled membrane. Am. J. Physiol. 260, C1315-C1324.

Blair, H.C., Teitelbaum, S.L., Grosso, L.E., Lacey, D.L., Tan, H., McCort, D.W., and Jeffrey, J.J. (1993). Extracellular matrix degradation at acid pH: Avian osteoclast acid collagenase isolation and characterization. Biochem. J. 290, 873-884.

Boskey, A.L. (1989). Noncollagenous matrix proteins and their role in mineralization. Bone and Mineral 6, 111-123.

Boyce, B.F., Wright, K., Reddy, S.V., Koop, B.A., Story, B., Devlin, R., Leach, R.J., Roodman, G.D., and Windle, J.J. (1995). Targeting simian virus 40 T antigen to the osteoclast in transgenic mice causes osteoclast tumors and transformation and apoptosis of osteoclasts. Endocrinol. 136, 5751-5759.

Cao, X., Ross, F.P., Zhang, L., MacDonald, P.N., Chappel, J., and Teitelbaum, S.L. (1993). Cloning of the promoter for the avian integrin β_3 subunit gene and its regulation by 1,25-dihydroxyvitamin D_3. J. Biol. Chem. 268, 27371-27380.

Cao, X., Teitelbaum, S.L., Zhu, H., Zhang, L., Feng, X., and Ross, F.P. (1996). Competition for a unique response element mediates retinoic acid inhibition of vitamin D_3-stimulated transcription. J. Biol. Chem. (In press).

Chambers, T.J. (1988). The regulation of osteoclastic development and function. Ciba Found. Symp. 136, 92-107.

Chiba, M., Teitelbaum, S.L., Cao, X., and Ross, F.P. (1996). Retinoic acid stimulates expression of the functional osteoclast integrin $\alpha_v\beta_3$. Transcriptional activation of the β_3 but not α_v gene. J. Cell Biochem. (In Press.)

Collins, C.A. (1994). Dynein-based organelle movement. In: Microtubules. (Anonymous Eds.), pp. 367-380. Wiley-Liss, New York.

Elferink, L.A. and Scheller, R.H. (1993). Synaptic vesicle proteins and regulated exocytosis. J. Cell Sci. 17, 75-79.

Fath, K.R., Mamajiwalla, S.N., and Burgess, D.R. (1993). The cytoskeleton in development of epithelial cell polarity. J. Cell Sci. (Suppl. 17), 65-73.

Ferro-Novick, S. and Novick, P. (1993). The role of GTP-binding proteins in transport along the exocytic pathway. Annu. Rev. Cell Biol. 9, 575-599.

Fisher von Mollard, G., Stahl, B., Li, C., Sudhof, T., and Jahn, R. (1994). Rab proteins in regulated exocytosis. TIBS 19, 164-168.

Fisher, J.E., Caulfield, M.P., Sato, M., Quartuccio, H.A., Gould, R.J., Garsky, V.M., Rodan, G.A., and Rosenblatt, M. (1993). Inhibition of osteoclastic bone resorption in vivo by echistatin, an "arginyl-glycyl-aspartyl" (RGD)-containing protein. Endocrinol. 132, 1411-1413.

Geppert, M., Bolshakow, V.Y., Siegelbaum, S.A., Takei, K., De Camilli, P., Hammer, R.E., and Sudhof, T.C. (1994). The role of Rab3A in neurotransmitter release. Nature 369, 493-497.

Glass, C.K. (1994). Differential recognition of target genes by nuclear receptor monomers, dimers, and heterodimers. Endocr. Revs. 15, 391-407.

Goto, T., Kiyoshima, T., Moroi, R., Tsukuba, T., Nishimura, Y., Himeno, M., Yamamoto, K., and Tanaka, T. (1994). Localizaiton of cathepsins B, D, and L in the rat osteoclast by immuno-light and -electron microscopy. Histochemistry 101, 33-40.

Gronowicz, G.A. and Derome, M.E. (1994). Synthetic peptide containing Arg-Gly-Asp inhibits bone formation and resorption in a mineralizing organ culture system of fetal rat parietal bones. J. Bone Miner. Res. 9, 193-201.

Gronthos, S., Graves, S.E., and Simmons, P.J. (1994). The STRO-1$^+$ fraction of adult human bone marrow contains the osteogenic precursors. Blood 84, 4164-4173.

Haylock, D.N., To, L.B., Dowse, T.L., Juttner, C.A., and Simmons, P.J. (1992). Ex vivo expansion and maturation of peripheral blood CD34$^+$ cells into the myeloid lineage. Blood 80, 1405-1412.

Horowitz, M.C. (1993). Cytokines and estrogen in bone: Antiosteoporotic effects. Science 260, 626-627.

Horton, M.A., Taylor, M.L., Arnett, T.R., and Helfrich, M.H. (1991). Arg-gly-asp (RGD) peptides and the antivitronectin receptor antibody 23C6 inhibit dentine resorption and cell spreading by osteoclasts. Exp. Cell Res. 195, 368-375.

Hough, S., Avioli, L.V., Muir, H., Gelderblom, D., Jenkins, G., Kurasi, H., Slatopolsky, E., Bergfeld, M.A., and Teitelbaum, S.L. (1988). Effects of hypervitaminosis A on the bone and mineral metabolism of the rat. Endocrinol. 122, 2933-2939.

Huang, S. and Terstappen, W.M.M. (1994). Lymphoid and myeloid differentiation of single human CD34$^+$, HLA-DR$^+$, CD38$^-$ hematopoietic stem cells. Blood 83, 1515-1526.

Hynes, R.O. (1992). Integrins: Versatility, modulation, and signaling in cell adhesion. [Review]. Cell 69, 11-25.

Ingber, D.E., Dike, L., Hansen, L., Karp, S., Liley, H., Maniotis, A., McNamee, H., Mooney, D., Plopper, G., and Sims, J. (1994). Cellular tensegrity: Exploring how mechanical changes in the cytoskeleton regulate cell growth, migration, and tissue pattern during morphogenesis. Int. Rev. Cytol. 150, 173-224.

Inoue, M., Teitelbaum, S.L., Hurter, L., Hruska, K., Seftor, E., Hendrix, M., and Ross, F.P. (1995). GM-CSF regulates expression of the functional integrins $\alpha_v\beta_3$ and $\alpha_v\beta_5$ in a reciprocal manner during osteoclastogenesis. J. Bone Miner. Res. 10, S163.

Kakegawa, H., Nikawa, T., Tagami, K., Kamioka, H., Sumitani, K., Kawata, T., Drobnic-Kosorok, M., Lenarcic, B., Turk, V., and Katunuma, N. (1993). Participation of cathepsin L on bone resorption. FEBS Lett 321, 247-250.

Kallio, D.M., Garant, P.R., and Minkin, C. (1971). Evidence of coated membranes in the ruffled border of osteoclasts. J. Ultrastruct. Res. 37, 169-177.

King, K.L., D'Anza, J.J., Bodary, S., Pitti, R., Siegel, M., Lazarus, R.A., Dennis, M.S., Hammonds, Jr., R.G., and Kukreja, S.C. (1994). Effects of kistrin on bone resorption in vitro and serum calcium in vivo. J. Bone Miner. Res. 9, 381-387.

Kitazawa, S., Ross, F.P., McHugh, K., and Teitelbaum, S.L. (1995). Interleukin-4 induces expression of the integrin $\alpha_v\beta_3$ via transactivation of the β_3 gene. J. Biol. Chem. 270, 4115-4120.

Kodama, H., Nose, M., Niida, S., and Yamasaki, A. (1991). Essential role of macrophage colony-stimulating factor in the osteoclast differentiation supported by stromal cells. J. Exp. Med. 173, 1291-1294.

Lacey, D.L., Grosso, L.E., Moser, S.A., Erdmann, J., Tan, H.L., Pacifici, R., and Villareal, D.T. (1993). IL-1-induced murine osteoblast IL-6 production is mediated by the type-1 IL-1 receptor and is increased by 1,25 dihydroxyvitamin D3. J. Clin. Invest. 91, 1731-1742.

Lacey, D.L., Erdmann, J.M., and Tan, H. (1994). Interleukin 4 increases type-5 acid phosphatase mRNA expression in murine bone marrow macrophages. J. Cell. Biochem. 54, 365-371.

Lacey, D.L., Erdmann, J.M., Teitelbaum, S.L., Tan, H., Ohara, J., and Shioi, A. (1995). Interleukin-4, interferon and prostaglandin E impact the osteoclast-forming potential of murine bone marrow macrophages in vitro. Endocrinol. 136, 2367-2376.

Lewis, D.B., Liggitt, H.D., Effmann, E.L., Motley, S.T., Teitelbaum, S.L., Jepsen, K.J., Goldstein, S.A., Bonadio, J., Carpenter, J., and Perlmutter, R.M. (1993). Osteoporosis-induced in mice by overproduction of interleukin 4. Proc. Natl. Acad. Sci. USA 90, 11618-11622.

Li, C., Ross, F.P., Cao, X., and Teitelbaum, S.L. (1995). Estrogen enhances $\alpha_v\beta_3$ integrin expression by avian osteoclast precursors via stabilization of β_3 integrin mRNA. Molec. Endocrinol. 9, 805-813.

Lledo, P.M., Vernier, P., Vincent, J.D., Mason, W.T., and Zorec, R. (1993). Inhibition of Rab3B expression attenuates Ca^{2+}-dependent exocytosis in rat anterior pituitary cells. Nature 364, 540-544.

Lomri, A. and Baron, R. (1992). 1α,25-dihydroxyvitamin D3 regulates the transcription of carbonic anhydrase II mRNA in avian myelomonocytes. Proc. Natl. Acad. Sci. USA 89, 4688-4692.

MacDonald, B.R., Mundy, G.R., Clark, S., Wang, E.A., Kuehl, T.J., Stanley, E.R., and Roodman, G.D. (1986). Effects of human recombinant CSF-GM and highly purified CSF-1 on the formation of multinucleated cells with osteoclast characteristics in long-term bone marrow cultures. J. Bone Miner. Res. 1, 227-233.

Manolagas, S.C. and Jilka, R.L. (1995). Bone marrow, cytokines, and bone remodeling. Emerging insights into the pathophysiology of osteoporosis. N. Engl. J. Med. 332, 305-311.

Mattsson, J.P., Vaananen, K., Wallmark, B., and Lorentzon, P. (1991). Omeprazole and bafilomycin, two proton pump inhibitors: Differentiation of their effects on gastric, kidney, and bone H^+-translocating ATPases. Biochim. Biophys. Acta 1065, 261-268.

Mattsson, J.P., Schlesinger, P.H., Keeling, D.J., Teitelbaum, S.L., Stone, D.K., and Xie, X. (1994). Isolation and reconstitution of a vacuolar-type proton pump of osteoclast membranes. J. Biol. Chem. 269, 24979-24982.

McHugh, K., Teitelbaum, S.L., Kitazawa, S., and Ross, F.P. (1994). Cloning and characterization of the murine integrin β_3 gene promoter. J. Bone Miner. Res. 9, S248 (Abstract).

McHugh, K., Teitelbaum, S.L., and Ross, F.P. (1995). Interleukin-13, like interleukin-4, modulates murine osteoclastogenesis and expression of the integrin $\alpha_v\beta_3$ on osteoclast precursors. J. Bone Miner. Res. 10, S487 (Abstract).

Medhora, M.M., Teitelbaum, S.L., Chappel, J., Alvarez, J., Mimura, H., Ross, F.P., and Hruska, K. (1993). 1α,25-dihydroxyvitamin D_3 upregulates expression of the osteoclast integrin $\alpha_v\beta_3$. J. Biol. Chem. 268, 1456-1461.

Mimura, H., Cao, X., Ross, F.P., Chiba, M., and Teitelbaum, S.L. (1994). $1,25(OH)_2D_3$ vitamin D_3 transcriptionally activates the β_3-integrin subunit gene in avian osteoclast precursors. Endocrinol. 134, 1061-1066.

Miyauchi, A., Alvarez, J., Greenfield, E.M., Teti, A., Grano, M., Colucci, S., Zambonin-Zallone, A., Ross, F.P., Teitelbaum, S.L., and Cheresh, D. (1991). Recognition of osteopontin and related peptides by an $\alpha_v\beta_3$ integrin stimulates immediate cell signals in osteoclasts. J. Biol. Chem. 266, 20369-20374.

Montgomery, A.M.P., Reisfeld, R.A., and Cheresh, D.A. (1994). Integrin $\alpha_v\beta_3$ rescues melanoma cells from apoptosis in three-dimensional dermal collagen. Proc. Natl. Acad. Sci. USA 91, 8856-8860.

Nickols, G.A., Settle, S.L., Engleman, V.W., Ruminski, P.G., Ross, F.P., and Teitelbaum, S.L. 1995). Prevention of oophorectomy-induced bone loss by a nonpeptide antagonist of the osteoclast integrin $\alpha_v\beta_3$. J. Bone Miner. Res. 10, S151 (Abstract).

Novick, P. and Garrett, M.D. (1994). No exchange without receipt. Nature 369, 18-19.

Nuoffer, C. and Balch, W.E. (1995). GTPases: Multifunctional molecular switches regulating vesicular traffic. Ann. Rev. Biochem. 63, 949-990.

Ohsawa, Y., Nitatori, T., Higuchi, S., Kominami, E., and Uchiyama, Y. (1993). Lysosomal cysteine and aspartic proteinases, acid phosphatase, and an endogenous cysteine proteinase inhibitor, cystatin-β, in rat osteoclasts. J. Histochem. Cytochem. 41, 1075-1083.

Pacifici, R. (1996). Postmenopausal osteoporosis. In: Osteoporosis. (Marcus, R., Feldman, D., and Kelsey, J. Eds.), pp. 727-743. Academic Press, San Diego.

Page, A.E., Warburton, M.J., Chambers, T.J., Pringle, J.A., and Hayman, A.R. (1992). Human osteoclastomas contain multiple forms of cathepsin B. Biochim. Biophys. Acta 1116, 57-66.

Parsons, J.T. (1996). Integrin-mediated signalling: Regulation by protein tyrosine kinases and small GTP-binding proteins. Curr. Op. Cell Biol. 8, 146-152.

Perkins, S.L., Link, D.C., Kling, S., Ley, T.J., and Teitelbaum, S.L. (1991). 1,25-dihydroxyvitamin D_3 induces monocytic differentiation of the PLB-985 leukemic line and promotes c-*fgr* mRNA expression. J. Leukocyte Biol. 50, 427-433.

Perkins, S.L. and Teitelbaum, S.L. (1991). 1,25-dihydroxyvitamin D_3 modulates colony stimulating factor-1 receptor binding by murine bone marrow macrophage precursors. Endocrinol. 128, 303-311.

Pfeffer, S.R. (1994). Rab GTPases: Master regulators of membrane trafficking. Curr. Op. Cell Biol. 6, 522-526.

Quelo, I., Kahlen, J., Rascle, A., Jurdic, P., and Carlberg, C. (1994). Identification and characterization of a vitamin D_3 response element of chicken carbonic anhydrase-II. DNA Cell Biol. 13, 1181-1187.

Raff, E.C. (1994). The role of multiple tubulin isoforms in cellular microtubule function. In: Microtubules., pp. 85-109. Wiley-Liss, Inc., New York.

Reinholt, F.P., Hultenby, K., Oldberg, A., and Heinegard, D. (1990). Osteopontin—a possible anchor of osteoclasts to bone. Proc. Natl. Acad. Sci. USA 87, 4473-4475.

Reponen, P., Sahlberg, C., Munaut, C., Thesleff, I., and Tryggvason, K. (1994). High expression of 92-kD type-IV collagenase (gelatinase B) in the osteoclast lineage during mouse development. J. Cell Biol. 124, 1091-1102.

Roodman, G.D., Ibbotson, K.J., MacDonald, B.R., Kuehl, T.J., and Mundy, G.R. (1985). 1,25-dihydroxyvitamin D_3 causes formation of multinucleated cells with several osteoclast characteristics in cultures of primate marrow. Proc. Natl. Acad. Sci. USA 82, 8213-8217.

Ross, F.P., Alvarez, J.I., Chappel, J., Sander, D., Butler, W.T., Farach-Carson, M.C., Mintz, K.A., Robey, P.G., Teitelbaum, S.L., and Cheresh, D.A. (1993). Interactions between the bone matrix proteins osteopontin and bone sialoprotein and the osteoclast integrin $\alpha_v\beta_3$ potentiate bone resorption. J. Biol. Chem. 268, 9901-9907.

Rothman, J.E. and Warren, G. (1994). Implications of the SNARE hypothesis for intracellular membrane topology and dynamics. Curr. Op. Cell Biol. 4, 220-232.

Sasaki, T. and Ueno-Matsuda, E. (1993). Cysteine-proteinase localization in osteoclasts: An immunocytochemical study. Cell Tissue Res. 271, 177-179.

Schekman, R. and Orci, L. (1996). Coat proteins and vesicle budding. Science 271, 1526-1538.

Scheven, B.A. and Hamilton, N.J. (1990). Retinoic acid and 1,25-dihydroxyvitamin D_3 stimulate osteoclast formation by different mechanisms. Bone 11, 53-59.

Shi, G.P., Chapman, H.A., Bhairi, S.M., DeLeeuw, C., Reddy, V.Y., and Weiss, S.J. (1995). Molecular cloning of human cathepsin O, a novel endoproteinase and homologue of rabbit OC2. FEBS Lett 357, 129-134.

Shima, M., Teitelbaum, S.L., Holers, M.V., Ruzicka, C., Osmack, P., and Ross, F.P. (1995). M-CSF regulates expression of the integrin $\alpha_4\beta_1$ and $\alpha_5\beta_1$ in murine bone marrow macrophages. Proc. Natl. Acad. Sci. USA 92, 5179-5183.

Shioi, A., Ross, F.P., and Teitelbaum, S.L. (1994). Enrichment of generated murine osteoclasts. Calcif. Tissue Int. 55, 387-394.

Silver, I.A., Murrills, R.J., and Etherington, D.J. (1988). Microelectrode studies on the acid microenvironment beneath adherent macrophages and osteoclasts. Exp. Cell Res. 175, 266-276.

Suda, T., Takahashi, N., and Martin, T.J. (1992). Modulation of osteoclast differentiation. Endocr. Revs. 13, 66-80.

Suda, T., Udagawa, N., Nakamura, I., Miyaura, C., and Takahashi, N. (1996). Modulation of osteoclast differentiation by local factors. Bone 17, 87S-91S.

Sudhof, T.C. (1995). The synaptic vesicle cycle: A cascade of protein-protein interactins. Nature 375, 645-653.

Takahashi, N., Akatsu, T., Udagawa, N., Sasaki, T., Yamaguchi, A., Moseley, J.M., Martin, T.J., and Suda, T. (1988). Osteoblastic cells are involved in osteoclast formation. Endocrinol. 123, 2600-2602.

Teitelbaum, S.L., Tondravi, M.M., and Ross, F.P. (1996). Osteoclast Biology. In: Osteoporosis. (Marcus, R., Feldman, D., and Kelsey, J. Eds.), pp. 61-94. Academic Press, San Diego.

Teti, A., Blair, H.C., Teitelbaum, S.L., Kahn, A.J., Koziol, C.M., Konsek, J., Zambonin-Zallone, A., and Schlesinger, P. (1989). Cytoplasmic pH regulation and chloride/bicarbonate exchange in avian osteoclasts. J. Clin. Invest. 83, 227-233.

Togari, A., Kondo, M., Arai, M., and Matsumoto, S. (1991). Effects of retinoic acid on bone formation and resorption in cultured mouse calvaria. Gen. Pharmacol. 22, 287-292.

Udagawa, N., Takahashi, N., Akatsu, T., Sasaki, T., Yamaguchi, A., Kodama, H., Martin, T.J., and Suda, T. (1989). The bone marrow-derived stromal cell lines MC3T3-G2/PA6 and ST2 support osteoclastlike cell differentiation in cocultures with mouse spleen cells. Endocrinol. 125, 1805-1813.

Udagawa, N., Takahashi, N., Akatsu, T., Tanaka, H., Sasaki, T., Nishihara, T., and Koga, T. (1990). Origin of osteoclasts: Mature monocytes and macrophages are capable of differentiating into osteoclasts under a suitable microenvironment prepared by bone marrow-derived stromal cells. Proc. Natl. Acad. Sci. USA 87, 7260-7264.

Udagawa, N., Takahashi, N., Katagiri, T., Tamura, T., Wada, S., Findlay, D.M., Martin, T.J., Hirota, H., Tada, T., Kishimoto, T., and Suda, T. (1995). Interleukin (IL-1)-6 induction of osteoclast differentiation depends on IL-6 receptors expressed on osteoblastic cells but not on osteoclast progenitors. J. Exp. Med. 182, 1461-1468.

Vaes, G. (1988). Cellular biology and biochemical mechanism of bone resorption. A review of recent developments on the formation, activation, and mode of action of osteoclasts. Clin. Orthop. Rel. Res. 231, 239-271.

Wahli, W. and Martinez, E. (1991). Superfamily of steroid nuclear receptors: Positive and negative regulators of gene expression. FASEB J. 5, 2243-2249.

Weber, E., Berta, G., Tousson, A., St. John, P., Green, M.W., Gopalokrishnan, U., Jilling, T., Sorscher, E.J., Elton, T.S., and Abrahamson, D.R. (1994). Expression and polarized targeting of a rab3 isoform in epithelial cells. J. Cell Biol. 125, 583-594.

Werb, Z., Tremble, P.M., Behrendtsen, O., Crowley, E., and Damsky, C.H. (1989). Signal transduction through the fibronectin receptor induces collagenase and stromelysin gene expression. J. Cell Biol. 109, 877-889.

Wiktor-Jedrzejczak, W., Bartocci, A., Ferrante, A.W.J., Ahmed-Ansari, A., Sell, K.W., Pollard, J.W., and Stanley, E.R. (1990). Total absence of colony-stimulating factor 1 in the macrophage-deficient osteopetrotic (op/op) mouse. Proc. Natl. Acad. Sci. USA 87, 4828-4832.

Wucherpfennig, A.L., Li, Y.P., Stetler-Stevenson, W.G., Rosenberg, A.E., and Stashenko, P. (1994). Expression of 92-kD, type-IV collagenase/gelatinase B in human osteoclasts. J. Bone Miner. Res. 9, 549-556.

Zerial, M. and Stenmark, H. (1993). Rab GTPases in vesicular transport. Curr. Op. Cell Biol. 5, 613-620.

Zhu, H., Ross, F.P., Cao, X., and Teitelbaum, S.L. (1996). Phorbol myristate acetate transactivates the avian β_3 integrin gene and induces $\alpha_v\beta_3$ integrin expression. J. Cell. Biochem. 61, 420-429.

Zimolo, Z., Wesolowski, G., Tanaka, H., Hyman, J.L., Hoyer, J.R., and Rodan, G.A. (1994). Soluble $\alpha_v\beta_3$-integrin ligands raise $[Ca^{2+}]_i$ in rat osteoclasts and mouse-derived osteoclastlike cells. Am. J. Physiol. 266, C376-C381.

THE OSTEOCLAST CYTOSKELETON

Alberta Zambonin Zallone, Maria Grano,
and Silvia Colucci

I. INTRODUCTION

The functions related to the cytoskeletal organization of the cytoplasm have become much better understood over recent years. The cytoskeleton can no longer be considered responsible only for the mantainance of cell shape. Locomotion, mitosis, and cytokinesis require the precisely orchestrated activity of proteins that regulate actin and tubulin assembly and disassembly, together with motor proteins

Advances in Organ Biology
Volume 5B, pages 347-357.

ISBN: 0-7623-0390-5

based on microtubules and actin filaments. A much more subtle expression of cytoskeletal function is the mantainance of the precise distribution of ions, metabolites, macromolecules, and organelles in time and space within the living cell. Moreover it is known that signal transduction initiated by growth factors results in rearrangement of all components of the cytoskeleton (Bockus and Stiles, 1984), but the molecular pathway connecting the two events remains to be determined (Ridley and Hall, 1994).

Osteoclasts are good models to study cytoskeletal organization because of their complex structure, but their studies are complicated by the fact they are highly dynamic and their status can change frequently and dramatically from migrating, to stationary, or from nonpolarized to polarized. These dynamic changes can be induced by the presence of systemic or local stimuli that can switch bone resorption on and off. Additionally, osteoclast are morphologically heterogeneous cells. In the same area of the skeleton, osteoclasts of different sizes and activity can easily be found. The life cycle of these cells is also very peculiar, due to the unique situation of a continuously renewing syncitium (Jaworski et al., 1981). Thus, osteoclasts are thought to receive new nuclei and lose old ones while the resorption lasts, and at the end of the resorption cycle the cells disappear very probably for apoptosis (Hughes et al., 1995).

In their life span, osteoclasts are highly motile and can alternate resorption phases with migration on the bone surfaces. Before the onset of bone resorption, however, osteoclasts become polarized and their inner organization dramatically changes. When resorption ends, they can move toward another bone surface, returning to the previous nonpolarized morphology. Moreover, as demonstrated with time lapse microcinematography, these changes are, at least *in vitro*, fast and frequent, often giving rise to multiple or closely located resorption lacunae (Kanehisa et al., 1990). Changes that we observe in the cytoskeletal organization support these changes.

Osteoclast cytoskeletal components have been extensively reviewed in the past (Teti and Zambonin-Zallone, 1992). In this chapter, we highlight the more recent studies on the molecular organization and functional role of the different components of the osteoclast cytoskeleton.

II. THE OSTEOCLAST CYTOSKELETON

Eukaryotic cells contain three major classes of cytoskeletal fibers: 7 nm diameter actin microfilaments, 24 nm diamenter microtubules (MTs), and 10 nm diameter intermediate filaments. All of these fibers are formed by finely regulated polymerization of protein subunits. These cytoskeletal elements play a fundamental role not only in determining the steady-state organization of the cytoplasm, but also in facilitating selective delivery between spatially segregated organelles. Notably, eukaryotic cells have developed highly regulated membrane trafficking pathways that function to mediate the exchange of lipid or protein between distinctive membrane-bound compartments or organelles. Transport intermediates that utilize these pathways must often travel significant intracellular distances to reach specific targets. This regulated mode of transport occurs within an architectural framework that it is likely to impose significant constraints on the diffusion of macromolecular components within the cytoplasm (Luby-Phelps, 1994). Recently a model for the organization of organelles and membrane transport pathways and their relation to centrosomally-arranged MTs has been presented. In this model, endoplasmic recticulum-to-Golgi transport, as the movement of the endocytic carrier vesicles from peripheral to late endosomes, is thought to involve MT minus end-directed movement (Cole and Lippincott-Schwartz, 1995). In polarized epithelial cells, post-Golgi vesicles are transported to the apical cortex by a dynein driven movement, while myosin-I provides the subsequent force for vesicle delivery to the apical membrane (Fath et al., 1994).

The actin and MT cytoskeletal systems are closely interrelated both structurally and functionally in cells. Both filament systems commonly coexist within domains of the cytoplasm, and factors that cause redistribution of one often lead to a change in the distribution of the other. The extents to which the two systems are redundant and the extent to which they function coordinately are areas of active investigation (for a review see Langford, 1995).

Cell motility is a cytoskeleton-dependent function. At the leading edge of migrating mouse fibroblasts, a biased assembly of actin is involved in the formation and stabilization of protrusions, while the perinuclear tail region exhibits a net disassembly of actin fibers (Giuliano and Taylor, 1994). The osteoclast can be in a secretory or in a migratory state depending on the phase of its activity. The more recent findings on cytoskeleton involvement in these processes can very possibly apply to this very peculiar cell. However, although there are many new results concerning osteoclast-matrix interaction and on related integrin receptors, there has been comparatively little work specifically addressed to the other cytoskeletal components.

A. Microtubules

MTs represent one of the fiber systems of the cytoskeleton. A MT consists of a core cylinder built of heterodimers of α and β-tubulin monomers. It is generally acknowledged that MTs and their associated motor proteins play fundamental roles in the organization of organelles and in the efficient transport of protein and lipid between different cell compartments. However, the way this is accomplished is far from clear.

Three main classes of proteins interact with tubulin. First, the MT-associated proteins, or MAPs, tend to stabilize and promote the assembly of MTs. Second, the motor proteins, kinesin and dynein, generate movement along MTs using chemical energy generated through ATP hydrolysis. Thirdly, a more heterogeneous class of proteins that interact with tubulin include glycolytic enzymes and kinases.

In many cell types, MTs are nucleated during interphase from a perinuclear MT-organizing center (MTOC) and radiate out toward the cell periphery to form an extensive network throughout the cytoplasm. The minus end (slow-growing) is buried in the MT organizing structure containing a γ-tubulin. The plus, fast-growing, end terminates with a crown of α-tubulin. Microtubules contain not-exchangeable GTP on α-tubulin and an exchangeable GTP on β-tubulin. An additional ATP-binding site is present on α-tubulin. The half-lives of most MTs are in the range of minutes and the predominant mechanism of redistribution appears to be dynamic instability (reviewed by Cassimeris, 1993).

The MT organization of osteoclasts do not differ from other types of cells. They form an extensive and elaborate array radiating from the perinuclear region throughout the cytoplasm toward the cell margin, but do not establish contact with the cell membrane (Figure 1). Notably, the following pattern has been observed. The intense staining around the nuclei resulting from many overlapping MTs in this thick region were often found to obscure the MTOC (Figure1B). However, one or up to four MTOCs could be observed in very spread osteoclasts (Zambonin-Zallone et al., 1983; Warshafsky et al., 1985; Turksen et al., 1988). The distribution of the centriole was found to differ from one cell to another. A single centriole, each one close to one nucleus, as well as aggregates of several centrioles localized near the nuclei, have been described (Turksen et al., 1988) (Figure 2).

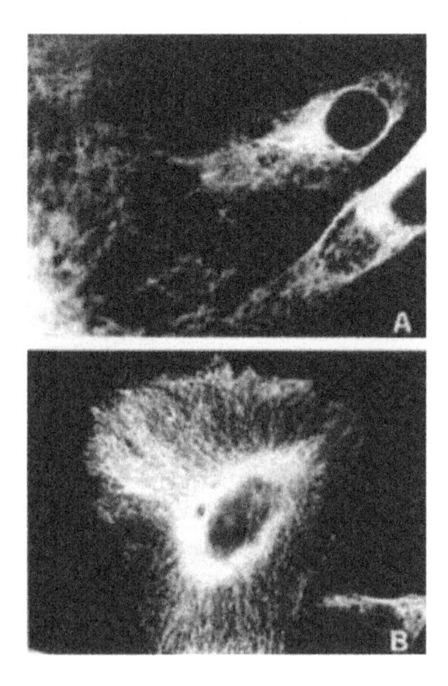

Figure 1. Immunofluorescence staining for α and β-tubulin. (**A**) Peripheral MTs in a large osteoclast seem to contact the MT array of two osteoclast precursors in the process of fusion with the mature cell. (**B**) Retinol-treated osteoclasts show a dense array of MTs radiating from the perinuclear area: The MT organizing center (MTOC) is not identifiable. Magnification: x 1000.

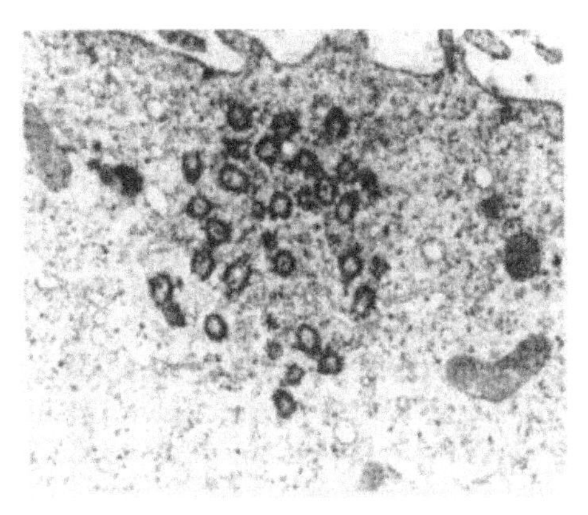

Figure 2. Transmission electron micrograph of an osteoclast. A group of centrioli close to the plasma membrane are evident. Magnification: x 6000.

The treatment of osteoclasts with calcitonin results in cell retraction, but does not significantly alter MT distribution. The MTs, however, appear more tightly compacted in the inner cell body (Warshafsky et al., 1985). Treatment with retinol or retinoic acid induces a reversible depolymerization of MTs in avian osteoclasts (Oreffo et al., 1988), but similar results could not be confirmed in human giant cell tumor-derived osteoclast-like cell lines (Colucci et al., in preparation). Retinoic acid-treated osteoclast-like cells showed, compared to controls, a thicker array of tubules and a granular distribution of tubulin in the cytoplasm that was absent in the controls (Figure 1B). The significance of this effect is unclear but can be related to the increased secretion of matrix metalloproteinases that we observe under similar circumstances (Colucci et al., in preparation). Resorbing osteoclasts on bone laminae interestingly show a noticeable concentration of MTs at cytoplasmic sites closest to the resorption lacunae (Lakkakorpi and Väänänen, 1991).

B. Intermediate Filaments

Five classes of intermediate filaments have been recognized to date on the basis of sequence similarity in the rod domain. These are referred as types I to V. Type I filaments are acidic keratin, type II are neutral to basic keratin, type III contain the vimentin-related group, type IV are neurofilaments, and type V the lamins. All intermediate filaments are similar structurally. A highly conserved 40 nm-long central rodlike domain is formed by coiling of α-helices of two polypeptides around each other. The N- and C-terminal domain are globular. Dimers form antiparallel tetramers which, in turn, form long protofilaments; eight of these form a 10 nm diameter intermediate filament.

The expression of intermediate filaments is strictly tissue specific, suggesting that the specific type present in a cell is fundamentally related to its function. It has been assumed that the progenitor of the IF protein was a nuclear lamin (Dodemont et al., 1994). Vimentin, desmin, peripherin, and glial fibrillary acidic protein (GFAP) are the most similar in terms of primary structure. *In vitro*, each can copolymerize with the other. They assemble hierarchically, first as dimers, then as dimer-dimers (tetramers), and then as higher order structures. The formation of tetramers appears to be driven by ionic interactions (Traub et al., 1993; Meng et al., 1994).

Osteoclasts display an array of vimentin filaments (Zambonin-Zallone et al., 1983). Vimentin is a 57,000 molecular weight phosphoprotein and its level of phosphorylation changes during the cell cycle. It has been proposed that vimentin acts as a "phosphate sink" buffering the cell against excess kinase activity (Lai et al., 1993). If this were a real function of vimentin, its effect on cell physiology could be quite subtle: by essentially protecting one kinase, the presence of a vimentin system could modulate the extent to which other kinase target molecules were phosphorylated which, in turn, could lead to changes in cellular responses (Klimkowski, 1995).

In osteoclasts, intermediate filaments are radially arranged and form a circular belt not far from the cell margin. In some cells they can also form a reinforcing ring around the nuclear cluster (Zambonin-Zallone et al., 1983; Marchisio et al., 1984). The role of vimentin filaments in osteoclasts as well as in other cell types, however, remains to be elucidated.

C. Actin Microfilaments

Actin is the more abundant protein in almost all the eukaryotic cells. Actin filaments (F-actin) are 8 nm wide and consist of monomers of G-actin (molecular weight 42,000). The tri-dimensional organization of G-actin has been deduced from x-ray diffraction studies (Kabsch et al., 1990; Kabsch and Vanderkeekhove, 1992). Each actin subunit has a defined polarity and the subunits polymerize head-to-tail. As a consequence, F-actin also has a defined polarity. The role of actin in cells goes from actin-based cell locomotion to various kinds of cytoplasmic motility or to the organization of an extracellular matrix-linked intracellular filamentous network.

Osteoclasts are, during bone resorption, polarized cells that display two new membrane domains: the sealing or clear zone, which is the specialized cell-extracellular matrix adhesion structure, and the ruffled border, which is the membrane specialization in the resorbing compartment facing the resorption lacuna. Migrating, nonpolarized osteoclasts have a homogeneous organization of the plasma membrane and can display a diffuse F-actin distribution. This is evident when cells are immunostained with fluorescent phlloidin. Actin is particularly concentrated at membrane rufflings (Figure 3A). *In vitro,* a partial polarization of stationary osteoclasts is possible; however, the cells show only a clear zone and lack ruffled borders. In these cells F-actin is organized in discrete brightly fluorescent dots (Figures 3B,C,D) (Marchisio et al., 1984). In transmission electron

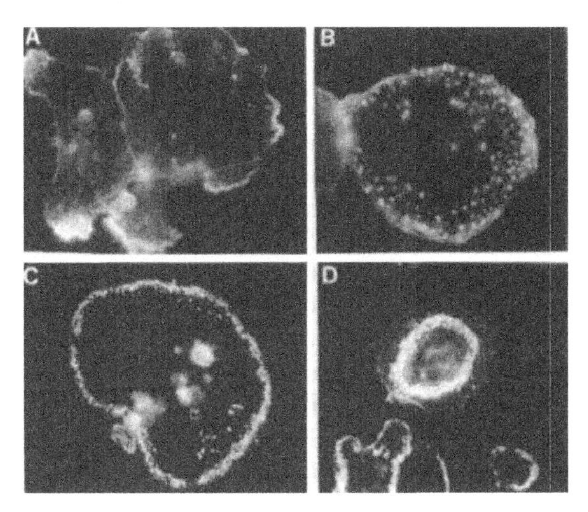

Figure *3.* Immunofluorescence micrographs of osteoclasts stained with rhodamine-conjugated phalloidin of F-actin. (**A**) Motile osteoclasts with membrane rufflings and diffuse microfilament network. (**B**) An osteoclast that is spread and polarized toward the culture surface. Podosomes are evident as single bright fluorescent dots at the cell margin. (**C, D**): The podosomes in these osteoclasts form multiple rows, giving rise to a broad actin band that encircles the central area of the cells. Magnification: A,B x 1000; C,D x 400.

micrographs of perpendicularly sectioned cells, the actin dots represent microfilaments oriented perpendicularly toward the ventral plasma membrane. They form discrete blunt protrusions called podosomes. Such structures have also been found in other cells that migrate and invade the extracellular matrix, as the Rous sarcoma transformed fibroblasts (Tarone et al., 1985) or in osteoclast-related cells, as monocytes and macrophages (Zambonin-Zallone et al., 1983).

III. THE PODOSOMES

Podosomes are adhesion organelles. They contain almost all the same proteins found in focal adhesions, but are arranged in a partially different way (Burridge et al, 1988; Zambonin-Zallone et al., 1988, 1989a,b; Aubin, 1992; Teti et al., 1993). Furthermore, while focal adhesions require the presence of serum and several hours to form, podosomes are highly dynamic and can assemble or disappear in minutes (Burridge et al., 1988). Studies with interference reflection microscopy (IRM) indicate that podosomes correspond to 30 nm grey reflections between the plasma membrane and the substratum. Their inner core is formed by an axis of actin containing microfilaments associated with the actin binding proteins, fimbrin and α-actinin, and with the regulatory protein gelsolin. The core is surrounded by vinculin and talin in a ringlike organization. The presence of more recently identified focal adhesion components such as paxillin and tensin has not yet been investigated.

Early work on osteoclast ultrastructure has shown the presence of actin microfilaments within the clear zone of resorbing osteoclasts. They were found to be oriented toward the bone facing plasma membrane (King and Holtrop, 1975). Despite these demonstrations, their structural and functional details have been lacking until recently. The fundamental role of podosomes in the process of recognition and adhesion of the osteoclast to the bone matrix has now been clearly demonstrated by several authors (Zambonin-Zallone et al., 1988, 1989a,b; Kanehisa et al., 1990; Lakkakorpi and Vaananen, 1991; Aubin, 1992; Teti et al., 1993).

Podosomes coincide with small indentations of the bone substrate both *in vivo* and *in vitro* (Zambonin-Zallone et al., 1988). An extracellular matrix-degrading proteolytic activity bound to podosome membranes has been demonstrated in transformed fibroblasts (Chen et al., 1984). Specific data in this regard are not available for osteoclasts. We could however detect by immunostaining the presence of urokinase that appeared to be colocalized with vinculin on human nontransformed osteoclasts obtained from surgical specimens (Zambonin-Zallone, unpublished observations). The significance of this remains unclear.

If osteoclasts are plated onto bone or dentine slices, the F-actin staining pattern in those cells that are engaged in resorption is very different from that seen with the nonresorbing ones. During activation, osteoclasts undergo rapid and dramatic changes in cell polarization. A very broad ring of podosomes outlines the boundary of resorbing lacunae. Remodeling and reorganization of the bands of podosomes precedes noticeable resorptive activity (Kanehisa et al., 1990; Lakkakorpi and Väänänen, 1991). Whether individual podosomes are still present during resorption, or whether actin and vinculin undergo a different kind of organization, is still a matter of debate. Ultrastructural evidence from resorbing osteoclasts demonstrates podosome-like protrusion on the plasma membrane of the clear zone. However, it is also possible that other kinds of actin containing adhesion structures exist in this region, with an ultrastructural morphology different from podosomes, but resulting in a close-contact type of adhesion. After a careful re-examination of published pictures it has been proposed by Aubin (1992) that podosomes are superimposed on a flat ring or disk-shaped area in close proximity to the substratum. A double ring of vinculin and talin, that may be partially formed by podosomes separated by a broad band of F-actin, has also been described during the resorptive phase (Lakkakorpi and Väänänen, 1991).

While the presence of podosomes during the resorptive phase is under debate, their role during bone matrix recognition and during the following polarization phase is not in doubt. Podosomes, as focal adhesions, cannot be considered only devices organized in order to stabilize cells on their substrate. They contain transmembrane proteins, the integrins, that are ligands for the extracellular matrix and that are capable of transmitting distinct signals to the cytoplasm and to the nucleus

(Clark and Brugge, 1995). The main integrin receptor in osteoclast membrane, involved in bone matrix recognition and following events, is the classical vitronectin receptor $\alpha_v\beta_3$ (Davies et al., 1989; Zambonin-Zallone et al., 1989). Osteoclast integrins, however, are reviewed in Chapter by Horton in Volume 5B and therefore will not be treated in detail here.

In vitro, several stimuli can change the cytoskeletal organization and the assembly and disassembly of podosomes in concert with a modification of osteoclast resorbing activity. In fact, metabolic acidosis and retinol, known to increase bone resorption rate in cultured chicken osteoclasts, stimulate cultured osteoclasts to polarize and to form podosome-rich clear zones (Oreffo et al., 1988; Teti et al., 1989). Conversely, alkalinization (Teti et al., 1989), increase of the extracellular Ca^{2+} concentration, high extracellular K^+ and treatment with the voltage-operated Ca^{2+} channel agonist BAY K8644 (Miyauchi et al., 1990), all of which reduce osteoclast bone resorption, induce the conversion of the osteoclasts into motile elements, devoid of the clear zone and lacking podosomes.

IV. SUMMARY

Osteoclasts contains all the typical component of the cytoskeleton, as MTs, intermediate filaments, and actin microfilaments. MTs can present one or more MTOCs, organized around the nuclear cluster. However, they are often not evident because of the thicker array of tubules in this area. A cluster of several nucleoli, close to the cell membrane or to the nuclei, can be observed upon transmission electron microscopy. Resorbing osteoclasts present a denser array of tubules. Vimentin containing intermediate filaments have been described, not different from what was observed in the other cells of mesodermal origin. Actin microfilaments have been particularly investigated. They seldom form stress fibers. Motile osteoclasts present a thin network of microfilaments concentrated at the cell edge, where membrane ruffling is also observed. Stationary osteoclasts organize F-actin in podosomes. These are modified focal adhesions in which the microfilaments are oriented perpendicularly to the cell ventral membrane, giving rise, when observed by immunofluorescence, to bright fluorescent dots. The F-actin at the membrane level is surrounded by vinculin and talin organized in a rosettelike structure, while the inner actin core contain fimbrin, α-actinin, and gelsolin. Podosomes are highly dynamic and can assemble and disassemble in minutes. They are also linked to integrin receptors, endowed with the capability of linking several proteins of the extracellular matrix. The number and distribution of podosomes is related to osteoclastic resorptive activity. During bone resorption, they are thickly packed in a ringlike structure that surrounds the resorption lacuna, so closely assembled to resemble, at low magnification, a continuous actin ring. In nonresorbing osteoclasts, they can be observed in small groups scattered on the ventral membrane, or in ringlike fashion, but singularly identifiable.

REFERENCES

Aubin, J.E., (1992). Osteoclast adhesion and resorption: the role of podosomes. J. Bone Min. Res. 7, 365-368.

Bockus, B.J. and Stiles, C.D. (1984). Regulation of cytoskeletal structures by platelet-derived growth factor, insulin,, and epidermal growth factor. Exp. Cell Res. 153, 186-197.

Burridge, K., Fath, K., Kelly, G. Nuckoll, G., and Turner, C. (1988). Focal adhesions: tranmembrane junctions between the extracellular matrix and the cytoskeleton. Annu .Rev. Cell Biol. 4, 487-525.

Cassimeris, L. (1993). Regulation of microtubules dynamic instability. Cell Moti. Cytoskeleton, 26, 275-281.

Chen, W.T., Olden, K., Bernard, B.A., and Chu, F. (1984). Expression of transformation-associated proteaseas that degrade fibronectin at cell contact sites. J. Cell Biol. 98, 1546-1555.

Clark, A.E. and Brugge, J.S. (1995). Integrins and signal transduction pathways: The road taken. Science (Washington, DC) 268, 233-239.

Cole, N.B. and Lippincott-Schwartz, J.,(1995). Organization of organelles and membrane traffic by microtubules. Curr.Opin. Cell Biol. 7, 55-64.

Colucci, S., Grano, M., and Zambonin Zallone, A. (1996). Modulation of metalloproteinase secretion and bone resorption by retinoic acid in human osteoclastlike cells. (Submitted.)

Davies, J., Warwick J, Totty, N., Philp, R., Helfrich, M, and Horton, M. (1989). The osteoclast functional antigen, implicated in the regulation of bone resorption, is biochemically related to the vitronectin receptor. J. Cell Biol. 109, 1817-1826.

Dodemont, H., Riemer, E., Ledger, N., and Weber, K. (1994). Eight genes and alternative RNA-processing pathways generate an unexpectedly large diversity of cytoplamic intermediate filament proteins in the nematode *Caenorhabditis elegans*. Embo J., 13, 2625-2638.

Fath, K., Trimbur, G.M., and Burgess, D.R .(1994). Molecular motors are differentially distributed on Golgi membranes from polarized epithelial cells. J. Cell Biol. 126, 661-675.

Giuliano, K.A. and Taylor, D.L (1994). Fluorescent actin analogs with a high affinity for profilin in vitro exhibit an enhanced gradient of assembly in living cells. J. Cell Biol. 124, 971-983.

Hughes, D.E., Wright, K.R., Uy, H.L., Sasaki, A., Yoneda, T., Roodiman, G.D., Mundy, G.R., and Boyce, B.F. (1995). Biphosphonates promote apoptosis in murine osteoclasts in vitro and in vivo. J. Bone Min. Res. 10, 1478-1487.

Jaworski, Z.F.G., Duck, B., and Sekali, G. (1981) Kinetics of osteoclasts and their nuclei in evolving secondary Haversian systems. J. Anat. 133, 397-405.

Kabsch, W., Mannhertz, H.G., Suck, D., Pai, E.F., and Holmes, K.C. (1990). Atomic structure of the actin: DNAse I complex. Nature 347, 37-44.

Kabsch, W. and Vanderkeekhove, J. (1992). Structure and function of actin. Annu. Rev. Biophys. Biomol. Struct. 21, 49-76.

Kanehisa, J., Yamanaka, T., Doi, S., Heerschhe, J.N.M., Aubin, J.E., and Takeuchi ,H., (1990). A band of F-actin-containing podosomes is involved in bone resorption by osteoclasts. Bone, 11, 287-293.

King, G.J., and Holtrop, M.E. (1975). Actinlike filaments in bone cells as demonstrated by binding to heavy meromiosin. J. Cell Biol. 66, 445-451.

Klimkowski, M.W. (1995). Intermediate filaments: New proteins, some answers, more questions. Curr. Opin. Cell Biol. 7, 46-54.

Lai, Y.K., Lee, W.C., and Che, K.D. (1993). Vimentin serves as a phosphate sink during the apparent activation of protein kinases by okadaic acid in mammalian cells. J. Cell. Biochem. 53, 161-168.

Lakkakorpi, P.T. and Väänänen, H.K. (1991). Kinetic of the osteoclast cytoskeleton during the resorption cycle in vitro. J. Bone Min. Res. 6, 817-826.

Langford, G.M. (1995). Actin- and microtubule-dependent organelle motors: Interrelationships between the two motility systems. Curr. Opin. Cell Biol. 7, 82-88.

Luby-Phelps, K. (1994). Physical properties of the cytoplasm. Curr. Opin. Cell Biol. 6, 3-9.

Marchisio, P.C., Cirillo, D., Naldini, L., Primavera, M.V., Teti, A., and Zambonin Zallone, A. (1984). Cell-substratum interaction of cultured avian osteoclasts is mediated by specific adhesion structures. J. Cell Biol. 99, 1696-1705.

Meng, J.J., Khan, S., and Ip, W. (1994). Charge interactions in the rod domain drive formation of tetramers during intermediate filaments assembly. J. Biol. Chem. 269, 18679-18685.

Miyauchi, A., Hruska, K.A., Greenfield, E.M., Duncan, R., Alvarez, J.,Barattolo, R., Colucci, S., Zambonin Zallone, A., Teitelbaum, S.L., and Teti, A. (1990). Osteoclast cytosolic calcium, regulated by voltage-gated calcium channels and extracellular calcium, controls podosome assembly and bone resorption. J. Cell Biol. 111, 2543-2552.

Oreffo, R.O.C., Teti, A., Triffitt, J.T., Franci, M.J.O., Carano, A., and Zambonin Zallone, A. (1988). Effect of Vitamin A on bone resorption: Evidence for a direct stimulation of isolated chicken osteoclasts by retinol and retinoic acid. J. Bone Min. Res. 3, 203-210.

Ridley, A.J. and Hall, A. (1994). Signal transduction pathways regulating rho-mediated stress fibers formation: Requirement for a tyrosine kinase. EMBO J. 13, 2600-2610.

Takaishi, K., Sasaki, T., Kato, M., Yamochi, W., Kuroda, S., Nakamura, T., Takeichi, M., and Takai, Y. (1994). Involvement of rho p21 small GTP-binding protein and its regulator in the HGF-induced cell motility. Oncogene 9, 273-279.

Tarone, G., Cirillo, D., Giancotti, F.G., Comoglio, P.M., and Marchisio, P.C (1985). Rous sarcoma virus transformed fibroblasts adhere primarily at discrete protrusions of the ventral membrane called podosomes. Exp. Cell Res. 159, 32-343.

Teti, A. and Zambonin-Zallone, A. (1992). Osteoclast cytoskeleton and attachment proteins. In: Biology and Physiology of the osteoclast. (Rifkin, B.R. and Gay, C.V., Eds.), pp. 245-258. CRC Press.

Teti, A., Blair, H.C., Schlesinger, P.H., Grano, M., Zambonin Zallone, A., Kahn, A.J., Teitelbaum, S.L., and Hruska, K.A. (1989). Extracellular protons acidify osteoclasts, reduce cytosolic calcium, and promote expression of cell matrix attachment structures. J. Clin. Invest. 84, 773-780.

Teti, A., Marchisio, P.C. and Zambonin Zallone, A., (1993). Clear zone in osteoclast function: Role of podosomes in regulation of bone-resorbing activity. Am. J. Phyiol. 261, C1-C7.

Traub, P., Kuhn, S., and Grub, S. (1993). Separation and characterization of homo- and hetero-oligomers of the intermediate filament protein desmin and vimentin. J. Mol. Biol. 230, 837-856.

Turksen, K., Kanehisa, J., Opas, M., Heersche, J.N.M., and Aubin, J.E. (1988). Adhesion patterns and cytoskeleton of rabbit osteoclasts on bone slices and glass. J. Bone Min .Res. 3, 389-400.

Warshafsky, J.B., Aubin, J.E., and Heersche, J.N.M. (1985) Cytoskeleton rearrangemet during calcitonin-induced changes in osteoclast motility in vitro. Bone 6, 179-185.

Zambonin-Zallone, A., Teti, A., Carano, A., and Marchisio, P.C. (1988). The distribution of podosomes in osteoclasts cultured on bone laminae: Effect of retinol. J. Bone Min. Res. 3, 517-523.

Zambonin-Zallone, A., Teti, A., Primavera, M.V., Naldini, L., and Marchisio, P.C (1983). Osteoclasts and monocytes have similar cytoskeletal structures and adhesion property in vitro. J. Anat. 136, 57-70.

Zambonin-Zallone, A, Teti, A., Gaboli, M., and Marchisio, P.C. (1989b). β3 subunit of vitronectin receptor is present in osteoclast adhesion structures and not in other monocyte-macrophage-derived cells. Connect. Tissue Res. 20, 143-149.

Zambonin-Zallone, A., Teti, A., Grano, M., Rubinacci, A., Abbadini, M., Gaboli, M., and Marchisio, P.C. (1989a). Immunicytochemical distribution of extracellular matrix receptors in human osteoclasts: $\alpha_v\beta_3$ integrin is colocalized with vinculin and talin in the podosomes of osteoclastoma giant cells. Exp. Cell Res. 182, 645-652.

ROLE OF PROTEASES IN OSTEOCLASTIC RESORPTION

Toshio Kokubo, Osamu Ishibashi, and
Masayoshi Kumegawa

I. INTRODUCTION

Major constituents of bone matrix are type I collagen and a basic calcium salt. Degradation of the bone collagen, as well as bone mineral solubilization, are the processes through which osteoclasts resorb bone. For the degradation of collagen, osteoclasts produce several kinds of proteases, including matrix metalloproteinases (MMPs) and cathepsins. Details of the proteolytic degradation, however, have not yet been well elucidated.

Advances in Organ Biology
Volume 5B, pages 359-370.

ISBN: 0-7623-0390-5

The native form of type I collagen has a characteristic tertiary structure of the triple helices which are quite resistant to degradation by many kinds of proteases. The initial hydrolytic cleavage of the native triple helical collagen is thus the rate-limiting step which requires a collagenase-type enzyme. Once this is achieved, collagen rapidly unfolds and becomes degradable by practically every lysosomal or neutral protease (Wooley, 1984). Identification of the collagenase-type enzyme in osteoclast proteases is a major objective of the study of osteoclast proteases. Furthermore, an interesting question to be answered by the study of osteoclast proteases is how several kinds of proteases interactively cooperate in the degradation of collagen. The two major classes of proteases produced by osteoclasts, i.e., the MMPs and the cathepsins, show distinct optimal pH values. MMPs are most active in the neutral pH. Cathepsin, which is a general term indicative of a lysosomal protease, shows an optimal acidic pH, reflecting the environment inside lysosomes. These two classes of proteases are naturally assumed to have different roles or to function at different events (or different sites) in the process of the collagen degradation by osteoclasts. We may alternatively ask the question why osteoclasts produce both of the neutral and lysosomal proteases.

Clear answers to these questions are not available yet, although many studies of each osteoclast protease have been conducted. There are specific issues that have prevented us from obtaining clear answers to these questions and providing a comprehensive view on the collagen degradation by osteoclast proteases. The issues include presence (or absence) of the interstitial collagenase (matrix metalloproteinase-1, MMP-1) in osteoclasts. Many studies that have been directed to this issue have repeatedly failed to provide a consensus. Another point at issue is which of the osteoclast cathepsins, if any, cleaves the native form of the collagen. The recent discovery of a novel cathepsin predominantly expressed in osteoclasts, i.e., cathepsin K, provides a somewhat new insight into the issue of the collagenase-like activity of cathepsins.

In this chapter, we first review recent progress in the studies of MMPs and cathepsins in osteoclasts. We then discuss models for the comprehensive process of the collagen degradation by osteoclast proteases.

II. MATRIX METALLOPROTEINASES IN OSTEOCLASTS

MMPs are proteases (or endopeptidases) that are dependent on ionic zinc (Zn^{2+}) and are distributed widely in cells such as fibroblasts, endothelial cells, and epithelial cells. MMPs form a family that consists of three major subgroups, i.e., the interstitial collagenase, gelatinases, and stromelysins. Each subgroup is characterized by substrate selectivity, but all MMPs share the common function to degrade organic components of the tissue matrix (or extracellular matrix) under the physiological conditions. The structural and biochemical properties of MMPs are summarized in Table 1.

Table 1. Characteristics f Major Matrix Metalloproteinases

MMP No.	*Enzyme Name (EC No.)*	*Molelcular Weight*		*Degradable Matrix Components*
		pro-Form	*Active Form*	
1	Interstitial collagenase (EC 3. 4. 24. 7)	56,000	45,000	Collagen (type I, II, III, VII, X), gelatin, proteoglycan
8	Neutrophile collagenase (EC 3. 4. 24. 34)	75,000	65,000	Collagen (type I, II, III)
2	72 kDa gelatinase, Gelatinase A (EC 3. 4. 24. 24)	72,000	67,000	Gelatine, collagen (type IV, V, VII, X, XI), fibronectin, laminin, elastin, proteoglycan
9	92 kDa gelatinase, Gelatinase B (EC 3, 4, 24, 35)	92,000	83,000 (66,000)	Gelatin, collagen (type IV, V), elastin proteoglycan
3	Stromelysin I (EC 3, 4, 24, 17)	57,000	45,000 (28,000)	Proteoglycan, fibronectin, collagen (type III, IV, IX), activation of *pro*-MMP-1, 2, 9
10	Stromelysin II (EC 3, 4, 24, 22)	57,000	44,000 (28,000)	Proteoglycan, fibronectin, collagen (type III, IV, IX), activation of *pro*-MMP-1, 2, 9
7	Matrilysin, Pump I (EC 3, 4, 24, 23)	28,000	19,000	Gelatin, proteoglycan, fibronectin, activation of *pro*-MMP-1

Among these MMPs, MMP-1 (interstitial collagenase, EC 3.4.24.7) is the only enzyme capable of degrading native type I collagen, the major organic component of bone (Murphy and Reynolds, 1985). Once cleaved into 1/4 and 3/4 size triple-helical fragments by MMP-1, type I collagen denatures at body temperature and can subsequently be degraded further by other proteases (Wooley, 1984). Thus, it appears reasonable that MMP-1 participates in bone resorption. Indeed, there is evidence that MMP-1 is produced by osteoblastic cells *in vivo* and in organ cultures of bone (Sakamoto and Sakamoto, 1984a,b). On the other hand, as for production of MMP-1 by osteoclasts, a consensus has not been reached despite extensive investigations using techniques including immunohistochemistry and *in situ* hybridization as well as studies using inhibitors.

Initially, there were a series of reports suggesting that osteoclasts do not produce MMP-1 and thus MMP-1 is not involved in osteoclastic bone resorption. For instance, Sakamoto and Sakamoto (1984a) by using fractionated bone cells, demonstrated that MMP-1 was localized primarily in osteoblastic cells, but not in the cell populations rich in osteoclastic cells. In addition, Delaissé et al. (1987) demonstrated that CI-1, a selective inhibitor of MMPs, had no inhibitory effect on pit formation with isolated osteoclasts on dentine slices. However, using organ culture of mouse calvaria as an *in vitro* model system, the inhibitor was found to suppress the secretion of hydroxyproline and calcium into culture medium (Hill et al., 1995). The different results are interpreted to be due to the difference of the substrates and the model systems. Dentine slices used in the pit formation assay are composed of only mineralized tissue, whereas the surface of bone in organ culture is covered with a demineralized component, namely periosteum. These observations are consistent with the well recognized hypothesis proposed by Chambers et al. (1985) that osteoblast-derived MMP-1 is responsible for degrading the nonmineralized osteoid layer covering bone surfaces, thereby exposing the underlying mineralized matrix to osteoclastic action. The results of the inhibitor study with the organ culture have not been considered to prove the direct involvement of MMP-1 in osteoclastic bone resorption.

These initial negative results were soon challenged. Delaissé et al. (1993) demonstrated by an immunohistochemical study the presence of (*pro*-)MMP-1 both in osteoclasts and in extracellular subosteoclastic bone-resorbing compartment. The intracellular collagenase was observed in osteoclasts whether the cells were plated on bone or cultured on glass coverslips. Okamura et al. (1993) reported that the mRNA for MMP-1 was detectable in odontoclasts, which are morphologically similar to osteoclasts, by *in situ* hybridization. Consistent with these reports, an inhibitor study performed by Hill et al. (1994b) showed that Ro 31-7467, a concentration dependent selective inhibitor of MMP-1, inhibited pit formation by isolated rat osteoclasts cultured on dentine slices, as well as reduced the release of hydroxyproline from mouse calvarial explants. However, Fuller and Chambers (1995) reported by *in situ* hybridization, that the mRNA for MMP-1 was not detected in osteoclasts, but was detected in chondrocytes and bone surface cells adjacent to osteoclasts. Thus, the issue as to whether or not MMP-1 is involved in osteoclastic bone resorption has not yet been resolved. To clarify the current confused views, more careful and quantitative studies, taking into account limits and specificity of the methods used for the detection of MMP-1, would be required. Experiments based on molecular genetics such as gene targeting would also be helpful in solving this problem.

In marked contrast to MMP-1, there have been several lines of reports that strongly suggest the direct involvement of MMP-9 (gelatinase B, EC 3.4.24.35) in osteoclastic bone resorption. The studies using the differential screening of cDNA identified MMP-9 as the cDNA clone or mRNA predominantly expressed in human and rabbit osteoclasts as well as in multinucleated giant cells in the human osteoclastoma (Tezuka et al., 1994a; Wucherpfennig et al., 1994). MMP-9, as well as MMP-2 (gelatinase A, EC 3. 4. 24. 24), can cleave the native form of type IV collagen, a specific component of the basement membranes (Fessler et al., 1984). MMP-9 has a high proteolytic activity against denatured collagen (gelatin) (Sorsa et al., 1989), whereas it is not able to cleave the native form of type I collagen. Thus, there are theoretically two possible functions of MMP-9 associated with osteoclasts, namely degradation of the basement membrane and degradation type I collagen in collaboration with other proteinases capable of denaturing the native collagen.

It has been shown that MMP-9 is secreted by cultured peripheral blood polymorphonuclear leukocytes and monocytes, as well as by cytotrophoblasts and alveolar macrophages *in vitro* (Mainardi et al., 1984; Hibbs et al., 1985; Wilhelm et al.,

1989; Murphy et al., 1990; Librach et al., 1991; Behrendtsen et al., 1992; Birkedal-Hansen et al., 1993). Since these cells are capable of migrating through connective tissue barriers, it has been speculated that they use the enzyme to degrade basement membranes. Since osteoclasts are recruited from hematopoietic stem cells in the bone marrow (Roodman et al., 1985), MMP-9 produced by osteoclasts or their progenitors may have a similar function of degrading the basement membrane to let them migrate from a blood vessel to the bone surface. Reponen et al. (1994), however, demonstrated by *in situ* hybridization that mature osteoclasts absorbing bone expressed MMP-9 at very high levels. Since the bone matrix does not contain type IV collagen, it is likely that MMP-9 is mainly used for the turnover of type I collagen together with other collagenase-type enzymes. The report by Hill et al. (1995), demonstrating inhibition of osteoclastic bone resorption with CT1166 and CT543, concentration dependent selective inhibitors of MMP-9 and -2, strongly supports this hypothesis. However, it should be taken into account that the denatured collagen is subjected to degradation not only by MMP-9 but also by many other proteinases including cathepsins. Nevertheless, the involvement of MMP-9 in osteoclastic bone resorption is much better understood than is the involvement of MMP-1.

Besides the two MMPs described above, MMP-2 and MMP-3 (stromelysin 1, EC 3.4.24.17) have been suggested to be involved in osteoclastic bone resorption (Case et al., 1989; Hill et al., 1994b). However, there is poor evidence for this function. Since MMPs are produced and secreted in the *pro*-form to be activated by selective cleavage with other proteinases, activation mechanisms of MMPs in osteoclasts remain to be investigated.

III. CATHEPSINS IN OSTEOCLASTS

Several cysteine proteases and a few aspartic and serine proteases are present ubiquitously in lysosomes of various mammalian cells and serve as scavengers that degrade denatured, unnecessary, or harmful proteins. Cysteine proteases, also known as lysosomal proteases, include cathepsins B, C, H, K, L, O, and S. Representative aspartic and serine proteases localized in lysosomes are cathepsin D and G, respectively. Most of these cathepsins are active under acidic conditions, reflecting the environment inside lysosomes. Cathepsins have endopeptidase (or proteinase) activities, but a few proteases such as cathepsins C and H are characterized as exopeptidases rather than endopeptidases. It is generally believed that these proteases are ubiquitously present in various tissues to efficiently degrade proteins sorted to lysosomes. However, it is also true that these enzymes are heterogeneous in tissue localization (Kominami et al., 1985; Bando et al., 1986). The heterogeneity in localization may reflect differences in metabolites of proteins among various types of cells and thus expression of cathepsins is closely related to cellular functions (Uchiyama et al., 1994).

Cathepsins B, C, D, G, H, K, and L have been demonstrated to be localized in osteoclasts by immunohistochemical studies (Goto et al., 1993, 1994; Ohsawa et al., 1993; Sasaki and Ueno-Matsuda, 1993). Osteoclastic bone resorption takes place within a tightly sealed zone beneath the ruffled border, called a resorption lacuna. The microenvironment of the segregated extracellular space is moderately acidified through the action of proton pumps and resembles a secondary lysosome. It has thus been suggested for some years that these cathepsins are secreted into the resorption lacuna and participate in the degradation of the bone organic matrix. Among the cathepsins detected in osteoclasts, those with endopeptidase activities, i.e., cathepsins B, D, G, K, and L, would be potential candidates of the protease that makes significant contribution to the collagen degradation.

The question as to which family of the lysosomal proteases—cysteine, serine or aspartic proteases—contributes most to the osteoclastic collagen degradation has been directed by several lines of *in vitro* studies. Evidence for the involvement of the cysteine proteases was provided by the finding that their inhibitors, such as leupeptin and E-64, prevented resorption of bone explants (Delaissé et al., 1980, 1984; Everts et al., 1988; Lerner and Grubb, 1992). Inhibition of cysteine proteases was also found to lead to decreased bone matrix degradation in the pit formation assay on bone or dentine slices with isolated osteoclasts (Delaissé et al., 1987; Rifkin et al., 1991). In other hands, representative aspartic and serine protease inhibitors, pepstatin A and aprotinin, showed no effect on the bone resorption induced by parathyroid hormone in cultured mouse calvaria (Barrett, 1977; Delaissé and Vaes 1992). It has thus been generally believed that cysteine proteases, including cathepsins B, K, and L, are mainly responsible for degradation of the bone matrix.

The involvement of cysteine proteases in bone resorption has been also suggested by *in vivo* studies. The intraperitoneal injection of leupeptin or E-64 in rats caused a fall in serum calcium and urinary excretion of hydroxyproline (Delaissé et al., 1984). Taken together, the *in vitro* and *in vivo* studies stress even more strongly that cysteine proteases play a key role in the osteoclastic bone resorption.

Since the two cysteine proteases, cathepsins B and L, have been known to exist in osteoclasts for a long time, their participation in osteoclastic bone resorption has been repeatedly investigated. An acid collagenase, later identified as cathepsin B, was purified from avian osteoclasts (Blair et al., 1993). The expression of multiple forms of cathepsin B mRNA was also detected at high levels in human osteoclastoma cells and avian osteoclasts (Page et al., 1992; Dong et al., 1995). However, Rifkin et al. (1991) suggested that the activity of cathepsin L appeared to be much higher than that of cathepsin B in avian osteoclasts by using two selective synthetic substrates. Consistent with this suggestion, it was reported that parathyroid hormone-induced bone resorption by rat osteoclasts was markedly reduced by inhibitors that strongly inhibit cathepsin L but not cathepsin B, such as pig leukocyte cysteine proteinase inhibitor (PLCPI) and chymostatin. On the other hand, CA-074, a derivative of E-64, designed as a specific inhibitor of cathepsin B, failed to show any inhibitory effect on the bone resorption, indicating that cathepsin L is the major cysteine protease responsible for bone collagen degradation (Kakegawa et al., 1993).

The clear and simple conclusion drawn from the inhibitor study, however, was soon challenged by a few lines of studies. Hill et al. (1994a) found that the methyl ester of CA-074 (CA-074Me), which is a permeable membrane precursor of CA-074, inhibited bone resorption in cultured neonatal mouse calvariae, as well as in the pit formation assay on bone slices. Because CA-074 is a negatively charged molecule, its ability to enter cells appears to be quite limited. CA-074Me, on the other hand, is diffusible into cells and inhibits cathepsin B, presumably following de-esterification (Murata et al., 1991). The result shows that cathepsin B cannot be excluded from the list of cysteine proteases involved in bone resorption. Cathepsin B may mediate its effect intracellularly, possibly through the activation of another protease participating in the bone collagen degradation within the resorption lacuna (Hill et al., 1994a).

The assumed major contribution of cathepsin L to the bone collagen degradation was also questioned by the failure to detect the mature form of cathepsin L in the culture medium of unfractionated bone cells (Kakegawa et al., 1995). By Western blotting, cathepsin L was detected only as the *pro*-enzyme and no signal corresponding to the mature enzyme was observed. Since the *pro*-enzyme is supposedly converted rapidly to the mature form on the secretion to the resorption lacunae, the absence of a detectable population of the mature enzyme poses a question to the involvement of cathepsin L as a major collagenolytic enzyme.

cDNA clones for a novel lysosomal cysteine protease, called cathepsin K, were cloned from human, rabbit and mouse (Tezuka et al., 1994b; Inaoka et al., 1995; Gelb et al., 1996a). By Northern blot analyses, it was demonstrated that cathepsin K is predominantly expressed in osteoclasts (Tezuka et al., 1994b) and in the giant cell tumor of bone (Inaoka et al., 1995). Cathepsin K was successfully purified from a tissue of human osteoclastoma in the form of the mature enzyme, which was also detected in a lysate of rabbit osteoclasts by Western blotting (Ishibashi et al., 1995). Cathepsin K immunoreactivity was detected at the ruffled border of bone-resorbing osteoclasts, indicating that it is secreted into the resorption lacunae (Ishibashi et al., 1995; Littlewood-Evans et al., 1997).

Interestingly, the purified enzyme protein was found to cleave the acid soluble type I collagen, even at the triple helical region, into small fragments (Ishibashi et al., 1996). Although cathepsins B and L are also recognized to be capable of digesting type I collagen, they cleave off only the terminal region that does not form the triple helix telopeptides if the reaction mixture is not warmed to accelerate the denaturation of collagen (Kirschke et al., 1982). In the strictest sense, therefore, only cathepsin K has the collagenase-like activity among the known lysosomal cysteine proteases.

The deduced amino acid sequence of cathepsin K is 25, 35, 42, and 48% identical with those of cathepsins B, H, L, and S, respectively (Inaoka et al., 1995). Since the homology in sequence with cathepsin L is significantly high, cathepsin K may have similar biochemical characteristics to cathepsin L and the inhibitors regarded as selective or specific to cathepsin L may inhibit cathepsin K as well. Chymostatin, one of the inhibitors formerly reported as selective for cathepsin L (Kakegawa et al., 1993), was indeed found to inhibit cathepsin K to the same extent as cathepsin L (unpublished data). Further, we recently determined by competitive reverse transcription-polymerase chain reaction the expression level of cathepsin K to be about 130 times higher than that of cathepsin L in osteoclast-like cells isolated from osteoclastoma (Ishibashi et al., 1996). Hence, it is very likely that contribution of cathepsin K to bone collagen degradation is much higher than that of cathepsin L and the suppression of the pit formation by chymostatin reported before is due to inhibition of the activity of cathepsin K.

Recently, evidence to indicate major contribution of cathepsin K to bone collagen degradation was provided by both in vivo and in vitro studies. Pycnodisostosis is a rare autosomal recessive trait characterized by osteosclerosis and short stature. By the linkage analysis and genomic DNA sequence analysis, this inherited disease was found to be caused by cathepsin K deficiency (Gelb et al., 1996b). It has been suggested that the phenotype results from impaired bone formation coupled with incomplete osteoclastic bone resorption. On the other hand, it was demonstrated in vitro that the selective suppression of the protein synthesis of cathepsin K with an antisense oligodeoxynucleotide resulted in inhibition of osteoclastic bone resorption. The maximum inhibitory effect of the antisense on the pit formation was almost equal to that of a cysteine protease inhibitor with broad selectivity, E-64 (Inui et al., 1997). These studies, taken together, strongly suggest that the contribution of cysteine proteases to bone collagen degradation is mediated primarily by the action of cathepsin K, whose function is not supplemented by other cysteine proteases.

IV. MODELS OF COLLAGEN DEGRADATION

As discussed above, both MMPs and lysosomal cysteine proteinases, most probably MMP-9 and cathepsin K, are likely to be responsible for bone matrix degradation. Everts et al. (1992) demonstrated that MMP and cysteine protease inhibitors CI-1 and E-64 show synergism in inhibiting bone degradation in a combination dosage. However, an implication for a role for MMPs in osteoclastic bone resorption still requires a reasonable explanation to the key question as to how the enzymes retain their activities under the acidified conditions within the subosteoclastic resorption lacunae.

It has been suggested that pH within the resorption lacunae depends on the activity of osteoclastic acid secretion and the buffering potential of the calcium (Ca^{2+}) and phosphate ions (PO_4^{2-}) released during mineral dissolution (Delaissé and Vaes, 1992). It is also known that bone resorbing osteoclasts show dynamic changes in their morphology and active migration, which is evident from shapes of the pits formed on bone slices. It would be quite natural to assume that periodical changes in the acid secretion, and thus changes in pH within the resorption lacuna, take place with the morphological changes and/or the migration. It is likely that MMPs and cysteine proteases function as bone matrix degrading enzymes alternately, depending upon pH in the resorption lacunae. A cysteine protease(s), most likely cathepsin K, may be secreted concurrently with the release of protons and cleave the type I collagen, leading to its denaturation at low pH. At a later stage in the resorption process, MMPs including MMP-9 may become active for the complete digestion of the denatured collagen after the secreted acid has been rapidly neutralized by the released bone salts.

The proteases closely associated with osteoclastic bone resorption would have pathogenic significance in human disorders related to bone, e.g., osteoporosis and osteoarthritis (Esser et al., 1994). For pathological and medicinal studies of the osteoclast proteases, a better understanding of their physiological functions is essential, especially on the interaction of proteases, including proteolytic processing of *pro*-enzymes to the active forms.

V. SUMMARY

There are a few issues that have prevented researchers from obtaining a comprehensive view on the proteolytic degradation of bone collagen by osteoclasts. The issues include presence (or absence) of collagenase or MMP-1 in osteoclasts, and identification of a lysosomal cysteine protease that is capable of cleaving the native form of the collagen. We have thus reviewed recent studies of MMPs and cathepsins produced by osteoclasts mainly to address these issues. As for production of MMP-1 by osteoclasts, a consensus has not yet been obtained even with immunohistochemistry, *in situ* hybridization, and inhibitor studies. In a marked contrast to MMP-1, there have been several lines of reports that strongly suggest the involvement of MMP-9. Direct evidence for the involvement of lysosomal cysteine proteases or cathepsins in osteoclastic collagen degradation has been provided by the finding that their inhibitors prevent bone resorption both *in vitro* and *in vivo*. Although cathepsin L was once proposed to be the major cysteine protease responsible for the collagen degradation, recent studies, especially those that have lead to the discovery of a novel cathepsin (cathepsin K) that is predominantly expressed in osteoclasts, are providing clues to the final identification of the functional collagenolytic cysteine protease. A new model of osteoclastic collagen degradation that assumes cooperative action of MMPs and cathepsins has been discussed.

REFERENCES

Bando, Y., Kominami, E., and Katsunuma, N. (1986). Purification and tissue distribution of rat cathepsin L. J. Biochem. 100, 35-42.

Barrett, A. J. (1977). In: Proteinases in Mammalian Cells and Tissues (Barrett, A.J., Ed.), pp. 1-55 and 181-208, North-Holland Publishing Co., Amsterdam.

Behrendtsen, O., Alexander, C.M., and Werb, Z. (1992). Metallo proteinases mediate extracellular matrix degradation by cells from mouse blastocyst outgrowths. Development 114, 447-456.

Birkedal-Hansen, H., Moore, W.G.I., Bodden, M.K., Windsor, L.J., Birkedal-Hansen, B., DeCarlo, A., and Engler, J.A. (1993). Matrix metalloproteinases: A review. Crit. Rev. Oral. Biol. Med. 4, 197-250.

Blair, H.C., Teitelbaum, S.L., Grosso, L.E., Lacey, D.L., Tan, H.-L., McCourt, D.W., and Jeffrey, J.J. (1993). Extracellular-matrix degradation at acid pH: Avian osteoclast acid collagenase isolation and characterization. Biochem. J. 290, 873-884.

Case, J. P., Sano, H., Lafyatis, R., Remmers, E. F., Kumkumian, G.K., and Wilder, R.L. (1989). Transin/stromelysin expression in the synovium of rats with experimental erosive arthritis. J. Clin. Invest. 84, 1731-1740.

Chambers, T.J., McSheehy, P.M.J., Thomson, B.M., and Fuller, K. (1985). The effect of calcium-regulating hormones and prostaglandins on bone resorption by osteoclasts disaggregated from neonatal rabbit bones. Endocrinology 116, 234-239.

Delaissé, J.M., Eeckhout, Y., and Vaes, G. (1980). Inhibition of bone resorption in culture by inhibitors of thiol proteinases. Biochem. J. 192, 365-368.

Delaissé, J.M., Eeckhout, Y., and Vaes, G. (1984). In vivo and in vitro evidence for the involvement of cysteine proteinases in bone resorption. Biochem. Biophys. Res. Commun. 125, 441-447.

Delaissé, J.M., Boyde, A., Maconnachie, E., Ali, N.N., Sear, C.H.J., Eeckhout, Y., Vaes, G., and Jones, S.J. (1987). The effects of inhibitors of cysteine proteinases and collagenase on the resorptive activity of isolated osteoclasts. Bone 8, 305-313.

Delaissé, J.M., and Vaes, G. (1992). Mechanism of Mineral Solubilization and Matrix Degradation in Osteoclastic Bone Resorption. In: Biology and Physiology of the Osteoclast (Rifkin, B.R. and Gay, C.V., Eds.), pp. 289-314, CRC Press, Boca Raton.

Delaissé, J.M., Eeckhout, Y., Nett, L., François-Gillet, C., Henriet, P., Su, Y., Vaes, G., and Baron, R. (1993). (Pro)collagenase (matrix metalloproteinase-1) is present in rodent osteoclasts and in the underlying bone-resorbing compartment. J. Cell Sci. 106, 1071-1082.

Dong, S.S., Stransky, G.I., Whitaker, C.H., Jordan, S.E., Schlesinger, P.H., Edwards, J.C., and Blair, H.C. (1995). Avian cathepsin B cDNA: Sequence and demonstration that mRNAs of two sizes are produced in cell types producing large quantities of the enzyme. Biochim. Biophys. Acta 1251, 69-73.

Esser, R.E., Angero, R.A., Murphey, M.D., Watts, L.M., Thornburg, L.P., Palmer, J.T., Talhouk, J.W., and Smith, R.E. (1994). Cysteine proteinase inhibitors decrease articular cartilage and bone destruction in chronic inflammatory arthritis. Arthritis Rheum. 37, 236-247.

Everts, V., Beertsen, W., and Schröder, R. (1988). Effects of the proteinase inhibitors leupeptin and E-64 on osteoclastic bone resorption. Calcif. Tissue Int. 43, 172-178.

Everts, V., Delaisse, J.M., Korper, W., Niehof, A., Vaes, G., and Beertsen, W. (1992). Degradation of collagen in the bone-resorbing compartment underlying the osteoclast involves both cysteine-proteinases and matrix metalloproteinases. J. Cell. Phys. 150, 221-231.

Fessler, L., Duncan, K., and Tryggrason, K. (1984). Identification of the procollagen IV cleavage products produced by a specific tumor collagenase. J. Biol. Chem. 259, 9783-9789.

Fuller, K. and Chambers, T. J. (1995). Localization of mRNA for collagenase in osteocytic, bone surface, and chondrocytic cells, but not osteoclasts. J. Cell Sci. 108, 2221-2230.

Gelb, B.D., Moissoglu, K., Zhang, J., Martignetti, J.A., Bromme, D., and Desnick, R.J. (1996a). Cathepsin K: Isolation and characterization of the murine cDNA and genomic sequence, the homologue of the human pycnodysostosis gene, Biochem, Mol. Med. 59, 200-206.

Gelb, B.D., Shi, G-P., Chapman, H.A., and Desnick, R.J. (1996b). Pycnodysostosis, a lysosomal disease caused by cathepsin K deficiency. Science 273, 1236-1238.

Goto, T., Tsukuba, T., Kiyoshima, T., Nishimura, Y., Kato, K., Yamamoto, K., and Tanaka, T. (1993). Immunohistochemical localization of cathepsins B, D, and L in the rat osteoclast. Histochemistry 99, 411-414.

Goto, T., Kiyoshima, T., Moroi, R., Tsukuba, T., Nishimura, Y., Himeno, M., Yamamoto, K., and Tanaka, T. (1994). Localization of cathepsins B, D, and L in the rat osteoclast by immuno-light and -electron microscopy. Histochemistry 101, 33-40.

Hibbs, M.S., Hasty, K.A., Seyer, J.M., Kang, A.H., and Mainardi, C.L. (1985). Biochemical and immunological characterization of the secreted forms of human neutrophil gelatinase. J. Biol. Chem. 260, 2493-2500.

Hill, P.A., Buttle, D.J., Reynolds, J.J., and Meikle, M.C. (1994a). Inhibition of bone resorption by selective inactivators of cysteine endopeptidases. Int. J. Exper. Pathol. 75, A12-A13.

Hill, P.A., Murphy, G., Docherty, A.J.P., Hembry, R.M., Millican, T.A., Reynolds, J.J., and Meikle, M.C. (1994b). The effects of selective inhibitors of matrix metalloproteinases (MMPs) on bone resorption and the identification of MMPs and TIMP-1 in isolated osteoclasts. J. Cell Sci. 107, 3055-3064.

Hill, P.A., Docherty, A.J.P., Bottomley, M.K., O'Connell, J.P., Morphy, J.R., Reynolds, J.J., and Meikle M.C. (1995). Inhibition of bone resorption in vitro by selective inhibitors of gelatinase and collagenase. Biochem. J. 308, 167-175.

Inaoka, T., Bilbe, G., Ishibashi, O., Tezuka, K., Kumegawa, M., and Kokubo, T. (1995). Molecular cloning of human cDNA for cathepsin K: Novel cysteine proteinase predominantly expressed in bone. Biochem. Biophys. Res. Commun. 206, 89-96.

Inui, T., Ishibashi, O., Inaoka, T., Origane, Y., Kumegawa, M., Kokubo, T., and Yamamura, T. (1997). Cathepsin K antisense oligodeoxynucleotide inhibits osteoclastic bone resorption. J. Biol. Chem. 272, 8109-8112.

Ishibashi, O., Inaoka, T., Togame, H., Bilbe, G., Nakamura, Y., Mori, Y., Honda, Y., Hakeda, Y., Ozawa, H., Kumegawa, M., and Kokubo, T. (1995). A novel cysteine proteinase localized at ruffled border: Cathepsin K. J. Bone Miner. Res. 10, S426.

Ishibashi, O., Togame, H., Mori, Y., Kumegawa, M., and Kokubo, T. (1996). Cathepsin K expressed at a high level in osteoclasts shows strong collagenolytic activity. J. Bone Min. Res., 11, S181.

Kakegawa, H., Nikawa, T., Tagami, K., Kamioka, H., Sumitani, K., Kawata, T., Drobnic-Kosorok, M., Lenarcic, B., Turk, V., and Katsunuma, N. (1993). Participation of cathepsin L on bone resorption. FEBS Lett. 321, 247-250.

Kakegawa, H., Tagami, K., Ohba, Y., Sumitani, K., Kawata, T., and Katsunuma, N. (1995). Secretion and processing mechanisms of procathepsin L in bone resorption. FEBS Lett. 370, 78-82.

Kominami, E., Tsukahara, T., Bando, Y., and Katsunuma, N. (1985). Distribution of cathepsins B and H in rat tissues and peripheral blood cells. J. Biochem. 98, 87-94.

Kirschke, H., Kembhavi, A.A., Boley, P., and Barrett, A.J. (1982). Action of rat liver cathepsin L (EC-3. 4. 22. 15) on collagen and other substrates. Biochem. J. 201, 367-372.

Lerner, U.H. and Grubb, A. (1992). Human cystatin C, a cysteine proteinase inhibitor, inhibits bone resorption in vitro stimulated by parathyroid hormone and parathyroid hormone-related peptide of malignancy. J. Bone Miner. Res. 7, 433-440.

Librach, C.L., Werb, Z., Fitzgerald, M.F., Chiu, K., Corwin, N.M., Esteues, R.A., Grobelny, D., Galardy, R., Damsky, C.H., and Fisher, S.J. (1991). 92-kDa type-IV collagenase mediates invasion of human cytotrophoblasts. J. Cell Biol. 113, 437-449.

Littlewood-Evans, A., Kokubo, T., Ishibashi, O., Inaoka, T., Wlodarski, B., Gallagher, J.A., and Bilbe, G. (1997). Localization of cathepsin K in human osteoclasts by in situ hybridization and immunohistochemistry. Bone 20, 81-86.

Mainardi, C.L., Hibbs, M.S., Hasty, K.A., and Seyer, J.M. (1984). Purification of a type-V collagen-degrading metalloproteinase from rabbit alveolar macrophages. Collagen Relat. Res. 4, 479-492.

Murata, M., Miyashita, S., Yokoo, C., Tamai, M., Hanada, K., Hatayama, K., Towatari, T., Nikawa, T., and Katsunuma, N. (1991). Novel epoxysuccinyl peptides: Selective inhibitors of cathepsin B. FEBS Lett. 280, 307-310.

Murphy, G. and Reynolds, J.J. (1985). Current views of collagen degradation. Progress towards understanding the resorption of connective tissues. Bio Essays 2, 55-60.

Murphy, G., Ward, R., Hembry, R.M., Reynolds, J.J., Kuhn, K., and Tryggvason, K. (1990). Characterization of gelatinase from pig polymorphonuclear leucocytes. Biochem. J. 258, 463-472.

Ohsawa, Y., Nitatori, T., Higuchi, S., Kominami, E., and Uchiyama, Y. (1993). Lysosomal cysteine and aspartic proteinases, acid phosphatase, and an endogenous cysteine proteinase inhibitor, cystatin-b, in rat osteoclasts. J. Histochem. Cytochem. 41, 1075-1083.

Okamura, T., Shimokawa, H., Takagi, Y., Ono, H., and Sasaki, S. (1993). Detection of collagenase mRNA in odontoclasts of bovine root-resorbing tissue by in situ hybridization. Calcif. Tissue Int. 52, 325-330.

Page, A.E., Warburton, M.J., Chambers, T.J., Pringle, J.A., and Hayman, A.R. (1992). Human osteoclastomas contain multiple forms of cathepsin B. Biochim. Biophys. Acta. 1116, 57-66.

Reponen, P., Sahlberg, C., Munaut, C., Theslett, I., and Tryggvason, K. (1994). High expression of 92-kDa type-IV collagenase (gelatinase B) in the osteoclast linage during mouse development. J. Cell Biol. 124, 1091-1102.

Rifkin, B.R., Vernillo, A.T., Kleckner, A.P., Auszmann, J.M., Rosenberg, L.R., and Zimmerman, M. (1991). Cathepsin B and L activities in isolated osteoclasts. Biochem. Biophys. Res. Commun. 179, 63-69.

Roodman, G.T., Ibbotson, K.J., MacDonald, B.R., Kuehl, T.J., and Mundy, G.R. (1985). 1,25$(OH)_2$ vitamin D_3 causes formation of multinucleated cells with osteoclast characteristics in cultures of primate marrow. Proc. Natl. Acad. Sci. U.S.A. 82, 8213-8217.

Sakamoto, M. and Sakamoto, S. (1984a). Immunocytochemical localization of collagenase in isolated mouse bone cells. Biomed. Res. 5, 29-38.

Sakamoto, S. and Sakamoto, M. (1984b). Isolation and characterization of collagenase synthesized by mouse bone cells in culture. Biomed. Res. 5, 39-46.

Sasaki, T. and Ueno-Matsuda, E. (1993). Cysteine-proteinase localization in osteoclasts, an immunocytochemical study. Cell Tissue Res. 271, 177-179.

Sorsa, T., Suomalainen, K., Konttinen, Y.T., Saari, H.T., Lindy, S., and Uitto, V.J. (1989). Identification of protease(s) capable of further degrading native 3/4- and 1/4-collagen fragments generated by collagenase from native type-I collagen in human neutrophils. Proc. Finn. Denn. Soc. 85, 3-11.

Tezuka, K., Nemoto, K., Tezuka, Y., Sato, T., Ikeda, Y., Kobori, M., Kawashima, H., Eguchi, H., Hakeda, Y., and Kumegawa, M. (1994a). Identification of matrix metalloproteinase 9 in rabbit osteoclasts. J. Biol. Chem. 269, 15006-15009.

Tezuka, K., Tezuka, Y., Maejima, A., Sato, T., Nemoto, K., Kamioka, H., Hakeda, Y., and Kumegawa, M. (1994b). Molecular cloning of a possible cysteine proteinase predominantly expressed in osteoclasts. J. Biol. Chem. 269, 1106-1109.

Uchiyama, Y., Waguri, S., Sato, N., Watanabe, T., Ishidoh, K., and Kominami, E. (1994). Review: Cell and tissue distribution of lysosomal cysteine proteinases, cathepsins B, H, and L, and their biological roles. Acta Histochem. Cytochem. 27, 287-308.

Wilhelm, S.M., Collier, I.E., Marmer, B.L., Eisen, A.Z., Grant, G.A., and Goldberg, G.I. (1989). SV40-transformed human lung fibroblasts secrete a 92-kDa type-IV collagenase, which is identical to that secreted by normal human macrophages. J. Biol. Chem. 264, 17213-17221.

Wooley, D.E. (1984). Mammalian collagenase. In: Extracellular Matrix Biochemistry (Piez, K.A., and Reddi, A.H., Eds.), pp. 119-157. Elsevier, New York.

Wucherpfennig, A.L., Yi-Ping, L., Stetler-Sterenson, W.G., Rosenberg, A.E., and Stashenko, P. (1994). Expression of 92-kDa type-IV collagenase/gelatinase B in human osteoclasts. J. Bone Min. Res. 9, 549-556.

EXTRACELLULAR CALCIUM ION SENSING IN OSTEOCLASTS

Olugbenga Adebanjo and Mone Zaidi

Advances in Organ Biology
Volume 5B, pages 371-383.

ISBN: 0-7623-0390-5

I. INTRODUCTION

It has recently become clear that not only is the critical role of Ca^{2+} in cell signaling subserved through classical intracellular mechanisms, but that Ca^{2+} is also an extracellular or first messenger (Zaidi et al., 1993a; Brown et al., 1995). Certain cell types, including the osteoclast, bear Ca^{2+} sensors on their plasma membrane (Zaidi et al., 1993a). The function of these cells is thus regulated through changes in their ambient Ca^{2+} concentration (Zaidi et al., 1993a; Brown et al., 1995). For example, parathyroid cells, renal cells, and thyroid C cells possess a G protein-coupled "seven-pass" membrane receptor for Ca^{2+} (hence, the widely used term, Ca^{2+}-sensing receptor) (Brown et al., 1993; Mithal et al., 1994; Ricardi et al., 1995). The human Ca^{2+}-sensing receptor gene has been cloned and sequenced; point mutations in it result in the abnormal Ca^{2+} sensing that is seen in certain familial hypercalcemic syndromes (Pollak et al., 1993; Brown et al., 1995). It should be noted however that Ca^{2+} sensors do not belong to a family of related molecules. Instead, a diverse set of unrelated structures appear to subserve a somewhat similar biological function. In the trophoblast, for example, Ca^{2+} sensing occurs through a low-density lipoprotein receptor-like molecule (Lundgren et al., 1994). In the osteoclast, the Ca^{2+} sensor is a plasma membrane-resident ryanodine receptor (Zaidi et al., 1995). The latter is otherwise a Ca^{2+} release channel found in microsomal membranes (Meissner, 1994).

Here, we will briefly describe recent work which led to the initial discovery of the phenomenon of extracellular Ca^{2+} sensing in the osteoclast, the subsequent prediction of a unique plasma membrane-resident Ca^{2+} sensor, and finally, a detailed characterization of the Ca^{2+} sensing mechanism. From our more recent studies, we have concluded that a ryanodine receptor, expressed in the osteoclast plasma membrane, functions as the Ca^{2+} sensor (Zaidi et al., 1995).

II. OSTEOCLAST REGULATION BY EXTRACELLULAR CA^{2+}

In 1989, Alberta Zambonin-Zallone's group in Italy (Malgaroli et al., 1989) and my own group (Zaidi et al., 1989) demonstrated that osteoclasts were sensitive to changes in their ambient Ca^{2+} concentration (Figure 1). Microelectrode studies revealed Ca^{2+}concentrations of between eight and 40 mM within the lacunae covered by resorbing osteoclasts (Silver et al., 1988). We found that when freshly isolated rat osteoclasts were settled onto bone slices and allowed to incubate in a high Ca^{2+} concentration (5 to 20 mM), there was a dramatic and concentration-dependent reduction in their bone resorbing activity (Zaidi et al., 1989; Datta et al., 1990b; Moonga et al., 1990). This inhibitory effect was mimicked by ionomycin, a Ca^{2+} ionophore (Zaidi et al., 1989; Moonga et al., 1990). In addition, we found that the rate of secretion of the osteoclastic phosphohydrolase, tartrate-resistant (band 5) acid phosphatase, fell sharply, in a concentration-dependent manner, within an hour of exposure to high Ca^{2+} (5 to 20 mM) or to ionomycin

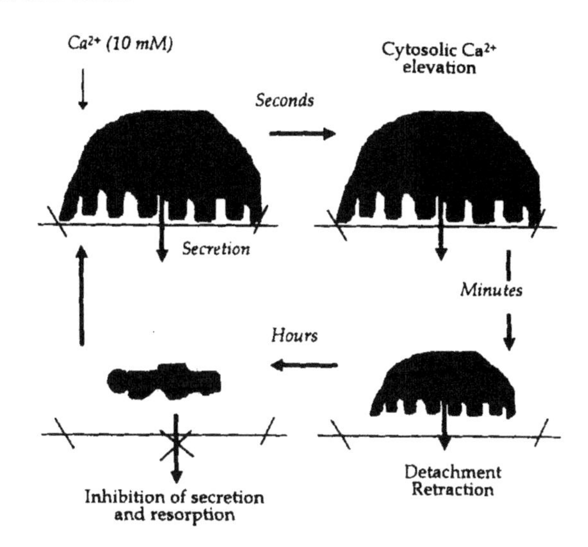

Figure 1. Effects of extracellular Ca^{2+} on osteoclast function.

(Moonga et al., 1990). Interestingly, resorbing osteoclasts produced lower quantities of acid phosphatase than those plated on plastic, suggesting that locally produced Ca^{2+} may inhibit secretion. In parallel, we found that osteoclasts retracted within minutes of exposure to a high Ca^{2+} level (Datta et al., 1990b; Adebanjo et al., 1994a). Their retracted margins remained motile (Zaidi, 1990), suggesting that the cells were still capable of moving to, and resorbing at, a different location. Similarly, both podosome formation (Malgaroli et al., 1989) and cell-matrix adhesion (Makgoba and Datta, 1992) were attenuated sharply upon exposure of osteoclasts to a high Ca^{2+} level. Finally, we found that an increase in extracellular Ca^{2+} triggered a rapid rise in cytosolic Ca^{2+} (Zaidi et al., 1989; Bax et al., 1992). Each such Ca^{2+} transient was biphasic in form, a feature reminiscent of hormone action. Notably, the triggered responses were not attenuated by the voltage-gated Ca^{2+} channel antagonists, verapamil, nifedipine, and diltiazem (Datta et al., 1990a). We thus hypothesized that Ca^{2+} generated locally could regulate osteoclastic bone resorption through a voltage-insensitive Ca^{2+} signaling pathway (Zaidi et al., 1989).

III. EVIDENCE FOR A CA2+ (DIVALENT CATION) SENSOR

A. Studies with Medullary Bone Osteoclasts

To meet with the increased requirement for Ca^{2+} during egg shell calcification, medullary bone of the Japanese quail is known to undergo unrestricted osteoclastic resorption (Dacke et al., 1993). Our hypothesis was that an anti-resorptive mechanism, such as Ca^{2+} sensing, should switch off during this time. We found that osteoclasts harvested during egg-lay displayed neither a cytosolic Ca^{2+} change nor a retraction response upon exposure to high Ca^{2+} (Bascal et al., 1992). Most surprisingly, however, their sensitivity to extracellular Ca^{2+} was promptly restored when the cells were cultured for up to seven days in a Ca^{2+}-free medium (Bascal et al., 1994). This suggested the existence of a putative sensor for Ca^{2+}, whose expression was being regulated, at least in part, by the required level of bone turnover.

B. Sensitivity to Divalent and Trivalent Cations

We next attempted to characterize the activation properties of the putative Ca^{2+} sensor by using divalent cations other than Ca^{2+}, as well as trivalent cations (Zaidi et al., 1991; Shankar et al., 1992a,b, 1993). This approach was used with success in the initial studies on the parathyroid cell Ca^{2+} receptor (Racke and Nemeth, 1993). The application of divalent or trivalent cations to resorbing osteoclasts resulted in concentration-dependent inhibitory effects on bone resorption, acid phosphatase secretion and cell spreading that were indistinguishable from those resulting from a high Ca^{2+} level (Zaidi et al., 1991). In this respect, the transition metal cations, Ni^{2+} and Cd^{2+}, as well as La^{3+}, were significantly more potent than the alkaline earth metals, Ca^{2+}, Ba^{2+}, Sr^{2+}, and Mg^{2+} (Shankar et al., 1992a,b, 1993). Our detailed studies with Ni^{2+} provided further evidence in favor of a divalent cation (Ca^{2+}) sensor. Firstly, in a way quite characteristic of agonist-receptor interactions, Ni^{2+} activated cytosolic Ca^{2+} release (Hill coefficient = 1). Secondly, like the desensitization of hormone receptors, there was clear evidence for inactivation in response to repeated cation application (Shankar et al., 1993). Finally, reminiscent of classical heterologous desensitization, a given cation attenuated the cytosolic Ca^{2+} response to the subsequent application of a different cation (Bax et al., 1993; Shankar et al., 1993).

C. Evidence for the Release of Intracellularly Stored Ca^{2+}

Our argument for the existence of a discrete surface entity became more compelling when we showed that its activation triggered Ca^{2+} release from intracellular stores (Shankar et al., 1993; Zaidi et al., 1993b). Thus, the rapid cytosolic Ca^{2+} transients elicited by Ni^{2+} in 1 mM-extracellular Ca^{2+} remained unchanged, both in form and magnitude, when extracellular Ca^{2+} was sequestered to nanomolar levels (Shankar et al., 1993). The responses were, however, abolished when potentially releasable Ca^{2+} stores were depleted either by ionomycin (Shankar et al., 1993) or by the highly selective microsomal membrane Ca^{2+}-ATPase inhibitor, thapsigargin (Putney, 1990; Zaidi et al., 1993b). These phenomena again paralleled the classic hormone-induced Ca^{2+} release responses.

D. The Ca^{2+} Sensor: A Pronase-Sensitive Protein

We next examined whether the putative sensor for Ca^{2+} was a cell surface protein. For this, we used pronase, a proteolytic enzyme that renders cells refractory to receptor-mediated hormone effects (Salgado et al., 1985). We found that pronase inhibited Ni^{2+}- and Cd^{2+}-induced elevations in cytosolic Ca^{2+} in a concentration- and time-dependent manner in the absence of membrane damage (Zaidi et al., 1993c). This study suggested that Ni^{2+}-sensitive site on the osteoclast was a cell surface-expressed pronase-sensitive protein.

IV. CONTROL OF CA^{2+} SENSING

We next determined whether any local or systemic influences to which the osteoclast is exposed can alter Ca^{2+} sensing (Figure 2). We found that calcitonin, at femtomolar concentrations, attenuated cation-induced Ca^{2+} release through protein kinase A phosphorylation (Zaidi et al., 1996). We inferred that calcitonin controls the extent to which an osteoclast can sense a change in its ambient Ca^{2+} concentration. We also found that a low extracellular pH enhanced the amplitude of the cytosolic Ca^{2+} response to Ca^{2+} or Ni^{2+} (Adebanjo et al., 1994b). That the pH beneath a resorbing osteoclast can fall to as low as 3.5 units is noteworthy (Silver et al., 1988). It seems logical if Ca^{2+} sensing was to increase in the face of a high H^+ extrusion rate as this would prevent further mineral dissolution. Finally, we have pursued in depth the effect of changing the osteoclast membrane potential on its divalent cation-sensitivity (Pazianas et al., 1993; Shankar et al., 1995a). Indeed, it is well known that the osteoclast membrane can spontaneously become hyperpolarized. We thus made use of the K^+ ionophore, valinomycin, that allowed the cell's membrane voltage to be clamped in a

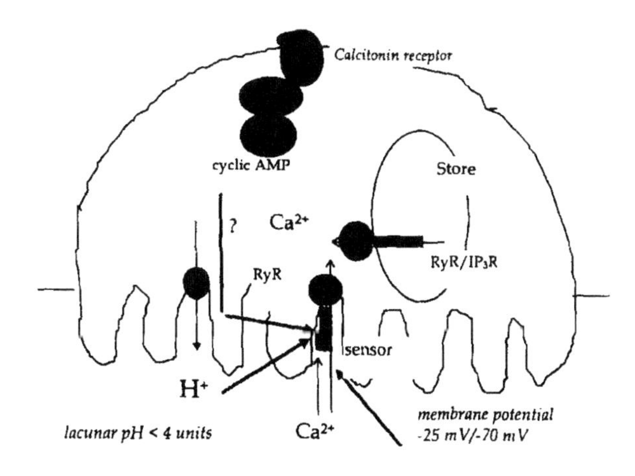

Figure 2. Control of Ca^{2+} sensing.

given direction. We noted that hyperpolarization resulted in a marked slowing of both the rising and decay phases of the Ni^{2+}-induced Ca^{2+} transient. This indicated that the osteoclast Ca^{2+} sensor could remain activated for longer times when the plasma membrane is hyperpolarized.

V. CA2+ SENSING THROUGH A CELL SURFACE RYANODINE RECEPTOR

A. Ryanodine receptor-bearing intracellular Ca2+ channels: an overview

In most eukaryotic cells, Ca^{2+} is released from intracellular stores into the cytosol through Ca^{2+} channels resident in the microsomal membrane (Berridge, 1993; Meissner, 1994). These fall into two distinct families: the IP_3 receptor family, and the ryanodine receptor family (ryanodine is a plant alkaloid) (Figure 3) (Berridge, 1993) The latter has three known isoforms, I, II, and III (Berridge, 1993; Meissner, 1994). Isoform I is expressed exclusively in skeletal muscle sarcoplasmic recticulum where it is coupled electrically to the plasma membrane voltage-sensing dihydropyridine receptor (Huang, 1993). Thus, it mediates Ca^{2+} release in response to

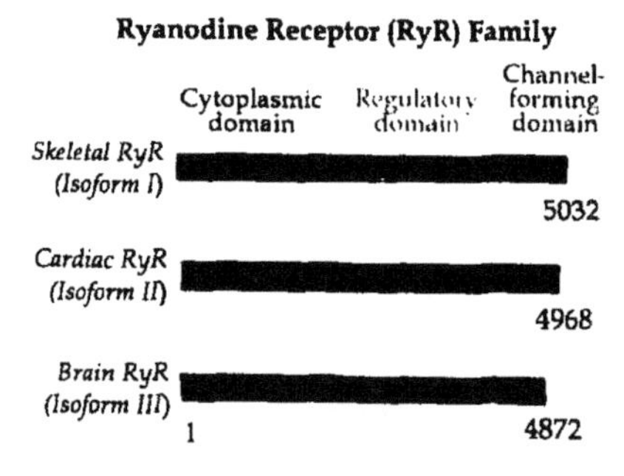

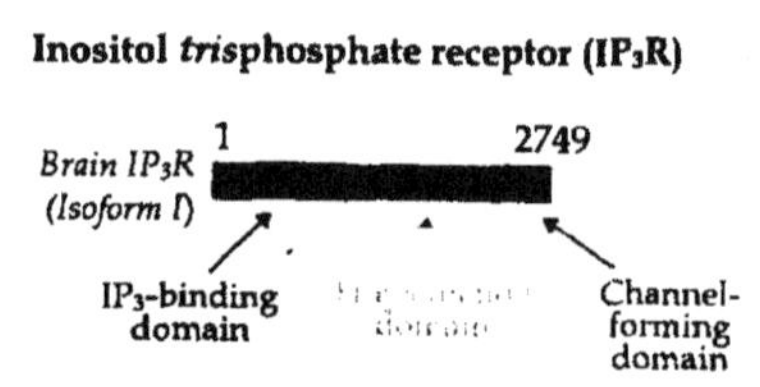

Figure 3. Ryanodine receptor (RyR) and IP_3 receptor isoforms.

tubular membrane depolarization (Huang, 1993). Isoforms II and III are found in the microsomal membranes of cardiac muscle cells, brain cells, and certain nonexcitable cells. They gate Ca^{2+} release in response to changes in cytosolic Ca^{2+} that are monitored by high-affinity cytosolic Ca^{2+}-binding sites (Berridge, 1993; Meissner, 1994). In the osteoclast, we have shown that, in addition to the microsomal receptor, the type II ryanodine receptor is expressed in the cell's plasma membrane (Zaidi et al., 1995). It is notable that in cardiac muscle this isoform monitors not only cytosolic Ca^{2+} level, but also the luminal Ca^{2+} concentration through a low-affinity (millimolar) Ca^{2+}-binding site (Anderson et al., 1989) In the osteoclast, this site should present at the cell's exterior surface, and at this location subserve the function of an extracellular Ca^{2+} sensor.

B. Analogies with Excitation-Contraction Coupling

We have noted that modulators of excitation-contraction coupling in muscle affect osteoclast Ca^{2+} sensing profoundly. First, changes in membrane potential that activate Ca^{2+} release in muscle also modulate Ca^{2+} sensor activation in the osteoclast (Huang, 1993). Second, Ni^{2+}, triggers Ca^{2+} release in the osteoclast, while in skeletal muscle, it renews voltage sensor activation (Huang, 1993). Third, a chaotropic anion, perchlorate, a charge movement activator in skeletal muscle (Huang, 1993), not only triggers Ca^{2+} release in the osteoclast, but also mimics the effects of a high Ca^{2+} level on cell function (Moonga et al., 1991). Recent receptor reconstitution studies have clarified that perchlorate interacts directly with ryanodine receptors (Anderson et al., 1993). Finally, dantrolene-Na^{+}, a Ca^{2+} release blocker in skeletal muscle, inhibits Ca^{2+}-induced Ca^{2+} release in the osteoclast (Malgaroli et al., 1989). Notably, dantrolene-Na^{+} is used for the management of malignant hyperthermia, a familial disorder that results from defective ryanodine receptor function (Wappler et al., 1994). Such striking analogies between the osteoclast and the muscle cell prompted us to look for a role for a ryanodine receptor in osteoclast Ca^{2+} sensing.

C. Pharmacomodulation

We first found that the alkaloid, ryanodine, inhibited Ni^{2+}-induced cytosolic Ca^{2+} release, suggesting an interaction between the Ca^{2+} sensor and a ryanodine-sensitive molecule (Zaidi et al., 1992). This was confirmed independently when we found that Ni^{2+} attenuated [^{3}H]-ryanodine binding to isolated osteoclasts (Zaidi et al., 1995). More importantly, we showed that the ryanodine effect on Ca^{2+} sensing was unusually sensitive to changes in the cell's membrane voltage, suggesting an action at or near the plasma membrane (Zaidi et al., 1992).

We next went on to examine the effect on Ca^{2+} sensing, of series of known ryanodine receptor modulators, including caffeine (Shankar et al., 1995b), ruthenium red (Adebanjo et al., 1996), and the more recently discovered molecule, cyclic ADP-ribose (Lee et al, 1994; Adebanjo et al., 1996) All three modulators not only attenuated Ni^{2+}-induced cytosolic Ca^{2+} release, but also triggered modest elevations in cytosolic Ca^{2+} (Shankar et al., 1995b; Adebanjo et al., 1996). Several features of caffeine action in the osteoclast, however, contrasted those in skeletal muscle (Huang, 1993). First, its concentrations, 5 μM to 250 μM, were substantially lower than those active in muscle (2 to 10 mM) (Shankar et al., 1995b). Second, at higher molar concentrations, 500 μM to 2 mM, caffeine action exhibited a use-dependent inactivation (Shankar et al., 1995b), a feature not seen in skeletal muscle (Huang 1993). Finally, unlike in muscle, the Ca^{2+} release triggered by caffeine itself was modulated by the extracellular Ca^{2+} concentration (Shankar et al., 1995b).

D. Isoform- and Epitope-Specific Anti-Ryanodine Receptor Antibodies

To explore the unprecedented possibility that a ryanodine receptor, expressed in the osteoclast plasma membrane, could function as a Ca^{2+} sensor, we raised several epitope-specific polyclonal antisera to the known ryanodine receptor isoforms. The strategy we used was to identify sequences unique to specific ryanodine receptor isoforms by sequence comparison (Zaidi et al., 1995). Figure 4, a similarity plot, depicts the percentage of similarity between the three ryanodine receptor isoforms, I, II, and III, following their optimal alignment (Zaidi et al., 1995). The overall sequence identity is around 70% (Meissner, 1994). However, there are several regions of marked sequence divergence shown as downward deflections. The largest region corresponding to the channel-forming domain at the C-terminal, one-fifth, end of the molecule (amino acids -4100 to 4800), was used to select unique sequences in isoforms II and III (Figure 4) (Zaidi et al., 1995). These sequences were used to scan the Swissprot and Leeds Databases to confirm their lack of similarity with any other known protein sequence (Zaidi et al., 1995). Synthetic peptides were then used to raise antisera Ab^{129} (anti-type II) and Ab^{180} (anti-type III) (Zaidi et al., 1995). Furthermore, an antiserum was raised to the intact purified type I ryanodine receptor, and yet another antiserum, Ab^{34}, was raised to a cytosolic calmodulin-binding consensus sequence (Zaidi et al., 1995).

We next confirmed the specificity of the antisera by immunoblotting with microsomal membranes prepared from skeletal muscle (lane 1), cardiac muscle (lane 2), or brain (lane 3) tissue (Zaidi et al., 1995). Figure 5 shows an immunoblot

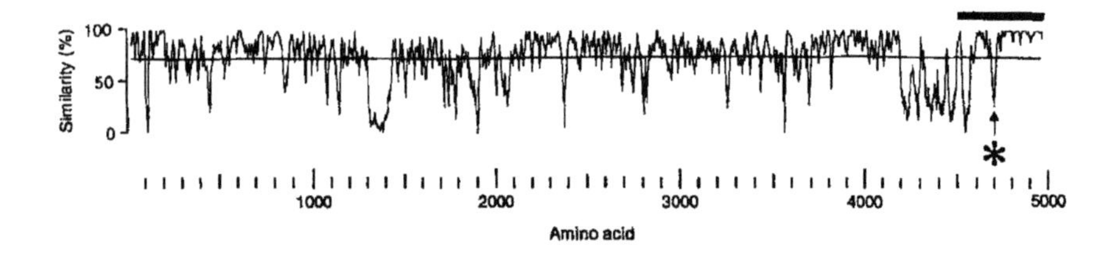

Figure 4. A similarity plot for the ryanodine receptor sequences.

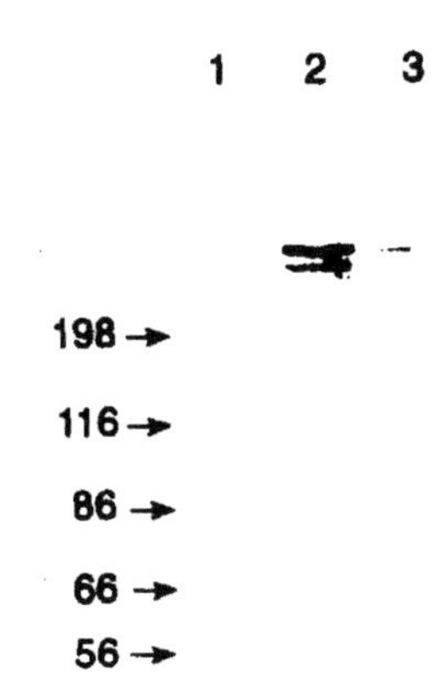

Figure 5. Western blot showing the specificity of antiserum Ab[129].

with antiserum Ab[129]. The antiserum recognized mainly the type II receptor protein found in cardiac muscle and, to a lesser extent, in brain microsomes (Zaidi et al., 1995). The other ryanodine receptor isoforms and members of the IP_3 receptor family were not detected by the antiserum (Zaidi et al., 1995). These antisera are critical for the subsequent experiment performed to characterize the osteoclast ryanodine receptor.

E. A Cell Surface Ryanodine Receptor

Our hypothesis was that should a microsomal membrane-resident molecule present at the plasma membrane, its intraluminal domains should present to the exterior, while its cytosolic domains should be retained intracellularly. Thus, if a ryanodine receptor was to be expressed in osteoclast plasma membrane, portions of its channel-forming domain should be accessible extracellularly to a relevant antiserum. With this working hypothesis, and through the use of antiserum Ab[129], we have succeeded in (a) unraveling the existence of a type II ryanodine receptor in the osteoclast plasma membrane, and (b) establishing its function as the divalent cation (Ca^{2+}) sensor (Zaidi et al., 1995).

Figure 6A is a confocal micrograph showing intense peripheral staining of intact, live, trypan blue-negative, osteoclasts with antiserum Ab[129]. This typical pattern, reminiscent of plasma membrane staining, was abolished upon the co-incubation of osteoclasts with the competing peptide, confirming detection of the Ab[129] epitope (Zaidi et al., 1995). Definitive evidence for a cell surface localization of the Ab[129] staining came from our scanning electron microscopic studies (Zaidi et al., 1995).

In contrast, permeabilized osteoclasts, stained with antiserum Ab[34], showed distinctive cytoplasmic staining. Note the clear nuclear outlining shown in Figure 6B. Intact osteoclasts did not stain with Ab[34], confirming a cytosolic localization of its

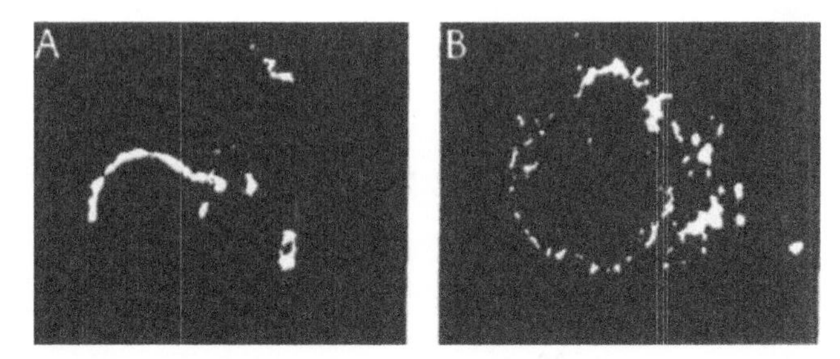

Figure 6. Panels A and B show confocal images of osteoclasts stained with antisera Ab^{129}. (**A**) Intact live cell, (**B**) fixed, permaebilized cell.

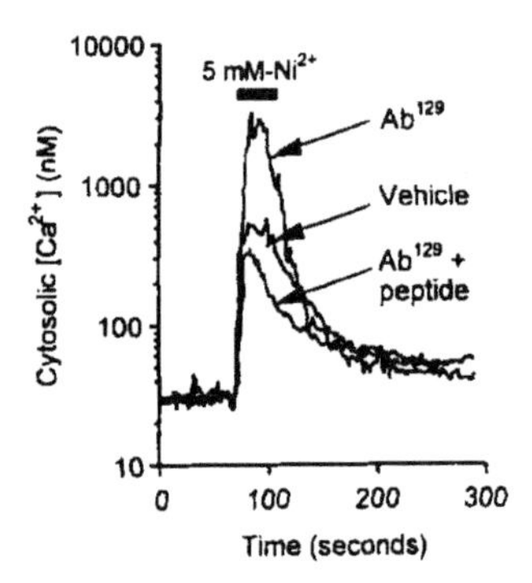

Figure 7. Potentiation of Ni^{2+}-induced cytosolic Ca^{2+} response after preincubation with antiserum Ab^{129}; the latter was abolished with the competing peptide.

epitope (Zaidi et al., 1995). Finally, we examined whether the plasma membrane type II ryanodine receptor (type II) could function as a Ca^{2+} sensor (Zaidi et al., 1995). For this, osteoclasts were treated with the surface-acting antiserum Ab^{129}, and then exposed to a Ni^{2+} pulse. We found that Ab^{129} potentiated the Ni^{2+}-induced cytosolic Ca^{2+} increase by 10-fold. As would be expected, the potentiation was abolished by co-incubation of the cells with the competing peptide (Figure 7) (Zaidi et al., 1995). These data, we believe, represent the strongest aspect of our evidence that the Ab^{129}-epitope is functional and, in view of its uniqueness, defines a novel location and function of a type II ryanodine receptor.

REFERENCES

Adebanjo, O.A., Pazianas, M., Zaidi, A., Shankar, V.S., Bascal, Z.A., Dacke, C.G., Huang, C.L.-H., and Zaidi, M. (1994a) Quantitative studies on the effect of prostacyclin on freshly isolated rat osteoclasts in culture. J. Endocrinol. 143, 375-381.

Adebanjo, O.A., Shankar, V.S., Pazianas, M., Zaidi, A., Huang, C.L.-H., and Zaidi, M. (1994b). Modulation of the osteoclast Ca^{2+} receptor by extracellular protons. Possible linkage between Ca^{2+} sensing and extracellular acidification. Biochem. Biophys. Res. Commun. 194, 742-747.

Adebanjo, O.A., Shankar, V.S., Pazianas, M., Simon, B., Lai, F.A., Huang, C.L.-H., Zaidi, M. (1996). Extracellularly applied ruthenium red and cyclic ADP-ribose elevate cytosolic Ca^{2+} in isolated rat osteoclasts. Am. J. Physiol. 270, F469-F475.

Anderson, K., Lai, F.A., Liu, Q.Y., Erickson, H.P., and Meissner, G. (1989) Structural and functional characterization of the purified cardiac ryanodine receptor-calcium release channel. J. Biol. Chem. 264, 1329-1335.

Anderson, M.J., Shirokov, R., Levis, R., Gonzalez, A., Karharnek, M, Hosey, M.M., Meissner, G., and Rios, E. (1993). Effects of perchlorate on the molecules of excitation-contraction coupling of skeletal and cardiac muscle. J. Gener. Physiol. 102, 423-448.

Bascal, Z.A., Alam, A.S.M.T., Zaidi, M., and Dacke, C.J. (1994) Effect of raised extracellular calcium on cell spread area of quail medullary bone osteoclasts in vitro. Exp. Physiol. 79, 15-24.

Bascal, Z.A., Moonga, B.S., Dacke, C.G., and Zaidi, M. (1992) Osteoclasts from medullary bone of egg-laying Japanese quail do not express the putative calcium 'receptor'. Exp. Physiol. 77, 501-504.

Bax, B.E., Shankar, V.S., Bax, C.M.R., Alam, A.S.M.T., Zara, S.J., Pazianas, M., Huang, C.L.-H., and Zaidi, M. (1993) Functional consequences of the interaction of Ni^{2+} with the osteoclast Ca^{2+} receptor. Exp. Physiol. 78, 517-529.

Bax, C.M.R., Shankar, V.S., Moonga, B.S., Huang, C.L.-H., and Zaidi, M. (1992). Is the osteoclast calcium "receptor" a receptor-operated calcium channel? Biochem. Biophys. Res. Commun. 183, 619-625.

Berridge, M.J. (1993) Inositol trisphosphate and calcium signaling. Nature 361, 315-325.

Brown, E.M., Gamba, G., Ricardi, I.D., Lombardi, M., Butters, R., Kifor, O., Sun, A., Hediger, M.A., Lytton, J., and Hebert, S.C. (1993). Cloning and characterization of an extracellular calcium-sensing receptor from bovine parathyroid. Nature (London) 366, 575-579.

Brown, E.M., Pollak, M., Seidman, C.E., Seidman, J.G., Chou, Y-H.W., Ricardi, D., and Hebert, S.C. (1995). Calcium-ion-sensing cell-surface receptors. New Eng. J. Med. 333, 234-240.

Dacke, C.G., Arkle, S.S., Cook, J., Wormstone, I.M., Jones, S., Zaidi, M., and Bascal, Z.A. (1993). Medullary bone and avian calcium regulation. J. Exper. Biol. 183, 63-88.

Datta, H.K., MacIntyre, I., and Zaidi, M. (1990a). Intracellular calcium in the control of osteoclast function. I. Voltage-insensitivity and lack of effects of nifedipine, BAYK8644, and diltiazem. Biochem. Biophys. Res. Commun. 167, 183-188.

Datta, H.K., MacIntyre, I., and Zaidi, M. (1990b). The effect of extracellular calcium elevation on morphology and function of isolated rat osteoclasts. Bioscience Reports 9, 747-751.

Huang, C.L.-H. (Ed.)(1993). Intramembrane charge movements in striated muscle. pp. 1-295. Clarendon Press, Oxford.

Lee, H.C., Galione, A., and Walseth, T.F. (1994). Cyclic ADP-Ribose: Metabolism and Calcium Mobilizing Function. In: *Vitamins and Hormones, Vol. 48.* (Litwack, G., Ed.), pp 199-258. Academic Press, Orlando.

Lundgren, S., Hjalm, G., Hellman, P.E.K.B., Juhlin, G, Rostad, J., Klares Kog, L., Akerstrom, G., and Rask, L. (1994). A protein involved in calcium sensing of the human parathyroid and placental cytotrophoblast cells belongs to the LDL-receptor protein superfamily. Exper. Cell Res. 212, 344-350.

Makgoba, M.W. and Datta, H.K. (1992). The critical role of magnesium ions in osteoclast-matrix interaction: Implications for the divalent cations in the study of osteoclast adhesion molecules and bone resorption. Eur. J. Clin. Invest. 22, 692-696.

Malgaroli, A., Meldolesi, J., Zambonin-Zallone, A., and Teti, A. (1989). Control of cytosolic-free calcium in rat and chicken osteoclasts. The role of extracellular calcium and calcitonin. J. Biol. Chem. 264, 14342-14347.

Meissner, G. (1994). The ryanodine receptor: Structure and function. Annl. Rev. Physiol. 56, 485-508.

Mithal, A., Kifor, O., Thun, R., Krapcho, K., Fuller, F., Hebert, S.C., Brown, E.M., and Tamir, H. (1994). Highly purified sheep C-cells express an extracellular Ca receptor similar to that present in parathyroid. J. Bone Miner. Res. 9 (Suppl. 1), B 209.

Moonga, B.S., Datta, H.K., Bevis, P.J.R., Huang, C.L-H., MacIntyre, I., and Zaidi, M. (1991). Correlates of osteoclast function in the presence of perchlorate ions in the rat. Exp. Physiol. 76, 923-933.

Moonga, B.S., Moss, D.W., Patchell, A., and Zaidi, M. (1990). Intracellular regulation of enzyme secretion from rat osteoclasts and evidence for a functional role in bone resorption. J. Physiol. 429, 29-45.

Pazianas, M., Zaidi, M., Huang, C.L.-H., Moonga, B.S., and Shankar, V.S. (1993). Voltage-sensitivity of the osteoclast calcium receptor. Biochem. Biophys. Res. Commun. 192, 1100-1105.

Pollak, M.R,. Brown, E.M., Chou, Y-H.W., Hebert, S.C., Marx, S.J., Steinmann, B., Levy, T., Seidman, C.E., and Seidman, J.G. (1993) Mutations in the human Ca^{2+}-sensing receptor gene cause familial hypocalciuric hypercalcemia and neonatal severe hyperparathyroidism. Cell 75, 1297-1303.

Putney, Jr., J.W. (1990). Capacitative calcium influx revisited. Cell Calcium 11, 611-624.

Racke, F.K. and Nemeth, E.F. (1993). Cytosolic calcium homeostasis in bovine parathyroid cells and its modulation by protein kinase C. J. Physiol. 464, 141-162.

Ricardi, D, Pak, J., Lee, W-S., Gamba, G., Brown, E.M., and Hebert, S.C. (1995). Cloning and functional expression of a rat kidney extracellular calcium/polyvalent cation-sensing receptor. Proc. Natl. Acad. Sci. USA 9, 131-135.

Salgado, V.L., Yeh, J.Z., and Narahashi, T. (1985). Voltage-dependent removal of sodium inactivation by N-bromoacetamide and pronase. Biophys. J. 47, 567-571.

Shankar, V.S., Alam, A.S.M.T., Bax, C.M.R., Bax, B.E., Pazianas, M., Huang, C.L.-H., and Zaidi, M. (1992a). Activation and inactivation of the osteoclast Ca^{2+} receptor by the trivalent cation, La^{3+}. Biochem.Biophys. Res. Commun. 187, 907-912.

Shankar, V.S., Bax, C.M.R., Alam, A.S.M.T., Bax, B.E., Huang, C.L.-H., and Zaidi, M. (1992b). The osteoclast Ca^{2+} receptor is highly sensitive to activation by transition metal cations. Biochem. Biophys. Res. Commun. 187, 913-918.

Shankar, V.S., Bax, C.M.R., Bax, B.E., Alam, A.S.M.T., Simon, B., Pazianas, M., Moonga, B.S., Huang, C.L.-H., and Zaidi, M. (1993). Activation of the Ca^{2+} 'receptor' on the osteoclast by Ni^{2+} elicits cytosolic Ca^{2+} signals: Evidence for receptor activation and inactivation, intracellular Ca^{2+} redistribution, and divalent cation modulation. J. Cell. Physiol. 155, 120-129.

Shankar, V.S., Huang, C.L.-H., Adebanjo, O.A., Simon, B.J., Alam, A.S.M.T., Moonga, B.S., Pazianas, M., Scott, R.H., and Zaidi, M. (1995a). The effect of membrane potential on surface Ca^{2+} receptor activation in rat osteoclasts. J. Cell. Physiol. 162, 1-8.

Shankar, V.S., Pazianas, M., Huang, C.L.-H., Simon, B., Adebanjo, O., and Zaidi, M. (1995b). Caffeine modulates Ca^{2+} receptor activation in isolated rat osteoclasts and induces intracellular Ca^{2+} release. Am. J. Physiol. 268, F447-F454.

Silver, I.A., Murrills, R.J., Etherington, D.J. (1988). Microelectrode studies on acid microenvironment beneath adherent macrophages and osteoclasts. Experimental Cell Research 175, 266-276.

Wappler, F., Roewer, N., Lenzen, C., Kochling, A., Scholz, J., Steinfath, M., and Schulte am Esch, J. (1994) High-purity ryanodine and 9,21-dehydroryanodine for in vitro diagnosis of malignant hyperthermia in man. Brit. J. Anesthesia 72, 240-242.

Zaidi, M. (1990). "Calcium receptors" on eukaryotic cells with special reference to the osteoclast. Bioscience Reports 10, 493-507.

Zaidi, M., Alam, A.S.M.T., Huang, C.L.-H., Pazianas, M., Bax, C.M.R., Bax, B.E., Moonga, B.S., Bevis, P.J.R., and Shankar, V.S. (1993a). Extracellular Ca^{2+} sensing by the osteoclast. Cell Calcium 14, 271-277.

Zaidi, M., Datta, H.K., Patchell, A., Moonga, B.S., and MacIntyre, I. (1989) "Calcium-activated" intracellular calcium elevation: A novel mechanism of osteoclast regulation. Biochem. Biophys. Res. Commun. 163, 1461-465.

Zaidi, M., Kerby, J., Huang, C.L.-H., Alam, A.S.M.T., Rathod, H., Chambers, T.J., and Moonga, B.S. (1991). Divalent cations mimic the inhibitory effects of extracellular ionized calcium on bone resorption by isolated rat osteoclasts: Further evidence for a "calcium receptor". J. Cell. Physiol. 149, 422-427.

Zaidi, M., Shankar, V.S., Adebanjo, O.A., Lai, F.A., Pazianas, M., Sunavala, G., Spielman, A.I., and Rifkin, B.R. (1996). Regulation of extracellular calcium sensing in rat osteoclasts by femtomolar calcitonin concentrations. Am. J. Physiol. 271, F637-F644.

Zaidi, M., Shankar, V.S., Alam, A.S.M.T., Moonga, B.S., Pazianas, M., and Huang, C.L.-H. (1992). Evidence that a ryanodine receptor triggers signal transduction in the osteoclast. Biochem. Biophys. Res. Commun. 188, 1332-1336.

Zaidi, M., Shankar, V.S., Bax, C.M.R., Bax, B.E., Bevis, P.J.R., Pazianas, M., Alam, A.S.M.T., and Huang, C.L.-H. (1993b). Linkage of extracellular and intracellular control of cytosolic Ca^{2+} in rat osteoclasts in the presence of thapsigargin. J. Bone Miner. Res. 8, 961-967.

Zaidi, M., Shankar, V.S., Latif, A.B., Adebanjo, O.A., Makinde, V., Huang, C.L.-H., and Pazianas, M. (1993c). Evidence that the osteoclast calcium receptor is a cell surface protein. Proceedings of the Second International Workshop on Osteobiology: Cell-Matrix Interactions in Health and Disease (Parma, 1-4 October, 1993). p. 26.

Zaidi, M., Shankar, V.S., Tunwell, R.E., Adebanjo, O.A., McKrill, J., Pazianas, M., O'Connell, D., Simon, B., Rifkin, B.R., Venkitaraman, A., Huang, C.L.-H., and Lai, F.A. (1995b). A ryanodine receptorlike molecule expressed in the osteoclast plasma membrane functions in extracellular Ca^{2+}sensing. J. Clin. Invest. 96, 1582-1590.

THE OSTEOCLAST MOLECULAR PHENOTYPE

Dennis Sakai and Cedric Minkin

Advances in Organ Biology
Volume 5B, pages 385-421.

ISBN: 0-7623-0390-5

I. INTRODUCTION

Osteoclasts perform many specialized functions during their life cycle. They fuse to form large polykaryons and develop a complex cytoskeletal architecture that allows them to attach tightly to bone surfaces and form a sealing zone which encloses the microenvironment in which resorption takes place. They synthesize a battery of enzymes responsible for the dissolution of the inorganic phase and digestion of the organic matrix of bone. They actively communicate by direct contact with matrix components and through soluble paracrine factors with osteoblasts and stromal cells in order to coordinate bone remodeling. They emerge on cue from a hemopoietic progenitor and may also perish on cue during hormonally-regulated cell death. In order to carry out these functions, cells of the osteoclast lineage must express a complex array of gene products that uniquely programs the phenotype of this highly specialized cell type (for a recent review, see Baron et al., 1993).

Osteoclasts have been difficult to study using standard biochemical and molecular biological methods due to their low abundance in bone and because they are terminally differentiated and cannot be propagated in culture. *In vitro* differentiation systems with hemopoietic progenitors have been used to generate cultures that are enriched for osteoclasts (reviewed by Nijweide and de Grooth, 1992; Suda et al., 1992), however osteoclast-like cells still comprise a minor fraction of these cultures. Histochemical methods utilizing enzyme and immunological markers have been traditionally used to identify osteoclasts in primary cultures and coculture systems and in tissue sections. Molecular (i.e., mRNA) markers would be advantageous for osteoclast studies as methods for their detection are more sensitive and specific. Unfortunately, few molecular markers for osteoclasts have been developed due, in part, to the difficulties in obtaining osteoclasts in large quantities necessary for mRNA isolation, and in specifically and unambiguously localizing expression of a gene to osteoclasts. It has also been difficult to ascertain the influence of growth factors, cytokines, and hormones upon osteoclast function *in vivo* or in mixed cell culture systems, as it is difficult to establish whether osteoclasts respond directly to a given treatment or to secondary signals elaborated by nonosteoclast cells.

In recent years, several new cell purification and molecular biology techniques have been applied to the study of osteoclasts that have made possible more sensitive and quantitative analysis of osteoclast gene expression. These methods are especially valuable for the study of cells that comprise only a minor population of a tissue such as osteoclasts in bone and have allowed a more precise definition of the osteoclast molecular phenotype. Some of these key developments are as follows. First, the development of anti-osteoclast monoclonal antibodies and cell separation techniques has permitted the isolation of highly enriched populations of osteoclast cells in quantities sufficient for standard biochemical and molecular biological analyses. Second, the use of *in situ* hybridization allows the localization of transcripts specifically to osteoclasts in their native environment. Third, micromanipulation techniques have permitted the isolation of homogeneous preparations of authentic osteoclasts. Fourth, the use of the highly sensitive and, in some schemes, quantitative reverse transcription polymerase chain reaction (RT-PCR) method can be used to detect mRNA molecules derived from individual cells or from small cell populations. Fifth, antisense inhibition and substrate inhibition experiments have been used to implicate specific gene products in osteoclastogenesis or osteoclast bone resorption. Finally, the molecular analysis of osteopetrotic mutant mouse models, including genetically engineered "knock-out" mice, has provided strong evidence for the necessity (or dispensability) of specific genes for osteoclast development or function.

In the following sections we discuss recent studies that have applied these methods towards the study of osteoclasts and summarize our current view of the osteoclast molecular phenotype.

II. THE LOCALIZATION AND QUANTITATION OF SPECIFIC GENE EXPRESSION IN OSTEOCLASTS

A. Application of New Methods for the Measurement of Specific Gene Expression in Osteoclasts

Traditional molecular and biochemical techniques such as enzyme assays, immunological assays, or nucleic acid hybridization require relatively large quantities of cells for analysis and so have, until recently, not been applicable to osteoclasts. In recent years, however, several methods have been developed for the large-scale purification of osteoclasts. Large numbers of osteoclasts are found in the chick limb and newborn rabbit long bones, and numerous osteoclast-like giant cells can be obtained from human osteoclastomas. Therefore, these sources have been frequently used as starting points for the preparation of enriched populations of osteoclasts and osteoclast-like cells.

Avian osteoclasts have been enriched by density gradient fractionation (Oursler et al., 1991b) and by immunomagnetic bead separation using osteoclast-selective monoclonal antibodies (Collin-Osdoby et al., 1991). Human osteoclastoma-derived giant cells and marrow-derived giant cells have also been purified by the immunomagnetic bead method (Oursler et al., 1994; Collin-Osdoby et al., 1995a). Rabbit osteoclasts have been purified by taking advantage of their tight attachment to tissue culture dishes. Treatment of rabbit osteoclast cultures with pronase E releases essentially all nonosteoclast cells leaving behind highly purified cultures of authentic osteoclasts (Tezuka et al., 1992). Recently, Wesolowski et al. (1995) have described a method utilizing echistatin-induced detachment for the enrichment of murine marrow cells that have the potential for differentiating into osteoclast-like cells. These "pre-osteoclasts" express many osteoclast markers and, when induced to differentiate into multinuclear cells by culturing in the presence of osteoblastic cells and 1,25-dihydroxyvitamin D_3 [$1,25(OH)_2D_3$], acquire the capability to resorb bone. An enriched population of mononuclear pre-osteoclastic cells has also been obtained from avian medullary bone by sequentially filtering the cell suspension through nylon mesh of progressively smaller pore sizes and recovering an intermediate-sized cell fraction (Prallet et al., 1992). This osteoclast-deficient fraction was subsequently cultured to generate tartrate-resistand acid phosphatase-(TRAP) positive bone resorbing cells.

Once an enriched cell preparation is generated, specific gene products can be analyzed by standard assays. Proteins can be detected by enzyme assays or immunological assays provided specific antibodies are available. Messenger RNAs can be detected by solution or blot hybridization procedures. Also, cDNA libraries can be prepared in order to screen for osteoclast transcripts that are selectively expressed in osteoclasts or are regulated by hormones or growth factors. Also, the enriched osteoclast preparations can be cultured in the presence or absence of growth factors, cytokines, or hormones to determine if osteoclast gene expression or bone resorptive function is directly regulated.

The detection of a specific gene sequence in RNA or cDNA libraries made from highly enriched preparations of osteoclasts is strong evidence that the gene is indeed expressed in osteoclasts. However, since these cell preparations are typically not homogeneous, there remains a finite probability that cells other than osteoclasts are responsible for production of the gene product in question. In order to provide conclusive evidence that osteoclasts do indeed express the gene, other methods must be used. *In situ* hybridization provides specific cell localization information within mixed cell preparations and within tissue sections. The expression of several genes has been localized to osteoclasts by this method (see Table 1). A limitation of *in situ* hybridization is that it is relatively insensitive so that genes that are expressed at low levels are undetectable by this method. An alternative to *in situ* hybridization is RT-PCR detection using microisolated cells (MI-RT-PCR; Tong et al., 1994). In this method, homogeneous preparations of osteoclasts are isolated by manual micromanipulation. RNA is extracted from pools of these cells and RT-PCR is performed using gene-specific primer pairs. The advantages of this method are the use of pure cell populations and, when optimized, the high sensitivity and specificity of the PCR reaction. Individual osteoclast cells have been analyzed by this method (Asotra et al., 1994; Tong et al., 1994). By using microisolated cell preparations for RT-PCR studies, one can unambiguously assign individual mRNAs or panels of mRNAs to specific cell types. For example, Tong et al. (1994) were able to distinguish osteoclasts from macrophage polykaryons on the basis of the expression pattern of 10 marker

Table 1. Osteoclast Markers: Gene Products That Are Preferentially Expressed in Osteoclasts (or Osteoclast-Like Cells) Relative to Other Cell Types

Gene	*Detection Method* [a]	*Reference(s)*
Expression restricted to osteoclasts in bone		
Tartrate-resistant acid phosphatase	Histochemistry	Minkin, 1982
(TRAP)	Northern blot,	
	RNase protection, ISH	Ek-Rylander et al., 1991
Carbonic anhydrase II (CA II)	Autoradiography, IC	Gay, 1992
	Northern blot, ISH	Zheng et al., 1993
	Northern blot, ISH, RT-PCR	Laitala and Väänänen, 1993
Calcitonin receptor	Autoradiography	Nicholson et al., 1986
	ISH	Lee et al., 1995b
121F antigen	Immunoassay, IC	Oursler et al., 1991a
Expression higher in osteoclasts compared to other cells		
c-src	IC	Horne et al., 1992
c-*fms*	IC	Weir et al., 1993
	Northern blot, ISH	Hofstetter et al., 1992
Cathepsin-OC2	Northern blot, ISH	Tezuka et al., 1994b
	ISH	Li et al., 1995b
	ISH	Shi et al., 1995
	ISH	Inaoka et al., 1995
Gelatinase B (MMP-9)	Zymogram, northern blot, ISH	Tezuka et al., 1994a
	IC, ISH	Wucherpfennig et al., 1994
	ISH	Reponen et al., 1994
Cystatin C	IC, ISH	Wucherpfennig et al., 1996
V-ATPase A subunit, HO68 isoform	Northern blot	van Hille et al., 1993, 1995
V-ATPase B subunit, HO57 isoform	Northern blot, ISH	van Hille et al., 1994
brain-type isoform	Northern blot, RT-PCR	Bartkiewicz et al., 1995
V-ATPase subunit D	Northern blot, RT-PCR	Durrin et al., unpublished
Integrin α_V subunit	Immunoassay, ISH	Nesbitt et al., 1993
	ISH	Shinar et al., 1993
integrin β_3 subunit	ISH	Shinar et al., 1993
osteopontin	Northern blot, ISH	Tezuka et al., 1992
	ISH	Merry et al., 1993
	Northern blot, ISH	Arai et al., 1993
creatine kinase, brain-type isozyme	Zymogram	Fukushima et al., 1994
	Northern blot	

Notes: [a] Abbreviations: IC, immunocytochemistry; ISH, *in situ* hybridization; RT-PCR, reverse transcription-polymerase chain reaction; MI-RT-PCR, microisolated cell RT-PCR; IS-RT-PCR, *in situ* RT-PCR; LDA-RT-PCR, limiting dilution assay RT-PCR; DD-PCR, differential-display PCR.

genes. Many genes have now been confirmed to be expressed in osteoclasts by use of MI-RT-PCR (Asotra et al., 1994; Arkett et al., 1994; Tong et al., 1994, 1995; Sakai et al., 1995; Yang et al., 1995a,b).

The ability to analyze gene expression in individual cells also permits us to examine the phenomenon of osteoclast cell heterogeneity. A single cell type (even cells in a clonal cell population) can exhibit heterogeneity in gene expression pattern due to temporal or environmental factors. Morphological and electrophysiologic evidence support the notion that osteoclasts can cycle between a resorptive phase and a nonresorptive phase (Lakkakorpi and Väänänen, 1991). Osteoclasts attached to bone surfaces display different electrophysiologic properties than those attached to plastic (Arkett et al., 1992). And cultured osteoclasts that are spread out with extensive podosomes differ in their gene expression pattern from "rounded" osteoclasts. Messenger RNAs for carbonic anhydrase II (CA II) and TRAP are elevated in activated (rounded) osteoclasts as compared to spread osteoclasts (Shibata et al., 1993; Asotra et al., 1994).

Because of the high sensitivity of MI-RT-PCR, it is also possible to determine with a high degree of confidence that a specific gene is not expressed in osteoclasts. Such a conclusion would be difficult to attain by use of any other methodological approach. As might be expected, osteoclasts do not express mRNAs for alkaline phosphatase, osteocalcin, a_1(I)procollagen, or epidermal growth factor (Tong et al., 1994).

A new technique, *in situ*-RT-PCR, combines the sensitivity of PCR with the localization capability of *in situ* hybridization. This method has been recently applied to the localization of vitamin D receptor mRNA in osteoclasts (Moore et al., 1995).

B. Osteoclast-Specific Gene Expression

There are several osteoclast markers which are known to be preferentially expressed in osteoclasts (Table 1) and which are often utilized to distinguish osteoclasts from other cell types. TRAP, CA II, the calcitonin receptor, and the 121F antigen each appear to be specifically expressed in osteoclasts in bone and are, therefore, quite useful for the identification of osteoclasts in that tissue (see Table 1). However, TRAP, CA II, and calcitonin receptor are also expressed in many nonbone cells so they are less useful as osteoclast markers in marrow culture systems or in tissues other than bone.

Histochemical and biochemical studies have demonstrated that TRAP enzyme activity in bone is specifically localized to osteoclast cells and their mononuclear precursors (Andersson et al., 1992; Minkin, 1982). This assessment was confirmed by the localization of TRAP mRNA to osteoclasts in rat bone by *in situ* hybridization (Ek-Rylander et al., 1991). However, further analysis of the tissue distribution of TRAP mRNA by a sensitive RNase protection assay demonstrated low but significant levels in all tissues examined. Thus, although TRAP appears to be osteoclast-specific in bone, it cannot be used as the sole criteria to distinguish osteoclast from nonbone cells. This is especially true for bone marrow cell cultures and cocultures of spleen and osteogenic cells as TRAP is expressed in macrophages and macrophage polykaryons under certain conditions of culture (Fuller and Chambers, 1989; Udagawa et al., 1990; Drexler and Gignac, 1994; Lacey et al., 1994; Shin et al., 1995). Despite its usefulness as a marker, the function of TRAP is unclear. The natural substrate for TRAP is unknown and its role in and necessity for bone resorption are undefined. However, TRAP can partially dephosphorylate the bone matrix phosphoproteins osteopontin and bone sialoprotein *in vitro*, and the dephosphorylated proteins fail to bind osteoclasts (Ek-Rylander et al., 1994). This suggests that the secretion of TRAP from osteoclasts into the resorption area could exert a regulatory influence on the attachment of the cells to the bone surface.

Another enzyme that has been used as an osteoclast-specific marker is CA II (see Gay, 1992). CA II is highly expressed in red cells, kidney, and pancreas where it functions in dissolving CO_2. High expression in osteoclasts is presumably required to generate hydrogen ions needed to acidify the resorption site. *In situ* hybridization studies demonstrated that CA II mRNA is specifically expressed in actively resorbing osteoclasts in trabecular bone (Laitala and Vnnen, 1993), and in multinucleated cells in osteoclastoma tissue (Zheng et al., 1993).

Calcitonin responsiveness has long been considered a specific marker for osteoclasts (Chambers and Magnus, 1982; Nicholson et al., 1986), despite the fact that many other cell types express the calcitonin receptor, including osteoblastic cells, myeloid cells, and macrophage polykaryons (Forrest et al., 1985; Vignery et al., 1991; Suzuki et al., 1995). Several alternatively spliced versions of the calcitonin receptor have been identified in various tissues so there remains an intriguing possibility that osteoclasts may express a unique isoform (see below).

The 121F monoclonal antibody detects a plasma membrane glycoprotein (possibly a Mn^{2+}-superoxide dismutase) which appears to be osteoclast-specific as it has not been detected immunologically in other cell types (Oursler et al., 1991a). However, proof that this marker is exclusively expressed in osteoclasts will require the cloning of its gene sequence in order to develop more sensitive methods for its detection. Nevertheless, 121F-reactivity has been shown to be useful for identifying osteoclasts in tissues and in cell culture (Collin-Osdoby et al., 1995a). The antibody has also been of great utility for the purification of osteoclasts and osteoclast-like cells (Collin-Osdoby et al., 1991; Oursler et al., 1994).

Several other genes have been shown to be preferentially expressed in osteoclasts though they may be widely expressed at lower levels in other bone and non-bone cell types (see Table 1). This is often sufficient for use as a marker provided that the method of detection is quantitative in nature so that high expressing cells can be distinguished from low expressing cells. In experiments utilizing immunological or histochemical detection methods or *in situ* hybridization, nonosteoclast cells may express the product of interest at levels below the limit of detection so that, operationally, the marker can be considered osteoclast-specific provided that they are defined as such with reference to a particular experimental protocol.

Because of the lack of completely osteoclast-specific markers, we have suggested that a panel of selectively expressed genes be used to uniquely describe the osteoclast molecular phenotype (Tong et al., 1994). Although no one member of such a panel of markers may be sufficient to distinguish an osteoclast, their simultaneous expression provides an unambiguous definition of an osteoclast. Also, important in this description is the quantitative information discussed above. For instance, although all cells express a vacuolar H^+-ATPase (V-ATPase), osteoclasts can be distinguished by their very high level of V-ATPase subunit expression (van Hille et al., 1993, 1994, 1995; Bartkiewicz, et al., 1995; Durrin et al., unpublished observations). The more markers that are included in this analysis, the more the cells gene expression profile will be uniquely defined. Thus, one of the challenges facing the osteoclast cell biologist is to identify panels of markers which can be used to define the osteoclast uniquely for a given experimental system. One step toward this objective is to identify additional known and novel genes that are preferentially expressed in osteoclasts and determine their tissue specificity.

C. The Search for Novel Osteoclast Markers

In efforts to discover new and possibly more specific osteoclast markers, cDNA libraries prepared from avian and rabbit osteoclasts and from human osteoclastoma have been screened in several ways for sequences that are selectively expressed in osteoclasts relative to other tissue sources. Tezuka and colleagues, (1992, 1994a) differentially screened a rabbit osteoclast cDNA library with osteoclast and spleen cDNA probes and isolated cDNA clones encoding osteopontin and cathepsin OC2. The latter is a novel cysteine protease related to cathepsins S and L (Bromme et al., 1996). The expression of these genes is not restricted to osteoclasts but they are among the more highly expressed osteoclast genes (see Table 2) and are only weakly expressed in spleen. This group has also isolated an osteoclast cDNA encoding the abundantly expressed gelatinase B (MMP-9) by differential screening with osteoclast and macrophage cDNA probes (Tezuka et al., 1994b).

In a similar differential screen protocol, Durrin et al. (unpublished observations) isolated rabbit osteoclast cDNAs encoding the B isozyme of creatine kinase. This isozyme is very highly expressed in osteoclasts and brain, but is ubiquitously expressed at lower levels in all other tissues. Interestingly, Fukushima et al. (1994) have found that serum levels of creatine kinase B are elevated in mice treated with bisphosphonates, which promote osteoclast cell death (Hughes et al., 1995b). Serum levels of the isoenzyme are also elevated in patients with bone resorption defects due to inheritance of autosomal dominant osteopetrosis type II (Yoneyama et al., 1992). These findings suggest that creatine kinase B may be a useful serum marker for osteoclast damage or some forms of osteopetrosis.

Wucherpfennig et al., (1994) have reported the isolation of MMP-9 and TRAP cDNAs during the differential screening of a human osteoclastoma cDNA library with giant cell and spleen cell probes. More recently these workers have described the cloning of cystatin C cDNA by screening for sequences that are preferentially expressed in multinucleated giant cells as compared to stromal cells; both cell preparations were derived from osteoclastoma tissue (Wucherpfennig et al., 1996). Localization of cystatin C and its mRNA to the giant cell subpopulation was confirmed by immunocytochemical staining and by *in situ* hybridization. Although, as exemplified by these results, differential screening has successfully identified several cDNAs that are preferentially expressed in osteoclasts relative to other cell

Table 2. cDNA Library Surveys of Abundantly Expressed Osteoclast Genes and Comparison With Their Frequency in the Entire dbEST Database

Gene	Rabbit OCL cDNA Library [a]	Rabbit OCL cDNA Library [b]	Human Osteoclastoma cDNA Library [c]	dbEST Database (v.030296) [d]
V-ATPase 16 kDa subunit	2.4–6.6			0.025
Cofilin	2.4–6.6			0.027
b-Actin	2.0–6.0			0.23
Creatine kinase B	1.2–4.9			0.023
c-*fms*	0.8–4.3			0.007
Ribosomal protein L18	0.8–4.3			0.029
Cathepsin-OC2	0.5–3.6	0.3	4.0	0.018
Cyclophilin A	0.5–3.6			0.052
d-Aminolevulinate synthase	0.5–3.6			0.020
Mitochondrial transcript	0.5–3.6			0.63
EST-4 (unidentified)	0.5–3.6			0.0005
EST-51 (unidentified)	0.5–3.6			0.0007
Osteopontin	0.0–1.5 [e]	0.7	11.5	0.016

Notes: [a]Percentage (95% confidence interval) of 194 rabbit osteoclast cDNAs randomly surveyed by sequencing (Sakai et al., 1995).
[b]Percentage of 5,000 rabbit osteoclast cDNAs screened by hybridization (Tezuka et al., 1992).
[c]Percentage of 9,300 human osteoclastoma cDNAs randomly surveyed by sequencing (Drake et al., 1995).
[d]Percentage of 438,652 expressed sequence tags (ESTs) (from dbEST ver. 030296) surveyed by sequencing. Includes ESTs from 52 species (80% from human). The majority of these ESTs have been generated by the I.M.A.G.E. consortium by random clone selection and sequencing (Lennon et al., 1996). Osteoclast ESTs (second column Sakai et al., 1995) are excluded from these calculations.
[e]No cDNA clones observed.

types, use of the method has failed to identify gene sequences that are uniquely expressed in osteoclasts. This is, in part, due to the use of only two tissue sources for comparative screening. Exhaustive screening of large numbers of osteoclast cDNAs with the use of probes derived from several different tissues is likely to be necessary for the identification of osteoclast-specific cDNA clones.

Another limitation of standard differential library screening protocols is that mRNA sequences that are expressed at low levels cannot be detected. An alternative approach to isolating differentially expressed gene sequences is differential display PCR (DD-PCR). This method entails the preparation of amplified mixtures of cDNA by RT-PCR using primers with arbitrary sequences (Liang and Pardee, 1992). It has been applied recently toward the isolation of cDNA sequences that are turned on during embryonic limb development (Cielinski et al., 1996), and the isolation of osteoclast mRNA sequences that are induced by PTH treatment (Tong, 1996). A disadvantage of the DD-PCR method is that, in contrast to the use of standard cDNA libraries, the sequences that are amplified by DD-PCR are not fully representative of the entire set of expressed gene sequences. Although DD-PCR cannot be considered a replacement for differential screening strategies as many differentially expressed genes will go undetected, it may complement differential screening by facilitating the isolation of some low abundance sequences.

cDNA libraries can also be screened by functional criteria. Takahashi et al. (1994) screened a cDNA expression library from human osteoclast-like multinuclear cells for clones that encode factors that stimulate osteoclast formation and bone resorbing activity. One cDNA encodes a protein with osteoclast-inductive activity and was identified as the calcium-dependent phospholipid-binding protein annexin II. This protein is expressed on the cell surface of osteoclastoma cells (Nesbitt and Horton, 1995). Inhibition of annexin II function with a neutralizing antibody partially inhibits bone resorption *in vitro*, indicating that it is required for full osteoclast activity (Nesbitt and Horton, 1995). Another cDNA clone encoding a potential osteoclast stimulating factor (OSF) displays no sequence homology to any known growth factors (Reddy et al., 1995).

Utilizing osteoclast-selective monoclonal antibodies to screen an osteoclastoma expression cDNA library, Roberts et al. (1995) have recently cloned a cDNA encoding an osteoclastic antigen. The sequence of the cDNA does not resemble any known gene suggesting that it may encode a highly specific osteoclast marker. Oursler et al. (1991a) have utilized a monoclonal antibody (121F) directed against an avian osteoclast cell surface antigen to purify a selectively-expressed protein with superoxide dismutase activity. Further characterization of this marker awaits the cloning and sequencing of gene sequences encoding this antigen.

D. The Relationship of Authentic Bone-Derived Osteoclasts to Spleen and Osteoclastoma Giant Cells and *in vitro* Generated Osteoclast-Like Cells

Much evidence indicates that osteoclasts emerge from the monocyte/macrophage lineage (for review, see Nijweide and de Grooth, 1992). Hemopoietic progenitors are capable of differentiating into osteoclasts *in vitro* (Shinar et al., 1990; Kerby et al., 1992), and mature monocytes and macrophages can be coaxed to differentiate into osteoclast-like cells (Udagawa et al., 1990). This suggests that macrophage polykaryons and osteoclasts may be closely related cell types, perhaps diverging in their particular phenotypes only because of their distinct environments. Many established cell lines have been shown to differentiate into cells that express one or more osteoclast markers. Macrophage-like cell lines can be induced to differentiate *in vitro* into osteoclast-like cells (Lacey et al., 1994; Shin et al., 1995). Vitamin D-induced HL-60 myeloid leukemia cell's express TRAP and CA II (Biskobing and Rubin, 1993; Suzuki et al., 1995), but don't express calcitonin receptors (Suzuki et al., 1995). Phorbol ester-induced HL-60 cells express integrin α_V and osteopontin (Somerman et al., 1995). Treatment of the human leukemic cell line FLG 29.1 with phorbol esters also results in the expression of osteoclastic characteristics, including multinuclearity, TRAP expression, calcitonin-responsiveness, and bone resorption activity (Gattei et al., 1992).

Primary cultures of cells of the macrophage/monocyte lineage can also be induced to express osteoclast markers. Alveolar macrophage giant cells express calcitonin receptor (Vignery et al., 1991). Foreign body giant cells express low levels of TRAP and the vitronectin receptor (Kadoya et al., 1994). Hemopoietic cells and macrophages cultured under certain conditions express variable levels of TRAP (Drexler and Gignac, 1994), which, ironically, is a marker that is often used to distinguish these cell types from osteoclasts. Although macrophages and macrophage polykaryons are capable of expressing at times certain osteoclast markers, they clearly have molecular phenotypes distinct from osteoclasts when an expanded number of markers are monitored simultaneously (Tong et al., 1994).

In summary, there are many instances where macrophages, macrophage-like cells, or myeloid cells can be coerced into expressing certain osteoclast-like characteristics. However, in most instances these cells are incapable of efficiently resorbing bone, and therefore lack the major functional criterion of an osteoclast. These results support the notion that osteoclasts and macrophage/monocytes arise from a common lineage and suggest that macrophages possess the *potential* to differentiate into osteoclasts. It should not be construed, however, that these results provide proof that macrophages can, or do, differentiate into osteoclast *in vivo*. There is as yet no definitive evidence that such a mechanism of osteoclast ontogeny exits. Cells in culture will often display "de-differentiated" phenotypes and their gene expression profile can be substantially manipulated by culture conditions. Rather, these results may simply indicate that the TRAP gene is subject to similar regulation in diverse cell types. Also, these findings further illustrate the limitations of the current set of known osteoclast markers and the necessity for discovering novel markers that can be used to distinguish among related cell types under different experimental conditions.

E. Gene Expression Profiles of Osteoclasts: the Osteoclast Molecular Phenotype

As part of the Human Genome Initiative, several laboratories have performed extensive sequencing surveys of cDNA libraries prepared from various cell lines and tissues in order to define their gene expression profile, find differences and similarities in cell type-specific and regulated expression patterns, and discover novel gene sequences (Adams et al., 1995; Lennon et al., 1996). Over 400,000 partial cDNA sequences in the form of expressed sequence tags (ESTs) have been deposited in the publicly-accessible dbEST database (Boguski et al., 1993; Boguski, 1995). ESTs from 52 species including over 37 different human tissues have been surveyed so far and several generalizations can be gleaned from the results. A total of approximately 88,000 genes (a likely majority of all human genes) have been identified by partial sequencing (Adams et al., 1995). Many of these genes have been found to be ubiquitously expressed and therefore meet the definition of a "housekeeping" gene. Only a small percentage of the genes ($\approx 0.05\%$) have been found to be expressed in only a single tissue source and can be classified as tissue-specific. About 88% of the putative gene sequences have not been associated with any known gene or protein product and are likely to represent novel gene sequences. This indicates that a random clone selection approach can be very effective at discovering novel genes and classifying known and novel genes according to abundance class and tissue expression pattern.

Osteoclast cDNA libraries from rabbit long bone and human osteoclastoma have also been surveyed by random clone selection and partial sequencing (Drake et al., 1995; Sakai et al., 1995). These analyses have identified the most abundantly expressed osteoclast genes, several of which have known osteoclast functions (Table 2). All the genes listed in Table 2 (with the exception of mitochondrial transcripts) are much more frequently represented (as much as 1,000-fold higher) in the osteoclast cDNA libraries (columns A, B, C) than in the entire dbEST database (column D), indicating that they are preferentially expressed in osteoclasts compared to other tissues. Among the abundantly expressed osteoclast mRNAs are the novel cathepsin OC2 and two additional novel gene sequences, EST-4 and EST-51. Also highly expressed are the V-ATPase proton channel subunit, cofilin, c-*fms*, and creatine kinase B. Interestingly, osteopontin was found to be the highest expressed gene in osteoclastoma-derived multinuclear cells, while it is expressed at a significantly lower level in rabbit osteoclast preparations. This may be due to differences in purity of the cell preparations, to a species difference, or to differences in the physiological state of osteoclasts isolated from these two environments.

Inspection of the types of functional classes of genes that are expressed by osteoclasts (Sakai et al., 1995) indicate that they synthesize large amounts of metabolic enzymes and protein synthesis components but, unlike cell types such as pancreatic islets (Takeda et al., 1993) or liver (Okubo et al., 1992), synthesize relatively small amounts of secretory products. Osteoclasts also synthesize high levels of cytoskeletal and cell surface and membrane proteins. This is consistent with their extensive membrane structure, their complex morphology, and with their high motility.

The use of large-scale EST surveys is an powerful method of characterizing gene expression profiles during cell differentiation and gene regulation and may become more routine with the development of improved methods of automated gene analysis (Adams et al., 1995; Lee et al., 1995a).

III. REGULATION OF OSTEOCLAST GENE EXPRESSION

A. Regulation of Gene Expression Upon Osteoclast Activation

Osteoclasts appear to cycle between a resorbing and a nonresorbing state, and genes associated with resorptive functions are predicted to be coordinately regulated during this cycling. Table 3 lists some gene products whose expression has been shown to be upregulated during osteoclast activation.

Table 3. Genes Whose Expression Level Correlates With Osteoclast Activation (Elevated Bone Resorption Activity)

Gene	*Quantitation Method* [a]	*Reference(s)*
TRAP	RT-PCR, ISH	Shibata et al., 1993
	LDA-RT-PCR	Gu et al., 1994
	Enzyme assay, Northern blot	Oursler et al., 1994
CA II	ISH	Laitala and Väänänen, 1993
	ISH	Zheng et al., 1994a
	MI-RT-PCR, ISH	Asotra et al., 1994
V-ATPase 16 kDa subunit	ISH	Laitala and Väänänen, 1993
121F antigen	IC	Collin-Osdoby et al., 1995a
c-*src*	Enzyme assay	Yoneda et al., 1993b
c-*fos*	ISH	Lee et al., 1994
	ISH	Hoyland and Sharpe, 1994
Osteopontin	Northern blot	Kaji et al., 1994
Urokinase receptor	ISH	Anderson et al., 1995
LEP-100	Northern blot	Oursler et al., 1993a
Lysozyme	Enzyme assay, Northern blot	Oursler et al., 1993a
Cathepsin B	Enzyme assay, Northern blot	Oursler et al., 1994
Cathepsin D	Northern blot	Oursler et al., 1994
Cathepsin C	DD-PCR, RNase protection	Tong, 1996

Note: [a]See Table 1 for definitions of abbreviations.

TRAP mRNA and enzyme activity appear to correlate with osteoclast maturation and activation. As the population of osteoclast lineage cells matures during fetal bone development, the level of TRAP mRNA increases, and this increase exceeds the accompanying increase in number of "TRAP-positive" cells (Gu et al., 1994). The mean TRAP mRNA levels in mature multinuclear osteoclasts is approximately one order of magnitude higher than the levels in mononuclear osteoclasts or TRAP-positive osteoclast precursors. In this study mRNA was quantitated by the novel method of limiting dilution assay-RT-PCR (Gu et al., 1994). With this method cDNA molecules are quantitated by determining the extent to which a solution of cDNA must be diluted for the DNA to no longer be detectable by a hemi-nested PCR reaction optimized to detect single DNA molecules. The number of mRNA molecules is subsequently calculated from the efficiency of reverse transcription of mRNA into cDNA.

Several other genes also appear to increase in expression during the maturation of osteoclasts into an actively bone resorbing state. The osteoclast-specific 121F antigen is upregulated during osteoclast-like cell formation induced by the treatment of avian marrow cells with osteoblast conditioned media (Collin-Osdoby et al., 1995a). Upregulation of the 121F antigen presumably accompanies the activation of pre-osteoclasts into resorbing osteoclasts. c-*fos* expression parallels the activation of osteoclasts which occurs following PTH treatment *in vivo* (Lee et al., 1994). The increase in c-*fos*, as measured by quantitative *in situ* hybridization, follows the appearance of mature osteoclasts and correlates with their activation into resorbing cells. c-*src* expression is enhanced in $1,25(OH)_2D_3$-treated cocultures of stromal and marrow cells in concert with the emergence of activated osteoclast-like cells (Yoneda et al., 1993b).

Actively resorbing osteoclasts express higher levels of TRAP than resting osteoclasts (Chambers et al., 1987). Osteoclasts cultured under low pH conditions have elevated resorption activity (Arnett and Dempster, 1986) and increased TRAP mRNA as determined by quantitative RT-PCR and *in situ* hybridization analyses (Shibata et al., 1993). In a similar study, Asotra et al. (1994) showed that CA II mRNA is also elevated in low pH-activated osteoclasts. Furthermore, by use of microisolation techniques, these authors were able to demonstrate that, among cells in the same culture, "rounded" osteoclasts expressed higher levels of CA II mRNA on average than "spread" osteoclasts. Osteoclasts with rounded morphologies may represent an osteoclast state with heightened bone resorption activity. Urokinase receptor mRNA also appears to be upregulated in osteoclasts by low pH activation (Anderson et al., 1995).

The V-ATPase plays a central role in bone resorption, the acidification of the extracellular space that forms the resorption lacunae. This is consistent with the finding that the expression of the 16 kDa V-ATPase proton channel subunit mRNA is elevated in actively resorbing osteoclasts as quantitated by *in situ* hybridization (Laitala and Väänänen, 1993). These authors also demonstrated that CA II mRNA is elevated in actively resorbing osteoclasts. Upregulation of carbonic anhydrase activity during osteoclast activation might be predicted as it is necessary for hydrogen ion generation.

Calcitonin treatment is known to inhibit osteoclast resorptive function (Kallio et al., 1972; Chambers and Magnus, 1982). The expression of several osteoclast gene products is downregulated by calcitonin and, therefore, correlates with the resorption process. Calcitonin treatment inhibits the expression of both CA II mRNA (Zheng et al., 1994a) and osteopontin mRNA (Kaji et al., 1994). Osteopontin is believed to function in the attachment of osteoclasts to bone surfaces, an early and essential step in the resorptive cycle. However, many other cell types in bone express osteopontin so it is unclear if this decrease in osteopontin expression by osteoclasts is physiologically important. Calcitonin treatment also results in the downregulation of its own receptor (Lee et al., 1995b; Rakopoulos et al., 1995; Takahashi et al., 1995; Wada et al., 1995), a mechanism that may be the basis for the "escape phenomenon" of calcitonin resistance.

Estrogen treatment of avian osteoclast and human giant cell cultures also inhibits bone resorbing activity, and simultaneously inhibits the expression of several genes (Oursler et al., 1991c, 1993a, 1994; Tobias and Chambers, 1991). The mRNAs encoding lysozyme, LEP-100, and cathepsins B and D are all downregulated by estrogen administration *in vitro*. These enzymes may function in the digestion of the matrix component of bone.

Utilizing the technique of DD-PCR, Tong (1996) has recently identified osteoclast cDNAs whose expression is enhanced in parathyroid hormone-(PTH) stimulated bone cell cultures. One cDNA was found to encode cathepsin C, another cysteine protease that may be involved in bone resorption.

One would expect that osteoclast activation would result in the coordinate expression of a large number of genes related to bone resorptive function. Additional studies along the lines described above would allow us to define the molecular phenotype of the actively resorbing osteoclast and compare it with that of nonresorbing cells. This information will give us a better perspective on the biochemical processes involved in osteoclast bone resorption. We note, however, that for most of the genes discussed above, it is not clear if their induction is in any way related to increased resorption activity or whether they are simply induced as a consequence of the general change in osteoclast phenotype that accompanies activation. Additional biochemical or genetic studies will be required to delineate the roles (if any) of these genes in the bone resorption process.

B. Direct Regulation of Osteoclast Gene Expression by Hormones and Growth Factors

Many hormones and growth factors have been shown to modulate osteoclast activity and development (reviewed by Heersche, 1992; Mundy, 1992; Manolagas, 1995). However, in most cases, it is unclear whether the agents act directly upon osteoclasts or through nonosteoclast target cells (or both). The availability of purified osteoclast preparations has permitted rigorous evaluation of the systemic and local factors that can directly impact osteoclast function. Proof that osteoclasts can act as target cells for a particular factor requires the demonstration that the relevant receptor is present and that physiological doses affect osteoclast bone resorbing activity or gene expression. Furthermore, the receptors biochemical properties (ligand affinity and number of binding sites per cell) should be consistent with the concentrations of ligand necessary to elicit a biological response. These criteria have been documented in only a few examples that we cite below and in Table 4.

Receptors for estrogens and androgens have been detected in purified avian osteoclasts and have been shown to display physiological affinity and specificity for steroid hormones and to mediate genomic effects (Oursler et al., 1991c, 1993b). Functional estrogen receptors have also been found in osteoclastoma-derived giant cells (Oursler et al., 1994) and a pre-osteoclastic cell line FLG 29.1 (Fiorelli et al., 1995). Immunological evidence has been presented for the localization of estrogen receptors in human osteoclasts and androgen receptor in mouse osteoclastic cells (Pensler et al., 1990; Mizuno et al., 1994). Brubaker and Gay (1994) have demonstrated saturable estradiol binding sites on the cell surface of avian osteoclasts. Treatment of osteoclasts with estradiol conjugates that are incapable of diffusion across the plasma membrane produce rapid cell shape changes and inhibition of acid production, demonstrating that these bindings sites are biologically functional. However, like steroid binding sites found on the cell surfaces of other cell types, the

Table 4. Growth Factor and Hormone Receptors Expressed by Osteoclasts and Osteoclast-Like Cells

Gene	*Detection Method* [a]	*Reference(s)*
CSF-1 receptor (c-*fms*)	ISH	Hofstetter et al., 1992
TGF-β receptor, type II	cDNA cloning	Dixon et al., 1994
	Northern blot, ISH	Zheng et al., 1994b
IGF receptor, type I	ISH	Middleton et al., 1995, 1996
IL-6 receptor	IC	Ohsaki et al., 1992
	ISH	Hoyland et al., 1994
IL-11 receptor	ISH	Romas et al., 1995
Calcitonin receptor	Autoradiography	Nicholson et al., 1986
	MI-RT-PCR	Tong et al., 1994
	Northern blot, RT-PCR	Ikegame et al., 1995
Parathyroid hormone receptor	Autoradiography	Teti et al., 1991
	Ligand binding, histochemistry	Agarwala and Gay, 1992
	MI-RT-PCR	Tong et al., 1995
Ryanodine receptor	Autoradiography, IC	Zaidi et al., 1995
P_{2U} purinoceptor	cDNA cloning	Bowler et al., 1995
Estrogen receptor	IC	Pensler et al., 1990
	Ligand binding, Western blot Northern blot	Oursler et al., 1991c
	Western blot, RT-PCR	Oursler et al., 1994
	Ligand binding, Western blot	Fiorelli et al., 1995
Androgen receptor	Ligand binding	Oursler et al., 1993b
	IC	Mizuno et al., 1994
Retinoic acid receptor β and retinoid X receptor α	Northern blot	Saneshige et al., 1995
Vitamin D receptor	IS-RT-PCR	Moore et al., 1995
Urokinase receptor isoform I	MI-RT-PCR, ISH	Yang et al., 1995a

Note: [a]See Table 1 for definitions of abbreviations.

osteoclast binding sites are of relatively low affinity and are unlikely to mediate responses to physiological ligand concentrations.

In cell cultures containing mixtures of osteoclast lineage cells and osteoblasts or osteogenic stromal cells, it appears that the major effects of sex steroids upon osteoclast numbers and activity are mediated indirectly by actions upon osteoblast and/or stromal cells (Poli et al., 1994; Bellido et al., 1995; Most et al., 1995; Zheng et al., 1995). Estrogens and androgens suppress the production of the osteoclastogenic cytokine IL-6 by these cells, causing reduced osteoclast development and bone resorption. However, these results do not rule out the possibility that estrogens and androgens can have significant direct effects upon the bone resorptive activity of mature osteoclasts under some physiological conditions. Estrogens have been shown to upregulate osteoclast expression of integrin β_3 mRNA in avian osteoclast precursors (Li et al., 1995a), to downregulate TRAP, cathepsin, and lysozyme mRNAs in human giant cells (Oursler et al., 1993a, 1994), and to upregulate progesterone receptor mRNA in FLG 29.1 human preosteoclastic cells (Fiorelli et al., 1995).

Glucocorticoids also modulate osteoclast function (Tobias and Chambers, 1989). Apart from their ability to inhibit bone resorptive activity, glucocorticoids may also induce osteoclast apoptosis (Hughes et al., 1995a). Wada et al. (1994) found that calcitonin receptor numbers are increased by glucocorticoid treatment of mouse marrow-derived osteoclast-like cells. However, because these studies did not utilize pure osteoclast preparations, it is unclear if these responses are due to direct action upon osteoclastic cells. Functional glucocorticoid receptors have not as yet been shown to reside in osteoclasts or osteoclast-like cells.

Osteoclast precursors and progenitors are $1,25(OH)_2D_3$ target cells. Vitamin D metabolites induce the differentiation of progenitor cells into osteoclasts in hemopoietic cell cultures (Kerby et al., 1992; Suda et al., 1992) as well as metatarsal organ cultures (Tao and Minkin, 1994). Functional receptors for $1,25(OH)_2D_3$ have been found in avian osteoclast precursors wherein vitamin D metabolites induce the transcription of genes encoding integrins α_V (Medhora et al., 1993) and β_3 (Mimura et al., 1994). In treated cells, expression of the vitronectin receptor is increased, enhancing the attachment of cells to vitronectin-coated dishes. Although it is clear that $1,25(OH)_2D_3$ directly regulates the differentiation of precursor cells into osteoclasts, it is less evident whether mature osteoclasts can act as $1,25(OH)_2D_3$ target cells. Moore et al. (1995) have recently detected the presence of $1,25(OH)_2D_3$ receptor mRNA in osteoclasts using the highly sensitive *in situ*-RT-PCR method.

Recently, receptor subunits for retinoic acid (retinoic acid receptor α and retinoid X receptor β subunits) have been located in rabbit osteoclasts (Saneshige et al., 1995). Administration of all-*trans*-retinoic acid to rabbit osteoclast cultures induces the expression of cathepsin OC2 and osteopontin mRNAs (Kaji et al., 1995; Saneshige et al., 1995).

In an effort to identify additional members of the steroid receptor gene family that may be expressed in osteoclasts, Smith-Oliver et al. (1995) performed RT-PCR experiments with degenerate oligonucleotide primers capable of amplifying many of the known receptors. Using RNA from enriched rabbit osteoclasts, they were able to identify mRNAs for the mineralocorticoid receptor, peroxisome proliferator receptor, and germ cell nuclear factor.

Calcitonin clearly has direct effects upon osteoclast attachment and bone resorption (Chambers and Magnus, 1982; Chambers et al., 1984). ^{125}I-calcitonin binding studies clearly demonstrate the existence of high affinity, saturable binding sites for calcitonin on the surface of rat osteoclasts and human giant cells (Nicholson et al., 1986). MI-RT-PCR experiments have provided indisputable evidence for the expression of calcitonin receptor mRNA in osteoclasts (Tong et al., 1994). Although one study has shown that rodent marrow-derived osteoclast-like cells express two different calcitonin receptor (CTR) mRNA alternatively-spliced isoforms (C1a and C1b type; Ikegame et al., 1995), a study with fetal mouse metatarsals detected only one (C1a) isoform (Gu et al., 1994). Since only the C1a isoform possesses high affinity for calcitonin, it is likely to be the physiologically relevant protein utilized by osteoclasts (Houssami et al., 1994). Semi-quantitative RT-PCR experiments indicate that CTR mRNA is autoregulated by calcitonin treatment (Lee et al., 1995b; Wada et al., 1995). This correlates with the downregulation of ^{125}I-calcitonin binding sites observed in calcitonin treated cultures (Wada et al., 1995).

Colony-stimulating factor (CSF-1, M-CSF) plays a key role in the maturation of macrophages and osteoclasts (Tanaka et al., 1993) and its receptor, the product of the c-*fms* gene, is expressed in osteoclast precursors. It was unexpected, however, to find that mature osteoclasts also express high levels of CSF-1 receptor (Hofstetter et al., 1992; Weir et al., 1993). In fact, the c-*fms* gene is among the most highly expressed genes in mature osteoclasts (Sakai et al., 1995). Although CSF-1 has been reported to promote osteoclast survival and resorptive function (Fuller et al., 1993; Felix et al., 1994; Amano et al., 1995b), it is unclear whether this is a direct effect upon mature cells and not an effect upon the precursors or an indirect effect mediated by nonosteoclast cells. Reminiscent of the autoregulation of the calcitonin receptor, CSF-1 treatment downregulates the expression of c-*fms* in osteoclasts (Amano et al., 1995a).

TGF-β has inhibitory effects upon bone resorption (Pfeilschifter et al., 1988) and in osteoclast development (Hattersley and Chambers, 1991). It may also accelerate osteoclast cell death (Hughes et al., 1994). TGF-β can also have stimulatory effects on the production of osteoclast-like cells in mouse bone marrow cultures, possibly by stimulating prostaglandin synthesis by stromal cells (Shinar and Rodan, 1990). Studies with isolated rat osteoclasts indicate that they can respond directly to treatment with TGF-β and that they express TGF-β receptor subunit mRNA (Dixon et al., 1994). TGF-β affects the motility of cultured osteoclasts, stimulating both cell locomotion and chemotaxis. TGF-β receptor is also expressed in giant cell tumors of bone where TGF-β may be involved in mediating recruitment of osteoclast-like cells (Zheng et al., 1994b).

In situ hybridization experiments have demonstrated expression of insulinlike growth factor (IGF-I) receptor mRNA in human osteophytic osteoclasts and osteoclastoma-derived giant cells (Middleton et al., 1995, 1996). Since stromal cells and osteoblasts secrete IGF-I, this suggests another mechanism by which bone resorption can be locally regulated. However, it has not yet been demonstrated if osteoclasts are capable of responding directly to IGF-I treatment. *In situ* hybridization experiments have also been used to show that osteoclasts express IL-6 receptor mRNA (Hoyland et al., 1994). Since IL-6 is important in osteoclast ontogeny, this result indicates that osteoclasts have the potential to respond directly to IL-6. Another cytokine that may act directly upon osteoclasts is IL-11. Romas et al. (1995) have recently detected the IL-11 receptor in both stromal and osteoclast cells by *in situ* hybridization. The responsiveness of osteoclasts to IL-6 and IL-11 requires the presence of osteoblasts or stromal cells (Ohsaki et al., 1992; Udagawa et al., 1995a). Since osteoblasts can respond directly to cytokines by production of osteoclastogenic factors (Suda et al., 1992; Greenfield et al., 1993), the major effects of IL-6 and IL-11 upon osteoclasts are likely to be indirect. Alternatively, the microenvironment generated by cytokine-primed osteoblasts or stromal cells may simply provide a "permissive" condition that osteoclasts require for cytokine responsiveness.

The stimulatory effects of PTH upon osteoclastic bone resorption appear to be largely mediated via osteoblasts (McSheehy and Chambers, 1986; Collin et al., 1992; Greenfield et al., 1993). Both osteoclast numbers and osteoclast cellular activity can be upregulated by paracrine factors secreted by osteoblasts in response to PTH. However, several studies have indicated that osteoclasts and their precursors express PTH receptors (Teti et al., 1991; Agarwala and Gay, 1992; Tong et al., 1995), indicating that they possess the potential to respond directly to the hormone.

Osteoclasts express a ryanodine receptor that appears to function as part of a plasma membrane-localized, extracellular calcium sensor (Zaidi et al., 1995). In muscle, ryanodine receptors are normally situated on intracellular sarcoplasmic membranes, so a plasma membrane localization is unique to osteoclasts. Elevation in extracellular calcium ion induces rapid increases in osteoclast cytosolic Ca^{2+} concentration. This subsequently triggers cell retraction and inhibition of bone resorption.

Osteoclasts also express a purinoceptor that binds extracellular ATP, another potential regulator of bone homeostasis (Bowler et al., 1995). Yang et al. (1995a) have recently demonstrated the expression of many components of the plasminogen activator/plasmin system (including urokinase and its receptor) in osteoclasts by using MI-RT-PCR and *in situ* hybridization. It is not known what role these gene products may play in osteoclast physiology.

C. Growth Factors and Cytokines Expressed by Osteoclasts

During bone remodeling osteoclasts must coordinate their activities with other cell types. This communication is mediated by cell-cell interactions and by the secretion of locally acting factors. Osteoclasts are now known to express a number of growth factors and cytokines (Table 5).

By use of either *in situ* hybridization or RT-PCR (or both), the mRNAs for TGF-β_1, IGF-I, IGF-II, IL-1α, IL-6, platelet derived growth factor A chain, and annexin II have been demonstrated to be expressed in osteoclasts or osteoclastoma-derived giant cells (see Table 5 for references). A cDNA encoding the cytokine inhibitor, IL-1 receptor antagonist, was cloned from an osteoclast cDNA library (Sakai et al., 1995). It should be noted that for many of these studies the actual expression of the growth factor or cytokine by osteoclasts has not been demonstrated (see Table 5).

Nitric oxide (NO) is a potent physiological regulator in many tissues and exerts rapid inhibitory effects upon osteoclast activity (Collin-Osdoby et al., 1995b). Brandi et al. (1995) have demonstrated that FLG 29.1 preosteoclastic cells generate NO in sufficient quantities to function as an autocrine regulator of bone resorption. Osteoclasts express both an inducible isoform and a calcium-sensitive isoform of NO synthase as assessed by RNA blot analysis and immunocytochemistry. Osteoclasts

Table 5. Growth Factors and Cytokines Expressed by Osteoclasts and Osteoclast-Like Cells

Gene	*Detection Method* [a]	*Reference(s)*
TGF-β_1	IC	Oursler, 1994
	MI-RT-PCR	Tong et al., 1994
	Northern blot, ISH	Zheng et al., 1994b
IGF-I	IC	Lazowski et al., 1994
	ISH	Middleton et al., 1995
IGF-II	ISH	Andrew et al., 1993
	ISH	Middleton et al., 1995
IL-1α	ISH	Okamura et al., 1993
	IC, ISH	Stashenko et al., 1994
IL-1 receptor antagonist	cDNA cloning	Sakai et al., 1995
IL-6	IC, RT-PCR, ISH	Ohsaki et al., 1992
	ISH	Hoyland et al., 1994
PDGF A chain	IC, ISH	Andrew et al., 1995
Annexin II	cDNA cloning, ISH	Takahashi et al., 1994
Nitric oxide synthase	IC, Northern blot	Brandi et al., 1995

Note: [a]See Table 1 for definitions of abbreviations.

attached to bone surfaces appear to express elevated levels of NO compared to nonresorbing osteoclasts.

The receptors for some of the growth factors and cytokines listed in Table 5 are also expressed in osteoclasts (see Table 4) indicating that they can potentially function in an autocrine manner. In the cases of IL-6 and annexin II, autoregulatory function has been reconstituted during suitable *in vitro* culture conditions (Ohsakai et al., 1992; Takahashi et al., 1994). However, due to the fact that osteoclasts are greatly outnumbered in their natural bone microenvironment, it would appear that any possible autocrine loop would be insignificant *in vivo* relative to the influence of paracrine and systemic factors generated by other cell types. In fact, studies with IL-6-deficient mice clearly indicate that IL-6 is dispensable for osteoclast maturation or function (Poli et al., 1994).

IV. GENES WHOSE EXPRESSION IS REQUIRED FOR OSTEOCLAST MATURATION OR ACTIVATION

Although the expression of many genes has been shown to be preferential in osteoclasts or correlated with osteoclast maturation or bone resorption activity, there are only a handful of genes that are known to be essential for osteoclast function. Such evidence has been obtained by the use of enzyme inhibitors, inhibitory peptides, neutralizing antibodies, or antisense nucleic acids to inhibit specific gene products, and by the analysis of osteopetrotic mouse genetic models (see Table 6).

Enzyme inhibitors have been used to implicate several enzymes in bone resorption. Bafilomycin A1 and WY 47766, specific inhibitors of the vacuolar-type

Table 6. Genes Whose Expression in Osteoclasts is Required for Osteoclast Maturation or Function

Gene	Experimental Evidence [a]	Reference(s)
c-fms	Osteopetrotic (*op*) mouse model	Yoshida et al., 1990
	CSF-1 requirement, neutralizing antibody	Tanaka et al., 1993
	Marrow/spleen cell coculture	Takahashi et al., 1991
mi	Osteopetrotic mouse model	Hodgkinson et al., 1993
PU.1	Osteopetrotic mouse model	Tondravi et al., 1995
oc	Osteosclerotic mouse model, marrow/spleen cell coculture	Udagawa et al., 1992
c-src	Osteopetrotic mouse model	Boyce et al., 1992
	Transplantation	Lowe et al., 1993
	Enzyme inhibition	Yoneda et al., 1993a
	Enzyme inhibition	Hall et al., 1994
	Enzyme inhibition	Feuerbach et al., 1995
c-fos	Osteopetrotic mouse model	Wang et al., 1992
	Transplantation	Grigoriadis et al., 1994
	Antisense inhibition	Udagawa et al., 1995b
CA II	Osteopetrotic human model	Roth et al., 1992
	Enzyme inhibition	Gay, 1992
	Antisense inhibition	Laitala and Väänänen, 1994
V-ATPase	Enzyme inhibition	Sundquist et al., 1990
V-ATPase 16 kDa subunit	Enzyme inhibition	Hall and Schaueblin, 1994
	Antisense inhibition	Laitala and Väänänen, 1994
Cathepsins B/L	Enzyme inhibitors	Hill et al., 1994a
Gelatinase B (MMP-9)	Enzyme inhibitors	Hill et al., 1994a
Phosphatidylinositol-3 Kinase	Enzyme inhibition	Hall et al., 1995
	Enzyme inhibition	Nakamura et al., 1995
RhoA	Enzyme inhibition	Zhang et al., 1995
FAK	Enzyme inibition, antisense inhibition	Tanaka et al., 1995
E-cadherin	Neutralizing antibody, peptide antagonists	Mbalaviele et al., 1995
Integrin subunits α_2, α_V, b_1, β_3	Neutralizing antibody, peptide antagonists	Horton et al., 1991
	Antisense inhibition	Townsend and Horton, 1995

Note: [a]See text for additional descriptions of methodologies.

H^+-ATPases, inhibit bone resorption in osteoclast-containing cell cultures (Sundquist et al., 1990; Hall and Schaueblin, 1994). Since osteoclasts express very high levels of the V-ATPase compared to other bone cell types, it is likely that these inhibitors are acting directly upon osteoclasts and not upon the nonosteoclast cells in these cultures. The V-ATPase is no doubt essential for acidification of the resorption compartment. More recently, Laitala and Väänänen (1994) demonstrated the requirement for V-ATPase activity by using antisense nucleic acids to inhibit V-ATPase subunit expression in osteoclast-containing cell cultures. Antisense oligodeoxynucleotides and antisense RNAs directed against the 16 kDa and 60 kDa subunits were found to inhibit bone resorption as well as the accumulation of the targeted mRNAs.

Another essential enzyme for osteoclast bone resorption is CA II. Both enzyme inhibitor studies (see Gay, 1992) and antisense inhibition experiments (Laitala and Väänänen, 1994) have demonstrated the requirement for this enzyme in bone resorption. The importance of CA II in bone resorption is further indicated by the osteopetrotic phenotype of patients with inherited CA II gene defects (Roth et al., 1992).

Wortmannin, a specific inhibitor of phosphatidylinositol-3 kinase, inhibits bone resorption *in vitro* (Hall et al., 1995; Nakamura et al., 1995) and *in vivo* (Nakamura et al., 1995). The inhibitor appears to function by disrupting formation of the osteoclast ruffled border, implicating phosphatidylinositol-3 kinase in that process.

The G protein RhoA regulates cytoskeletal organization. Treatment of bone marrow cell cultures with C3 exoenzyme (an ADP-ribosyltransferase that specifically modifies and inactivates RhoA) disrupts the ringed podosome structure of osteoclasts and inhibits bone resorption (Zhang et al., 1995). Apparently, sufficient C3 exoenzyme is able to enter osteoclasts for it to effectively ADP-ribosylate RhoA. Inhibition was more effective when the C3 exoenzyme was microinjected into osteoclasts.

Osteoclasts secrete many proteases that are necessary for the degradation of bone matrix (for review, see Delaissé and Vaes, 1992). Numerous inhibitor studies have shown the involvement of several classes of proteases in bone resorption including lysosomal cysteine proteinases and matrix metalloproteinases. Osteoclasts express high levels of cathepsins and gelatinase B (MMP-9) and these appear to be required for full resorption activity (Hill et al., 1994a,b). However, because enzyme inhibitors are not entirely specific in their actions, additional lines of experimentation will be necessary to prove that any one particular osteoclast enzyme is necessary for bone resorption and not an isozyme or an unrelated enzyme with similar specificity.

Cadherins are Ca^{2+}-dependent cell surface proteins that mediate cell-cell adhesion. Mbalaviele et al. (1995) studied the effect of inhibiting cadherin function upon osteoclastic cell development and fusion in mouse bone marrow cultures. Neutralizing antibodies to E-cadherin (but not N-cadherin or P-cadherin) decreased osteoclastic cell formation and bone resorption by inhibiting the fusion of mononuclear osteoclast precursors. Furthermore, synthetic peptides containing the cell adhesion recognition sequence of cadherins also decreased osteoclastic cell formation. These findings suggest that E-cadherins may be involved in the fusion of hemopoietic osteoclast precursors into mature multinucleated osteoclasts.

Integrins are heterodimeric cell surface glycoproteins that mediate cell-substratum interactions. Osteoclasts express integrin subunits α_V, α_1, α_2, β_1, and β_3 and perhaps others (Clover et al., 1992; Hughes et al., 1993; Nesbitt et al., 1993;

Shinar et al., 1993; Grano et al., 1994) (see also Chapters 13 and 14). The $\alpha_V\beta_3$ dimer constitutes the major form of the osteoclasts vitronectin receptor. Anti-vitronectin receptor antibodies and peptides containing the Arg-Gly-Asp sequence (the consensus sequence found in substrates that are recognized by the vitronectin receptor) are effective inhibitors of osteoclast cell spreading and bone resorption (Horton et al., 1991). Recently, Townsend and Horton (1995) have described the inhibition of osteoclast attachment to dentine or serum- or collagen-coated glass using antisense oligonucleotides targeted against integrin subunits. Oligonucleotides that targeted integrins $\alpha 2$, α_V, β_1, and β_3 inhibited adhesion whereas control "sense" oligonucleotides and antisense oligonucleotides directed against β_2 and β_5 integrin subunits had no significant effect.

Antisense oligonucleotide inhibition experiments have also provided evidence that IL-6 can stimulate osteoclast bone resorptive activity (Reddy et al., 1994). Antisense IL-6 oligonucleotides inhibited the secretion of IL-6 and the formation of resorption pits by cultures enriched ($\approx 50\%$) in human giant cells. It is unclear, however, whether osteoclastic cells or stromal cells in these cultures contributed more to IL-6 production. In any event, IL-6 appears to be nonessential for osteoclast maturation and normal bone resorption (Poli et al., 1994).

Protein tyrosine kinase inhibitors decrease bone resorption in osteoclast-containing cell cultures (Hall et al., 1994), in cocultures of osteoblastic cells and bone marrow cells (Yoneda et al., 1993a; Tanaka et al., 1995), and in rat long bone tissue explants (Feuerbach et al., 1995). Herbimycin A and mycotrienins were found to suppress the activity of *src* protein tyrosine kinase (Yoneda et al., 1993a; Feuerbach et al., 1995). Herbimycin A was also found to inhibit FAK kinase (Tanaka et al., 1995). *Src* is highly expressed in actively resorbing osteoclasts and presumably functions as a constituent of the signal transduction pathways that modulate osteoclast bone resorption activity (Horne et al., 1992; Tanaka *et al.*, 1992). FAK is localized in focal adhesion plaques where it is thought to regulate cell attachment and podosome formation. Transgenic mouse strains with gene knock-outs can be generated by homologous recombination in embryonic stem cells and the construction of embryo chimeras. Strains constructed with loss-of-function mutations in the c-*src* gene develop osteopetrosis but are otherwise phenotypically normal (Soriano et al., 1991). Considering the fact that *src* is ubiquitously expressed in all tissues, this observation emphasizes the key role of *src* in osteoclast function. Alternatively, cells other than osteoclasts may simply express redundant enzyme activities that substitute for normal *src* functions. The osteopetrotic defect due to *src* deficiency appears not to be due to diminished numbers of mature osteoclasts, but rather to a defect in their resorption activity (Boyce et al., 1992, 1993). Transplantation experiments have demonstrated that *src* expression in osteoclasts, but not osteoblasts, is essential for bone resorption (Lowe et al., 1993).

Knock-out mouse strains containing null mutations in the c-*fos* gene develop osteopetrosis because of a general defect in hemopoietic cell development (Johnson et al., 1992; Wang et al., 1992). FOS is a transcriptional regulatory protein of the basic-leucine zipper family that is constitutively expressed in osteoclasts. Bone marrow transplantation experiments show that the osteopetrotic defect can be overcome by the transplantation of hemopoietic progenitors from normal mice (Grigoriadis et al., 1994), strongly suggesting that FOS expression (and expression of genes that are regulated by FOS) within osteoclast progenitors is necessary for osteoclast maturation. Using a osteoblast/spleen cell coculture system, Udagawa et al. (1995b) found that the proliferation of osteoclast progenitors (and the development of mature osteoclasts) was inhibited by the presence of FOS antisense oligodeoxynucleotides (and not by control sense-strand oligodeoxynucleotides). The antisense oligodeoxynucleotide did not inhibit the differentiation of postmitotic osteoclast precursors; nor did it inhibit bone resorption by mature osteoclasts. Other observations, however, suggest that FOS expression (at appropriate levels) may modulate the activity of mature osteoclasts. Transgenic mice that express high levels of FOS display abnormal bone remodeling defects, presumably due to dysfunctional osteoclasts (Rther et al., 1987). Overexpression of FOS in avian osteoclast precursors by transfection of c-*fos* appears to enhance osteoclastic bone resorption activity (Miyauchi *et al.*, 1994). Also, FOS expression is elevated in Pagetic osteoclasts which have high bone resorbing activity (Hoyland and Sharpe, 1994).

Osteoprotegerin (OPG) deficient mice develop a skeletal phenotype associated with early onset osteoporosis (Bucay et al., 1998), while hepatic overexpression of OPG in transgenic mice leads to osteopetrosis (Simonet et al., 1997). OPG, also known as osteoclastogenesis inhibitory factor (OCIF), is a recently discovered soluble members of the tumor-necrosis factor (TNF) receptor superfamily that inhibits osteoclastogenesis (Simonet et al., 1997; Yasuda et al., 1998). The OPG ligand (OPGL) has been identified as a membrane-associated, TNF-related cytokine (Lacey et al., 1998; Yasuda et al., 1998). This cytokine induces osteoclast formation as well as differentiation and appears to bind to a unique hematopoietic progenitor committed to the osteoclast lineage (Lacey et al., 1998), perhaps the colony-forming unit-osteoclast (CFU-O) recently identified by Maguruma and Lee (1998).

Another transgenic mouse strain has recently been constructed with a null mutation in the gene encoding the monocyte/lymphoid transcription factor PU.1 (Tondravi et al., 1995). These mice have severe osteopetrosis due to a complete lack of osteoclasts, and because of a simultaneous deficiency in macrophage and lymphocyte development, succumb to opportunistic infections shortly after birth. PU.1 was shown to be expressed in normal mature osteoclasts suggesting that PU.1-regulated gene expression in the osteoclast lineage is required for osteoclast maturation. Nevertheless, until genetic or transplantation or coculture studies are performed, we cannot know for certain exactly which cell types must express PU.1 for osteoclast development.

Several naturally occurring mutant mouse models have been described that show clear genetic influences upon bone resorption. The *op* mutation disrupts the reading frame of the gene encoding CSF-1 (M-CSF) and results in osteopetrosis

(Yoshida et al., 1990). Maturation of osteoclasts is greatly reduced (but not eliminated) in these strains. Osteoclast development and bone resorption can be partially rescued by CSF-1 replacement (Felix et al., 1989; Tanaka et al., 1993; Sundquist et al., 1995). Moreover, coculture studies demonstrate that the inability of osteoclasts to form in osteopetrotic mice is due to a defect in the local microenvironment provided by osteoblastic cells and not in the osteoclast lineage (Takahashi et al., 1991). When antibodies against CSF-1 or the CSF-1 receptor were added to the osteoblast-spleen cell coculture system, osteoclast-like multinuclear cell formation was abrogated (Tanaka et al., 1993). These results are consistent with the model that osteoclast precursors and/or progenitors express a receptor for CSF-1 (the product of the c-*fms* gene) and that occupation of this receptor by osteoblast-produced (or exogenously-derived) CSF-1 is required for osteoclast maturation. As stated above, osteoclasts and their precursors express abundant CSF-1 receptors.

The *mi* mutation causes osteopetrosis due to deficiency of mature osteoclasts and also impairs melanocyte and mast cell development. The product of the *mi* gene is a transcriptional regulatory protein of the helix-loop-helix-basic-leucine zipper family (Hodgkinson et al., 1993). Since *mi* is thought to mediate the genomic effects of CSF-1, and it has been demonstrated to be expressed within osteoclasts (Yang et al., 1995b), it is probable that *mi* expression is a requirement for normal osteoclast maturation. However, because *mi* is also expressed abundantly in other bone cell types (chondrocytes, osteoblasts) it is uncertain whether its expression in osteoclasts alone is sufficient for normal osteoclast development or that expression in other cell types in addition (or instead) is necessary.

The osteosclerotic (*oc/oc*) mouse has bone remodeling defects that appear to be due to a bone resorption dysfunction of mature osteoclasts. Cocultures of *oc* osteoblastic cells and normal spleen cells form osteoclast-like multinuclear cells that can resorb bone; cocultures of normal osteoblastic cells and *oc* spleen cells form multinuclear cells that don't resorb bone (Udagawa et al., 1992). The molecular defect of the *oc* mutation is unknown, but it is likely to effect a gene product that is important in osteoclast bone resorption.

V. CONCLUSIONS

Recent advances in osteoclast cell and molecular biology methodology have permitted increasingly detailed and sensitive measurements of osteoclast gene expression profiles. These techniques have led to discoveries of several novel gene products of apparent importance to osteoclast physiology and have permitted more quantitative analysis of individual mRNA levels and their regulation by hormones and growth factors. They have also permitted the unambiguous localization of gene products to osteoclast cells. However, these advances have also complicated our understanding of the osteoclast molecular phenotype. Gene products that were once thought to be uniquely expressed in osteoclasts are now known to be expressed, albeit at lower levels, in other cell types. Several regulatory mechanisms that were once believed to function only in nonosteoclast cells have now been found to exist within osteoclastic cells. The complexity of osteoclast molecular biology is further increased by the revelation that many of its gene products are encoded by multi-gene families or by alternatively spliced mRNA isoforms.

Osteoclast gene expression is clearly dependent upon the environment that surrounds the cell. Because of the dynamic nature of osteoclast gene expression, it is difficult, if not impossible, to summarize a single osteoclast "molecular phenotype." Rather, there may exist multiple osteoclast phenotypes that describe the osteoclast under various conditions. It would be less confusing if the "cell type" of the cells under investigation were defined by appropriate physiological criteria such as their morphological properties (e.g., multinucleation) and tissue of origin (e.g., bone), and not by the genes that they may be expressing at the time. It is evident from our current understanding that the gene expression profile of osteoclasts can vary considerably with its physiological state. And so, identifying cells as osteoclasts on the basis of the expression of one or a few "osteoclast" markers can result in incorrect or misleading classification. Thus, the use of the term osteoclast is best reserved for bone resorbing cells found in or isolated from bone tissue.

Although we have learned a great deal about gene expression in osteoclasts and osteoclastic cells in recent years, we still have a poor understanding of how the osteoclast phenotype is regulated or what gene products are important for the various osteoclast functions. Future studies will be needed to determine which signal transduction mechanisms function in osteoclasts, and how its genetic repertoire and bone resorption activities are modulated by hormones, growth factors, cytokines, and cell-cell and cell-substratum interactions. One approach to understanding how osteoclasts are regulated by these factors is to determine the osteoclast gene expression profile while under stimulation by different factors. We also need to improve our understanding of osteoclast ontogeny. Little is known about the molecular phenotype of osteoclast precursors and progenitors, and the precise lineage from which osteoclasts develop remains to be defined. Finally, this molecular information will be useful for future studies that address the function of specific gene products in osteoclast biology. Genetic studies utilizing naturally occurring or transgenic mutant mouse models will be important in addressing the *in vivo* function of gene products during osteoclast ontogeny, while cell culture studies employing inhibitors of gene expression or protein function will enable determination of the role of specific gene products in osteoclast formation as well as bone resorption.

VI. SUMMARY

Recent advances in osteoclast cell and molecular biology methodology have significantly broadened our knowledge of the osteoclast molecular phenotype. New strategies for osteoclast cell isolation and purification, and for the quantitation and tissue-localization of mRNA have permitted more accurate and sensitive analysis of osteoclast-specific gene expression and its regulation by endocrine and paracrine factors. Studies utilizing enzyme inhibitors, inhibitory peptides, neutralizing antibodies, or antisense nucleic acids have implicated many gene products in osteoclast formation or bone resorptive function. The molecular analysis of osteopetrotic mutant mouse models, including genetically engineered knock-out mice has identified genes that are essential for osteoclast development or function. In this chapter, we have discussed recent studies that have utilized these strategies to study genes expressed in osteoclasts and summarize our current view of the osteoclast molecular phenotype.

REFERENCES

Adams, M.D., Kerlavage, A.R., Fleischmann, R.D., Fuldner, R.A., Bult, C.J., Lee, N.H., Kirkness, E.F. et al. (1995). Initial assessment of human gene diversity and expression patterns based upon 83 million nucleotides of cDNA sequence. Nature 377 (Suppl.), 3-174.

Agarwala, N. and Gay, C. V. (1992). Specific binding of parathyroid hormone to living osteoclasts. J. Bone Miner. Res. 7, 531-539.

Amano, H., Hofstetter, W., Cecchini, M. G., Fleisch, H., and Felix, R. (1995a). Downregulation of colony-stimulating factor-I (CSF-1) binding by CSF-1 in isolated osteoclasts. Calcif. Tissue Int. 57, 367-370.

Amano, H., Teramoto, T., Yamada, S., Hofstetter, W., Fleisch, H., and Felix, R. (1995b). Colony stimulating factor-1 activates mature osteoclasts both morphologically and functionally. Bone 17 (Suppl. S), 577.

Anderson, G.I., Allan, E.H., Zhou, H., Heersche, J.N.M., and Martin, T.J. (1995). Osteoclasts activated by lowered pH express urokinase receptor message and induce greater plaminogen activation. J. Bone Miner. Res. 10 (Suppl. 1), S222.

Andersson, G., Ek-Rylander, B., and Minkin, C. (1992). Acid phosphatases. In: Biology and Physiology of the Osteoclast. (Rifkin, B.R. and Gay, C.V., Eds.), pp. 55-80. CRC Press, Boca Raton, FL.

Andrew, J.G., Hoyland, J., Freemont, A.J., and Marsh, D. (1993). Insulinlike growth factor gene expression in human fracture callus. Calcif.Tissue Int. 53, 97-102.

Andrew, J.G., Hoyland, J.A., Freemont, A.J., and Marsh, D.R. (1995). Platelet-derived growth factor expression in normally healing human fractures. Bone 16, 455-460.

Arai, N., Ohya, K., and Ogura, H. (1993). Osteopontin mRNA expression during bone resorption: An in situ hybridization study of induced ectopic bone in the rat. Bone Miner. 22, 129-145.

Arkett, S. A., Dixon, S. J., and Sims, S. M. (1992). Substrate influences rat osteoclast morphology and expression of potassium conductances. J. Physiol. 458, 633-653.

Arkett, S.A., Dixon, S.J., Yang, J.-N., Sakai, D.D., Minkin, C., and Sims, S.M. (1994). Mammalian osteoclasts express a transient potassium channel with properties of Kv1.3. Receptors Channels 2, 281-293.

Arnett, T.R. and Dempster, D.W. (1986). Effect of pH on bone resorption by rat osteoclasts in vitro. Endocrinology 119, 119-124.

Asotra, S., Gupta, A.K., Sodek, J., Aubin, J.E., and Heersche, J.N. (1994). Carbonic anhydrase II mRNA expression in individual osteoclasts under resorbing and nonresorbing conditions. J. Bone Miner. Res. 9, 1115-1122.

Baron, R., Ravesloot, J.-H., Neff, L., Chakraborty, M., Chatterjee, D., Lomri, A., and Horne, W. (1993). Cellular and molecular biology of the osteoclast. In: Cellular and Molecular Biology of Bone. (Noda, M., Ed.), pp. 445-495. Academic Press, New York.

Bartkiewicz, M., Hernando, N., Reddy, S.V., Roodman, G.D., and Baron, R. (1995). Characterization of the osteoclast vacuolar H^+-ATPase B subunit. Gene 160, 157-164.

Bellido, T., Jilka, R. L., Boyce, B.F., Girasole, G., Broxmeyer, H., Dalrymple, S.A., Murray, R., and Manolagas, S.C. (1995). Regulation of interleukin-6, osteoclastogenesis, and bone mass by androgens: The role of the androgen receptor. J. Clin. Invest. 95, 2886-2895.

Biskobing, D.M. and Rubin, J. (1993). 1,25-dihydroxyvitamin D_3 and phorbol-myristate acetate produce divergent phenotypes in a monomyelocytic cell line. Endocrinology 132, 862-866.

Boguski, M. (1995). The turning point in genome research. Trends Biochem. Sci. 20, 295-296.

Boguski, M.S., Lowe, T.M.J., and Tolstoshev, C.M. (1993). dbESTdatabase for "expressed sequence tags". Nature Genet. 4, 332-333.

Bowler, W.B., Birch, M.A., Gallagher, J.A., and Bilbe, G. (1995). Identification and cloning of human P_{2U} purinoceptor present in osteoclastoma, bone, and osteoblasts. J. Bone Miner. Res. 10, 1137-1145.

Boyce, B.F., Yoneda, T., Lowe, C., Soriano, P., and Mundy, G.R. (1992). Requirement of $pp60^{c\text{-}src}$ expression for osteoclasts to form ruffled borders and resorb bone in mice. J. Clin. Invest. 90, 1622-1627.

Boyce, B.F., Chen, H., Soriano, P., and Mundy, G. R. (1993). Histomorphometric and immunocytochemical studies of *src*-related osteopetrosis. Bone 14, 335-340.

Brandi, M.L., Hukkanen, M., Umeda, T., Moradi-Bidhendi, N., Bianchi, S., Gross, S.S., Polak, J.M., and MacIntyre, I. (1995). Bidirectional regulation of osteoclast function by nitric oxide synthase isoforms. Proc. Natl. Acad. Sci. U.S.A. 92, 2954-2958.

Brömme, D., Okamoto, K., Wang, B.B., and Biroc, S. (1996). Human cathepsin O2, a matrix protein-degrading cysteine protease expressed in osteoclasts. J. Biol. Chem. 271, 2126-2132.

Brubaker, K.D and Gay, C.V. (1994). Specific binding of estrogen to osteoclast surfaces. Biochem. Biophys. Res. Commun. 202, 643-644.

Bucay, N., Sarosi, I., Dunstan, C.R., Morony, S., Tarpley, J., Capparelli, C., Scully, S., Tan, H.L., Xu, W., Lacey, D.L., Boyle, W.J., and Simonet, W.S. (1998). Osteoprotegerin-deficient mice develop early onset osteoporosis and arterial calfication. Genes Dev. 12, 1260-1268.

Chambers, T.J. and Magnus, C.J. (1982). Calcitonin alters behaviour of isolated osteoclasts. J. Pathol. 136, 27-39.

Chambers, T.J., Revell, P.A., Fuller, K., and Athanasou, N.A. (1984). Resorption of bone by isolated rabbit osteoclasts. J. Cell Sci. 66, 383-399.

Chambers, T.J., Fuller, K., and Darby, J.A. (1987). Hormonal regulation of acid phosphatase release by osteoclasts disaggregated from neonatal rat bone. J. Cell. Physiol. 132, 90-96.

Cielinski, M.J., Kwong, Y., and Stashenko, P. (1996). Identification of gene products related to osteoclast recruitment and development. J. Dent. Res. 75 (Suppl. A), 351.

Clover, J., Dodds, R.A., and Gowen, M. (1992). Integrin subunit expression by human osteoblasts and osteoclasts in situ and in culture. J. Cell Sci. 103, 267-271.

Collin, P., Guenther, H.L., and Fleisch, H. (1992). Constitutive expression of osteoclast-stimulating activity by normal clonal osteoblastlike cells. Effects of parathyroid hormone and 1,25-dihydroxyvitamin D_3. Endocrinology 131, 1181-1187.

Collin-Osdoby, P., Oursler, M. J., Webber, D., and Osdoby, P. (1991). Osteoclast-specific monoclonal antibodies coupled to magnetic beads provide a rapid and efficient method of purifying avian osteoclasts. J. Bone Miner. Res. 6, 1353-1365.

Collin-Osdoby, P., Oursler, M.J., Rothe, L., Webber, D., Anderson, F., and Osdoby, P. (1995a). Osteoclast 121F antigen expression during osteoblast conditioned medium induction of osteoclastlike cells in vitro: Relationship to calcitonin responsiveness, tartrate-resistant acid-phosphatase levels, and bone resorptive activity. J. Bone Miner. Res. 10, 45-58.

Collin-Osdoby, P., Nickols, G.A., and Osdoby, P. (1995b). Bone cell function, regulation, and communication: A role for nitric oxide. J. Cell. Biochem. 57, 399-408.

Delaissé, J.-M., and Vaes, G. (1992). Mechanism of mineral solubilization and matrix degradation in osteoclastic bone resorption. In: Biology and Physiology of the Osteoclast. (Rifkin, B.R. and Gay, C.V., Eds.), pp. 289-314. CRC Press, Boca Raton.

Dixon, S. J., Hapak, L. K., Wiebe, S. H., Sakai, D. D., Minkin, C., and Sims, S.M. (1994). Function and expression of transforming growth factor-b receptors in mammalian osteoclasts. J. Bone Miner. Res. 9 (Suppl. 1), S242.

Drake, F., Dodds, R.A., James, I., Connor, J., Lee-Rykaczewski, E., Hastings, G., Rosen, C., and Gowen, M. (1995). Large scale sequencing of ESTs from human osteoclast cDNA library: Electronic northern blot. Bone 17 (Suppl. S), 579.

Drexler, H.G. and Gignac, S.M. (1994). Characterization and expression of tartrate-resistant acid phosphatase (TRAP) in hematopoietic cells. Leukemia 8, 359-368.

Durrin, L.K., Minkin, C., and Sakai, D. (1996). The rabbit vacuolar H^+-ATPase D subunit: High expression in osteoclasts and alternative splicing. J. Bone Miner. Res. 11(Suppl.1), S187.

Ek-Rylander, B., Bill, P., Norgard, M., Nilsson, S., and Andersson, G. (1991). Cloning, sequence, and developmental expression of a type-5, tartrate-resistant, acid phosphatase of rat bone. J. Biol. Chem. 266, 24684-24689.

Ek-Rylander, B., Flores, M., Wendel, M., Heinegård, D. , and Andersson, G. (1994). Dephosphorylation of osteopontin and bone sialoprotein by osteoclastic tartrate-resistant acid phosphatase. Modulation of osteoclast adhesion in vitro. J. Biol. Chem. 269, 14853-14856.

Felix, R., Cecchini, M.C., and Fleisch, H. (1989). Macrophage colony-stimulating factor restores in vivo bone resorption in the *op/op* osteopetrotic mouse. Endocrinology 127, 2592-2594.

Felix, R., Hofstetter, W., Wetterwald, A., Cecchini, M.G., and Fleisch, H. (1994). Role of colony stimulating factor-I in bone metabolism. J. Cell. Biochem. 55, 340-349.

Feuerbach, D., Waelchli, R., Fehr, T., and Feyen, J.H.M. (1995). Mycotrienins. A new class of potent inhibitors of osteoclastic bone resorption. J. Biol. Chem. 270, 25949-25955.

Fiorelli, G., Gori, F., Petilli, M., Tanini, A., Benvenuti, S., Serio, M., Bernabei, P., and Brandi, M.L. (1995). Functional estrogen receptors in a human preosteoclastic cell line. Proc. Nat. Acad. Sci. U.S.A. 92, 2672-2676.

Forrest, S.M., Ng, K.W., Findlay, D.M., Michelangeli, V.P., Livesey, S.A., Partidge, N.C., Zajac, J.D., and Martin, T.J. (1985). Characterization of an osteoblastlike clonal cell line, which responds to both parathyroid hormone and calcitonin. Calif. Tissue Int. 37, 51-56.

Fukushima, S., Nagao, Y., Niida, S., Kodama, H., and Kawashima, H. (1994). Bisphosphonates increase serum creatine kinase BB isozyme via inhibition of osteoclast activity. J. Bone Miner. Res. 9 (Suppl. 1), S370.

Fuller, K. and Chambers, T.J. (1989). Bone matrix stimulates osteoclastic differentiation in cultures of rabbit bone marrow cells. Endocrinology 124, 1689-1696.

Fuller, K., Owens, J.M., Jagger, C.J., Wilson, A., Moss, R., and Chambers, T.J. (1993). Macrophage colony-stimulating factor stimulates survival and chemotactic behavior in isolated osteoclasts. J. Exp. Med. 178, 1733-1744.

Gattei, V., Bernabei, P.A., Pinto, A., Bezzini, R., Ringressi, A., Formigli, L., Tanini, A., Attadia, V., and Brandi, M.L. (1992). Phorbol ester-induced osteoclastlike differentiation of a novel human leukemic cell line (FLG 29.1). J. Cell Biol. 116, 437-447.

Gay, C. (1992). Osteoclast ultrastructure and enzyme histochemistry: Functional implications. In: Biology and Physiology of the Osteoclast. (Rifkin, B.R. and Gay, C. V., Eds.), pp. 135-137. CRC Press, Boca Raton.

Grano, M., Zigrino, P., Colucci, S., Zambonin, G., Trusolino, L., Serra, M., Baldini, N., Teti, A., Carlo Marchisio, P., and Zambonin Zallone, A. (1994). Adhesion properties and integrin expression of cultured human osteoclastlike cells. Exp. Cell Res. 212, 209-218.

Greenfield, E.M., Gornik, S.A., Horowitz, M.C., Donahue, H.J., and Shaw, S.M. (1993). Regulation of cytokine expression in osteoblasts by parathyroid hormone: Rapid stimulation of interleukin-6 and leukemia inhibitory factor messenger RNA. J. Bone Miner. Res. 8, 1163-1171.

Grigoriadis, A.E., Wang, Z.-Q., Cecchini, M.G., Hofstetter, W., Felix, R., Fleisch, H.A., and Wagner, E.F. (1994). c-Fos: A key regulator of osteoclast-macrophage lineage determination and bone remodeling. Science 266, 443-448.

Gu, Y., Sakai, D.D., and Minkin, C. (1994). Expression of messenger RNAs for calcitonin receptor and TRAP in fetal limb development. J. Bone Miner. Res. 9 (Suppl. 1), S174.

Hall, T.J. and Schaueblin, M. (1994) A pharmacological assessment of the mammalian osteoclast vacuolar H^+-ATPase. Bone Miner. 27, 159-166.

Hall, T.J., Schaeublin, M., and Missbach, M. (1994). Evidence that c-*src* is involved in the process of osteoclastic bone resorption. Biochem. Biophys. Res. Commun. 199, 1237-1244.

Hall, T.J., Jeker, H., and Schaueblin, M. (1995). Wortmannin, a potent inhibitor of phosphatidylinositol 3-kinase, inhibits osteoclastic bone resorption in vitro. Calcif. Tissue Int. 56, 336-338.

Hattersley, G. and Chambers, T.J. (1991). Effects of transforming growth factor β1 on the regulation of osteoclastic development and function. J. Bone Miner. Res. 6, 165-172.

Heersche, J.N.M. (1992). Systemic factors regulating osteoclast function. In: Biology and Physiology of the Osteoclast. (Rifkin, B.R. and Gay, C.V., Eds.), pp. 151-169. CRC Press, Boca Raton, FL.

Hill, P.A., Buttle, D.J., Jones, S.J., Boyde, A., Murata, M., Reynolds, J.J., and Meikle, M.C. (1994a). Inhibition of bone resorption by selective inactivators of cysteine proteases. J. Cell. Biochem. 56, 118-130.

Hill, P.A., Murphy, G., Docherty, A.J., Hembry, R.M., Millican, T.A., Reynolds, J.J., and Meikle, M.C. (1994b). The effects of selective inhibitors of matrix metalloproteinases (MMPs) on bone resorption and the identification of MMPs and TIMP-1 in isolated osteoclasts. J. Cell Sci. 107, 3055-3064.

Hodgkinson, C.A., Moore, K.J., Nakayama, A., Steingrimsson, E., Copeland, N.G., Jenkins, N.A., and Arnheiter, H. (1993). Mutations at the mouse microphthalmia locus are associated with defects in a gene encoding a novel basic-helix-loop-helix-zipper protein. Cell 74, 395-404.

Hofstetter, W., Wetterwald, A., Cecchini, M.C., Felix, R., Fleisch, H., and Mueller, C. (1992). Detection of transcripts for the receptor for macrophage colony-stimulating factor, c-*fms*, in murine osteoclasts. Proc. Natl. Acad. Sci. U.S.A. 89, 9637-9642.

Horne, W.C., Neff, L., Chatterjee, D., Lomri, A., Levy, J.B., and Baron, R. (1992). Osteoclasts express high levels of $pp60^{c\text{-}src}$ in association with intracellular membranes. J. Cell Biol. 119, 1003-1013.

Horton, M.A., Taylor, M.L., Arnett, T.R., and Helfrich, M.H. (1991). Arginine-gly-asp (RGD) peptides and the anti-vitronectin receptor antibody 23C6 inhibit cell spreading and dentine resorption by osteoclasts. Exp. Cell. Res. 135, 368-375.

Houssami, S., Findlay, D.M., Brady, C.L., Myers, D.E., Martin, T.J. (1994). Isoforms of the rat calcitonin receptor: Consequences for ligand binding and signal transduction. Endocrinology 135, 183-190.

Hoyland, J. and Sharpe, P.T. (1994). Upregulation of c-*fos* protooncogene expression in Pagetic osteoclasts. J. Bone Miner. Res. 9, 1191-1194.

Hoyland, J.A., Freemont, A.J., and Sharpe, P.T. (1994). Interleukin-6, IL-6 receptor, and IL-6 nuclear factor gene expression in Pagets disease. J. Bone Miner. Res. 9, 75-80.

Hughes, D.E., Salter, D.M., Dedhar, S., and Simpson, R. (1993). Integrin expression in human bone. J. Bone Miner. Res. 8, 527-533.

Hughes, D.E., Wright, K.R., Mundy, G.R., and Boyce, B.F. (1994). TGFb1 induces osteoclast apoptosis in vitro. J. Bone Miner. Res. 9 (Suppl. 1), S138.

Hughes, D.E., Jilka, R., Manolagas, S., Dallas, S.L., Bonewald, L.F., Mundy, G.R., and Boyce, B.F. (1995a). Sex steroids promote osteoclast apoptosis in vitro and in vivo. J. Bone Miner. Res. 10 (Suppl. 1), S150.

Hughes, D.E., Wright, K.R., Uy, H.L., Sasaki, A., Yoneda, T., Roodman, G.D., Mundy, G. R., and Boyce, B.F. (1995b). Bisphosphonates promote apoptosis in murine osteoclasts in vitro and in vivo. J. Bone Miner. Res. 10, 1478-1487.

Ikegame, M., Rakopoulos, M., Zhou, H., Houssami, S., Martin, T.J., Moseley, J.M., and Findlay, D.M. (1995). Calcitonin receptor isoforms in mouse and rat osteoclasts. J. Bone Miner. Res. 10, 59-65.

Inaoka, T., Bilbe, G., Ishibashi, O., Tezuka, K., Kumegawa, M., and Kokubo, T. (1995). Molecular cloning of human cDNA for cathepsin K: Novel cysteine proteinase predominantly expressed in bone. Biochem. Biophys. Res. Commun. 206, 89-96.

Johnson, R.S., Spiegelman, B.M., and Papaioannou, V. (1992). Pleiotropic effects of a null mutation in the c-*fos* protooncogene. Cell 71, 577-586.

Kadoya, Y., Al-Saffar, N., Kobayashi, A., and Revell, P.A. (1994). The expression of osteoclast markers on foreign body giant cells. Bone Miner. 27, 85-96.

Kaji, H., Sugimoto, T., Miyauchi, A., Fukase, M., Tezuka, K, Hakeda, Y., Kumegawa, M., and Chihara, K. (1994). Calcitonin inhibits osteopontin mRNA expression in isolated rabbit osteoclasts. Endocrinology 135, 484-487.

Kaji, H., Sugimoto, T., Kanatani, M., Fukase, M., Kumegawa, M., and Chihara, K. (1995). Retinoic acid induces osteoclastlike cell formation by directly acting on hemopoietic blast cells and stimulates osteopontin mRNA expression in isolated osteoclasts. Life Sci. 56, 1903-1913.

Kallio, D.M., Garant, P.R., and Minkin, C. (1972). Ultrastructural effects of calcitonin on osteoclasts in tissue culture. J. Ultrastruct. Res. 39, 205-216.

Kerby, J.A., Hattersley, G., Collins, D.A., and Chambers, T.J. (1992). Derivation of osteoclasts from hematopoietic colony forming cells in culture. J. Bone Miner. Res. 7, 353-362.

Lacey, D.L., Erdmann, J.M., and Tan, H.L. (1994). Interleukin-4 increases type 5 acid phosphatase messenger RNA expression in murine bone marrow macrophages. J. Cell. Biochem. 54, 365-371.

Lacey, D.L., Timms, E., Tan, H.L., Kelley, M.J., Dunstan, C.R., Burgess, T., Elliott, R., Colombero, A., Elliott, G., Scully, S., Hsu, H., Sullivan, J., Hawkins, N., Davy, E., Capparelli, C., Eli, A., Qian, Y.X., Kaufman, S., Sarosi, I., Shalhoub, V., Senaldi, G., Guo, J., Delaney, J., and Boyle, W.J. (1998). Osteoprotegerin ligand is a cytokine that regulates osteoclast differentiation and activation. Cell 93, 165-176.

Laitala, T. and Vnnen, K (1993). Proton channel part of vacuolar H^+-ATPase and carbonic anhydrase II expression is stimulated in resorbing osteoclasts. J. Bone Miner. Res. 8, 119-126.

Laitala, T. and Vnnen, K (1994). Inhibition of bone resorption in vitro by antisense RNA and DNA molecules targeted against carbonic anhydrase II or two subunits of vacuolar H+-ATPase. J. Clin. Invest. 93, 2311-2318.

Lakkakorpi, P.T. and Vnnen, H.K. (1991). Kinetics of the osteoclast cytoskeleton during the resorption cycle in vitro. J. Bone Miner. Res. 6, 817-826.

Lazowski, D.A., Fraher, L.J., Hodsman, A., Steer, B., Modrowski, D., and Han, V.K.M. (1994). Regional variation of insulinlike growth factor-I gene expression in mature rat bone and cartilage. Bone 15, 563-576.

Lee, K., Deeds, J.D., Chiba, S., Unno, M., Bond, A.T., and Segre, G.V. (1994). Parathyroid hormone induces sequential c-*fos* expression in bone cells in vivo: In situ localization of its receptor and c-*fos* messenger ribonucleic acids. Endocrinology 134, 441-450.

Lee, N.H., Weinstock, K.G., Kierkness, E.F., Earle-Hughes, J.A., Fuldner, R.A., Marmaros, S., Glodek, A., Gocayne, J.D., Adams, M.D., Kerlavage, A.R., Fraser, C.M., and Venter, J.C. (1995a). Comparative expressed-sequence-tag analysis of differential gene expression profiles in PC-12 cells before and after nerve growth factor treatment. Proc. Natl. Acad. Sci. U.S.A. 92, 8303-8307.

Lee, S.K., Goldring, S.R., and Lorenzo, J.A. (1995b). Expression of the calcitonin receptor in bone marrow cell cultures and in bone: A specific marker of the differentiated osteoclast that is regulated by calcitonin. Endocrinology 136, 4572-4581.

Lennon, G.G., Auffray, C., Polymeropoulos, M., and Soares, M.B. (1996) The I.M.A.G.E. consortium: An integrated molecular analysis of genomes and their expression. Genomics 33, 151-152.

Li, C.F., Ross, P., Cao, X., and Teitelbaum, S.L. (1995a). Estrogen enhances $\alpha_v\beta_3$ integrin expression by avian osteoclast precursors via stabilization of β_3 integrin messenger RNA. Molec. Endocrinol. 9, 805-813.

Li, Y.-P., Alexander, M., Wucherpfennig, A.L., Yelick, P., Chen, W., and Stashenko, P. (1995b). Cloning and complete coding sequence of a novel human cathepsin expressed in giant cells of osteoclastomas. J. Bone Miner. Res. 10, 1197-1202.

Liang, P. and Pardee, A.B. (1992). Differential display of eukaryotic messenger RNA by means of the polymerase chain reaction. Science 257, 967-971.

Lowe, C., Yoneda, T., Boyce, B.F., Chen, H., Mundy, G.R., and Soriano, P. (1993). Osteopetrosis in Src-deficient mice is due to an autonomous defect of osteoclasts. Proc. Natl. Acad. Sci. U.S.A. 90, 4485-4489.

Manolagas, S.C. (1995). Role of cytokines in bone resorption. Bone 17 (Suppl.), 63S-67S.

Mbalaviele, G., Chen, H., Boyce, B.F., Mundy, G.R., and Yoneda, T. (1995). The role of cadherin in the generation of multinucleated osteoclasts from mononuclear precursors in murine marrow. J. Clin. Invest. 95, 2757-2765.

McSheehy, P.M.J. and Chambers, T.J. (1986). Osteoblastic cells mediate osteoclast responsiveness to parathyroid hormone. Endocrinology 118, 824-828.

Medhora, M.M., Teitelbaum, S., Chappel, J., Alvarez, J., Mimura, H., Ross, F.P., and Hruska, K. (1993). 1a,25-dihydroxyvitamin-D_3 upregulates expression of the osteoclast integrin $\alpha_v\beta_3$. J. Biol. Chem. 268, 1456-1461.

Merry, K., Dodds, R., Littlewood, A., and Gowen, M. (1993). Expression of osteopontin messenger RNA by osteoclasts and osteoblasts in modeling adult human bone. J. Cell Sci. 104, 1013-1020.

Middleton, J., Arnott, N., Walsh, S., and Beresford, J. (1995). Osteoblasts and osteoclasts in adult human osteophyte tissue express the mRNAs for insulinlike growth factors I and II and the type-1 IGF receptor. Bone 16, 287-293.

Middleton, J., Arnott, N., Walsh, S., and Beresford, J. (1996). The expression of mRNA for insulinlike growth factors and their receptor in giant cell tumors of human bone. Clin. Ortho. Rel. Res. 322, 224-231.

Mimura, H., Cao, X., Ross, F.P., Chiba, M., and Teitelbaum, S.L. (1994). 1,25-dihydroxyvitamin D_3 transcriptionally activates the β_3-integrin subunit gene in avian osteoclast precursors. Endocrinology 134, 1061-1066.

Minkin, C. (1982). Bone acid phosphatase: Tartrate-resistant acid phosphatase as a marker of osteoclast function. Calcif. Tissue Int. 34, 285-290.

Miyauchi, A., Kuroki, Y., Fukase, M., Fujita, T., Chihara, K., and Shiozawa, S. (1994). Persistent expression of proto-oncogene c-fos stimulates osteoclast differentiation. Biochem. Biophys. Res. Commun. 205, 1547-1555.

Mizuno, Y., Hosoi, T., Inoue, S., Ikegami, A., Kaneki, M., Akedo, Y., Nakamura, T., Ouchi, Y., Chang, C., and Orimo, H. (1994). Immunocytochemical identification of androgen receptor in mouse osteoclastlike multinucleated cells. Calcif. Tissue Int. 54, 325-326.

Moore, P.R., Mee, A.P., Mawer, E.B., Freemont, A.J., Gokal, R., and Hoyland, J.A. (1995). Demonstration of vitamin D receptor (VDR) messenger RNA expression by osteoclasts using in situ reverse-transcriptase PCR (IS-RT-PCR). J. Am. Soc. Nephrol. 6, 938.

Most, W., Schot, L., Ederveen, A., van der Wee-Pals, L., Papapoulos, S., and Löwik, C. (1995). In vitro and ex vivo evidence that estrogens suppress increased bone resorption induced by ovariectomy or PTH stimulation through an effect on osteoclastogenesis. J. Bone Miner. Res. 10, 1523-1530.

Muguruma, Y., and Lee, M.Y. (1998). Isolation and characterization of murine clonogenic osteoclast progenitors by cell surface phenotype analysis. Blood 91, 1272-1279.

Mundy, G.R. (1992). Local factors regulating osteoclast function. In: Biology and Physiology of the Osteoclast. (Rifkin, B.R. and Gay, C.V., Eds.) pp. 171-185. CRC Press, Boca Raton, FL.

Nakamura, I., Takahashi, N., Sasaki, T., Tanaka, S., Udagawa, N., Murakami, H., Kimura, K., Kabuyama, Y., Kurokawa, T., and Suda, T. (1995). Wortmannin, a specific inhibitor of phosphatidylinositol-3 kinase, blocks osteoclastic bone resorption. FEBS Lett. 361, 79-84.

Nesbitt, S. and Horton, M.A. (1995). Osteoclast annexins bind collagen and may play a role in bone resorption. J. Bone Miner. Res. 10 (Suppl. 1), S221.

Nesbitt, S., Nesbit, A., Helfrich, M., and Horton, M. (1993). Biochemical characterization of human osteoclast integrins. Osteoclasts express $\alpha_v\beta_3$, $\alpha_2\beta_1$, and $\alpha_v\beta_1$ integrins. J. Biol. Chem. 268, 16737-16745.

Nicholson, G.C., Moseley, J.M., Sexton, P.M., Mendelsohn, F.A.O., and Martin, T.J. (1986). Abundant calcitonin receptors in isolated rat osteoclasts. Biochemical and autoradiographic characterization. J. Clin. Invest. 78, 355-360.

Nijweide, P.J. and de Grooth, R. (1992). Ontogeny of the osteoclast. In: Biology and Physiology of the Osteoclast. (Rifkin, B.R. and Gay, C.V., Eds.), pp. 81-104. CRC Press, Boca Raton, FL.

Ohsaki, Y., Takahashi, S., Scarcez, T., Demulder, A., Nishihara, T., Williams, R., and Roodman, G.D. (1992). Evidence for an autocrine/paracrine role for interleukin-6 in bone resorption by giant cells from giant cell tumors of bone. Endocrinology 131, 2229-2234.

Okamura, T., Shimokawa, H., Takagi, Y., Ono, H., and Sasaki, S. (1993). Detection of collagenase mRNA in odontoclasts of bovine root resorbing tissue by in situ hybridization. Calcif. Tissue Int. 52, 325-330.

Okubo, K., Hori, N., Matoba, R., Niiyama, T., Fukushima, A., Kojima, Y., and Matsubara, K. (1992). Large scale cDNA sequencing for analysis of quantitative and qualitative aspects of gene expression. Nature Genet. 2, 173-179.

Oursler, M., Li, L., and Osdoby, P. (1991a). Purification and characterization of an osteoclast membrane glycoprotein with homology to manganese superoxide dismutase. J. Cell. Biochem. 46, 219-233.

Oursler, M., Li, L., Collin-Osdoby, P., Webber, D., Anderson, F., and Osdoby, P. (1991b). Isolation of avian osteoclasts: Improved techniques to preferentially purify viable cells. J. Bone Miner. Res. 6, 375-385.

Oursler, M.J., Osdoby, P., Pyfferoen, J., Riggs, B. L., and Spelsberg, T.C. (1991c). Avian osteoclasts as estrogen target cells. Proc. Natl. Acad. Sci. U.S.A. 88, 6613-6617.

Oursler, M.J., Pederson, L., Pyfferoen, J., Osdoby, P., Fitzpatrick, L., and Spelsberg, T.C. (1993a). Estrogen modulation of avian osteoclast lysosomal gene expression. Endocrinology 132, 1373-1380.

Oursler, M.J., Riggs, B.L., and Spelsberg, T.C. (1993b). Androgen action on avian osteoclasts and human giant cell tumors of the bone. J. Bone Miner. Res. 8 (Suppl. 1), S387.

Oursler, M.J. (1994). Osteoclast synthesis and secretion and activation of latent transforming growth factor β. J. Bone Miner. Res. 9, 443-452.

Oursler, M.J., Pederson, L., Fitzpatrick, L., Riggs, B.L., and Spelsberg, T. (1994) Human giant cell tumors of the bone (osteoclastomas) are estrogen target cells. Proc. Natl. Acad. Sci. U.S.A. 91, 5227-5231.

Pensler, J.M., Radosevich, J.A., Higbee, R., and Langman, C.B. (1990). Osteoclasts isolated from membranous bone in children exhibit nuclear estrogen and progesterone receptors. J. Bone Miner. Res. 5, 797-802.

Pfeilschifter, J.P., Seyedin, S., and Mundy, G.R. (1988). Transforming growth factor β inhibits bone resorption in fetal rat long bones. J. Clin. Invest. 82, 680-685.

Poli, V., Balena, R., Fattori, E., Markatos, A., Yamamoto, M., Tanaka, H., Ciliberto, G., Rodan, G. A., and Costantini, F. (1994). Interleukin-6 deficient mice are protected from bone loss caused by estrogen depletion. EMBO J. 13, 1189-1196.

Prallet, B., Male, P., Neff, L., and Baron, R. (1992). Identification of a functional mononuclear precursor of the osteoclast in chicken medullary bone marrow cultures. J. Bone Miner. Res. 7, 405-414.

Rakopoulos, M., Ikegame, M., Findlay, D.M., Martin, T.J., and Moseley, J.M. (1995). Short treatment of osteoclasts in bone marrow culture with calcitonin causes prolonged suppression of calcitonin receptor messenger RNA. Bone 17, 447-453.

Reddy, S.V., Takahashi, S., Dallas, M., Williams, R.E., Neckers, L., and Roodman, G.G. (1994). Interleukin-6 antisense deoxyoligonucleotides inhibit bone resorption by giant cells from human giant cell tumors of bone. J. Bone Miner. Res. 9, 753-757.

Reddy, S.V., Devlin, R., and Roodman, G.D. (1995). Cloning and characterization of a novel autocrine osteoclast (OCL) stimulating factor (OSF). J. Bone Miner. Res. 10 (Suppl. 1), S325.

Reponen, P., Sahlberg, C., Munaut, C., Thesleff, I., and Tryggvason, K. (1994). High Expression of 92-kDa type-IV collagenase (gelatinase-B) in the osteoclast lineage during mouse development. J. Cell Biol. 124, 1091-1102.

Roberts, E., Wagstaff, S.C., Birch, M.A., Bilbe, G., and Gallagher, J.A. (1995). Isolation of a clone coding for an osteoclastic antigen. Bone 17 (Suppl. S), 585.

Romas, E., Zhou, H., Udagawa, N., Ikegame, M., Hilton, D.J., Martin, T.J., and Ng, K.W. (1995). In-situ localization of interleukin-11 receptor mRNA in both stromal-osteoblastic cells and osteoclasts. Bone 17 (Suppl. S), 585.

Roth, D.E., Venta, P.J., Tashian, R.E., and Sly, W.W. (1992). Molecular basis of human carbonic anhydrase II deficiency. Proc. Natl. Acad. Sci. U.S.A. 89, 1804-1808.

Rüther, U., Garber, C., Komitowski, D., Müller, R., and Wagner, E.F. (1987). Deregulated *c-fos* expression interferes with normal bone development in transgenic mice. Nature 325, 412-416.

Sakai, D., Tong, H., and Minkin, C. (1995). Osteoclast molecular phenotyping by random cDNA sequencing. Bone 17, 111-119.

Saneshige, S., Mano, H., Tezuka, K., Kakudo, S., Mori, Y., Honda, Y., Itabashi, A., Yamada, T., Miyata, K., Hakeda, Y., Ishii, J., and Kumegawa, M. (1995). Retinoic acid directly stimulates osteoclastic bone resorption and gene expression of cathepsin K/OC-2. Biochem. J. 309, 721-724.

Shi, G.-P., Chapman, H.A., Bhairi, S.M., Deleeuw, C., Reddy, V.Y., and Weiss, S.J. (1995). Molecular cloning of human cathepsin O, a novel endoproteinase and homolog of rabbit OC2. FEBS Lett. 357, 129-134.

Shibata, Y., Asotra, S., Anderson, G., and Heersche, J.N.M. (1993). Tartrate resistant acid phosphatase (TRAP) mRNA expression in resorptive and nonresorptive rabbit osteoclasts. J. Bone Miner. Res. 8 (Suppl. 1), S379.

Shin, J.H., Kukita, A., Ohki, K., Katsuki, T., and Kohashi, O. (1995). In vitro differentiation of the murine macrophage cell line BDM-1 into osteoclastlike cells. Endocrinology 136, 4285-4292.

Shinar, D.M., and Rodan, G.A. (1990). Biphasic effects of transforming growth factor β on the production of osteoclastlike cells in mouse bone marrow cultures: The role of prostaglandins in the generation of these cells. Endocrinology 126, 3153-3158.

Shinar, D.M., Sato, M., and Rodan, G.A. (1990). The effect of hemopoietic growth factors on the generation of osteoclastlike cells in mouse bone marrow cultures. Endocrinology 126, 1728-1735.

Shinar, D.M., Schmidt, A., Halperin, D., Rodan, G.A., and Weinreb, M. (1993). Expression of α_v and β_3 integrin subunits in rat osteoclasts in situ. J. Bone Miner. Res. 8, 403-414.

Simonet, W.S., Lacey, D.L., Dunstan, C.R., Kelley, M., Chang, M.S., Luthy, R., Nguyen, H.Q., Wooden, S., Bennett, L., Boone, T., Shimamoto, G., DeRose, M., Elliott, R., Colombero, A., Tan, H.L., Trail, G., Sullivan, J., Davy, E., Bucay, N., Renshaw-Grett, L., Hughes, T.M., Hill, D., Pattison, W., Campbell, P., Boyle, W.J. et al. (1997). Osteoprotegerin: A novel secreted protein involved in regulation of bone density. Cell 89, 309-319.

Smith-Oliver, T., Han, B., and Connolly, K. (1995). Identification of nuclear receptors in purified rabbit osteoclasts. J. Bone Miner. Res. 10 (Suppl. 1), S322.

Somerman, M.J., Berry, J.E., Khalkhali-Ellis, Z., Osdoby, P., and Simpson, R.U. (1995). Enhanced expression of α_v integrin subunit and osteopontin during differentiation of HL-60 cells along the monocytic pathway. Exp. Cell Res. 216, 335-341.

Soriano, P., Montgomery, C., Geske, R., and Bradley, A. (1991). Targeted disruption of the c-src proto-oncogene leads to osteopetrosis in mice. Cell 64, 693-702.

Stashenko, P., Wang, C.Y., Taniishii, N., and Yu, S.M. (1994). Pathogenesis of induced rat periapical lesions. Oral Surg. Oral Med. Oral Pathol. 78, 494-502.

Suda, T., Takahashi, N., and Martin, T. (1992). Modulation of osteoclast differentiation. Endocrine Rev. 13, 66-80.

Sundquist, K., Lakkakorpi, P., Wallmark, B., and Vnnen, K. (1990). Inhibition of osteoclast proton transport by bafilomycin A_1 abolishes bone resorption. Biochem. Biophys. Res. Commun. 168, 309-313.

Sundquist, K.T., Cecchini, M.G., and Marks, S.C. (1995). Colony stimulating factor-1 injections improve but do not cure skeletal sclerosis in osteopetrotic (*op*) mice. Bone 16, 39-46.

Suzuki, K., Uchii, M., and Nozawa, R. (1995). Expression of calcitonin receptors on human myeloid leukemia cells. J. Biochem. 118, 448-452.

Takahashi, N., Udagawa, N., Akatsu, T., Tanaka, H., Isogai, Y., and Suda, T. (1991). Deficiency of osteoclasts in osteopetrotic mice is due to a defect in the local microenvironment provided by osteoblastic cells. Endocrinology 128, 1792-1796.

Takahashi, S., Reddy, S.V., Chirgwin, J.M., Devlin, R., Haipek, C., Anderson, J., and Roodman, G.D. (1994). Cloning and identification of annexin-II as an autocrine/paracrine factor that increases osteoclast formation and bone resorption. J. Biol. Chem. 269, 28696-28701.

Takahashi, S., Goldring, S., Katz, M., Hilsenbeck, S., Williams, R., and Roodman, G.D. (1995). Downregulation of calcitonin receptor mRNA expression by calcitonin during human osteoclastlike cell differentiation. J. Clin. Invest. 95, 167-171.

Takeda, J., Yano, H., Eng, S., Zeng, Y., and Bell, G.I. (1993). A molecular inventory of human pancreatic islets: Sequence analysis of 1000 cDNA clones. Hum. Molec. Genet. 2, 1793-1798.

Tanaka, S., Takahashi, N., Udagawa, N., Sasaki, T., Fukui, Y., Kurokawa, T., and Suda, T. (1992). Osteoclasts express high levels of $p60^{c\text{-}src}$, preferentially on ruffled border membranes. FEBS Lett. 313, 85-89.

Tanaka, S., Takahashi, N., Udagawa, N., Tamura, T., Akatsu, T., Stanley, E.R., Kurokawa, T., and Suda, T. (1993) Macrophage colony-stimulating factor is indispensable for both prolifereation and differentiation of osteoclast progenitors. J. Clin. Invest. 91, 257-263.

Tanaka, S., Takahashi, N., Udagawa, N., Murakami, H., Nakamura, I., Kurokawa, T., and Suda, T. (1995). Possible involvement of focal adhesion kinase, $p125^{FAK}$, in osteoclastic bone resorption. J. Cell. Biochem. 58, 424-435.

Tao, H. and Minkin, C. (1994). The effects of 1,25-dihydroxyvitamin D_3 on osteoclast formation in fetal mouse metatarsal organ cultures. Bone 15, 217-223.

Teti, A., Rizzoli, R., and Zambonin-Zallone, A. (1991). Parathyroid hormone binding to cultured avian osteoclasts. Biochem. Biophys. Res. Commun. 174, 1217-1222.

Tezuka, K., Sato, T., Kamioka, H., Nijweide, P. J., Tanaka, K., Matsuo, T., Ohta, M., Kurihara, N., Hakeda, Y., and Kumegawa, M. (1992). Identification of osteopontin in isolated rabbit osteoclasts. Biochem. Biophys. Res. Commun. 186, 911-917.

Tezuka, K., Tezuka, Y., Maejima, A., Sato, T., Nemoto, K., Kamioka, H., Hakeda, Y., and Kumegawa, M. (1994a). Molecular cloning of a possible cysteine proteinase predominantly expressed in osteoclasts. J. Biol. Chem. 269, 1106-1109.

Tezuka, K., Nemoto, K., Tezuka, Y., Sato, T., Ikeda, Y., Kobori, M., Kawashima, H., Eguchi, H., Hakeda, Y., and Kumegawa, M. (1994b). Identification of matrix metalloproteinase 9 in rabbit osteoclasts. J. Biol. Chem. 269, 15006-15009.

Tobias, J.H. and Chambers, T.J. (1989). Glucocorticoids impair bone resorptive activity and viability of osteoclasts disaggregated from neonatal rat long bones. Endocrinology 125, 1290-1296.

Tobias, J.H. and Chambers, T.J. (1991). The effect of sex hormones on bone resorption by rat osteoclasts. Acta. Endocrinol. 124, 121-127.

Tondravi, M.M., Erdmann, J.M., McKercher, S., Anderson, K., Maki, R., and Teitelbaum, S. L. (1995). Novel osteopetrosis mutation caused by the knock out of the hematopoietic transcription factor PU.1 in transgenic mice. J. Bone Miner. Res. 10 (Suppl. 1), S175.

Tong, H.-S., Sakai, D.D., Sims, S.M., Dixon, S.J., Yamin, M., Goldring, S.R., Snead, M.L., and Minkin, C. (1994). Murine osteoclasts and spleen cell polykaryons are distinguished by messenger RNA phenotyping. J. Bone Miner. Res. 9, 577-584.

Tong, H., Lin, H., Wang, H., Sakai, D., and Minkin, C. (1995) Mouse osteoclasts respond to parathyroid hormone and express mRNA for its receptor. Bone 16 (Suppl. S), 172.

Tong, H.-S. (1996). Osteoclast molecular phenotyping. Ph.D. thesis, University of Southern California School of Dentistry.

Townsend, P.A. and Horton, M.A. (1995). The use of antisense oligonucleotides in studying the role of integrins in rabbit osteoclasts. Bone 17 (Suppl. S), 586.

Udagawa, N., Takahashi, N., Akatsu, T., Tanaka, H., Sasaki, T., Nishihara, T., Koga, T., Martin, T. J., and Suda, T. (1990). Origin of osteoclasts: Mature monocytes and macrophages are capable of differentiating into osteoclasts under a suitable microenvironment prepared by bone marrow-derived stromal cells. Proc. Natl. Acad. Sci. U.S.A. 87, 7260-7264.

Udagawa, N., Sasaki, T., Akatsu, T., Tanaka, S., Tamura, T., Tanaka, H., and Suda, T. (1992). Lack of bone resorption in osteosclerotic (oc/oc) mice is due to a defect in osteoclast progenitors rather than the local microenvironment provided by osteoblastic cells. Biochem. Biophys. Res. Commun. 184, 67-72.

Udagawa, N., Takahashi, N., Katagiri, T., Tamura, T., Wada, S., Findlay, D. M., Martin, T. J., Hirota, H., Taga, T., Kishimoto, T., and Suda, T. (1995a). Interleukin (IL)-6 induction of osteoclast differentiation depends on IL-6 receptors expressed on osteoblastic cells but not on osteoclast progenitors. J. Exp. Med. 182, 1461-1468.

Udagawa, N., Wada, S., Chan, J., Hamilton, J., Findlay, D. M., and Martin, T.J. (1995b). c-*fos* antisense DNA inhibits proliferation of osteoclast progenitors in osteoclast development but not macrophage differentiation in vitro. J. Bone Miner. Res. 10 (Suppl. 1), S177.

van Hille, B., Richener, H., Evans, D.B., Green, J.R., and Bilbe, G. (1993). Identification of two subunit A isoforms of the vacuolar H^+-ATPase in human osteoclastoma. J. Biol. Chem. 268, 7075-7080.

van Hille, B., Richener, H., Schmid, P., Puettner, I., Green, J.R., and Bilbe, G. (1994). Heterogeneity of vacuolar H^+-ATPase: Differential expression of 2 human subunit B isoforms. Biochem. J. 303, 191-198.

van Hille, B., Richener, H., Green, J.R., and Bilbe, G. (1995). The ubiquitous VA68 isoform of subunit-A of the vacuolar H^+-ATPase is highly expressed in human osteoclasts. Biochem. Biophys. Res. Commun. 214, 1108-1113.

Vignery, A., Raymond, M.J., Qian, H.Y., Wang, F., and Rosenzweig, S.A. (1991). Multinucleated rat alveolar macrophages express functional receptors for calcitonin. Am. J. Physiol. 261, F1026-F1032.

Wada, S., Akatsu, T., Tamura, T., Takahashi, N., Suda, T., and Nagata, N. (1994). Glucocorticoid regulation of calcitonin receptor in mouse osteoclastlike multinucleated cells. J. Bone Miner. Res. 9, 1705-1712.

Wada, S., Martin, T.J., and Findlay, D.M. (1995). Homologous regulation of the calcitonin receptor in mouse osteoclastlike cells and human breast cancer T47D cells. Endocrinology 136, 2611-2621.

Wang, Z.-Q., Ovitt, C., Grigoriadis, A.E., Möhle-Steinlein, U., Rüther, U., and Wagner, E.F. (1992). Bone and haematopoietic defects in mice lacking c-fos. Nature 360, 741-745.

Weir, E.C., Horowitz, M.C., Baron, R., Centrella, M., Kacinski, B.M., and Insogna, K.L. (1993). Macrophage colony-stimulating factor release and receptor expression in bone cells. J. Bone Miner. Res. 8, 1507-1518.

Wesolowski, G., Duong, L.T., Lakkakorpi, P.T., Nagy, R.M., Tezuka, K., Tanaka, H., Rodan, G.A., and Rodan, S.B. (1995). Isolation and characterization of highly enriched prefusion mouse osteoclastic cells. Exp. Cell Res. 219, 679-686.

Wucherpfennig, A.L., Li, Y.P., Stetler-Stevenson, W.G., Rosenberg, A.E., and Stashenko, P. (1994). Expression of 92 kDa type-IV collagenase/gelatinase B in human osteoclasts. J. Bone Miner. Res. 9, 549-556.

Wucherpfennig, A.L., Li, Y.-P., and Stashenko, P. (1996). Expression of cystatin C by human osteoclasts. J. Dent. Res. 75 (Suppl. A), 308.

Yang, J.-N., Allan, E.A., Anderson, G.I., Martin, T.J., and Minkin, C. (1995a). The plasminogen activator/plasmin system in osteoclasts. J. Bone Miner. Res. 10 (Suppl. 1), S227.

Yang, J.-N., Sakai, D., and Minkin, C. (1995b). Expression of the microphthalmia gene in bone cells. J. Bone Miner. Res. 10 (Suppl. 1), S227.

Yasuda, H., Shima, N., Nakagawa, N., Mochizuki, S.I., Yano, K., Fujise, N., Sato, Y., Goto, M., Yamaguchi, K., Kuriyama, M., Kanno, T., Murakami, A., Tsuda, E., Morinaga, T., and Higashio, K. (1998a). Identity of osteoclastogenesis inhibitory factor (OCIF) and osteoprotegerin (OPG): A mechanism by which OPG/OCIF inhibits osteoclastogenesis in vitro. Endocrinology 139, 1329-1337.

Yasuda, H., Shima, N., Nakagawa, N., Yamaguchi, K., Kinosaki, M., Mochizuki, S., Tomoyasu, A., Yano, K., Goto, M., Murakami, A., Tsuda, E., Morinaga, T., Higashio, K., Udagawa, N., Takahashi, N., and Suda, T. (1998b). Osteoclast differentiation factor is a ligand for osteoprotegerin/osteoclastogenesis–inhibitory factor and is identical to TRANCE/RANKL.Proc. Natl. Acad. Sci. U.S.A. 95, 3597-3602.

Yoneda, T., Lowe, C., Lee, C.H., Gutierrez, G., Niewolna, M., Williams, P.J., Izbicka, E., Uehara, Y., and Mundy, G.R. (1993a). Herbimycin-A, a $pp60^{c\text{-}src}$ tyrosine kinase inhibitor, inhibits osteoclastic bone resorption in vitro and hypercalcemia in vivo. J. Clin. Invest. 91, 2791-2795.

Yoneda, T., Niewolna, M., Lowe, C., Izbicka, E., and Mundy, G.R. (1993b). Hormonal regulation of $pp60^{c\text{-}src}$ expression during osteoclast formation in vitro. Molec. Endocrinol. 7, 1313-1318.

Yoneyama, T., Fowler, H.L., Pendleton, J.W., Sforza, P.P., Gerard, R.D., Lui, C.Y., Eldridge, T.H., and Iranmanesh, A. (1992). Elevated serum levels of creatine kinase BB in autosomal dominant osteopetrosis type II - a family study. Clin. Genet. 42, 39-42.

Yoshida, H., Hayashi, S., Kunisada, T., Ogawa, M., Nishikawa, S., Okamura, H., Suda, T., Shultz, L.D., and Nishikawa, S. (1990). The murine mutation osteoperosis is in the coding region of the macrophage colony stimulating factor gene. Nature 345, 442-444.

Zaidi, M., Shankar, V.S., Tunwell, R., Adebanjo, O.A., Mackrill, J., Pazianas, M., O'Connell, D., Simon, B.J., Rifkin, B.R., Venkitaraman, A.R., Huang, C.L.-H., and Lai, F.A. (1995). A ryanodine receptorlike molecule expressed in the osteoclast plasma membrane functions in extracellular Ca^{2+}-sensing. J. Clin. Invest. 96, 1582-1590.

Zhang, D., Udagawa, N., Nakamura, I., Murakami, H., Saito, S., Yamasaki, K., Shibasaki, Y., Morii, N., Narumiya, S., Takahashi, N., and Suda, T. (1995). The small GTP-binding protein, rho-p21, is involved in bone resorption by regulating cytoskeletal organization in osteoclasts. J. Cell Sci. 108, 2285-2292.

Zheng, M.H., Fan, Y., Wysocki, S., Wood, D.J., and Papadimitriou, J.M. (1993). Detection of messenger RNA for carbonic anhydrase II in human osteoclastlike cells by in situ hybridization. J. Bone Miner. Res. 8, 113-118.

Zheng, M.H., Fan, Y., Wysocki, S., Wood, D.J., and Papadimitriou, J.M. (1994a). Carbonic anhydrase II gene transcript in cultured osteoclasts from neonatal rats: Effect of calcitonin. Cell Tissue Res. 276, 7-13.

Zheng, M.H., Fan, Y., Wysocki, S.J., Lau, A.T.-T., Robertson, T., Beilharz, M., Wood, D.J., and Papadimitriou, J.M. (1994b). Gene expression of transforming growth factor-β1 and its type-II receptor in giant cell tumors of bone. Possible involvement in osteoclastlike cell migration. Am. J. Pathol. 145, 1095-1104.

Zheng, M.H., Holloway, W., Fan, T., Collier, F., Criddle, A., Prince, R., Wood, D.J., and Nicholson, G.C. (1995). Evidence that human osteoclastlike cells are not the major estrogen target cells. Bone 16 (Suppl. S), 93.

ION CHANNELS IN OSTEOCLASTS

A. Frederik Weidema, S. Jeffrey Dixon,
and Stephen M. Sims

Advances in Organ Biology
Volume 5B, pages 423-442.

ISBN: 0-7623-0390-5

I. INTRODUCTION

Ion channels play key roles in many cellular functions, including the generation of action potentials in nerve, initiation of muscle contraction and stimulation of secretion. Just as knowledge of ion channels is fundamental for understanding the physiology and pathology of nerve and muscle, such information is instrumental for understanding osteoclasts and resorption of bone. Isolation of osteoclasts in quantities sufficient for biochemical studies has proven difficult. Therefore, single cell electrophysiology is a valuable approach for investigating the function and regulation of osteoclasts. The earliest studies by Mears used microelectrodes to record membrane potential of osteoclasts (Mears, 1971). As for other cell types, the introduction of patch clamp recording techniques has revolutionized the characterization of ion channels in osteoclasts, as reviewed previously (Sims et al., 1992; Ypey et al., 1992; Dixon et al., 1993).

When actively resorbing bone and other mineralized tissues, osteoclasts transport H^+ across the ruffled border membrane, acidifying the resorption lacuna. In this acidic compartment, the mineral dissolves and the organic components of the bone matrix are degraded. Acidification is mediated by a vacuolar-type of ATP-driven H^+ pump located in the ruffled border membrane. This class of H^+ pump is electrogenic (Forgac, 1989) and therefore contributes to the total transmembrane current. Thus, transport of H^+ by this pump should modify, and in turn be regulated by, the membrane potential of the osteoclast. Several types of ion channels and electrogenic transporters determine the membrane potential. The purpose of this chapter is to review the types of ion channels in osteoclasts, their biophysical and pharmacological properties, and their possible roles in the function and regulation of osteoclasts. A summary of ion channels that have been identified in osteoclasts using electrophysiological methods is presented in Table 1.

II. POTASSIUM CHANNELS

As in many other cell types, the predominant conductances in the membrane of osteoclasts are due to K^+ channels and, accordingly, they were the first to be characterized (Ravesloot et al., 1989; Sims and Dixon, 1989). There is now evidence for at least three classes of K^+ channels in osteoclasts.

A. Inward Rectifier K^+ Channel

Description

A common feature of both avian and mammalian osteoclasts is the presence of inwardly rectifying K^+ currents (Ravesloot et al., 1989; Sims and Dixon, 1989; Sims et al., 1991; Kelly et al., 1992). This current is readily identified under voltage clamp based upon its characteristic voltage-activation properties (Figure 1) and

Table 1. Ion Channels in Osteoclasts

Channel	*Species*	*Channel conductance (pS)*	*Blockers*	*References*
Potassium				
Inwardly rectifying	chicken	30	Ba^{2+}, Cs^{+}	Ravesloot et al., 1989 Weidema, 1995
	mouse		Ba^{2+}, Cs^{+}	Arkett et al., 1994c
	rabbit	31	Ba^{2+}, Cs^{+}	Kelly et al., 1992 Hammerland et al., 1994 Yamashita et al., 1994
	rat	25	Ba^{2+}, Cs^{+}	Sims and Dixon, 1989 Sims et al., 1991
Transient, outwardly rectifying	chicken		4-AP, TEA, verapamil	Ravesloot et al., 1989; Weidema, 1995
	mouse		4-AP, CTX	Arkett et al., 1994c
	rabbit		4-AP, CTX	Hammerland et al., 1994
	rat	13	4-AP, CTX, Ni^{2+}, MgTX, quinine	Arkett et al., 1992 Arkett et al., 1994c
Ca^{2+}-activated	chicken	150	Ba^{2+}, TEA	Weidema et al., 1993
	rat			Weidema et al., 1996
Proton	rabbit		Zn^{2+}	Nordstrom et al., 1995
Sodium	chicken		TTX	Gaspar et al., 1995
Nonselective cation				
Stretch-activated	chicken	60		Ypey et al., 1992 Wiltink et al., 1995
ATP-activated	mouse			Modderman et al., 1994
	rat			Weidema et al., 1996
Chloride	rat		DIDS, DNDS, SITS	Sims et al., 1991
	rabbit	20	DIDS, DNDS, SITS, niflumic acid	Kelly et al., 1994
	chicken	350 to 400	SITS	Weidema, 1995

Note: Summary of types of ion channels identified in osteoclasts. Blanks are present where no data are available. Abbreviations: 4-AP, 4-aminopyridine; CTX, charybdotoxin; MgTX, margatoxin; TEA, tetraethylammonium; DIDS, 4,4′-diisothiocyanatostilbene-2,2′disulfonic acid; DNDS, 4,4-dinitrostilbene-2,2′-disulfonic acid; SITS, 4-acetamido-4′-isothiocyanatostilbene-2,2′-disulfonic acid; TTX, tetrodotoxin.

blockade by Ba^{2+} and Cs^{+}. Hyperpolarizing commands to voltages negative to the K^{+} equilibrium potential (E_K) elicit large inward currents, indicating a large conductance at negative potentials (Figure 1). The current decreases rapidly at voltages positive to E_K, so that only small outward currents are evident. This nonlinear current voltage relationship is characteristic of an inwardly rectifying conductance. The basis of the inward rectification in osteoclasts has not been studied, but rectification of these channels in other cell types is due to blockade by polyamines or Mg^{2+} (Johnson, 1996).

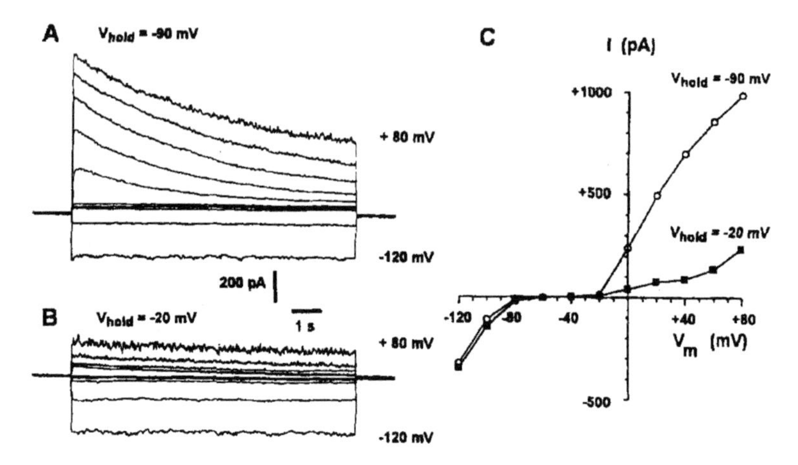

Figure 1. Three types of voltage-dependent K^+ conductances in chicken osteoclasts on glass substrate. (**A, B**) Current traces showing activation and inactivation of currents by voltage commands from -120 to +80 mV in 20 mV increments from holding potentials indicated. Inward rectifying K^+ current (downward deflections) is activated at potentials negative to -80 mV and is not dependent on holding potential. At potentials more positive than 0 mV an outwardly rectifying K^+ current is activated, consisting of both transient and sustained components. Because of voltage-dependent inactivation, the transient outward K^+ current is reduced with a holding potential of -20 mV, revealing the sustained outward Ca^{2+}-activated K^+ current (**B**). The Ca^{2+}-activated K^+ current is most evident at potentials more positive than +60 mV as increased current noise. (**C**) Current-voltage relations of the peak outward currents in **A** (○) and **B** (■).

Single-channel currents underlying the inward rectifier K^+ conductance have been recorded using the cell-attached patch configuration. With high K^+-containing solutions in the pipette, the single-channel conductance ranges from 25 pS in rat osteoclasts (Sims et al., 1991) to 31 pS in rabbit osteoclasts (Kelly et al., 1992). This is in the range reported for inward rectifier K^+ channels in many other tissues, and is close to that reported for the inward rectifier K^+ channel (IRK1), which has been cloned from the J774 macrophage cell line (Kubo et al., 1993). Messenger RNA encoding IRK1 has been identified in isolated murine osteoclasts using reverse transcription polymerase chain reaction (Arkett et al., 1994c).

Regulation

During bone resorption, the concentration of Ca^{2+} around osteoclasts is likely increased, a change which may affect channel properties. Indeed, high concentrations of extracellular divalent cations reduce the K^+ conductance in rat and rabbit osteoclasts (Arkett et al., 1994a; Hammerland et al., 1994; Yamashita et al., 1994). However, the basis of this effect is not resolved. Elevation of extracellular Ca^{2+} appears to block the channel, reducing the conductance at all voltages (Arkett et al., 1994a; Hammerland et al., 1994). It has also been suggested that binding of Ca^{2+} to a cell surface receptor activates G proteins, which in turn inhibit the inward rectifier K^+ channel (Yamashita et al., 1994). In this regard, the inward rectifier in osteoclasts has been shown to be regulated by G proteins, since inclusion of GTPγS or fluoroaluminate in the recording pipette inhibits channel activity within 12 minutes after breaking into the cell (Arkett et al., 1994b; Yamashita et al., 1994). However, this observation is somewhat puzzling, because IRK1 K^+ channels are not considered to be regulated by G proteins in other systems. Further studies are required to define the subclass of inward rectifier K^+ channels in osteoclasts, as well as details of the signaling pathways involved in their regulation.

Channel Function

The inward rectifier plays an important role in determining the membrane potential of mammalian osteoclasts. Ba^{2+}, which blocks the inward rectifier, depolarizes rat osteoclasts from -70 mV to around 0 mV (Sims and Dixon, 1989). Another role for the inward rectifier K^+ channel may involve dissipation of charge arising from activity of the electrogenic H^+-ATPase in osteoclasts. Transport of H^+ would be expected to cause hyperpolarization of the membrane potential, which would ultimately prevent further H^+ efflux. The inwardly rectifying K^+ conductance allows inward movement of K^+ at hyperpolarized potentials, making it ideally suited to counteract the change in membrane potential arising from activity of the H^+ pump, preventing excessive hyperpolarization, while minimizing efflux of K^+ at depolarized potentials. However, the inward rectifier may not be essential for H^+ pumping, since preliminary data indicate that Cs^+ and Ba^{2+} do not prevent resorption by chicken osteoclasts in an *in vitro* pit formation assay (Weidema, 1995).

It has recently been shown that inward rectifier K^+ channels participate in differentiation of hematopoietic progenitor cells. Inhibition of expression of the inward rectifier IRK1 with antisense oligonucleotides or channel blockade with Ba^{2+} or Cs^+ prevents cytokine-induced expansion of progenitor cells (Shirihai et al., 1996). Similarly, preliminary findings indicate that blockade of the inward rectifier with Cs^+ prevents differentiation of osteoclast-like cells in a mouse coculture system (Ypey et al., 1995). It is possible that setting the membrane potential at negative values is crucial for membrane signaling events that regulate the proliferation and differentiation of osteoclast precursors.

B. Transient Outward Rectifier K^+ Channel

Description

In addition to the inward rectifier described above, chicken and mammalian osteoclasts exhibit a time- and voltage-dependent outward rectifier K^+ current (Ravesloot et al., 1989; Arkett et al., 1992; Hammerland et al., 1994). The outward rectifier K^+ current activates with depolarizations beyond -40 mV (Figure 1A), giving rise to the characteristic outward rectification of the current-voltage relationship (Figure 1C). This transient current peaks then declines, indicating time-dependent inactivation (Figure 1A). The current also exhibits voltage-dependent inactivation, and is almost completely inactivated at a holding potential of -20 mV (Figure 1B). Transient outward K^+ currents are found in a wide variety of cells including lymphocytes, macrophages, and osteoblasts (Ypey et al., 1988; Dixon et al., 1993; Lewis and Cahalan, 1995). The K^+ channel underlying the whole-cell current in rat osteoclasts has a single channel conductance of 13 pS and exhibits the expected time- and voltage-dependent activation and inactivation (Arkett et al., 1994c).

There are marked differences between the pharmacological sensitivity of transient K^+ channels in chicken and mammalian osteoclasts. Whereas 4-aminopyridine blocks transient K^+ channels in osteoclasts from both chicken and mammals, tetraethylammonium blocks only the chicken K^+ current (Ravesloot et al., 1989; Arkett et al., 1992; Hammerland et al., 1994; Weidema, 1995). In contrast, the high affinity scorpion toxin charybdotoxin inhibits transient K^+ current in mammalian, but not chicken, osteoclasts. Similarly, margatoxin, a highly selective blocker of the Kv1.3 subclass of K^+ channels, blocks the transient K^+ current of rat and mouse, but not chicken, osteoclasts (Arkett et al., 1994c; Weidema, 1995). Messenger RNA encoding Kv1.3 has been identified in murine osteoclasts (Arkett et al., 1994c), but further studies are clearly warranted to determine the molecular basis for the pharmacological differences between species.

Regulation

Expression of the transient K^+ current is regulated by interaction of isolated osteoclasts with the substrate. Most rat osteoclasts plated on glass or plastic are "spread" in appearance and exhibit inward rectifier K^+ channels. When plated on collagen or dentin, a larger proportion of osteoclasts are "rounded" in appearance and exhibit transient outward K^+ channels (Arkett et al., 1992). This correlation between cell morphology and the dominant type of K^+ current has also been reported in rabbit osteoclasts (Hammerland et al., 1994), but not chicken osteoclasts (Weidema, 1995). It has been suggested that in mammals, the rounded osteoclasts are actively resorbing cells, whereas spread osteoclasts are motile cells (Arkett et al., 1992). The mechanisms underlying the selective expression of these channels remain to be identified. A number of possibilities exist, including insertion of Kv1.3 channels during formation of the ruffled border, or posttranslational modifications leading to activation or suppression of channel activity. In this regard, the primary sequence of Kv1.3 contains consensus sites for phosphorylation by protein kinases A and C, which can lead to inhibition or activation of this channel in lymphocytes (Lewis and Cahalan, 1995).

Alterations in extracellular concentrations of divalent cations or H^+ have marked effects on the transient outward K^+ current (Figure 2; Arkett et al., 1994a;

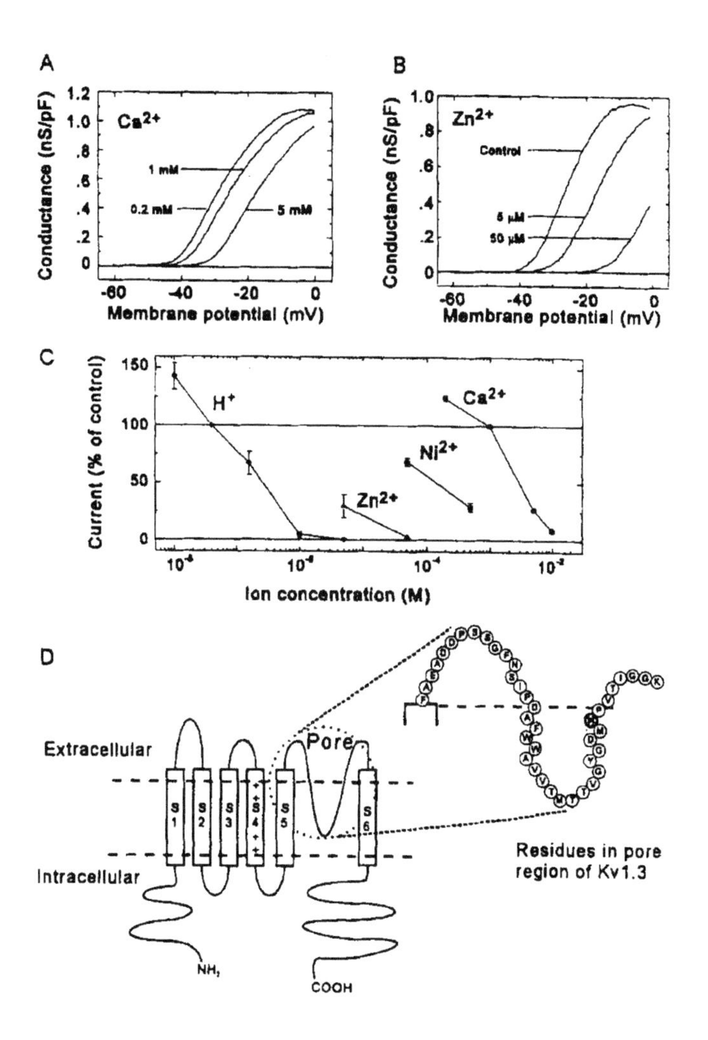

Figure 2. Effects of protons and divalent cations on transient outward rectifier K^+ channels of rat osteoclasts. (**A**) Increasing extracellular $[Ca^{2+}]$ from 0.2 to 5 mM shifted the K^+ conductance-voltage curve to more positive potentials. Effects were rapid and reversible. (**B**) Similar shifts in channel activation were obtained with Zn^{2+} at lower concentrations. The parallel shift in the voltage dependence of the channel indicates that divalent ions screen surface charge in or near the channel. (**C**) Concentration-dependence of the effects of cations on K^+ current. Reduction of current was measured at a fixed voltage and reflects shifts in the voltage dependence of channel activation. (**D**) Cartoon showing proposed structure of Kv1.3 K^+ channel, consisting of six membrane spanning domains (S1–S6). The pore region of Kv1.3 between S5 and S6 is shown expanded at right. Several residues, such as the shaded histidine (single letter amino acid code is used) may be sites for interaction with divalent cations and H^+. Data in **A–C** modified with permission from Arkett et al., 1994a.

Hammerland et al., 1994). The voltage activation range of the K^+ conductance is shifted to more positive potentials upon elevation of H^+ or divalent cations (Ca^{2+}, Zn^{2+}, Ni^{2+}). This effect is consistent with these cations altering the voltage-dependence of channel gating, similar to surface charge effects on most other types of voltage-dependent ion channels (Hille, 1992). In addition, some cations, such as Ni^{2+}, appear to block K^+ channels. The concentration-dependent shift in the current-voltage relation caused by cations is consistent with a binding site, perhaps a histidine residue, in or near the pore region of the channel, which affects voltage-dependent gating (Figure 2D).

Channel Function

The outward rectifier K^+ current could contribute to the resting membrane potential of osteoclasts expressing this conductance. The overlap of activation and inactivation results in a K^+ "window current" that could set the membrane potential between -45 and -25 mV. However, it is unlikely that the outward rectifier maintains the resting potential at more hyperpolarized levels, because these channels are not active at potentials close to -60 mV. In hyperpolarized cells, the membrane potential could be set by an additional outward current arising from the activity of electrogenic ion pumps, such as the Na^+/K^+-ATPase and H^+-ATPase.

The outward rectifier Kv1.3 may be involved in other cell functions. For example, exposure of rabbit osteoclasts to hypoosmotic medium activates an outwardly rectifying Cl^- current, which could play a role in regulatory volume decrease (Kelly et al., 1994; see section V below). If Cl^- current contributes to salt loss, then it must be accompanied by cation efflux, which may involve Kv1.3, as reported for T lymphocytes (Deutsch and Chen, 1993). In lymphocytes, the K^+ channel blocker charybdotoxin inhibits mitogen-induced proliferation and interleukin 2 production (Price et al., 1989). Whether the outward rectifier also plays a role in osteoclast formation or activation is yet to be resolved.

C. Ca^{2+}-Activated K^+ Channels

In addition to the transient outward rectifier K^+ current, chicken osteoclasts express a sustained outward rectifier K^+ current (Ravesloot et al., 1989; Weidema, 1995), which has not been identified in mammalian osteoclasts. This sustained K^+ current activates at potentials more positive than +60 mV and does not exhibit time- or voltage-dependent inactivation (Figure 1). The sustained K^+ current is blocked by extracellular tetraethylammonium and Ba^{2+} (Weidema, 1995). The channels underlying this outward K^+ current have been identified in cell-attached and excised patches and shown to be Ca^{2+}-activated with a single channel conductance of 150 pS (Weidema et al., 1993). The large single channel conductance and the opening and closure of these channels results in noisy current traces at potentials more positive than +60 mV (Figure 1). Surprisingly, charybdotoxin, a potent blocker of many large conductance Ca^{2+}-activated K^+ channels

(Miller et al., 1985), does not inhibit this sustained outward K^+ current in chicken osteoclasts (Weidema, 1995). There is also a report of a K^+ current in rat osteoclasts, which is activated by increases in cytosolic free Ca^{2+} concentration ($[Ca^{2+}]_i$). In contrast to the sustained current in chicken osteoclasts, the mammalian Ca^{2+}-activated K^+ current is not voltage-dependent, so that once opened by elevation of $[Ca^{2+}]_i$, it remains open at all voltages (Weidema et al., 1996; see note added in proof).

Factors that elevate $[Ca^{2+}]_i$ activate the large conductance K^+ channel in chicken osteoclasts. For example, mechanical stimulation of isolated osteoclasts increases $[Ca^{2+}]_i$ with concomitant activation of the K^+ channel. This has been proposed to cause hyperpolarization of the osteoclast, which would increase Ca^{2+} influx, giving rise to positive feedback (Wiltink et al., 1995). The roles of the Ca^{2+}-activated K^+ conductances in osteoclasts deserve further attention.

III. OTHER CATION CHANNELS

A. Proton Channel

Protons (H^+) are generated inside the osteoclast as the result of several metabolic processes. H^+ are exported into the resorption lacuna by H^+-ATPases, while cytosolic pH levels are regulated by Na^+/H^+ and HCO_3^-/Cl^- exchangers (Teti et al., 1989; Ravesloot et al., 1995). Recently, proton channels have been identified in a number of cell types (DeCoursey and Cherny, 1994), including rabbit osteoclasts (Nordstrom et al., 1995). This conductance becomes apparent when all other permeant ions are removed from the electrode and bath solutions, effectively eliminating other ionic currents. Voltage clamp commands to positive potentials reveal an outwardly rectifying H^+ current that activates much more slowly than the K^+ channels described above, with activation time constants of 0.2–1.0 seconds. In the presence of physiological pH gradients, this H^+ conductance activates at potentials more positive than 20 mV, but acidification of the cytosol shifts the activation range to more negative potentials. Although specific blockers of this channel have not been identified, Zn^{2+}, which is a potent inhibitor of bone resorption (Moonga and Dempster, 1995), does reduce the current (Nordstrom et al., 1995). It is interesting that the effect of Zn^{2+} on the H^+ current is similar to its effect on the transient K^+ current described above. In both cases, Zn^{2+} appears to shift the voltage-activation range to more positive potentials, consistent with changes in surface charge. In addition to the thorough characterization of H^+ current in rabbit osteoclasts (Nordstrom et al., 1995; see note added in proof), there is a preliminary report of a similar current in chicken osteoclasts (Weidema, 1995).

Although the role of this H^+ current in osteoclast function is presently uncertain, several possibilities exist. For example, oxygen free radicals, which are thought to be necessary for bone resorption, are produced by an NADPH oxidase (Darden et al., 1996). It has been suggested that, as in macrophages and neutrophils, H^+ channels in osssteoclasts are required to extrude H^+ released during an oxidative burst.

Activation of NADPH oxidase would be expected to produce strong depolarization and local acidification, conditions which would support H^+ efflux through the channels (Nordstrom et al., 1995). However, in a resting cell, it is unlikely that this conductance would play a role in constitutive H^+ extrusion, since, under these conditions, the electrochemical driving force on H^+ is inwardly directed. Similarly, in the actively resorbing osteoclast, it is unlikely that H^+ channels could mediate acidification of the resorption lacuna. In contrast to active transport of H^+ by the vacuolar H^+-ATPase, which establishes a large concentration gradient across the ruffled border, movement of H^+ through channels is passive.

B. Sodium Channel

Voltage-dependent Na^+ channels are found in many excitable cells, including neuronal, muscle, and secretory cells (Hille, 1992). A rapidly activating Na^+ conductance has been identified in chicken osteoclasts (Gaspar et al., 1995), but has not been seen in mammalian osteoclasts. This inward current activates promptly at membrane potentials more positive than -30 mV (peak amplitude within 1–2 ms) then inactivates rapidly (within 5–7 ms) (Figure 3), kinetics which are more rapid than those of other voltage-gated channels in osteoclasts. This Na^+ current is inhibited by nanomolar concentrations of tetrodotoxin (TTX), a blocker of certain voltage-gated Na^+ channels (Hille, 1992).

In many excitable cells, Na^+ channels are responsible for the depolarizing phase of action potentials. In chicken osteoclasts, voltage-activated Na^+ conductance at the whole-cell level is relatively small compared to other conductances. This low density of channels, and their transient activation, does not allow generation of action potentials. In other cells, Na^+ conductances play a role in secretion (Stein, 1990) and possibly in regulation of cell proliferation (Wen et al., 1994). It is conceivable that a larger Na^+ conductance plays a role in the proliferation of pre-osteoclasts and that only a remnant remains in differentiated chicken osteoclasts.

C. Calcium Channels

Calcium plays an important role in regulating the activity of virtually all cells. Studies using fluorescent indicator dyes reveal that $[Ca^{2+}]_i$ in osteoclasts is influenced by both release from intracellular stores and influx across the plasma membrane. However, using electrophysiological techniques, Ca^{2+} influx pathways have not yet been identified in osteoclasts. In particular, no evidence has been found for voltage-activated Ca^{2+} channels (Ravesloot et al., 1989; Sims et al., 1991; Gaspar et al., 1995), but other Ca^{2+} entry pathways, such as ligand-gated channels, are likely present in the plasma membrane. For example, certain stretch-activated and P2X channels (see below) are thought to be permeable to Ca^{2+}. Another Ca^{2+} influx pathway is the calcium-release activated calcium (CRAC) channel which in many cells is activated by depletion of intracellular Ca^{2+} stores (Hoth and Penner, 1993). In this

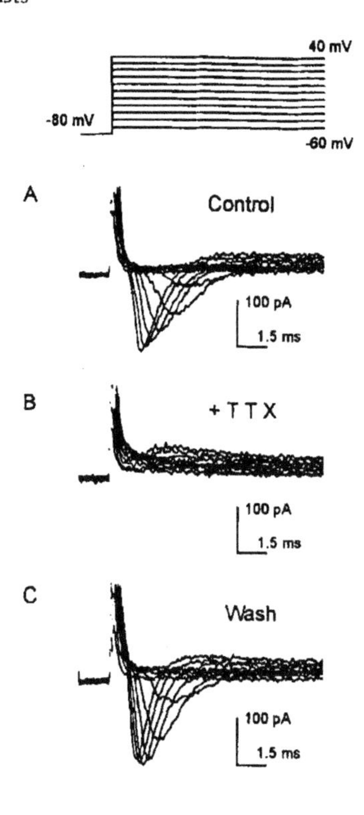

Figure 3. Voltage-activated sodium channels in chicken osteoclasts. Cell was held under voltage clamp at -80 mV and stepped from -60 to +40 mV in 10 mV increments. The bath medium contained Ba^{2+} (5 mM) and tetraethylammonium (10 mM) to block inward and outward potassium channels. Under control conditions (**A**) inward current was seen as transient downward deflection at potentials positive to -20 mV. Maximal activation occurred within 1.5 ms and inactivation was complete in 6 ms. (**B**) Inward current was inhibited by tetrodotoxin (TTX, 100 nM), with recovery after wash shown in (**C**). Traces taken with permission from Gaspar et al., 1995.

regard, depletion of Ca^{2+} stores by thapsigargin activates capacitative Ca^{2+} influx in rat osteoclasts (Zaidi et al., 1993). Electrophysiological identification and characterization of osteoclast Ca^{2+} channels are important areas for future studies (see Chapter by Adebanjo in volume 5B).

D. Nonselective Cation Channels

Stretch-Activated Channel

Bones remodel in response to mechanical stimuli, however the cellular basis of this phenomenon is poorly understood (see Chapter by Burge in volume 5A).

Stretch-activated ion channels have been described in osteoblastic cells (Duncan and Misler, 1989; Davidson et al., 1990) and chicken osteoclasts (Ypey et al., 1992; Wiltink et al., 1995), but have not yet been described in mammalian osteoclasts. In many cell types, a variety of channels are activated in patches of membrane when the membrane is stretched by applying suction to the recording pipette. The channel identified in chicken osteoclasts shows inward rectification (giving rise largely to inward currents) and has a single channel conductance of 60 pS with high potassium in the pipette. The channel permits passage of Na^+, K^+ and Ca^{2+} ions, and is therefore considered a nonselective cation channel. Influx of Ca^{2+} ions through this channel is sufficient to elevate $[Ca^{2+}]_i$ in osteoclasts (Wiltink et al., 1995). Stretch-activated channels are thought to play a role in cell volume regulation, and it is tempting to speculate that these channels may also play a role in the transduction of mechanical stimuli in bone.

ATP-Activated Channels

Extracellular nucleotides such as ATP activate ion channels via the P2X class of purinoceptors, which have been demonstrated in several cell types, including smooth muscle cells, neurons, and neuroendocrine cells (Surprenant et al., 1995). P2X purinoceptors are multimeric ion channels that are directly gated by the binding of extracellular nucleotides. A recent report has shown that ATP activates a large, voltage-independent, nonselective conductance in mouse osteoclasts (Modderman et al., 1994). In the absence of extracellular Mg^{2+}, activation of this receptor by ATP forms pores in the membrane permeable to molecules up to 900 Da. It was proposed that the ligand for this channel was ATP^{4-}, since addition of Mg^{2+} to the bath rapidly decreased the current, as described in other systems for P2Z purinoceptors. In this regard, the P2Z receptor has recently been found to be a member of the P2X class of purinoceptors (referred to as $P2X_7$; Surprenant et al., 1996). In the presence of Mg^{2+}, the $P2X_7$ channel behaves like other P2X receptors as a nonselective cation channel (i.e., permeable only to small cations). However, in Mg^{2+}-free conditions, activation of $P2X_7$ receptors by nucleotides results in the formation of pores permeable to large molecules (Surprenant et al., 1996). It is possible that the $P2X_7$ receptor underlies the nucleotide-activated pores described by Modderman and co-workers (1994). However, whether the effect of Mg^{2+} is to complex with ATP^{4-} or to regulate behavior of the osteoclast channel remains to be established.

There is also a report of a nonselective cation current activated by ATP in rat osteoclasts (Weidema et al., 1996; see Note Added in Proof). This inwardly rectifying current shows rapid activation and inactivation kinetics. The channel is selective for small cations such as Na^+ and K^+ in solutions containing physiological levels of Mg^{2+}. It remains to be determined whether this conductance is mediated by $P2X_7$ receptors or other members of the P2X family.

There are a number of conditions under which nucleotides may be released into the extracellular fluid to influence osteoclast function. Nucleotides are released locally during trauma and inflammation (Dubyak and El-Moatassim, 1993). In addition, it has been suggested that efflux of ATP is mediated by the ATP-binding cassette family of transport proteins, which includes the multidrug resistance gene product, known as P-glycoprotein (Al-Awqati, 1995). Interestingly, P-glycoprotein has recently been shown to be expressed by chondrocytes and osteoblasts in mineralizing regions of the skeleton (Mangham et al., 1996). Efflux of ATP may locally influence osteoclast function through activation of the P2X receptors described above, or other types of purinoceptors present on osteoclasts (Yu and Ferrier, 1993; Bowler et al., 1995).

The P2X class of purinoceptors serves several functions in other cell types. For example, they play a role in fast synaptic transmission in the central and peripheral nervous systems, cytolytic activity of T lymphocytes, and induction of apoptosis (Surprenant et al., 1995). It is conceivable that activation of these receptors also induces apoptosis in osteoclasts. Suramin acts as an antagonist at many P2X purinoceptors (Surprenant et al., 1995). The previously reported effects of suramin on bone resorption (Walther et al., 1992; Farsoudi et al., 1993; Yoneda et al., 1995) may be mediated in part through inhibition of purinoceptor function in bone cells.

IV. CHLORIDE CHANNELS

Description

In many cell types, Cl^- channels contribute to a number of important functions, including setting the membrane potential, regulating cell volume and secreting electrolytes and fluid. Patch-clamp studies have revealed the presence of Cl^- channels in mammalian and chicken osteoclasts (Sims et al., 1991; Kelly et al., 1994; Weidema, 1995). In rat and rabbit osteoclasts, the current shows outward rectification, activates and deactivates rapidly, and shows little time-dependent inactivation. The Cl^- current is inhibited by stilbene disulfonates (SITS, DIDS, DNDS—defined in legend to Table 1) and niflumic acid (see Table 1). Studies carried out in the cell-attached patch configuration reveal that this channel has a unitary conductance of 20 pS at positive potentials (Kelly et al., 1994). A second type of SITS-sensitive Cl^- channel has been observed in excised patches from chicken osteoclasts (Weidema, 1995). This channel has several subconductance states, but its large unitary conductance of 350 to 400 pS clearly distinguishes it from the smaller channels described above. Since this channel is not observed in cell-attached or whole-cell configurations, its relevance remains uncertain.

The purification of a stilbene-sensitive Cl^--selective channel from chicken osteoclast cell membranes has been reported by Blair and Schlesinger (1990). A recent preliminary report has described reconstitution of a purified osteoclast Cl^- channel in planar lipid membranes, and the cloning of two putative Cl^- channels (Edwards and Schlesinger, 1995; see Note Added in Proof). It is likely that several types of Cl^- channels exist in osteoclasts, and their molecular identification and electrophysiological characterization remain important areas for future studies.

Regulation

Under control conditions, few osteoclasts exhibit the small conductance Cl^- channel (Sims et al., 1991). However, this current can be reproducibly activated by cell swelling induced by exposure to hypotonic bathing solution (Figure 4). The swelling activated conductance gives rise to large outwardly rectifying currents. When studied under current clamp conditions, swelling of osteoclasts is accompanied by reversible depolarization from resting potentials of -75 mV to -5 mV (Kelly et al., 1994). It remains to be determined as to whether this channel is

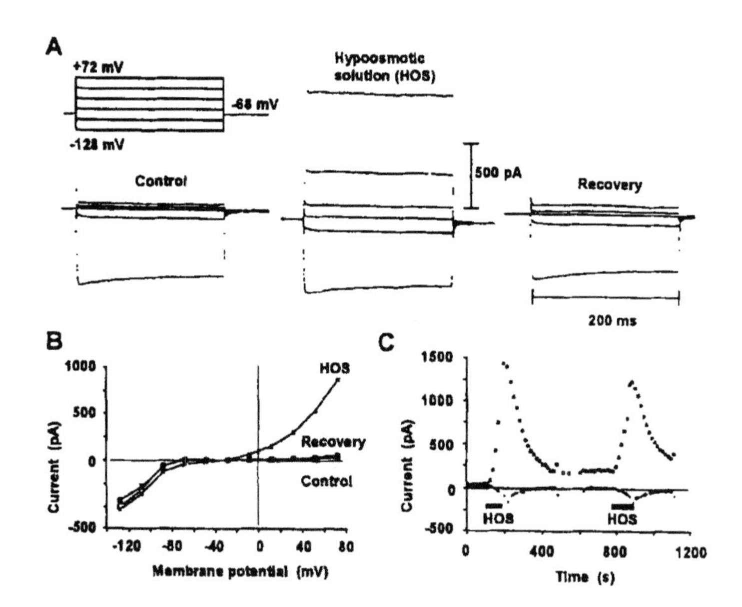

Figure 4. Hypoosmotic activation of outwardly rectifying chloride channels in rabbit osteoclasts. (**A**) Whole-cell currents recorded under control conditions show only inwardly rectifying K^+ current, seen as downward deflections upon hyperpolarization. Application of hypoosmotic solution to the cells (HOS, 205 mOsm, compared to control of 290 mOsm) caused reversible activation of Cl^- current (middle traces, with recovery at right). (**B**) Current-voltage relations of the peak currents in **A** reveal outward rectification of the current activated by hypoosmotic solution with little change in the inward current. The induced current reversed direction at -37 mV, close to the Nernst equilibrium potential for Cl^-, indicating that Cl^- is the main permeant ion. (**C**) Time course of the Cl^- current activated by hypoosmotic solution. Steady-state current levels were recorded at +52 mV and -68 mV, and reveal reversible and repetitive activation of Cl^- current. Traces taken with permission from Kelly et al., 1994.

activated directly by stretch of the membrane, or by other signaling mechanisms. In this regard, there is a preliminary report that extracellular Ca^{2+} activates an outwardly rectifying Cl^- current in rabbit osteoclasts (Yamashita et al., 1995). It has also been reported that extracellular RGD-containing peptides inhibit swelling-induced Cl^- current in rat osteoclasts (Shankar et al., 1994). Clearly, the regulation of Cl^- currents in osteoclasts is complex and warrants further studies.

Channel Function

As described above for the inward rectifier K^+ channel, the Cl^- channel in osteoclasts may provide a conductive pathway to dissipate charge arising from electrogenic H^+ pumping in the actively resorbing cell (Blair et al., 1991). It is likely that swelling-induced Cl^- channels also play a role in regulatory volume decrease in osteoclasts, as proposed for other cell types (e.g., Nilius et al., 1994). Cl^- channels may also participate in regulating cytoplasmic pH of osteoclasts by maintaining the inward Cl^- concentration gradient that drives HCO_3^-/Cl^- exchange. A number of findings provide evidence that Cl^- channels play an essential role in bone resorption. The Cl^- channel blockers DIDS and SITS reduce bone resorption (Hall and Chambers, 1989; Klein-Nulend and Raisz, 1989), an effect attributed to inhibition of the HCO_3^-/Cl^- exchanger. The possibility exists that DIDS and SITS inhibit resorption through blockade of osteoclast Cl^- channels. Similarly, tamoxifen, which inhibits bone resorption (Williams et al., 1996), has been shown to block Cl^- channels in other systems (Zhang et al., 1994).

V. FUTURE DIRECTIONS

Undoubtedly, there are other types of ion channels in osteoclasts that remain to be identified using electrophysiological and molecular approaches. Techniques have been developed to study gene expression in osteoclast preparations. These have already proven useful for characterizing classes of K^+ channels in mammalian osteoclasts (Arkett et al., 1994c), and it will be informative to apply such approaches to other types of ion channels. Furthermore, it is now possible to explore the role of ion channel mutations in hereditary disorders of bone.

Many questions remain concerning the localization, regulation, and function of ion channels in osteoclasts. It is well established that the vacuolar H^+-ATPase is located in internal membranes and the ruffled border membrane. It is important to determine whether specific ion channels are targeted to discrete membrane domains of the osteoclast. Localization can be addressed using antibody labeling and single channel electrophysiological techniques. It is also important to elucidate the signaling pathways responsible for regulation of channel expression and activity. A valuable approach will be to combine Ca^{2+} fluorescence imaging and patch-clamp techniques to investigate the role of $[Ca^{2+}]_i$ in regulation of channel activity. Further insights into ion channel function may eventually be obtained using targeted gene knock-out and transgenic approaches.

An ongoing area of investigation is the identification of pharmacological blockers and activators of ion channels in osteoclasts. These agents could be used to explore the role of specific channels in bone remodeling. Furthermore, as in other systems, membrane ion channels are potential targets for drug therapy. It is conceivable that agents will be identified that interact selectively with channels to modulate the function of osteoclasts in metabolic and inflammatory bone diseases.

ACKNOWLEDGMENTS

The work from the authors' laboratories reviewed in this chapter was supported by The Arthritis Society of Canada. S.J.D. was supported by a Development Grant, and S.M.S. by a Scientist award from the Medical Research Council of Canada.

NOTE ADDED IN PROOF

Since preparation of this chapter the following relevant papers have been published.

Arnett, T.R., and King, B.F. (1997). ATP as an osteoclast regulator? J. Physiol. (London) 503, 236.

Fujita, H., Matsumoto, T., Kawashima, H., Ogata, E., Fujita, T., and Yamashita, N. (1996). Activation of Cl^- channels by extracellular Ca^{2+} in freshly isolated rabbit osteoclasts. J. Cell. Physiol. 169, 217-225.

Kai, Y., Ikemoto, Y., Abe, K., and Oka, M. (1996). Two types of K^+ currents underlying inward rectification of rat osteoclast membrane: A single-channel analysis. Jpn. J. Physiol. 46, 231-241.

Nordstrom, T., Shrode, L.D., Rotstein, O.D., Romanek, R., Goto, T., Heersche, J.N., Manolson, M.F., Brisseau, G.F., and Grinstein, S. (1997). Chronic extracellular acidosis induces plasmalemmal vacuolar type H+ ATPase activity in osteoclasts. J. Biol. Chem. 272, 6354-6360.

Schlesinger, P.H., Blair, H.C., Teitelbaum, S.L., and Edwards, J.C. (1997). Characterization of the osteoclast ruffled border chloride channel and its role in bone resorption. J. Biol. Chem. 272, 18636-18643.

Shibata, T., Sakai, H., and Nakamura, F. (1996). Membrane currents of murine osteoclasts generated from bone marrow/stromal cell co-culture. Osaka City Med. J. 42, 93-107.

Shibata, T. Sakai, H., Nakamura, F., Shioi, A., and Kuno, M. (1997). Differential effect of high extracellular Ca^{2+} on K^+ and Cl^- conductances in murine osteoclasts. J. Membr. Biol. 158, 59-67.

Weideman, A.F., Barbera, J., Dixon, S.J., and Sims, S.M. (1997). Extracellular nucleotides activate nonselective cation and Ca^{2+}-dependent K^+ channels in rat osteoclasts. J. Physiol. (London) 503, 303-315.

Yoshida, N., Sato, T., Kobayashi, K., and Okada, Y. (1998). High extracellular Ca^{2+} and Ca^{2+}-sensing receptor agonists activate nonselective cation conductance in freshly isolated rat osteoclasts. Bone. 22, 495-501.

REFERENCES

Al-Awqati, Q. (1995). Regulation of ion channels by ABC transporters that secrete ATP. Science 269, 805-806.

Arkett, S.A., Dixon, S.J., and Sims, S.M. (1992). Substrate influences rat osteoclast morphology and expression of potassium conductances. J. Physiol. (London) 458, 633-653.

Arkett, S.A., Dixon, S.J., and Sims, S.M. (1994a). Effects of extracellular calcium and protons on osteoclast potassium currents. J. Membrane Biol. 140, 163-171.

Arkett, S.A., Dixon, S.J., and Sims, S.M. (1994b). Lamellipod extension and K^+ current in osteoclasts are regulated by different types of G proteins. J. Cell Sci. 107, 517-526.

Arkett, S.A., Dixon, S.J., Yang, J.N., Sakai, D.D., Minkin, C., and Sims, S.M. (1994c). Mammalian osteoclasts express a transient potassium channel with properties of Kv1.3. Receptors and Channels 2, 281-293.

Blair, H.C. and Schlesinger, P.H. (1990). Purification of a stilbene sensitive chloride channel and reconstruction of chloride conductivity into phospholipid vesicles. Biochem. Biophys. Res. Comm. 171, 920-925.

Blair, H.C., Teitelbaum, S.L., Tan, H.L., Koziol, C.M., and Schlesinger, P.H. (1991). Passive chloride permeability charge coupled to H^+-ATPase of avian osteoclast ruffled membrane. Am. J. Physiol. 260, C1315-C1324.

Bowler, W.B., Birch, M.A., Gallagher, J.A., and Bilbe, G. (1995). Identification and cloning of human $P2_U$ purinoceptor present in osteoclastoma, bone and osteoblasts. J. Bone Miner. Res. 10, 1137-1145.

Darden, A.G., Ries, W.L., Wolf, W.C., Rodriguiz, R.M., and Key, Jr., L.L. (1996). Osteoclastic superoxide production and bone resorption: Stimulation and inhibition by modulators of NADPH oxidase. J. Bone Miner. Res. 11, 671-675.

Davidson, R.M., Tatakis, D.W. and Auerbach, A.L. (1990). Multiple forms of mechanosensitive ion channels in osteoblastlike cells. Pflügers Arch. 416, 646-651.

DeCoursey, T.E. and Cherny, V.V. (1994). Voltage-activated hydrogen ion currents. J. Membrane Biol. 141, 203-223.

Deutsch, C. and Chen, L.Q. (1993). Heterologous expression of specific K^+ channels in T lymphocytes: Functional consequences for volume regulation. Proc. Natl. Acad. Sci. U.S.A. 90, 10036-10040.

Dixon, S.J., Arkett, S.A., and Sims, S.M. (1993). Electrophysiology of osteoclasts and macrophages. In: Blood Cell Biochemistry, Vol. 5: Macrophages and Related Cells. (Horton, M.A., Ed.), pp. 203-222, Plenum Press, New York.

Dubyak, G.R. and El-Moatassim, C. (1993). Signal transduction via P_2-purinergic receptors for extracellular ATP and other nucleotides. Am. J. Physiol. 265, C577-C606.

Duncan, R. and Misler, S. (1989). Voltage-activated and stretch-activated Ba^{2+} conducting channels in an osteoblastlike tumor cell line (UMR 106) FEBS Lett. 251, 17-21.

Edwards, J.C. and Schlesinger, P.H. (1995). The ruffled membrane chloride channel: Regulation of proton secretion in osteoclasts. J. Bone Miner. Res. 10 (Suppl. 1), S430 (Abstract.)

Farsoudi, K.H., Pietschmann, P., Cross, H.S., and Peterlik, M. (1993). Suramin is a potent inhibitor of calcemic hormone- and growth factor-induced bone resorption in vitro. J. Pharmacol. Exp. Ther. 264, 579-583.

Forgac, M. (1989). Structure and function of vacuolar class of ATP-driven proton pumps. Physiol. Rev. 69, 765-796.

Gaspar, Jr., R., Weidema, A.F., Krasznai, Z., Nijweide, P.J., and Ypey, D.L. (1995). Tetrodotoxin-sensitive fast Na^+ current in embryonic chicken osteoclasts. Pflügers Arch. 430, 596-598.

Hammerland, L.G., Parihar, A.S., Nemeth, E.F., and Sanguinetti, M.C. (1994). Voltage-activated potassium currents of rabbit osteoclasts: effects of extracellular calcium. Am. J. Physiol. 267, C1103-C1111.

Hall, T.J. and Chambers, T.J. (1989). Optimal bone resorption by isolated rat osteoclasts requires chloride bicarbonate exchange. Calcif. Tissue Int. 45, 378-380.

Hille, B. (1992). Ionic Channels in Excitable Membranes, Sinauer Associates Inc., Sunderland, Massachusetts.

Hoth, M. and Penner, R. (1993). Calcium release-activated calcium current in rat mast cells. J. Physiol. (London) 465, 359-386.

Johnson, T.D. (1996). Modulation of channel function by polyamines. Trends Pharmacol. Sci. 17, 22-27.

Kelly, M.E., Dixon, S.J., and Sims, S.M. (1992). Inwardly rectifying potassium current in rabbit osteoclasts: A whole-cell and single-channel study. J. Membrane Biol. 126, 171-181.

Kelly, M.E., Dixon, S.J., and Sims, S.M. (1994). Outwardly rectifying chloride current in rabbit osteoclasts is activated by hyposmotic stimulation. J. Physiol. (London) 475, 377-389.

Klein-Nulend, J. and Raisz, L.G. (1989). Effects of two inhibitors of anion transport on bone resorption in organ culture. Endocrinology 125, 1019-1024.

Kubo, Y., Baldwin, T.J., Jan, Y.N., and Jan, L.Y. (1993). Primary structure and functional expression of a mouse inward rectifier potassium channel. Nature (London) 362, 127-132.

Lewis, R.S. and Cahalan, M.D. (1995). Potassium and calcium channels in lymphocytes. Annu. Rev. Immunol. 13, 623-653.

Mangham, D.C., Cannon, A., Komiya, S., Gendron, R.L., Dunussi, K., Gebhardt, M.C., Mankin, H. J., and Arceci, R.J. (1996). P-Glycoprotein is expressed in the mineralizing regions of the skeleton. Calcif. Tissue Int. 58, 186-191.

Mears, D.C. (1971). Effects of parathyroid hormone and thyrocalcitonin on the membrane potential of osteoclasts. Endocrinology 88, 1021-1028.

Miller, C., Moczydlowski, E., Latorre, R., and Phillips, M. (1985). Charybdotoxin, a potent inhibitor of single Ca^{2+}-activated K^{+} channels from skeletal muscle. Nature (London) 313, 316-318.

Modderman, W.E., Weidema, A.F., Vrijheids Lammers, T., Wassenaar, A.M., and Nijweide, P.J. (1994). Permeabilization of cells of hemopoietic origin by extracellular ATP^{4-}: Elimination of osteoclasts, macrophages, and their precursors from isolated bone cell populations and fetal bone rudiments. Calcif. Tissue Int. 55, 141-150.

Moonga, B.S. and Dempster, D.W. (1995). Zinc is a potent inhibitor of osteoclastic bone resorption in vitro. J. Bone Miner. Res. 10, 453-457.

Nilius, B., Schrer, J., De-Greef, C., Raeymaekers, L., Eggermont, J., and Droogmans, G. (1994). Volume-regulated Cl^{-} currents in different mammalian nonexcitable cell types. Pflügers Arch. 428, 364-371.

Nordstrom, T., Rotstein, O.D., Romanek, R., Asotra, S., Heersche, J.N.M., Manolson, M.F., Brisseau, G.F., and Grinstein, S. (1995). Regulation of cytoplasmic pH in osteoclasts. Contribution of proton pumps and a proton-selective conductance. J. Biol. Chem. 270, 2203-2212.

Price, M., Lee, S.C., and Deutsch, C. (1989). Charybdotoxin inhibits proliferation and interleukin 2 production in human peripheral blood lymphocytes. Proc. Natl. Acad. Sci. U.S.A. 86, 10171-10175.

Ravesloot, J.H., Eisen, T., Baron, R., and Boron, W.F. (1995). Role of Na-H exchangers and vacuolar H^{+} pumps in pH regulation in neonatal rat osteoclasts. J. Gen. Physiol. 105, 177-208.

Ravesloot, J.H., Ypey, D.L., Vrijheid-Lammers, T., and Nijweide, P.J. (1989). Voltage-activated K^{+} conductances in freshly isolated embryonic chicken osteoclasts. Proc. Natl. Acad. Sci. U.S.A. 86, 6821-6825.

Shankar, G., Coetzee, W.A., and Horton, M.A. (1994). Rapid intracellular acidification and inhibition of stretch-activated chloride currents by integrin ligands in rat osteoclasts. J. Bone Miner. Res. 9 (Suppl. 1), S155 (Abstract.)

Shirihai, O., Merchav, S., Attali, B., and Dagan, D. (1996). K^{+}-channel antisense oligodeoxynucleotides inhibit cytokine-induced expansion of human hemopoietic progenitors. Pflügers Arch. 431, 632-638.

Sims, S.M. and Dixon, S.J. (1989). Inwardly rectifying K^+ current in osteoclasts. Am. J. Physiol. 256, C1277-C1282.

Sims, S.M., Kelly, M.E.M., Arkett, S.A., and Dixon, S.J. (1992). Electrophysiology of osteoclasts. In: Biology and Physiology of the Osteoclast. (Rifkin, B.R. and Gay, C.V., Ed.), pp. 223-244. CRC Press, Boca Raton, FL.

Sims, S.M., Kelly, M.E.M., and Dixon, S.J. (1991). K^+ and Cl^- currents in freshly isolated rat osteoclasts. Pflügers Arch. 419, 358-370.

Stein, W.D. (1990). Channels, carriers, and pumps: An introduction to membrane transport. Academic Press, San Diego.

Surprenant, A. Buell, G., and North, R.A. (1995). P_{2X} receptors bring new structure to ligand-gated ion channels. Trends in Neurosciences 18, 224-229.

Surprenant, A., Rassendren, F., Kawashima, E., North, R.A., and Buell, G. (1996). The cytolytic P_{2Z} receptor for extracellular ATP identified as a P_{2X} receptor ($P2X_7$). Science 272, 735-738.

Teti, A., Blair, H.C., Teitelbaum, S.L., Kahn, A.J., Koziol, C., Konsek, J., Zambonin-Zallone, A., and Schlesinger, P.H. (1989). Cytoplasmic pH regulation and chloride/bicarbonate exchange in avian osteoclasts. J. Clin. Invest. 83, 227-233.

Walther, M.M., Kragel, P.J., Trahan, E., Venzon, D., Blair, H.C., Schlesinger, P.H., Jamai-Dow, C., Ewing, M.W., Myers, C.E., and Linehan, W.M. (1992). Suramin inhibits bone resorption and reduces osteoblast number in a neonatal mouse calvarial bone resorption assay. Endocrinology 131, 2263-2270.

Weidema, A.F. (1995). Ion channels and ion transport in chicken osteoclasts. Ph.D. Thesis. Rijks Universiteit Leiden, The Netherlands.

Weidema, A.F., Barbera, J., Dixon, S.J., and Sims, S.M. (1996). Extracellular nucleotides activate nonselective cation and K^+ conductances in rat osteoclasts. Biophys. J. 70, A97 (Abstract.)

Weidema, A.F., Ravesloot, J.H., Panyi, G., Nijweide, P.J., and Ypey, D.L. (1993). A Ca^{2+}-dependent K^+-channel in freshly isolated and cultured chick osteoclasts. Biochim. Biophys. Acta 1149, 63-72.

Wen, R., Ming, G., and Steinberg, R.H. (1994). Expression of a tetrodotoxin-sensitive Na^+-current in cultured human retinal pigment epithelial cells. J. Physiol. (London) 476, 187-196.

Williams, J.P., Blair, H.C., McKenna, M.A., Jordan, S.E., and McDonald, J.M. (1996). Regulation of avian osteoclast H^+-ATPase and bone resorption by tamoxifen and calmodulin antagonists. J. Biol. Chem. 271, 12488-12495.

Wiltink, A., Nijweide, P.J., Scheenen, W.J., Ypey, D.L., and Van, D.B. (1995). Cell membrane stretch in osteoclasts triggers a self-reinforcing Ca^{2+} entry pathway. Pflügers Arch 429, 663-671.

Yamashita, N., Fujita, H., Ogata, E., Kawashima, H., and Matsumoto, T. (1995). Activation of Cl^- channels by extracellular Ca^{2+} in freshly isolated rabbit osteoclasts. J. Physiol. (London) 487, 200P (Abstract.)

Yamashita, N., Ishii, T., Ogata, E., and Matsumoto, T. (1994). Inhibition of inwardly rectifying K^+ current by external Ca^{2+} ions in freshly isolated rabbit osteoclasts. J. Physiol. (London) 480, 217-224.

Yoneda, T., Williams, P., Rhine, C., Boyce, B.F., Dunstan, C., and Mundy, G.R. (1995). Suramin suppresses hypercalcemia and osteoclastic bone resorption in nude mice bearing a human squamous cancer. Cancer Res. 55, 1989-1993.

Ypey, D.L., Folander, K., Weselowsky, G., Tanaka, H., Nagy, R., Rodan, G., Swanson, R., and Duong, L. (1995). On the origin and role of cloned inwardly rectifying K^+ channels in mouse bone marrow osteoclast/osteoblast cocultures. Biophys. J. 68, A268 (Abstract.)

Ypey, D.L., Ravesloot, J.H., Buisman, H.P., and Nijweide, P.J. (1988). Voltage-activated ionic channels and conductances in embryonic chick osteoblast cultures. J. Membrane Biol. 101, 141-150.

Ypey, D.L., Weidema, A. F., Hold, K.M., VanderLaarse, A., Ravesloot, J.H., VanderPlas, A. and Nijweide, P.J. (1992). Voltage, calcium, and stretch activated ion channels and intracellular calcium in bone cells. J. Bone Miner. Res. 7(Suppl. 2), S377-S387.

Yu, H. and Ferrier, J. (1993). ATP induces an intracellular calcium pulse in osteoclasts. Biochem. Biophys. Res. Commun. 191, 357-363.

Zaidi, M., Shankar, V.S., Bax, C.M.R., Bax, B.E., Bevis, P.J.R., Pazianas, M., Alam, A.S.M.T., Moonga, B.S., and Huang, C.L.-H. (1993). Linkage of extracellular and intracellular control of cytosolic Ca^{2+} in rat osteoclasts in the presence of thapsigargin. J. Bone Miner. Res. 8, 961-967.

Zhang, J.J., Jacob, T.J.C., Valverde, M.A., Hardy, S.P., Minting, G.M., Sepulveda, F.V., Gill, D.R., Hyde, S.C., Trezise, A.E.O., and Higgins, C.F. (1994). Tamoxifen blocks chloride channels. J. Clin. Invest. 94, 1690-1697.

SECTION III

BONE FORMATION

OSTEOBLASTS AND BONE FORMATION

Pierre J. Marie

Advances in Organ Biology
Volume 5B, pages 445-473.

ISBN: 0-7623-0390-5

I. INTRODUCTION

Scientific advances in the field of skeletal biology over the last decade have markedly improved our understanding of the osteoblast and its role in bone formation. Notably, in vitro and in vivo studies have led to a better understanding of the mechanisms controlling bone formation during bone development, bone remodeling, and repair. This has allowed delineation of the sequential events involved in osteoblast recruitment and differentiation, and identification of the critical factors acting at the different steps of cell proliferation and differentiation. In parallel, in vitro/ex vivo studies have demonstrated the importance of cell proliferation in the control of bone formation. In addition, some of the complex cell-matrix interactions and the cellular and molecular mechanisms controlling osteoblast differentiation have been elucidated. Thus our overall understanding of mechanisms controlling bone formation at the tissue, cellular, and molecular levels has improved substantially.

This chapter focuses on our current understanding of osteoblast biology and bone formation that is based upon recent data obtained in our and other laboratories. I will also discuss critical factors and mechanisms controlling bone formation, will raise yet unanswered questions, and outline future research directions.

II. BONE FORMATION

A. Advances in Methodological Approaches

Classical approaches used to study the function of osteoblasts and bone formation include *in vivo* studies, organ culture of bone, and osteoblastic cell cultures. Histological studies of bone have provided critical information on the physiological regulation of endosteal bone formation in humans (Parfitt, 1990) as well as animals (Marie et al., 1994). The development of several *in vitro* models using different types of osteoblastic cells has allowed the elucidation of the cellular and molecular mechanisms controlling the osteoblast function. Transformed osteoblasts and osteosarcoma cells have been widely used to study gene expression and molecular mechanisms (Rodan and Noda, 1991). However, interest in such cultures is in part limited because of the dysregulation of differentiation and gene expression in these cells (Stein and Lian, 1993).

Cultures of normal periosteal osteoblastic cells in rats have provided a powerful tool to study normal osteoblast biology (Bellows et al., 1991; Stein and Lian, 1993). Human neonatal calvaria cells also proved of interest to study the regulation of osteogenesis (de Pollak et al., 1996; 1997; Debiais et al., 1998). In addition, the regulation of endosteal osteoblastic cells has been studied using osteoblastic cells derived from trabecular bone in animals (Halstead et al., 1992; Modrowski and Marie, 1993) and humans (Beresford, 1989; Marie et al., 1989c). The recent development of osteoblast precursor cell cultures derived from the marrow stroma in animals (Maniatopoulos et al., 1988) and humans (Cheng et al., 1994; Fromigué et al., 1998) appears to be of use in the study of osteoblast progenitor cells. Finally, studies comparing the behavior of human endosteal osteoblastic cells *in vitro* and bone formation *ex vivo* has proven to be a powerful tool in allowing an evaluation of the osteoblastic abnormalities noted in metabolic bone diseases and in experimentally-induced osteopenia (Marie and de Vernejoul, 1993a; Marie, 1994).

The more recent development of methods to analyze gene expression in osteoblastic cells *in vitro* and *in vivo* has greatly improved our understanding of osteogenesis. Gene expression in osteoblasts can be studied *in vivo* at the periosteal level (Turner and Spelsberg, 1991; Jackson et al., 1994) or in endosteal bone (Zhang et al., 1995) both in normal and pathologic conditions. The use of *in situ* hybridization and *in situ* reverse transcriptase-polymerase chain reaction (RT-PCR) has allowed provision of information on the spatial expression of specific genes (Shinar et al., 1993). The evaluation of gene expression by RT-PCR has permitted studies on the differentiation of a subgroup of osteoprogenitor cells (Liu et al., 1994). Specific immunocytochemical techniques have allowed the identificaton of the expression of specific proteins in osteoblastic cells during *in vitro* osteogenesis (Malaval et al., 1994). The molecular analysis of gene transcription in vitro and in situ led to the identification of the main regulatory factors acting at the different steps of osteoblastic cell proliferation and differentiation (Rodan and Noda, 1991). The recent application of molecular approaches such as mRNA differential display analyses (Mason et al., 1997; Yotov et al., 1998) are also promising tools to identify novel genes involved in osteoblast differentiation. The production of genetically manipulated mice that either overexpress or lack gene(s) of interest has allowed evaluation of the role of genes believed to play a crucial role in bone formation (Karaplis, 1996). Finally, recent studies on the molecular mechanisms underlying genetic disorders of the skeleton have provided major insights into the role of particular genes in cartilage and bone during skeletal development (Jacenko et al., 1994; Erlebacher et al., 1995; Marie, 1998).

Thus, while most of our knowledge of bone formation came initially from histological and cellular studies, the development of in vitro models of osteogenesis in animals and humans, the analysis of gene expression in bone tissue in vivo and in vitro, and the identification of genes involved in skeletal abnormalities have led to greatly improve our understanding of the function of osteoblasts. This remarkable complementation of studies at the tissue, cellular, and molecular levels has also provided important information on the role of growth factors, cell-matrix interactions, intracellular signaling molecules and transcription factors that are involved in the regulation of the osteoblastic bone formation.

B. Bone Formation During Development

Bone formation may be classically defined as the process by which osteoblasts synthesize and mineralize an extracellular bone matrix during bone development, bone remodeling, or bone repair. The sequence of events and the genes and proteins involved in bone formation differ during the fetal and adult life, during repair of fractures and during bone remodeling. These differences come from the different envelopes and cells involved, the type of matrix synthesized, the heterogeneity of the cells and factors involved, and the variable local conditions such as the presence of marrow, cartilage, bone matrix, and the extent of vascularization. Most of our present knowledge concerns bone formation during bone remodeling (Rodan and Noda, 1991) and repair (Sanberg et al., 1993). Detailed information on the sequential events characterizing bone development is lacking, mainly because of the complexity of the bone structure and the multiple interactions between the cells and the matrix.

Recent studies in the biology of developmental genes, i.e., genes involved in embryonic and fetal skeletal development, indicate that multiples genes are involved in the sequential development of the skeleton (Johnson and Tabin, 1997). Notably, specific homeobox-containing genes control mesenchymal cells and the pattern of the skeleton in early development (Mavilio, 1993; Morgan and Tabin, 1993; Krumlauf, 1994). The condensation of mesenchyme and the development of skeletal elements in a temporally and spatially controlled manner are dependent on the actions of locally secreted inductive factors. The progression of cell populations along specific differentiation pathways not only requires the coordinate expression of different members of the transforming growth factor β (TGFβ) family, such as $TGFβ_1$, $TGFβ_2$, and bone morphogenetic protein-2 (BMP- 2), but also multiple interactions between these factors (Lyons et al., 1989). For example, that BMPs are expressed in developing cartilage and bone suggests a potential role in the local control of mesenchymal condensation (Lyons et al., 1991; Wozney, 1992). Interestingly, BMP-2 and -4 were found to be signals for the induction of homeobox-containing genes, such as the Msx-1 and Msx-2 genes, as well as for odontoblast differentiation during early tooth development (Vainio et al., 1993). The potential role of homeobox genes in bone formation has been recently emphasized by the finding that Msx1-deficient mice exhibit craniofacial abnormalities (Satokata and Maas, 1994). In addition, Msx-2 was found to be expressed by osteoblastic cells (Harris et al., 1993; Hodginkson et al., 1993) and is a transcriptional regulator of the osteocalcin promoter (Towler et al., 1994). Specific transcription factors may play a role in the determination of precursor cells toward osteoblasts. Recent data indicate that Osf2/Cbf1, a specific factor inducing osteoblast differentiation, is expressed in osteoblast precursors and mature osteoblasts and that local inducing factors such as BMP-7 induce Osf2 (Ducy et al., 1996; 1997).

During growth, the development of the skeleton involves the coordinated action of multiple systemic and local mediators (Canalis et al., 1991). Some of these factors may have autocrine or paracrine actions. For example, the parathyroid hormone-related protein (PTHrP) and parathyroid hormone (PTH)/PTHrP receptor were found to be distinctly expressed in growth plate cartilage (Lee et al., 1995), suggesting a paracrine-autocrine role for PTHrP on cartilage growth. The targeted ablation of PTHrP gene was found to result in severe chondrodysplasia (Amizuka et al., 1994), and an activating PTHrP receptor mutation has been reported to cause an abnormal endochondral bone formation in humans (Schipani et al., 1995). Taken together, these findings suggest that PTHrP may play a crucial role in cartilage ossification and skeletal development.

Furthermore, the growth of distinct sites of cartilage and bone may be controlled by specific growth factors which are expressed differently in distinct zones of the skeleton. For example, insulin-like growth factors (IGFs) are expressed distinctly in cartilage and bone in rats (Shinar et al., 1993; Lazowski et al., 1994). Growth factors may also act on different cells due to a distinct expression of their receptors. For example, fibroblast growth factor receptors (FGFRs) have been shown to have a distinct expression pattern in the skeleton (Partanen et al., 1992), and FGFs appear to have a distinct role in the development of the skull and cartilage. Mutations in the FGFR3 induce achondroplasia (Rousseau et al., 1994; Shiang et al., 1994), whereas mutations in FGFR1 and FGFR2 cause premature cranial suture ossification (Wilkie et al., 1995). Our recent data indicate that FGFR2 mutations enhance osteoblast differentiation in calvaria of fetus and neonates with Apert syndrome and craniosynostosis (Lomri et al., 1998). The restricted expression pattern of growth factor receptors, such as FGFRs, or of growth factors, such as IGFs, may account for, in part, the regulation of bone cells at distinct sites in the skeleton. Bone formation during development and growth therefore appears to be a complex process which is controlled by a multitude of factors acting in a coordinate fashion during the fetal and postnatal life.

C. Bone Formation During Remodeling

Bone formation during remodeling is characterized by the deposition of a collagenous matrix at the site of previously resorbed bone. Several mechanisms can account for the initiation of bone formation at specific sites. For example, agents capable of promoting osteoblastic cell recruitment may be left at the cement line at the end of the resorbing phase. These agents may be chemoattractant compounds (Pfeilschifter et al., 1990), bone matrix components with adhesive properties due to RGD (Arginin-glycin-aspartate) sequences (Gehron-Robey, 1989), or growth factors acting as coupling factors. One of these factors may be TGFβ which is abundant in the bone matrix, is released from the matrix during bone resorption, and stimulates bone formation (Bonewald and Dallas, 1994). It is, however, possible that several factors are required for the coordinate induction of bone formation at the right time and site along the bone surface during bone remodeling.

During the formation phase, the osteoblast differentiation pathway is characterized by a sequence of events involving the proliferation of osteoprogenitor cells, the progressive differentiation of preosteoblasts and the sequential expression of genes of the osteoblast phenotype and the synthesis, organization,

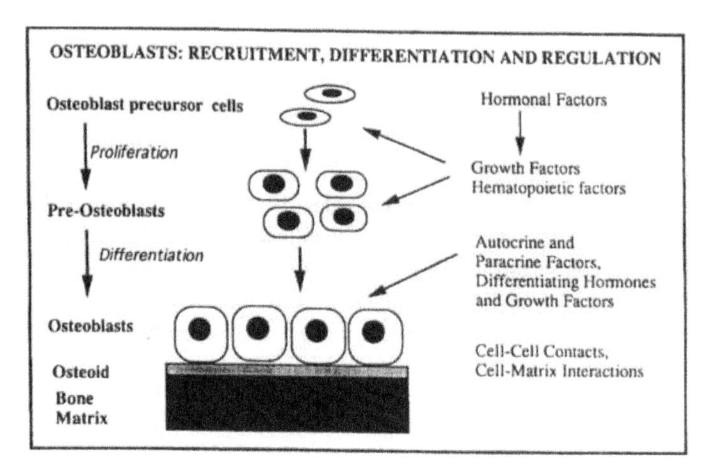

Figure 1. Schematic representation of the developmental sequence of bone formation and regulatory factors controlling osteoblast precursor cells, preosteoblasts, and differentiated osteoblasts.

deposition and mineralization of a bone matrix by post-mitotic mature osteoblasts. The recruitment of osteoblast precursors is in part controlled by mitogenic factors acting on osteoblast precursor cells. The differentiation of preosteoblasts into mature osteoblasts, and the synthesis of the bone matrix by osteoblasts are modulated by cell-cell contact and cell-matrix interactions, and are regulated by hormonal and local factors affecting cell differentiation (Figure 1). The normal deposition of new bone matrix in the remodeling unit during the formation phase requires the coordinated succession of these events and the appropriate induction by systemic and local factors during the sequence of osteoblastic cell differentiation.

III. OSTEOBLAST RECRUITMENT

A. Cells of the Osteoblast Lineage

The osteoblastic lineage comprises several cells at different stages of differentiation, starting from osteoprogenitor cells to fully differentiated osteoblasts. The term osteoblast should therefore be restricted to the more differentiated cells that express characteristics of mature osteoblasts, including the osteogenic capacity. Intermediate cells can be named osteoprogenitor cells, preosteoblasts or osteoblastic cells according to the progressive expression of markers of the osteoblast phenotype.

Osteoprogenitor cells are believed to originate from mesenchymal stem cells. Embryonic undifferentiated mesenchymal cells are multipotential cells that give rise to cartilage, bone, muscle cells, or adipocytes under the influence of hormones and growth factors (Grigoriadis et al., 1988; Kellermann et al., 1990). In the endosteum, osteoblasts originate from osteoprogenitor cells located in the bone marrow stroma (Owen, 1988). The precise identification of osteoprogenitor cells in the marrow stroma is still lacking, mainly because the frequency of occurrence of osteoblast precursors in the stromal compartment is very low (Falla et al., 1993). The latter compartment contains undifferentiated stem cells which may give rise to the various lineages of the stroma under appropriate stimulation (Beresford et al., 1992). For example, osteoprogenitor cells present in the marrow stroma in young adult rats can be induced to differentiate into osteoblasts and to produce a mineralized bonelike tissue in the presence of glucocorticoids (Maniatopoulos et al., 1988; Leboy et al., 1991).

The development of specific markers is an important issue for the identification of osteoblastic cells at different stages of differentiation in the endosteum. Some attempts have been made to identify markers on cells of the osteoblastic lineage along the differentiation pathway. Cellular and molecular analyses of cloned cells derived from rat calvaria indicate that preosteoblastic cells express alkaline phosphatase, the PTH receptor, and collagen type I, whereas differentiated osteoblasts express osteocalcin and bone sialoprotein (Turksen and Aubin, 1991; Aubin and Liu, 1996). In humans, studies of endosteal cells (Marie et al., 1989c), calvaria cells (de Pollak et al., 1996, 1997; Lomri et al., 1997; Debiais et al., 1998), and marrow stromal cells (Fromigué et al., 1997, 1998) led us to depict a general scheme of expression of phenotypic markers expressed during human osteoblastic cell differentiation (Marie, 1998). Preosteoblasts express markers such as alkaline phosphatase, osteopontin, and collagen type I whereas mature postmitotic osteoblasts express osteocalcin and contribute to the synthesis, organization, deposition, and mineralization of the bone matrix. Although antibodies recognizing cells of the osteoblastic lineage have been developed (Bruder and Caplan, 1989; Turksen et al., 1992; Walsh et al., 1994b), specific markers for osteoprogenitor cells at their early stages of differentiation are still lacking. The identification of these cells may help to identify the factors that are acting specifically on cells of the osteoblastic lineage at the early steps of osteoblast differentiation.

B. Control of Osteoblastic Cell Proliferation

Histomorphometric studies have shown that the rate of bone formation at the tissue level is dependent on the number or activity of mature osteoblasts (Parfitt, 1990; Marie et al., 1994). Recently, the cellular mechanisms directing the rate of bone formation have been determined. In a series of *ex vivo/in vitro* comparative experiments, we have demonstrated that the decreased trabecular bone formation in osteoporosis results from a lower than normal proliferative capacity of endosteal osteoblastic cells rather than from alteration of osteoblast differentiation (Marie et al., 1989a, 1991). In accordance with these findings, we found that the enhancement of osteoblastic cell proliferation by a mitogenic agent leads to increased bone formation in osteoporotic patients (Marie et al., 1992). These findings indicate that an insufficient number of osteoblasts may be the pathogenic basis for osteoporosis (Marie, 1995).

Recent experimental models of osteopenia have also shown that bone formation depends mainly on the number of osteoblasts. Age-related bone loss in animals is associated with a decreased number of stromal osteogenic cells (Kahn et al., 1995) whereas estrogen deficiency results in increased osteoblastic cell pool (Modrowski et al., 1993). In addition, osteopenia induced by unloading was found to result from a reduction in the proliferative capacity (Machwate et al., 1993) and reduced osteogenic capacity (Keila et al., 1994) of osteoblast precursor cells in the marrow stroma. Accordingly, treatment of osteopenic animals with mitogenic agents (Modrowski et al., 1992; Machwate et al., 1994, 1995a) was found to increase the rate of proliferation of osteoblast precursor cells and to improve bone formation. The recruitment of osteoblasts from progenitor cells appears to be the more important limiting step controlling bone formation at the tissue level (Marie, 1994), which stresses the importance of stimulating osteoblastic cell recruitment to improve bone formation in osteopenic disorders (Marie, 1995).

The recruitment of cells of the osteoblastic lineage is known to be controlled by several systemic and local agents (Goldring and Goldring, 1990; Canalis et al., 1991; Marie and de Vernejoul, 1993b; Martin et al., 1993; 1994; Mundy, 1995). These factors control osteoblastic cell proliferation by acting on cells expressing their receptors at a particular stage of maturation. Hormones, such as PTH, sex hormones, and growth hormone (GH), have been found to directly affect the growth of osteoblastic cells. In addition, glucocorticoids, PTH, PTHrP, estrogens, progesterone, and GH have been found to indirectly affect cell proliferation through changes in growth factor(s) production and activity (Goldring and Goldring, 1990). Among the multiple growth factors controlling bone cell proliferation, the most important are probably those that are present locally or that are present in the matrix and are released locally (Canalis et al., 1991; Mundy, 1995). *In situ* hybridization studies have shown that bone development, bone repair, and skeletal growth are associated with a high expression of IGFs, TGFβ and FGFs (Wanaka et al., 1991; Sandberg et al., 1993; Lazowski et al., 1994), suggesting that these factors are important local modulators of osteoblastic cell recruitment.

Indeed, FGF, IGF, and TGFβ are synthesized by osteoblastic cells and may be acting in an autocrine or paracrine manner to stimulate osteoblast precursor cell proliferation (Canalis et al., 1991; Centrella et al., 1994). In addition, normal osteoblastic cells produce some interleukin- (IL) 1, tumor necrosis factor α (TNFα), IL-6, and granulocyte macrophage colony-stimulating factor (GM-CSF) under basal conditions or through stimulation. The cells also possess receptors for these cytokines (Horowitz et al., 1989; Keeting et al., 1991; Littlewood et al., 1991; Gowen, 1992; Chaudhary et al., 1992). We recently reported that IL-1, TNFα (Modrowski et al., 1995), and GM-CSF (Modrowski et al., 1997) may act as autocrine growth factors for human osteoblastic cells, suggesting that these factors may be involved in the control of osteoblastic proliferation.

Some factors produced by osteoblastic cells may stimulate other cell types in the local environment. For example, IL-6, TNFα, and IL-1 produced by osteoblasts were postulated to play a role in the increased bone resorption induced by estrogen deficiency. However, these cytokines do not appear to be regulated by estrogens in human osteoblastic cells (Chaudhary et al., 1992; Gowen, 1992), and the production of these cytokines by osteoblastic cells is not increased in osteoporotic postmenopausal women with high bone turnover (Marie et al., 1993). Cytokines produced by other cell types (stromal cells, hematopoietic cells) may be involved in the increased bone resorption in estrogen deficiency. In fact, cytokines (Martin and Ng, 1995) and polypeptides produced by stromal cells (Bab and Einhorn, 1993) are probably important local factors regulating osteoclastogenesis and hematopoiesis as well as osteogenesis and bone marrow repair.

Because of the large number of growth factors and the complexity of their interaction, it is likely that, as in other tissues, the cellular action of growth factors on osteoblastic cells is controlled by complex mechanisms. At the extracellular level, growth factor action may be modulated by an equilibrium between latent and inactive forms (Bonewald and Dallas, 1994), and by the production of binding proteins, soluble receptors, and local inhibitors or agonists in the local environment. At the intracellular level, the action of growth factors may be modulated by cell membrane and bone matrix components. For example, the biological activity of FGFs depends on their binding to heparan sulfate (Rapraeger et al., 1991) and TGFβ effects are in part controlled by interactions with proteoglycans (Gehron-Robey, 1989). Recently, we found that the activity of GM-CSF is dependent on its binding to glycosaminoglycans present on human osteoblastic cell surface and on the extracellular matrix, a mechanism of action similar to that of FGF-2 (Modrowski et al., 1998). Finally, the action of growth factors may be controlled intracellularly by several signaling pathways. The intracellular mediators of some growth factors have been identified (Merriman et al., 1990; Machwate et al., 1995b), and the transduction signals involved in the transcriptional regulation of osteoblastic cells are starting to be understood (Siddhanti and Quarles, 1994). It appears, therefore, that multiple mechanisms acting at the extracellular, cellular, and molecular levels modulate the biological activity of growth factors that stimulate osteoblast proliferation.

Although in vitro studies have clarified the role of growth factors in the control of osteoblast proliferation, the physiological role of these factors in the regulation of bone formation in vivo remains uncertain. Some data point to a possible role of growth factors in disorders of bone formation. Aging is associated with diminished osteoblastic cell proliferation (Evans et al., 1990; Fedarko et al., 1992; Kahn et al., 1995), reduced cell responsiveness to growth factors (Pfeilschifter et al., 1993; Kato et al., 1995), and decreased IGF concentrations in bone in humans (Nicolas et al., 1994). Taken together, the findings suggest that a reduced local production of growth factor(s) may be involved in the decreased bone formation seen with aging.

A lower than normal expression of local growth factors was also reported in unloaded long bones (Zhang et al., 1995), indicating that an insufficient local production of growth factors may be involved in the reduced recruitment of osteoblast precursor cells and inhibition of bone formation induced by unloading. These recent data raise the interesting possibility that treatment with growth factors may stimulate bone formation in vivo in osteopenic disorders (Marie, 1997b).

In normal rodents, periosteal bone formation can be stimulated by local injections of TGFβ (Noda and Camillière, 1989; Marcelli et al., 1990), and endosteal bone formation can be increased by systemic administration of epidermal growth factor (EGF) (Marie et al., 1989b). In aged ovariectomized rats, the administration of IGF-I improves bone formation and bone mass (Mueller et al., 1994). In unloaded rats, the administration of IGF-I (Machwate et al., 1994) or TGFβ (Machwate et al., 1995a) stimulates osteoblast recruitment and prevents the decreased bone formation and osteopenia. These recent findings suggest that growth factors, by increasing the number of osteoblast precursors, may be of therapeutic value in osteopenic disorders characterized by insufficient bone formation (Marie, 1997b). It is however uncertain whether the administration of growth factors in humans could stimulate osteoprogenitor cell proliferation and differentiation, or could induce *de novo* bone matrix formation within the endosteal area without causing nonspecific or toxic effects on soft tissues or on bone growth.

C. Transition Between Cell Proliferation and Differentiation

The mechanisms involved in the transition from osteoblast precursor cell proliferation to the induction of osteoblast differentiation remain largely unknown. Recent data indicate that some transcription factors may be involved in the induction of osteoblast differentiation. Members of helix-loop-helix (HLH) DNA binding proteins appear to have a role in the downregulation of undifferentiated osteoblastic cell proliferation and the upregulation of osteoblast differentiation (Ogata and Noda, 1991; Siddhanti and Quarles, 1994; Tamura and Noda, 1994). Although HLH proteins may be involved in the onset of osteoblast phenotype, their role in the transcriptional control of osteoblast gene expression remains to be determined.

Early genes may also be involved in the switch between cell proliferation and differentiation (Siddhanti and Quarles, 1994). The recent molecular analysis of the development of osteogenesis by rat calvaria cells points to a role for the protooncogene, *c-fos*, in the transition from proliferation to differentiation in osteogenic cells. The induction of *c-fos* gene by mitogenic factors such as IGF-I (Merriman et al., 1990), $TGF\beta_2$ (Machwate et al., 1995b; Subramaniam et al., 1995), and PTH (Kano et al., 1994; Lee et al., 1994) indicates that the induction of osteoblast replication may result in part from the induction of immediate-early genes (Angel and Karin, 1991). *C-fos* is expressed by proliferating cells during the early phase of osteogenesis *in vitro,* and its expression precedes osteogenic differentiation (Closs et al., 1990). Gene expression is downregulated at the onset of differentiation (Owen et al., 1990a). *In vivo, c-fos* expression in the rat is high in osteoblast precursor cells. Its expression decreases with time, but shows a transient increase that precedes the expression of osteocalcin and the marked increase in bone formation that occurs during postnatal bone development (Machwate et al., 1995c). This suggests that *c-fos* may play an essential role in the transition from cell proliferation to differentiation. The downregulation of *c-fos* during the onset of differentiation *in vitro* is associated with the expression of genes whose promoters contain the consensus AP-1 binding site which binds the Fos/Jun heterodimer (Lian et al., 1991; Stein and Lian, 1993). The increased AP-1 activity in proliferating cells may repress the upregulation of osteocalcin by 1,25-dihydroxyvitamin D, and this phenotype suppression involving *c-fos* may partially explain the reciprocal relationship between proliferation and differentiation during osteogenesis *in vitro* (Owen et al., 1990b).

The pattern of *c-fos* expression during osteogenesis *in vitro* and *in vivo,* and studies in transgenic mice (Ruther et al., 1987; Johnson et al., 1992; Wang et al., 1992; Grigoriadis et al., 1993) support role for *c-fos* in bone formation in general. This role is also supported by the recent finding that *c-fos* (Candelière et al., 1995) is overexpressed in fibrous dysplasia, which may be related to an activating mutation of the $G_s\alpha$ protein in osteoblastic cells, inducing an overactive cAMP signaling pathway (Shenker et al., 1995). We showed that, in osteoblastic cells isolated from polyostotic and monostotic lesions that express missense mutations in the $G_s\alpha$ gene with substitution of His or Cys for Arg in position 201 (Shenker et al., 1995), intracellular basal cAMP production is increased and is associated with increased cell growth and decreased osteocalcin production. This indicates that the activating mutation of $G_s\alpha$ increases the proliferation of mesenchymal osteoprogenitor cells and results in accelerated matrix deposition in fibrous dysplastic lesions (Marie et al., 1997). Although *c-fos* may control in part the onset of differentiation, it is likely that multiple classes of transcription factors are involved in the induction of specific genes at the onset and during the development of osteoblast differentiation (Lian et al., 1996).

IV. OSTEOBLAST DIFFERENTIATION

A. Initiation of Osteoblast Differentiation

In the postnatal life, the osteoblast differentiation pathway is characterized by a sequence of events involving the adherence of preosteoblasts to the extracellular matrix, the expression of osteoblast markers, and the synthesis, organization, deposition, and mineralization of bone matrix. During the early step of cell differentiation, cell-cell contacts take place between preosteoblasts, inducing the development of multiple cell-to-cell communications. Mature osteoblasts located along the extracellular bone matrix are also in contact with each other through intercellular communications. Cell-cell contacts between osteoblastic cells are probably important for the expression of the osteoblast phenotype since increasing cell-cell contacts inhibits cell growth (Lomri et al., 1987; Owen et al., 1990b) and promotes the expression of osteoblast markers and genes of the osteoblastic phenotype. Furthermore, different patterns of gap junction proteins, named connexins, were recently found in human and rat osteoblastic cells. These are regulated by cAMP-enhancing agents (Civitelli et al., 1993). The intercellular transmission of signals through these functional mechanisms may be an important mechanism by which mature osteoblasts function in a coordinate fashion.

Numerous interactions can occur between osteoblastic cells and the extracellular bone matrix. Following cell contact with the bone surface, preosteoblastic cells adhere to the matrix through focal adhesion contacts, and this adhesion is primarily mediated by cell surface receptors, the integrins, which form connections with the cytoskeleton and components of the extracellular bone matrix (Ruoslahti, 1991). Cells of the osteoblastic lineage express various integrins (Hughes et al., 1993) which are involved in the association with the cytoskeleton (Adams and Watt, 1993). Collagenous and noncollagenous proteins in the extracellular matrix provide the main attachment sites of osteoblasts on the bone matrix. The RGD sequence present in some bone matrix proteins promotes osteoblast attachment or spreading (Gehron-Robey, 1989) and modulates the differentiation of bone cells (Puleo and Bizios, 1991; Andrianarivo et al., 1992). Recent data indicate, however, that osteoblastic cells may bind to collagen and fibronectin using RGD-independent receptors (Grzesik and Gehron-Robey, 1994). We have shown that a glycyl-histidyl-lysine (GHK) sequence present in the α_2 (I) chain of human collagen, thrombospondin and osteonectin (Lane et al., 1994), promotes cell attachment and modulates the osteoblast phenotype in rat and human osteoblastic cells (Godet and Marie, 1995), suggesting that osteoblastic cells may interact with bone matrix proteins using multiple receptor systems.

The complete osteogenic differentiation is determined by the induction of multiple cell-cell communications and cell-matrix interactions. Osteoblast precursor cells derived from rodent mesenchyme (Kellermann et al., 1990), postnatal calvaria (Owen et al., 1990a; Bellows et al., 1991), endosteal bone (Lomri et al., 1988; Modrowski and Marie, 1993), or marrow stroma (Maniatopoulos et al., 1988; Kasugai et al., 1991; Leboy et al., 1991) are able to form a mineralized bonelike matrix only when a three-dimensional structural organization is created. In contrast, diffuse calcification of the matrix occurs in two-dimentional structures formed by human trabecular (Schulz, 1995) or marrow-derived osteoblastic cells (Cheng et al., 1994; Fromigué et al., 1997; 1998). During *in vitro* osteogenesis, cell contacts with the collagenenous matrix are essential for the induction of the expression of osteoblast marker genes, such as alkaline phosphatase and osteocalcin (Owen et al., 1990a; Harada et al., 1991; Lian et al., 1991; Franceschi and Iyer, 1992; Lynch et al., 1995). Cytoskeletal proteins appear to be involved in the signal transduction induced by cell contact with the matrix, and in the transfer of second messengers to the nucleus, processes that generate changes in gene expression. These complex intracellular mechanisms may implicate interactions between integrins, cytoskeletal proteins and components of signal transduction pathways, such as the tyrosine phosphorylation of cytoplasmic kinases pp125^{FAK} (focal adhesion kinase) (Burridge et al., 1992; Damski and Werb, 1992; Adams and Watt, 1993) (Figure 2). Additionally, hormones, such as PTH and steroids (Lomri and Marie, 1990a,b), and growth factors such as TGFβ (Lomri and Marie, 1990c), have important effects on cytoskeletal proteins and cell shape through second messengers (intracellular calcium and cAMP) that modulate the assembly and biosynthesis of cytoskeletal proteins, or through effects on integrin synthesis (Lomri and Marie, 1996). The transmission of signals to the nucleus in response to hormones or growth factors results in transcription of genes involved in cell differentiation (Figure 2).

Recent reports indicate that cell-matrix interactions may also play an important role as mediators of mechanical forces on osteoblasts. The extracellular matrix is the site where mechanical forces are transmitted to the osteoblasts, and cells lining the bone surface and osteocytes are suitable cells that may sense the changes in the strain induced by mechanical loading (Jones and Bingmann,

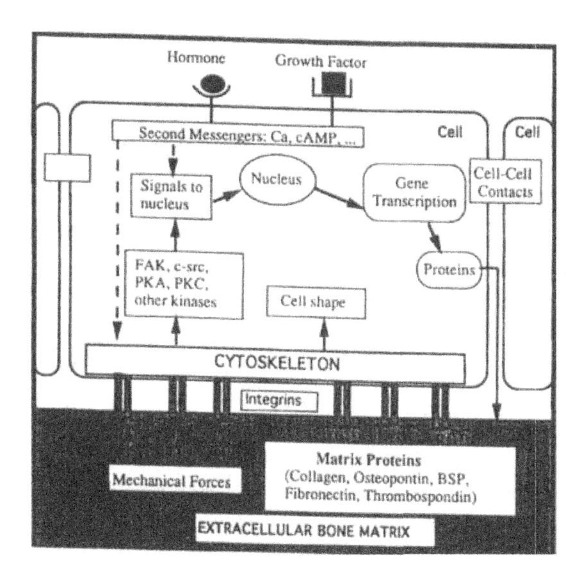

Figure 2. Mechanisms regulating osteoblast differentiation. After binding to their specific receptors, hormones and growth factors induce gene expression through second messengers. Changes in osteoblast differentiation induced by cell attachment or mechanical forces are mediated by complex integrin-cytoskeleton interactions. Cytoskeletal proteins are involved in the signal transduction and in the transfer of second messengers to the nucleus. These intracellular mechanisms implicate interactions between integrins, cytoskeletal proteins, and components of signal transduction pathways, such as the tyrosine phosphorylation of cytoplasmic kinases. Reproduced with kind permission from Lomei, A., and Marie, R.J. (1996). JAI Press, Stamford, CT.

1991). Loading within physiological range stimulates bone formation, and bone cells respond to strain *in vitro* by causing changes in cell alignment, increased rate of cell division, and increased synthesis of noncollagenous proteins. The cellular mechanisms of action by which strain stimulates bone cells is not fully understood. Mechanical forces applied to the bone matrix may transduce signals via integrins that interact with the cytoskeleton (Ingber, 1991; Jones et al., 1991). Transduction of mechanical forces into a biochemical information may be mediated by several second messengers.

In vitro, mechanical strain on bone cells induces signaling events that involve phospholipase C, phospholipase A_2 and protein kinase C activation, elevation of intracellular calcium, cAMP and cGMP levels, and prostaglandin release (Jones and Bingmann, 1991). Prostaglandins are produced in response to mechanical stimulus and may act as mediators of mechanical stimulus on osteoblasts (Rawlinson et al., 1991; Kawaguchi et al., 1995), although peptide growth factors secreted locally have also been implicated. The release of prostaglandins may induce osteoblasts to produce growth factors, such as IGF-I, which may in turn be responsible for the initiation of new bone formation. It is, however, likely that the anabolic effect of mechanical forces on bone formation involves several growth factors acting in a coordinate way on osteoblastic cells at different stages of their differentiation.

B. Development of Osteogenesis

The cellular and molecular analysis of osteoblastic cell differentiation *in vitro* has permitted the determination of the temporal and spatial expression of the particular genes and bone-related proteins during osteogenesis (Aubin and Liu, 1996; Chentoufi et al., 1993; Stein and Lian, 1993; Suva et al., 1993; Malaval et al., 1994; Yao et al., 1994; Stein et al., 1996). The sequence of osteogenic differentiation in these models is characterized by the expression of alkaline phosphatase, an early marker of the osteoblast phenotype, the synthesis of type I collagen, the deposition of an extracellular matrix, and an increase in the expression of osteocalcin and bone sialoprotein at the onset of mineralization. A similar sequence of events has been found to occur during neonatal rat bone development (Machwate et al., 1995c). In human postnatal calvaria, the sequential events involved in the progression of osteogenesis are also characterized by a decline in osteoblastic cell growth followed by a progressive expression of markers of differentiation (de Pollak et al., 1996). This indicates that postnatal development of the calvarium is characterized by an inverse relationship between osteoblast proliferation and differentiation of cells of the osteoblastic lineage (Figure 3). The sequence of gene expression characterizing the development of osteogenesis in human bone, however, remains to be determined.

The final step of osteogenesis involves matrix calcification, a process that requires appropriate mineral supply. Calcium transport from the extracellular fluid to the nucleation sites can be ensured both by passive diffusion and transcellular transport. Calcium pumps have been recently described in osteoblasts and these may play a role in calcium transfer to the bone matrix. Calcium can be extruded by a sodium/calcium exchange transport in osteoblast-like cells (Krieger, 1992). Matrix vesicles and calcium-binding proteins have been found to be associated with cartilage calcification (Balmain et al., 1995) and with the rapid calcification of woven bone, although such vesicles do not appear to be essential for the mineralization of lamellar bone in adults.

The availability of phosphate ions at sites of mineralization is mainly dependent upon their passive transfer from the systemic fluid to the bone extracellular compartment. Recent data indicate, however, that there is an active phosphate transfer in osteoblasts, which is a carrier-mediated saturable process with characteristics of a sodium-dependent cotransport system (Caversazio et al., 1988). This sodium-dependent phosphate transport system in osteoblastic cells is stimulated by PTH, cAMP, and IGF-I; all these molecules increase the number of such carriers in the plasma membrane. Fluoride also stimulates the sodium-dependent phosphate transport, and enhances the stimulatory effect of serum, insulin and IGF-I, independently of their mitogenic action, whereas 1,25$(OH)_2$vitamin D_3 inhibits phosphate uptake via a decreased number of carriers, and causes intracellular phosphate depletion (Green et al., 1993). This precisely regulated system may contribute to the transport of phosphate from the cell to the matrix.

The high alkaline phosphatase activity localized in the lateral domain of osteoblasts indicates that this enzyme may be involved in bone matrix mineralization. *In vitro,* alkaline phosphatase was found to hydrolyze phosphate esters into phosphate and to hydrolyze pyrophosphate, an inhibitor of mineralization, which may lead to increase the local phosphate concentrations. Osteoblastic cells display high alkaline phosphatase activity which is regulated by numerous hormones (Rodan and Noda, 1991). Osteogenesis *in vitro* requires high phosphate concentrations or a substrate for alkaline phosphatase (β glycero-phosphate) (Bellows et al., 1992). Note that alkaline phosphatase expression is increased during initial mineralization of the extracellular matrix *in vitro* (Owen et al., 1990a). These *in vitro* studies, together with *in vivo* studies showing that partial suppression of the bone type alkaline phosphatase results in inhibtion of bone mineralization (Garba et al., 1986), supports a potential, but not essential, role for the enzyme in bone matrix calcification.

In the matrix, collagen and noncollagenous proteins may contribute to the initiation of bone mineralization through their calcium-binding properties (Gehron-Robey, 1989; Glimcher, 1989). In particular, osteopontin and bone sialoprotein have been shown to be expressed at the late stage of osteoblast differentiation and at the onset of calcification during osteogenesis (Bianco et al., 1989; Kasugai et al., 1991). Although osteocalcin expresion is also increased prior to calcification *in vitro* (Collin et al., 1992), this protein is not clearly associated with bone spots of mineralization *in vivo,* which precludes a clear understanding of the exact contribution of this protein in bone matrix calcification.

C. Control of Osteoblast Differentiation

A number of hormones have been found to modulate the differentiation of osteoblastic cells. Osteoblastic responsiveness has been found to depend on the stage of cell differentiation (Rodan and Noda, 1991; Martin et al., 1993), which suggests that these hormones induce the osteoblast phenotype at some restricted stages of osteoblast differentiation (Owen et al., 1991). For example, glucocorticoids promote the differentiation of early osteoblast precursors in the marrow stroma (Kamalia et al., 1992) and inhibit collagen synthesis in more differentiated osteoblasts. In contrast, 1,25-dihydoxyvitamin D and retinoic acid induce several markers of the osteoblast phenotype in more differentiated cells and reduce cell proliferation (Chentoufi and Marie, 1994). The different responsiveness to these factors along the osteoblast differentiation pathway may be due to changes in receptor levels. Receptors for PTH, vitamin D, estrogens, progesterone, thyroid hormone, retinoic acid, and glucocorticoids have been found in osteoblastic cells (Rodan and Noda, 1991). Rat osteoblastic cells have been recently found to express receptors for mineralocorticoids. Aldosterone regulates cell proliferation and alkaline phosphatase activity in these cells, suggesting that these steroids may also modulate osteoblast differentiation (Agarwal et al., 1996).

Most of these hormones may directly induce the expression of genes of the osteoblast phenotype through transcriptional effects (Rodan and Noda, 1991; Lian and Stein, 1992). The intracellular mechanisms involved in the actions of these hormones are now better understood (Partridge et al., 1994; Siddhanti and Quarles, 1994). In addition to inducing direct effects, several hormones, such as sex hormones, growth hormone, PTH, glucocorticoids, and vitamin D, indirectly modulate the differentiation of osteoblastic cells through changes in the synthesis of IGF and TGFβ (Ernst et al., 1989; Tremollières et al., 1992; Bodine et al., 1995). Only a few growth factors have been shown to promote osteoblast differentiation directly. The latter include IGFs, TGFβs, and BMPs, which are potent stimulators of bone matrix protein synthesis in osteoblastic cell cultures (Schmid and Ernst, 1990; Canalis et al., 1991; Mundy, 1995). *In vitro*, TGFβ appears to act at different levels, since it regulates human bone marrow derived osteoprogenitor cells (Long et al., 1995), increases the synthesis of bone matrix proteins, and reduces the rate of matrix degradation (Centrella et al., 1991).

Other factors may, however, be involved in the induction or regulation of osteoblast differentiation. For example, PTHrP was found to be expressed and produced by human osteoblastic cells (Walsh et al., 1994a; Lomri et al., 1997) and may be implicated as an autocrine modulator of osteoblast differentiation. The finding that vascular endothelial cell growth factor (VEGF) is produced by rat osteoblasts and is regulated by glucocorticoids suggests that VEGF may have a role in the local regulation of osteoblast differentiation and proliferation (Harada et al., 1994).

Several studies indicate that BMPs may play a crucial role in the promotion of osteoblast differentiation and osteogenesis. Namely, BMP-2, -3, and -7 were shown to induce the expression of osteoblastic markers in uncommitted mesenchymal cells *in vitro* and to promote cell differentiation in more mature osteoblastic cells (Rosen and Thies, 1992; Wozney, 1992; Rosen et al., 1996; Marie, 1997a). TGFβs and BMPs may act sequentially on osteoblastic cells to induce successive stages of differentiation during osteogenesis *in vitro* (Ghosh-Choudhury et al., 1994; Fromigué et al., 1998). TGFβ appears to downregulate the expression of BMP-2 *in vitro,* which can account for the lack of stimulation of osteogenesis in rat calvaria cells (Harris et al., 1994). Thus, it is likely that BMPs are important factors controlling not only skeletal development, but also osteoblast differentiation during postnatal osteogenesis. The mechanism of action, receptor mediation, and regulation of BMP expression in osteoblasts during osteogenesis are not fully understood. Recent data indicate that transmission of signals through different BMP receptors play a critical role in the differentiation of osteoblasts or adipocytes (Chen et al., 1998). Thus, from our present knowledge, it appears that the development of osteogenesis in a spatially and temporally organized manner results from the actions of multiple local and systemic factors acting in a coordinate manner on target cells.

V. CONCLUSIONS AND PERSPECTIVES

Recent advances in the cellular and molecular biology of the skeleton in normal and pathologic conditions have led to important progress in our understanding of osteoblast biology and of the mechanisms controlling bone formation. The development of endosteal bone-forming cell cultures resulted in the demonstration that the proliferation of osteoprogenitor cells is an important limiting step controlling the rate of bone formation. Moreover, the main factors involved in the recruitment and proliferation of osteoblast precursor cells have been identified. Some of the complex mechanisms involved in the induction of the osteoblast phenotype, the sequence of genes expressed during the differentiation of osteoblasts, and the development of osteogenesis have been clarified. In addition, the actions of hormones and growth factors at the different steps of cell proliferation and differentiation are now better understood.

However, a number of questions concerning osteoblast recruitment and differentiation still remain unanswered. First, the identification of osteoprogenitor cells and the development of specific markers of early osteoblast precursor cells is still lacking. A better understanding of stem cells for osteoblasts may permit the generation of new strategies for increasing the pool of osteoblast precursor cells, and hence, for stimulating bone formation. On the other hand, the precise mechanisms controlling the amount of bone matrix deposited by osteoblasts in resorptive lacunae during the formation phase remains unknown. An answer to this question may help in defining possible ways to prolong the duration of the formation phase. In addition, we are far from understanding the cellular and molecular mechanisms involved in the induction of bone formation during bone development or remodeling

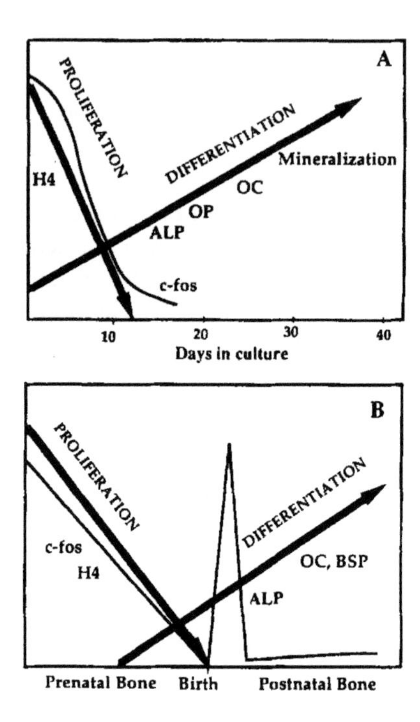

Figure 3. Expression of the osteoblast phenotype during osteogenesis *in vitro* (**A**) (Owen et al., 1990a) and during neonatal osteogenesis *in vivo* (**B**) (Machwate et al., 1995c). The early developement of osteogenesis *in vitro* and *in vivo* is characterized by a high rate of the proliferation of osteoblast precursor cells which declines progressively with time, and this is associated with decreased expression of H4 and *c-fos* gene expression. *In vivo, c-fos* is transiently expressed before the postnatal development of bone formation. Both *in vitro* and *in vivo,* osteoblast differentiation is characterized by the early induction of alkaline phosphatase activity (ALP), followed by expression of osteopontin (OP), bone sialoprotein (BSP), and osteocalcin (OC) during the late stages of differentiation and osteogenesis.

in vivo. The identification of the molecular mechanisms controlling the transition from cell proliferation to differentiation may clarify this issue. It is also likely that the determination of the expression of genes and phenotypic abnormalities in genetic skeletal disorders affecting bone formation will provide new information on the molecular process controlling bone formation in vivo (Marie, 1998). Finally, there is a need to determine the role and mechanisms of action of local growth factors on bone formation *in vivo.* More research on osteoblast cell biology is therefore required to determine the cellular and molecular mechanisms involved in pathological bone formation in local and metabolic bone diseases. This may permit the identification of the local factors involved in the age-related decrease in bone formation and study of the actions of specific growth factors able to stimulate endosteal bone formation *in vivo*. This may lead, in the long term, to the development of new strategies to increase the number of osteoblasts and to optimally induce new bone formation in osteopenic disorders.

VI. SUMMARY

Models of osteoblast differentiation have shown that osteogenesis *in vitro* and *in vivo* is a time and spatially regulated process characterized by the succession of coordinated events involving the proliferation of osteoprogenitor cells, the differentiation of preosteoblasts into differentiated osteoblasts, and the apposition of a calcified bone matrix. Evidence has accumulated from human and animal studies that osteoprogenitor cell proliferation is the more important limiting step controlling the rate of bone formation. *In vitro* models have provided valuable information on the role of growth factors in osteoblast recruitment. Current available literature suggests that the transition from cell proliferation to differentiation involves transcription factors, but the nature of the local factors involved in the induction of osteoblast differentiation is not yet known. Recent data also indicate that the differentiation of osteoblasts is induced by cell-cell communication and cell-matrix interactions involving complex cytoskeleton-integrins interactions. In addition, numerous cell-matrix interactions and local factors appear to be involved as mediators of the stimulatory effect of mechanical forces on bone formation. Finally, information has been gleaned on the effect of hormones and local factors in controlling those genes that appear to regulate osteoblast differentiation. The cellular mechanisms of action and signal transduction pathways of most of these factors are now better understood. It is likely that an improved understanding of the cellular and molecular mechanisms involved in bone formation in skeletal genetic disorders will provide further insights into the control of osteoblast generation and function and may ultimately help conceive new therapeutic strategies to stimulate bone formation in ossteopenic disorders.

ACKNOWLEDGMENTS

The author's research is supported by grants from INSERM and CNES. The author wishes to acknowledge A. Lomri, M. Machwate, J. Chentoufi, M. Hott, D. Modrowski, D. Godet, F. Debiais, J. Lemonnier, E. Hay, and C. dePollak for their valuable contribution to the work reported in this review.

REFERENCES

Adams, J.C. and Watt, F.M. (1993). Regulation of development and differentiation by the extracellular matrix. Development 117, 1183-1198.

Agarwal M.K., Mirshahi F., Mirshahi M., Bracq S., Chentoufi J., Hott M., Jullienne A., Marie P.J. (1996). Evidence for receptor-mediated mineralocorticoid action in rat osteoblastic cells. Am. J. Physiol. (In press.)

Amizuka, N., Warshawsky, H., Henderson, J.E., Goltzman, D., and Karaplis, A.C. (1994). Parathyroid hormone-related peptide-depleted mice show abnormal epiphyseal cartilage development and altered endochondral bone formation. J. Cell Biol. 126, 1611-1623.

Andrianarivo, A.G., Robinson, J.A., Mann, K.G., and Tracy, R.P. (1992). Growth on type-I collagen promotes expression of the osteoblastic phenotype in human osteosarcoma MG-63 cells. J. Cell. Physiol. 153, 256-265.

Angel, P. and Karin, M. (1991) The role of Jun, Fos, and the AP-1 complex in cell-proliferation and transformation. Biochim. Biophys. Acta 1072, 129-157.

Aubin, J.E., and Liu, F. (1996). The osteoblast lineage. In: *Principles of Bone Biology*, (Bilezikia, Raisz, and Rodan, Eds.), pp. 51-67. Academic Press, New York.

Bab, I. and Einhorn, T.A. (1993). Regulatory role of osteogenic growth polypeptides in bone formation and hemopoiesis. Crit. Rev. Eukaryot Gene Expression 3, 31-46.

Balmain, N., Von Eichel, B., Toury, R., Belquasmi, F., Hauchecorne, M., Klaus, G., Mehls, O., Ritz, E. (1995). Calbindin-D28K and -D9K and 1,25(OH)2 vitamin D3 receptor immunolocalization and mineralization induction in long term primary cultures of rat epiphyseal chondrocytes. Bone 17, 37-45.

Bellows, C.G., Aubin, J.E., and Heersche J.N.M. (1991). Initiation and progression of mineralization of bone nodules formed in vitro: The role of alkaline phosphatase and organic phosphate. Bone Miner. 14, 27-40.

Bellows, C.G., Heersche, J.N.M., and Aubin, J.E. (1992). Inorganic phosphate added exogenously or released from β-glycerophosphate initiates mineralization of osteoid nodules in vitro. Bone and Mineral 17, 15-29.

Beresford, J.N. (1989). Osteogenic stem cells and the stromal system of bone and marrow. Clin. Orthop. 240, 270-280.

Beresford, J.N., Bennett, J.H., Devlin, C., Leboy, P.S., and Owen, M.E. (1992). Evidence for an inverse relationship between the differentiation of adipocytic and osteogenic cells in rat marrow stromal cell cultures. J. Cell Sci. 102, 341-351.

Bodine, P.V.N., Riggs, L.B., and Spelsberg, T.C. (1995). Regulation of c-fos expression and TGF-β production by gonadal and adrenal androgens in normal human osteoblastic cells. J. Steroid Biochem. Molec. Biol. 52, 149-155.

Bianco, P., Fishert, L.W., and Young, M.F. (1989). Expression of bone sialoprotein in human developing bone as revealed by immunostaining and in situ hybridization. J. Bone Miner. Res. 4, 246-249.

Bonewald, L.F. and Dallas, S.L. (1994). Role of active and latent transforming growth factor in bone formation. J. Cell. Biochem. 55, 350-360.

Bruder, S. and Caplan, A. (1989). A monoclonal antibody against the surface of osteoblasts recognises alkaline phosphatase isoenzymes in bone, liver, kidney, and intestine. Bone 11, 133-139.

Burridge, K., Petch, L.A., and Romer, L.H. (1992). Signals from focal adhesions. Current Opinion Cell Biol. 2, 537-539.

Canalis, E., McCarthy, T.L., Centrella, M. (1991). Growth factors and cytokines in bone cell metabolism. Annu. Rev. Med. 42, 17-24.

Candelière, G.A., Glorieux, F.H., Prud'homme, J., and St. Arnaud, R. (1995). Increased expression of the c-fos proto-oncogene in bone from patients with fibrous dysplasia. N. Eng. J. Med. 332, 1546-1578.

Caversazio, J., Seltz, T., and Bonjour, J-P. (1988). Characteristics of phosphate transport in osteoblastlike cells. Cacif. Tissue Int. 43, 83-87.

Centrella, M., McCarthy, T., and Canalis, E. (1991). Current concepts review: Transforming growth factor β and remodeling of bone. J. Bone Joint. Surg. 73A, 1418-1428.

Centrella, M., Horowitz, M.C., Wozney, J.M., and McCarthy, T.L. (1994). Transforming growth factor-β gene family members and bone. Endocrine Rev. 15, 27-39.

Chaudhary, L., Spelsberg, T.C., and Riggs, B.L. (1992). Production of various cytokines by normal human osteoblastlike cells in response to interleukin-1β and tumor necrosis factor-α: Lack of regulation by 17β-estradiol. Endocrinology 130, 2528-2534.

Chen, D., Ji, X., Harris, M.A., Feng, J.K., Karsenty, G., Celeste, A.J., Rosen, V., Mundy, G.R., and Harris, S.E. (1998). Differential roles for bone morphogenetic protein (BMP) receptor type IB and IA in differentiation and specification of mesenchymal precursor cells to osteoblast and adipocyte lineages. J. Cell Biol. 142(1), 295-305.

Cheng, S.L., Yang, J.W., Rifas, L., Zhang, S.F., and Avioli, L.V. (1994). Differentiation of human bone marrow osteogenic stromal cells in vitro: Induction of the osteoblast phenotype by dexamethasone. Endocrinology 134, 277-286.

Chentoufi, J., Hott, M., Lamblin, D., Buc-Caron, M.H., Marie, P.J., and Kellermann, O. (1993). Kinetics of in vitro mineralization by an osteogenic clonal cell line (C1) derived from mouse teratocarcinoma. Differentiation 53, 181-189.

Chentoufi, J. and Marie, P.J. (1994). Interactions between retinoic acid and 1,25(OH)2D in mouse immortalized osteoblastic C1 cells. Am. J. Physiol. 266 (Cell Physiol. 35), 1247-1256.

Civitelli, R., Beyer, E.C., Warlow, P.M., Robertson, A.J., Geist, S.T., and Steinberg, T.H. (1993). Connexin 43 mediates direct intercellular communication in human osteoblastic cell networks. J. Clin. Invest. 91, 1888-1896.

Closs, E.I., Murray, B.A., Schmidt, J., Schon, A., Erfle, V., and Strauss, P.G. (1990). C-fos expression precedes osteogenic differentiation of cartilage cells in vitro. J. Cell Biol. 111, 1313-1323.

Collin, P., Nefussi, J.R., Wetterwald, A., Nicolas, V., Boy-Lefevre, M.L., Fleisch, H., and Forest, N. (1992). Expression of collagen, osteocalcin, and bone alkaline phosphatase in a mineralizing rat osteoblastic cell culture. Calcif. Tissue Int. 50, 175-183.

Damsky, C.H. and Werb, Z. (1992). Signal transduction by integrin receptors for extracellular matrix: Cooperative processing of extracellular information. Curr. Opin. Cell Biol. 4, 772-781.

Debiais, F., Hott, M., Graulet, A.M., and Marie, P.J. (1998). Fibroblast growth factor-2 differently affects human neonatal calvaria osteoblastic cells depending on the stage of cell differentiation. J. Bone Miner. Res. 13(4), 645-654.

De Pollak, C., Renier, D., Hott, M., and Marie, P.J. (1996). Increased bone formation and osteoblastic cell phenotype in premature cranial ossification (craniosynostosis). J. Bone Min. Res. 11, 401-407.

De Pollak, C., Arnaud, E., Renier, D., and Marie, P.J. (1997). Age-related changes in bone formation, osteoblastic cell proliferation, and differentiation during postnatal osteogenesis in human calvaria. J. Cell. Biochem. 64, 128-139.

Ducy, P., Desbois, C., Boyce, B., Pinero, G., Story, B., Dunstan, C., Smith, E., Bonadio, J., Goldstein, S., Gundberg, C., Bradley, A., and Karsenty, G. (1996). Increased bone formation in osteocalcin-deficient mice. Nature 382, 448-452.

Ducy, P., Zhang, R., Geoffroy, V., Ridall, A.L., and Karsenty, G. (1997). Osf2/Cbf1: A transcriptional activator of osteoblast differentiation. Cell 89, 747-754.

Erlebacher, A., Filvaroff, E.H., Gitelman, S.E., and Derynck, R. (1995). Toward a molecular understanding of skeletal development. Cell 80, 371-378.

Ernst, M., Heath, J.K., Schmid, C., Froesch, R.E., and Rodan, G.A. (1989). Evidence for a direct effect of estrogen on bone cells in vitro. J. Steroid Biochem. 34, 279-284.

Evans, C.E., Galasko, C.S.D., and Ward, C. (1990). Effect of donor age on the growth in vitro of cells obtained from human trabecular bone. J. Orthopaedic Res. 8, 234-237.

Falla, N., van Vlasselaer, P.., Bierkens, J., Borremans, B., Schoeters, G., and van Gorp, U. (1993). Characterization of a 5-fluorouracil-enriched osteoprogenitor population of the murine bone marrow. Blood 82, 3580-3591.

Fedarko, N.S., Vetter, U.K., Weibstein, S., and Gehron-Robey, P. (1992). Age-related changes in hyaluronan, proteoglycan, collagen, and osteonectin synthesis by human bone cells. J. Cell. Physiol. 151, 215-227.

Franceschi, R.T. and Iyer, B.S. (1992). Relationship between collagen synthesis and expression of the osteoblast phenotype in MC3T3-E1 cells. J. Bone Min. Res. 7, 235-240.

Fromigué, O., Marie, P.J., and Lomri, A. (1997). Differential effects of transforming growth factor-β, 1,25-dihydroxyvitamin D, and dexamethasone on human bone marrow stromal cells. Cytokine 9(8), 613-623.

Fromigué, O., Marie, P.J., and Lomri, A. (1998). Bone morphogenetic protein-2 and transforming growth factor β2 interact to modulate human bone marrow stromal cell proliferation and differentiation. J. Cell Biochem. 68, 411-426.

Garba, M.T. and Marie, P.J. (1986). Alkaline phosphatase inhibition by levamisole prevents 1,25-dihydroxyvitamin D3-stimulated bone mineralization in the mouse. Calcif. Tissue Int. 38, 296-302.

Gehron-Robey, P. (1989). The biochemistry of bone. Endocrinol. Metab. Clinics of North America 18, 859-902.

Ghosh-Choudhury, N., Harris, M.A., Feng, J.Q., Mundy, G.R., and Harris, S.E. (1994). Expression of the *BMP2* gene during bone cell differentiation. Crit. Rev. Euk. Gene Exp. 4, 345-355.

Glimcher, M.J. (1989). Mechanism of calcification: Role of collagen fibrils and collagen-phosphoprotein complexes in vitro and in vivo. Anat. Rec. 224, 139-153.

Godet, D. and Marie, P.J. (1995). Effects of the tripeptide glycyl-L-Histidyl-L-Lysine copper complex on osteoblastic cell spreading, attachment, and phenotype (1995). Cell. Mol. Biol. 41, 1081-1091.

Goldring, M.B. and Goldring, S.R. (1990). Skeletal tissue response to cytokines. Clin Orthop 258, 245-275.

Gowen, M. (Ed.) (1992). *Cytokines and Bone Metabolism*. pp. 1-417. CRC Press, Ann Arbor, Michigan.

Green, J., Luong, K.V.Q., Kleeman, C.R., Ye, L-H., and Chaimovitz, C. (1993). 1,25-dihydroxyvitamin D3 inhibits Na+-dependent phosphate transport in osteoblastic cells. Am. J. Physiol. 264 (Cell Physiol. 33), C287-C295.

Grigoriadis, A.E., Schellander, K., Wang, Z.Q., and Wagner, E.F. (1993). Osteoblasts are target cells for transformation in c-fos transgenic mice. J. Cell Biol. 122, 685-701.

Grigoriadis, A.E., Heersche, J.N.M., and Aubin, J.E. (1988). Differentiation of muscle, fat, cartilage and bone progenitor cells present in a bone-derived clonal cell population: Effect of dexamethasone. J. Cell Biol. 106, 2139-2151.

Grzesik, W.J. and Gehron-Robey, P. (1994). Bone matrix RGD glycoproteins: Immunolocalization and interaction with human primary osteoblastic bone cells in vitro. J. Bone Min. Res. 9, 487-495.

Halstead, L.R., Scott, M.J., Rifas, L., Avioli, L.V. (1992). Characterization of osteoblastlike cells from normal adult rat femoral trabecular bone. Calcif. Tissue Int. 50, 93-95.

Harada, S-I., Matsumoto, T., and Ogata, E. (1991). Role of ascorbic acid in the regulation of proliferation in osteoblastlike MC3T3-E1 cells. J. Bone Min. Res. 6, 903-908.

Harada, S., Nagy, J.A., Sullivan, K.A., Thomas, K.A., Endo, N., Rodan, G.A., and Rodan, S.B. (1994). Induction of vascular endothelial growth factor expression by prostaglandin E2 and E1 in osteoblasts. J. Clin. Invest. 93, 2490-2496.

Harris, S.E., Bonewald, L.F., Harris, M.A., Sabatini, M., Dallas, S., Feng, J.Q., Ghosh-Choudhury, N., Wozney, J., and Mundy, G.R. (1994). Effects of transforming growth factor β on bone nodule formation and expression of bone morphogenetic protein 2, osteocalcin, osteopontin, alkaline phosphatase, and type-I collagen mRNA in long-term cultures of fetal rat calvarial osteoblasts. J. Bone Miner. Res. 9, 855-863

Harris, S.E., Bonewald, L.F., Harris, M.A., Sabatini, M., Dallas, S., Feng, J., Ghosh-Choudhury, Hodgkinson, J.E., Davidson, C.L., Beresford, J., and Sharpe, P.T. (1993). Expression of a human homeobox-containing gene is regulated by $1{,}25(OH)_2D_3$ in bone cells. Biochim. Biophys. Acta 1174, 11-16.

Hodgkinson, J.E., Davidson, C.L., Beresford, J., Sharpe, P.T. (1993). Expression of a human homeobox-containing gene is regulated by 1,25(OH)2D3 in bone cells. Biochim. Biophys. Acta 1174, 11-16.

Horowitz, M.C., Coleman, D.L., Flood, P.M., Kupper, T.S., and Jilka, R.L. (1989). Parathyroid hormone and lipopolysaccharide induce murine osteoblastlike cells to secrete a cytokine indistinguishable from granulocyte-macrophage colony-stimulating factor. J. Clin. Invest. 83, 149-157.

Hughes, D.E., Salter, D.M., Dedphar, S., and Simpson, R. (1993). Integrin expression in human bone. J. Bone Miner. Res. 8, 527-533.

Ingber, D. (1991). Integrins as mechanochemical transducers. Curr. Opin. Cell Biol. 3, 841-848.

Jacenko, O., Olsen, B.R., and Warman, M.L. (1994). Of mice and men: Heritable skeletal disorders. Am. J. Hum. Genet. 54, 163-168.

Jackson, M.E., Shalhoub, V., Lian, J.B., Stein, G.S., and Marks, S.C. (1994). Aberrant gene expression in cultured mamallian bone cells demonstrates an osteoblast defect in osteopetrosis. J. Cell. Biochem. 55, 366-372.

Johnson, R.S., Spiegelman, B.M., and Papaioannou, V. (1992). Pleiotropic effects of a null mutation in the c-fos proto-oncogene. Cell 71, 577-586.

Johnston, R.L., and Tabin, C.J. (1997). Molelcular models for vertebrate limb development. Cell 90, 979-990.

Jones, D.B. and Bingmann, D. (1991). How do osteoblasts respond to mechanical stimulation? Cells Materials 1, 329-340.

Jones, D.B., Nolte, H., Scholbbers, J.G., Turner, E., and Veltel, D. (1991). Biochemical signal transduction of mechanical strain in osteoblastlike cells. Biomaterials 12, 101-110.

Kahn, A., Gibbons, R., Perkins, S., and Gazit D. (1995). Age-related bone loss. A hypothesis and initial assessment in mice. Clin. Oth. Rel. Res. 313, 69-75.

Kamalia, N., McCulloch, C.A.G., Tenenbaum, H.C., and Limeback, H. (1992). Dexamethasone recruitment of self-renewing osteoprogenitor cells in chick bone marrow stromal cell culture. Blood 79, 320-326.

Kano, J., Sugimoto, T., Kanatani, M., Kuroki, Y., Tsukamoto, T., Fukase, M., and Chihara, K. (1994). Second messenger signaling of *c-fos* gene induction by parathyroid hormone (PTH) and PTH-related peptide in osteoblastic ostesarcoma cells: Its role in osteoblast proliferation and osteoclastlike cell formation. J. Cell. Physiol. 161, 358-366.

Karaplis, A.C., (1996). Gene targeting. In: *Principles of Bone Biology*. (Bilezikian, J.P., Raisz, L.G., Rodan, G.A., (Eds.), pp. 1189-1202. Academic Press, New York.

Kato, H., Matsuo, R., Komiyama, O., Tanaka, T., Inazu, M., Kitagawa, H., and Yoneda, T. (1995). Decreased mitogenic and osteogenic responsiveness of calvarial osteoblasts isolated from aged rats to basic fibroblast growth factor. Geront. 41, 20-27.

Kawaguchi, H., Pilbeam, C.C., Harrison, J.R., and Raisz, L.G. (1995). The role of prostaglandins in the regulation of bone metabolism. Clin. Oth. Rel. Res. 313, 36-46.

Kasugai, S.,Todescan, R., Nagata, T., Yao, K.L., Butler, W.T., and Sodek, J. (1991). Expression of bone matrix proteins associated with mineralized tissue formation by adult rat bone marow cells in vitro: Inductive effects of dexamethasone on the osteoblastic phenotype. J. Cell. Physiol. 147, 111-120.

Keeting, P.E., Rifas, L., Harris, S.A., Colvard, D.S., Spelsberg, T.C., Peck, W.A., and Riggs, B.L. (1991). Evidence of interleukin-1 production by cultured normal human osteoblastlike cells. J. Bone Mineral Res. 6, 827-833.

Keila, S., Pitaru, S., Grosskopf, A., and Weinreb, M. (1994). Bone marow from mechanically unloaded rat bones express reduced osteogenic capacity in vitro. J. Bone Miner. Res. 9, 321-327.

Kellermann, O., Buc-Caron, M.H., Marie, P.J., Lamblin, D., and Jacob, F. (1990). An immortalized osteogenic cell line derived from mouse teratocarcinoma is able to mineralize in vivo and in vitro. J. Cell Biol. 110, 123-132.

Krieger, N.S. (1991). Evidence for sodium-calcium exchange in rodent osteoblasts. Ann. NY. Acad. Sci. 639, 660-662.

Krumlauf, R. (1994). Hox genes in vertebrate development. Cell 78, 191-201.

Lane, T.F., Iruela-Arispe, M.L., Johnson, R.S., and Sage, E.H. (1994). SPARC is a source of copper-binding peptides that stimulate osteogenesis. J. Cell Biol. 125, 929-943.

Lazowski, D.A., Fraher, L/J., Hodsman, A., Steer, B., Modrowski, D., and Han, V.K. (1994). Regional variation of insulinlike growth hormone factor-I gene expression in mature rat bone and cartilage. Bone 15, 563-576.

Leboy, P.S., Beresford, J.N., Devlin, C., and Owen, M.E. (1991). Dexamethasone induction of osteoblast mRNAs in rat marrow stromal cell cultures. J. Cell. Physiol. 146, 370-378.

Lee, K., Deeds, J.D., Chiba, S., Un-No, M., Bond, A.T., and Segre, G.V. (1994). Parathyroid hormone induces sequential c-fos expression in bone cells in vivo: In situ localization of its receptor and c-fos messenger ribonucleic acids. Endocrinology 134, 441-450.

Lee, K., Deeds, J.D., and Segre, G.V. (1995). Expression of parathyroid hormone-related peptide and its receptor messenger ribonucleic acids during fetal development of rats. Endocrinology 136, 453-463.

Lian, J.B. and Stein, G.S. (1992). Transcriptional control of vitamin D-regulated proteins. J. Cell. Biochem. 49, 37-45.

Lian, J.B., Stein, G.S., Bortell, R., and Owen, T.A. (1991). Phenotype suppression: A postulated molecular mechanism for mediating the relationship of proliferation and differentiation by fos/jun interactions at AP-1 sites in steroid responsive promoters elements of tissue-specific genes. J. Cell. Biochem. 45, 9-14.

Liu, F., Malaval, L., Gupta, A.K., and Aubin, J.E. (1994). Simultaneous detection of multiple bone-related mRNAs and protein expression during osteoblast differentiation: Polymerase chain reaction and immunocytochemical studies at the single cell level. Dev. Biol. 166, 220-234.

Littlewood, A.J., Aarden, L.A., Evans, D.B., Russell, R.G.G., and Gowen, M. (1991). Human osteoblastlike cells do not respond to interleukin-6. J. Bone Miner. Res. 6, 141-148.

Lomri, A., Marie, P.J., Escurat, M., and Portier, M.M. (1987). Cytoskeletal protein synthesis and organization in cultured mouse osteoblastic cells: Effects of cell density. Febs Letters 222, 311-316.

Lomri, A., Marie, P.J., Tran, P.V., and Hott, M. (1988). Characterization of endosteal osteoblastic cells isolated from mouse caudal vertebrae. Bone 9, 165-175.

Lomri, A. and Marie, P.J. (1990a). Changes in cytoskeletal proteins in response to PTH and $1,25(OH)_2D$ in human osteoblastic cells. Bone and Mineral 10, 1-12.

Lomri, A. and Marie, P.J. (1990b). Distinct effects of calcium- and cAMP-enhancing factors on cytoskeletal assembly and synthesis in mouse osteoblastic cells. Biochim. Biophys. Acta 1052, 179-186.

Lomri, A. and Marie, P.J. (1990c). Effects of TGFβ on expression of cytoskeletal proteins in endosteal mouse osteoblastic cells. Bone 11, 445-451.

Lomri, A. and Marie, P.J. (1996). Cytoskeleton in bone cell biology. In: *The Cytoskeleton.* (Hesketh, J.E. and Pyme, I., Eds.), pp. 229-264. JAI Press, Stamford, CT.

Lomri, A., de Pollak, C., Goltzman, D., Kremer, R., and Marie, P.J. (1997). Expression of PTHrP and PTH/PTHrP receptor in newborn human calvaria osteoblastic cells. Eur. J. Endocrinol. 136(8), 640-648.

Lomri, A., Lemonnier, J., Hott, M., de Perseval, N., Lajeunie, E., Munnich, A., Renier, D., and Marie, P.J. (1998). Increased calvaria cell differentiation and bone matrix formation induced by fibroblast growth factor receptor-2 mutations in Apert syndrome. J. Clin. Invest. 101, 1310-1317.

Long, M.W., Robinson, L.A., Ashcraft, E.A., and Mann, K.G. (1995). Regulation of human bone marrow-derived osteoprogenitor cells by osteogenic growth factors. J. Clin. Invest. 95, 881-887.

Lynch, M.P., Stein, J.L., Stein G.S., and Lian, J.B. (1995). The influence of type-I collagen on the development and maintenance of the osteoblast phenotype in primary and passaged rat calvarial osteoblasts: Modification of expression of genes supporting cell growth, adhesion, and extracellular matrix mineralization. Exp. Cell Res. 216, 35-45.

Lyons, K.M., Pelton, R.W., and Hogan, B.L.M. (1989). Patterns of expression of murine Vgr-1 and BMP-2a RNA suggest that transforming growth factor-β-like genes coordinately regulate aspects of embryonic development. Genes Dev. 3, 1657-1668.

Lyons, K.M., Michael Jones, C., and Hogan, B.L.M. (1991) The DVR gene family in embryonic development. Trends Gene 7, 408.

Machwate, M., Zerath, E., Holy, X., Hott, M., Modrowski, D., Malouvier, A., and Marie, P.J. (1993). Skeletal unloading in rat decreases proliferation of rat bone and marrow-derived osteoblastic cells. Am. J. Physiol. 264 (Endocrinol. Metab. 27), E790-E799.

Machwate, M., Zerath, E., Holy, X., Pastoureau, P., Marie, P.J. (1994). Insulinlike growth factor-I increases trabecular bone formation and osteoblastic cell proliferation in unloaded rats. Endocrinology 134, 1031-1038.

Machwate, M., Zerath, E., Holy, X., Hott, M., Godet, D., Lomri, A., and Marie, P.J. (1995a). Systemic administration of transforming growth factor-β 2 prevents the impaired bone formation and osteopenia by unloading in rats. J. Clin. Invest. 96, 1245-1259.

Machwate, M., Jullienne A., Moukhtar M., Lomri A., and Marie P.J. (1995b). c-fos proto-oncogene is involved in the mitogenic effect of transforming growth factor-β in osteoblastic cells. Molec. Endo. 9, 187-198.

Machwate, M., Jullienne, A., Moukhtar, M., and Marie, P.J. (1995c). Temporal variation of c-fos proto-oncogene expression during osteoblast differentiation and osteogenesis in developing bone. J. Cell. Biochem. 57, 62-70.

Malaval, L., Modrowski, D., Gupta, A., Aubin, J.E. (1994). Cellular expression of bone-related proteins in vitro osteogenesis in rat bone marrow stromal cell cultures. J. Cell. Physiol. 158, 555-572.

Maniatopoulos, C., Sodek, J., and Melcher, A.H. (1988). Bone formation in vitro by stromal cells obtained from bone marrow of young adult rats. Cell Tissue Res. 254, 317-330.

Marcelli, C., Yates, A.J.P., and Mundy, G.R. (1990). In vivo effects of human recombinant transforming growth factor- on bone turnover in normal mice. J. Bone Miner. Res. 5, 1087-1096.

Marie, P.J., Sabbagh, A., de Vernejoul, M.C., and Lomri, A. (1989a). Osteocalcin and deoxyribonucleic acid synthesis in vitro and histomorphometric indices of bone formation in postmenopausal osteoporosis. J. Clin. Endocrinol. Metab. 69, 272-279.

Marie, P.J., Hott, M., and Perheentupa, J. (1989b). Effects of epidermal growth factor on bone formation and resorption in vivo. Am. J.Physiol. 258, E275-E281.

Marie, P.J., Lomri, A., Sabbagh, A., and Basle, M. (1989c). Culture and behavior of osteoblastic cells isolated from normal trabecular bone surfaces. In Vitro Cell. Biol. Dev. 25, 373-380.

Marie, P.J., de Vernejoul, M.C., Connes, D., and Hott, M. (1991). Decreased DNA synthesis by cultured osteoblastic cells in eugonadal osteoporotic men with defective bone formation. J. Clin. Invest. 88, 1167-1172.

Marie, P.J., de Vernejoul, M.C., and Lomri, A. (1992). Stimulation of bone formation in osteoporosis patients treated with fluoride associated with increased DNA synthesis by osteoblastic cells in vitro. J. Bone Miner. Res. 7, 103-113.

Marie, P.J., Hott, M., Launay, J.M., Graulet, A.M., and Gueris, J. (1993). In vitro production of cytokines by bone surface-derived osteoblastic cells in normal and osteoporotic postmenopausal women: Relationship with cell proliferation. J. Clin. Endocrinol. Metab. 77, 824-830.

Marie, P.J. and de Vernejoul, M.C. (1993a). Proliferation of bone surface-derived osteoblastic cells and control of bone formation. Bone 14, 463-468.

Marie, P.J. and de Vernejoul, M.C. (1993b). Local factors influencing bone remodeling. Rev. Rhum. 60, 1, 55-63.

Marie, P.J., Hott, M., and Lomri, A. (1994). Regulation of endosteal bone formation and osteblasts in rodent vertebrae. Cells Mater. 4, 143-154.

Marie, P.J. (1994). Human osteoblastic cells: A potential tool to assess the etiology of pathologic bone formation. J. Bone Miner. Res. 9, 1847-1850.

Marie, P.J. (1995). Human osteoblastic cells: Relationship with bone formation. Calcif. Tissue Int. 56S, 13-16.

Marie, P.J. (1997a). Effects of bone morphogenetic proteins on cells of the osteoblastic lineage. J. Cell. Engin. 2(3), 92-99.

Marie, P.J. (1997b). Growth factors and bone formation in osteoporosis: Roles for IGF-I and TGF-β. Rev. Rhum. 64(1), 44-53.

Marie, P.J., de Pollak, C., Chanson, P., and Lomri, A. (1997). Increased osteoblastic cell proliferation associated with activating $G_s\alpha$ mutation in monostotic and polyostotic fibrous dysplasia. Am. J. Pathol. 150(3), 1059-1069.

Marie, P.J. (1998). Cellular and molecular alterations of osteoblasts in human disorders of bone formation. Histology and Histopathology (In press.).

Martin, T.J. and Ng, K.W. (1994). Mechanisms by which cells of the osteoblast lineage control osteoclast formation and activity. J. Cell. Bioch. 56, 357-366.

Martin, T.J., Findlay, D.M., Heath, J.K., and Ng, K.W. (1993). Osteoblasts: Differentiation and function. In: Physiology and Pharmacology of Bone. (Mundy, G.R. and Martin, T.J., Eds.), pp. 149-183. Springer-Verlag, New York.

Mavilio, F. (1993). Regulation of vertebrate homeobox-containing genes by morphogens. Eur. J. Biochem. 212, 273-288.

Merriman, H.L., LaTour, D., Likhart, T.A., Mohan, S., Baylink, D.J., and Strong, D.D. (1990). Insulinlike growth factor-I and insulinlike growth factor-II induce c-fos in mouse osteoblastic cells. Calcif. Tissue Int. 46, 258-262.

Modrowski, D., Miravet, L., Feuga, M., Bannie, F., and Marie, P.J. (1992). Effect of fluoride on bone and bone cells in ovariectomized rats. J. Bone Miner. Res.7, 961-969.

Modrowski, D. and Marie, P.J. (1993). Cells isolated from the endosteal bone surface in adult rats express differentiated osteoblastic characteristics in vitro. Cell and tissue Res. 271, 499-505.

Modrowski, D., Miravet, L., Feuga, M., and Marie, P.J. (1993). Increased proliferation of osteoblast precursor cells in estrogen-deficient rats. Am. J. Physiol. 264 (Endocrinol. Metab. 27), E190-E196.

Modrowski, D., Godet, D., and Marie, P.J. (1995). Involvement of interleukin-1 and tumour necrosis factor a as endogenous growth factors in human osteoblastic cells. Cytokine 7, 7.

Modrowski, D., Lomri, A., and Marie, P.J. (1997). Endogenous GM-CSF is involved as an autocrine growth factor for human osteoblastic cells. J. Cell. Physiol. 170, 35-46.

Modrowski, D., Lomri, A., and Marie, P.J. (1998). Glycosaminoglycans bind granulocyte-macrophage colony-stimulating factor and modulate its mitogenic activity and signaling in human osteoblastic cells. J. Cell. Physiol. (In press.).

Morgan, B.A. and Tabin, C.J. (1993). The role of homeobox genes in limb development. Curr. Opin. Genet. Dev. 3, 668-674.

Mueller, K., Cortesi, R., Modrowski, D., and Marie, P.J. (1994). Stimulation of trabecular bone formation by insulinlike growth factor I in adult ovariectomized rats. Am. J. Physiol. 267 (Endocrinol. Metab. 30), E1-E6.

Mundy, G.R. (1995). Local control of bone formation by osteoblasts. Clin. Orthop. Rel. Res. 313, 19-26.

Nicolas, V., Prewett, A., Bettica, P., Mohan, S., Finkelman, R.D., Baylink, D.J., and Farley, J.R. (1994). Age-related decreases in insulinlike growth factor-I and transforming growth factor-β in feromal cortical bone from both men and women: Implications for bone loss with aging. J. Clin. Endocrinol. Metab. 78, 1011-1016.

Noda, M. and Camilliere, J.J. (1989). In vivo stimulation of bone formation by transforming growth factor- . Endocrinology 124, 2991-2994.

Ogata, T. and Noda, M. (1991). Expression of Id, a member of HLH protein family, is downregulated at confluence and enhanced by dexamethasone in a mouse osteoblastic cell line, MC3T3E1. Biochem. Biophys. Res. Commun. 1991, 1194-1199.

Owen, M.E. (1988). Marrow stromal stem cells. J. Cell Sci. (Suppl.) 10, 63-76.
Owen, T.A., Aronow, M., Shalhoub, V., Lian, J.B., and Stein, G.S. (1990a). Progressive development of the rat osteoblast phenotype in vitro—reciprocal relationships in expression of genes associated with osteoblast proliferation and differentiation during formation of the bone extracellular matrix. J. Cell Physiol. 143, 420-430.
Owen, T.A., Bortell, R., Yocum, S.A., Smock, S.L., Zhang, M., Abate, C., Shalhoub, V., Aronin, N., Wright, K., Van Wijnen, A.J., Stein, J.L., Curran, T., Lian, J.B., and Stein, G.S. (1990b). Coordinate occupancy of AP-1 sites in the vitamin D-responsive and CCAAT box elements by Fos-Jun in the osteocalcin gene: Model for phenotype suppression of transcription. Proc. Natl. Acad. Sci. USA 87, 9990-9994.
Owen, T.A., Aronow, M., Barone, L.M. Lian, J.B., and Stein, G.S.(1991). Pleiotropic effects of vitamin D on osteoblast gene expression are related to the proliferative and differentiated state of the bone cell phenotype-dependency upon basal levels of gene expression, duration of exposure, and bone matrix competency in normal rat osteoblast cultures. Endocrinology 128, 1496-1504.
Parfitt, A.M. (1990). Bone forming cells in clinical conditions. In: Bone, A Treatise, Vol. 1: The Osteoblast and Osteocyte. (Hall, B.K., Ed.), pp 351-429. The Telford Press, Caldwell, NJ.
Partanen, J., Vainikka, S., Korhonen, J., Armstrong, E., and Alitalo, K. (1992). Diverse receptors for fibroblast growth factors. Progress Growth Factor Res., 4, 69-83.
Partridge, N.C., Bloch, S.R., and Pearman, A.T. (1994). Signal transduction pathways mediating parathyroid hormone regulation of osteoblastic gene expression. J. Cell. Biochem. 55, 321-327.
Pfeilschifter, J., Wolf, O., Naumann A., and Mundy, G.R. (1990). Chemotactic response of osteoblastlike cells to transforming growth factor β. J. Bone Miner. Res. 5, 825-830.
Pfeilschifter, J., Diel, I., Pilz, U., Brunotte, K., Naumann, A., and Ziegler, R. (1993). Mitogenic responsiveness of human bone cells in vitro to hormones and growth factors decreases with age. J. Bone Miner. Res. 8, 707-717.
Puleo, D.A. and Bizios, R. (1991). RGDS tetrapeptide binds to osteoblasts and inhibits fibronectin-mediated adhesion. Bone 1991, 12, 271-276.
Rapraeger, A., Krufka, A., and Olwin, B.B. (1991). Requirement of heparan sulfate for bFGF-mediated fibroblast growth and myoblast differentiation. Science 252, 1705-1708.
Rawlinson, S.C.F., El-Haj, A.J., Minter, S.L., Tavares, I.A., Bennett, A., and Lanyon, L.E. (1991). Loading-related increases in prostaglandin production in cores of adult canine cancerous bone in vitro: A role for prostacyclin in adaptive bone remodeling? J. Bone Miner. Res. 6, 1345-1351.
Rodan, G.A. and Noda, M. (1991). Gene expression in osteoblastic cells. Crit. Rev. Eukariotic Gene Expr.1, 85-98.
Rosen, V. and Thies, R.S. (1992). The BMP proteins in bone formation and repair. Trends Genet. 8, 97-102.
Rousseau, F., Bonaventure, J., Legeai-Mallet, L., Pelet, A., Rozet, J.M., Maroteaux, P., Le Merrer, M., and Munnich, A. (1994). Mutations in the gene encoding fibroblast growth factor receptor-3 in achondroplasia. Nature 371, 252-254.
Ruoslahti, E. (1991). Integrins. J. Clin. Invest. 87, 1-5.
Ruther, U., Garber, C., Komitowski, D., Müller, R., and Wagner, E.F. (1987). Deregulated c-fos expression interferes with normal bone development in transgenic mice. Nature 325, 412-416.
Sandberg, M.M., Aro, H.T., and Vuorio, E.I. (1993). Gene expression during bone repair. Clin. Orthop. 289, 292-312.
Satokata, I. and Maas, R. (1994). Msx1-deficient mice exhibit cleft palate and abnormalities of craniofacial and tooth development. Nature Genet. 6, 348-356.
Schipani, E., Kruse, K., and Jüppner, H. (1995). A constitutively active mutant PTH-PTHrP receptor in Jansen-type metaphyseal chondrodysplasia. Science 268, 98-100.
Schmid, C. and Ernst, M. (1990). Insulinlike growth factors. In: Cytokines and Bone Metabolism. (Gowen, M., Ed.), pp 229-265. CRC Press, Ann Arbor, Michigan.

Schulz, A. (1995). "True bone" in vitro? Virchow Archiv. 426, 103-105.

Shenker, A., Chanson, P., Weinstein, L.S., Spiegel, A.M., Lomri, A., and Marie, P.J. (1995). Osteoblastic cells from monostotic fibrous dysplasia contain ARG ^{201} mutation of Gs$_{\alpha}$ Hum. Mol. Gen. 4, 1675-1676.

Shiang, R., Thompson, L.M., Zhu, Y.Z., Church, D.M., Fielder, T.J., Bocian, M., Winokur, SS.T., and Wasmuth, J.J. (1994). Mutations in the transmembrane domain of FGFR3 cause the most common genetic form of dwarfism, achondroplasia. Cell 78, 335-342.

Shinar, D.M., Endo, N., Halperin, D., Rodan, G.A., and Weinreb, M. (1993). Differential expression of insulinlike growth factor-I (IGF-I) and IGF-II messenger ribonucleic acid in growing rat bone. Endocrinology 132, 1158-1167.

Siddhanti, S.R. and Quarles, L.D. (1994). Molecular to pharmacologic control of osteoblast proliferation and differentiation. J. Cell. Biochem. 55, 310-320.

Stein, G.S. and Lian, J.B. (1993). Molecular mechanisms mediating proliferation/ differentiation interrelationships during progressive development of the osteoblast phenotype. Endocr. Rev. 14, 424-442.

Stein, G.S., Lian, J.B., Stein, J.L., van Wijnen, A.J., Frenkel, B., and Montecino, M. (1996). Mechanisms regulating osteoblast proliferation and differentiation. In: *Principles of Bone Biology*. (Bilezikian, J.P., Raisz, L.G., Rodan, G.A., Eds.), pp. 69-86. Academic Press, New York.

Subramaniam, M., Oursler, M.J., Rasmussen, K., Riggs, B.L., and Spelsberg, T.C. (1995). TGF-β regulation of nuclear proto-oncogenes and TGF-β gene expression in normal human osteoblastlike cells. J. Cell. Biochem. 57, 52-61.

Suva, L.J., Seedor, J.G., Endo, N., Quartuccio, H.A., Thompson, D.D., Bab, I., and Rodan, G.A. (1993). The pattern of gene expression following rat tibial marrow ablation. J. Bone Min. Res. 8, 379-388.

Tamura, M. and Noda, M. (1994). Identification of a DNA sequence involved in osteoblast-specific gene expression via interaction with Helix-Loop-Helix- (HLH)-type transcription factors. J. Cell Biol. 126, 773-783.

Turner, R. and Spelsberg, T. (1991). Correlation between mRNA levels for bone cell proteins and bone formation in long bones of maturing rats. Am. J. Physiol. 261 (Endocrinol. Metab. 24), E348-E353.

Towler, D.A., Rutledge, S.J., and Rodan, G.A. (1994). Msx-2/Hox 8.1: A transcriptional regulator of the rat osteocalcin promoter. Mol. Endocrin. 8, 1484-1493.

Trémollières, F.A., Strong, D.D., Baylink, D.J., and Mohan, S. (1992). Progesterone and promegestone stimulate human bone cell proliferation and insulinlike growth factor-2 production. Acta Endocrin. 126, 329-337.

Turksen, K. and Aubin, J.E. (1991). Positive and negative immunoselection for enrichment of two classes of osteoprogenitor cells. J. Cell Biol. 114, 373-384.

Turksen, K., Bhargava, H., Moe, H., and Aubin, J.E. (1992). Isolation of monoclonal antibodies recognizing rat bone-associated molecules in vitro and in vivo. J. Histochem. 40, 1339-1352.

Vainio, S., Karavanova, I., Jowett, A., and Thesleff, I. (1993). Identification of BMP-4 a signal mediating secondary induction between epithelial and mesenchymal tissues during early tooth development. Cell 75, 45-58.

Walsh, C.A., Birch, M.A., Fraser, W.D., Robinson, J., Lawton, R., Dorgan, J., Klenerman, L., and Gallagher, J.A. (1994a). Primary cultures of human bone-derived cells produce parathyroid hormone-related peptide protein: A study of 40 patients of varying age and pathology. Bone Min. 27, 43-50.

Walsh, S., Dodds, R.A., James, I.E., Bradbeer, J.N., and Gowen, M. (1994b). Mononuclear antibodies with selective reactivity against osteoblasts and osteocytes in human bone. J. Bone Miner. Res. 9, 1687-1696.

Wanaka, A., Milbrandt, J., and Johnson, E. (1991). Expression of FGF receptor gene in rat development. Development, 111, 455-468.

Wang, Z., Ovit, Q.C., Grigoriadis, A.E., Mohle-Steinlein, U., Ruther, U., and Wagner, E.F. (1992). Bone and haematopoietic defects in mice lacking c-fos. Nature (London) 360, 741-745.

Wilkie, A.O.M., Morriss-Kay, G.M., Jones, E.Y., and Heath, J.K. (1995). Functions of fibroblast growth factors and their receptors. Curr. Biol. 5, 1-9.

Wozney, J.M. (1992). The bone morphogenetic protein family and osteogenesis. Mol. Reprod. Dev. 32, 160-167.

Yao, K.L., Todescan, R., and Sodek, J. (1994). Temporal changes in matrix protein synthesis and mRNA expression during mineralized tissue formation by adult rat bone marrow cells in culture. J. Bone Miner. Res. 9, 231-240.

Yotov, W.V., Moreau, A., and St-Arnaud, R. (1998). The α-chain of the nascent polypeptide-associated complex functions as a transcriptional coactivator. Mol. Cell. Biol. 18(3), 1303-1311.

Zhang, R., Supowit, S.C., Klein, G.L., Lu, Z., Christensen, M., Lozano, R., and Simmons, D. (1995). Rat tail suspension reduces messenger RNA level for growth factors and osteopontin and decreases the osteoblastic differentiation of bone marrow stromal cells. J. Bone Miner. Res. 10, 415-423.

OSTEOBLAST LINEAGE

James T. Triffitt and Richard O.C. Oreffo

I. INTRODUCTION

Bone deposition is spatially directed during embryological development and normally occurs in specific and characteristic sites in the adult organism. However, the capacity

Advances in Organ Biology
Volume 5B, pages 475-498.

ISBN: 0-7623-0390-5

to form bone exists in the adult in many nonskeletal tissues throughout the body, in such unlikely situations as the brain and the lung, as well as in skeletal muscle (Triffitt, 1987a). Residual primitive connective tissue cells within all these tissues retain the propensity for extensive osteogenesis under the appropriate stimuli. The cells with high potential for bone formation that are derived from the bone surfaces of the normally distributed skeletal tissues have been considered in the past to be different from those that are induced extra-skeletally and which result in pathological osteogenesis (Friedenstein, 1973). Nevertheless, evidence suggests that, in both types of situation, the activation of early progenitors or stem cells results in the eventual formation of large quantities of bone.

II. MESENCHYMAL STEM CELLS

A stem cell can be defined as "a cell type which, in the adult organism, can maintain its own numbers in spite of physiological or artificial removal of cells from the population" (Lajtha, 1982). Division of a stem cell *in vivo*, therefore, results in the production of a new stem cell together with a daughter cell, which has more limited potential but which supplies all the committed progenitors and end cells of the particular cell lineage, (Figure 1). Many stem cell concepts were initially formulated by reference to the hemopoietic system (Siminovitch et al., 1963), of which there is a great deal of knowledge, concerning the capacity of these cells for

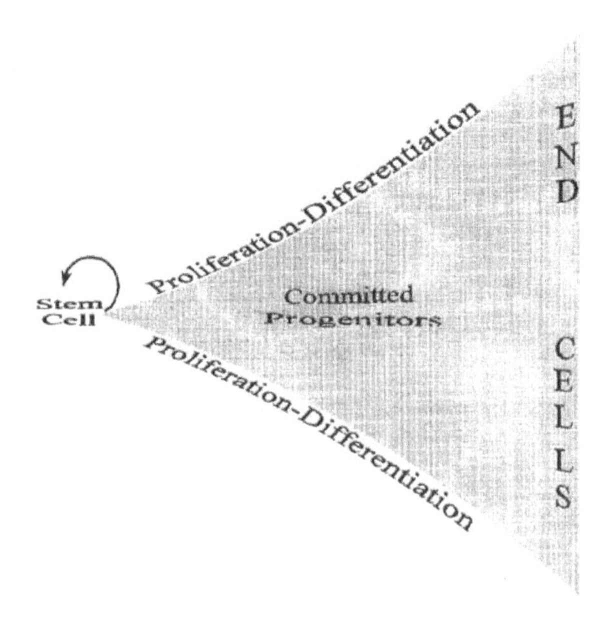

Figure 1. Diagram illustrating the production of cell populations from stem cells.

regeneration of blood cell lines. This was aided to a large extent by the readily recognizable morphological characteristics of the hemopoietic cell progenitors, committed cells, and end cells which lead to identification of the particular stage and lineage of the cells under study. Even so, the hemopoietic stem cell remains uncharacterised and the evidence for the existence of the hematopoietic stem cell is still indirect. As in the hemopoietic system, experiments involving transplantation of osteoblast progenitors have indicated indirectly the existence of stem cells for the osteoblast (Owen and Friedenstein, 1988) that also exhibit pluripotentiality by their inherent capacity to spawn a variety of related cell types including osteoblasts, chondroblasts, adipoblasts, myoblasts, and fibroblasts (Figure 2) (Grigoriadis et al., 1988; Bennett et al., 1991; Yamaguchi and Kahn, 1991; Beresford et al., 1992; Caplan, 1995; Saito et al., 1995; Triffitt, 1996). Despite the existence of the hemopoietic stem cells in close proximity to those stem cells generating bone there appears to be no interconversion between these two lines postfetally and there is no evidence that there is a single common progenitor even within fetal bone marrow (Waller et al., 1995). Effectively, therefore, the principal bone-forming cells, the osteoblasts, and bone- destroying cells, the osteoclasts, are derived from separate stem cell systems.

As bone grows by accretion, the progenitor cells exist in close proximity to bone surfaces, although possibly being less differentiated further from this surface. This means that such cells can be isolated from all tissues adjacent to bone, including endosteal marrow and periosteum (Triffitt, 1987a). Those isolated from marrow have been termed "marrow stromal stem cells," or from other sites "stromal stem cells" (Owen, 1985) or "stromal fibroblastic stem cells" (Triffitt, 1987a). Alternatively they have been named by the currently popular and nonspecific phrase "mesenchymal stem cells" (Caplan, 1995). None of these phrases are precise and adequate, but the latter will be used here.

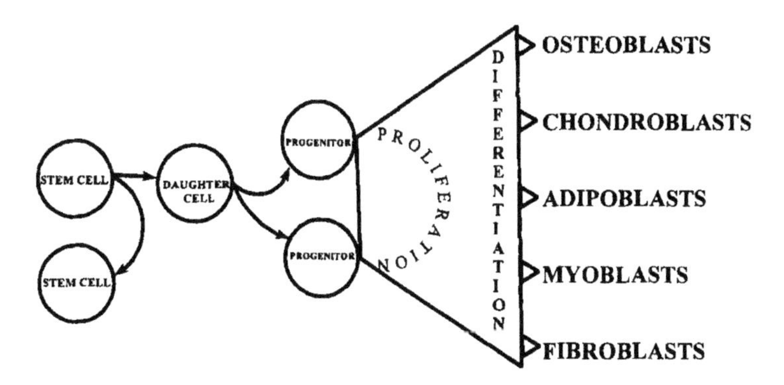

Figure 2. Diagram illustrating production of mesenchymal lineages from self-renewing stem cells and a daughter cell, which proliferates and has the capacity to produce all mesenchymal cell lineages.

III. MESENCHYMAL LINEAGES

The derivation of the mesenchyme tissues from the mesoderm in the embryo results from the division of primitive mesenchymal cell precursors with characteristics of stem cells. The mesenchymal tissues by definition include bone, muscle, and other connective tissues including the blood (Ham, 1969). As mentioned above, hemopoietic cells do not derive from a common progenitor, at least in the late embryo and postfetally, and are distinct histogenically. The mesenchymal lineages directly related to the osteoblast lineage include the adipogenic, myogenic, fibrogenic, and chondrogenic lineages (Figure 2). While the terminal cell stages of differentiation of all these lineages are readily recognizable by morphological and biochemical criteria, identification of stem cells and the earliest progenitors is almost impossible at the present time. However, the simultaneous reports from a number of groups on the identification of a transcription factor, core-binding factor A1 (CBFA1) which is essential for osteoblast differentiation and bone formation, has proven a major advance in our understanding of osteoblast differentiation and opened new avenues of research into the mechanisms regulating bone differentiation and formation (Ducy et al., 1997; Komori et al., 1997; Mundlos et al., 1997; Otto et al., 1997). CBFA1, the MyoD family of myogenic transcription factors which is essential for muscle differentiation and the peroxisome proliferator activated receptor $\gamma 2$ (PPAR $\gamma 2$) that is essential for fat cell differentiation, are a new class of "master genes" key in the differentiation of particular tissue types (reviewed in Triffit et al., 1998).

IV. OSTEOBLAST LINEAGE.

A. Stages of Differentiation

Progenitor cells of the osteoblast lineage progressively differentiate in a continuum of development from stem cells through osteoprogenitors, preosteoblasts, and osteoblasts to, effectively, two types of end cells termed osteocytes and lining cells (Triffitt et al., 1998) (Figure 3). The former are the most abundant cells in bone tissue and are those osteoblasts which surround themselves with calcified matrix during bone formation. They are interconnected by many cytoplasmic processes through canaliculi with their neighbors and subsequently with surface osteoblastic cells and other cell types through these syncytial membranous connections. They

Figure 3. Named cells of the osteoblastic lineage.

are thus in dynamic communication with cells that can modify bone architecture and have been suggested to be the major sensors of mechanical loading on the skeleton (Lanyon 1993; Burger et al., 1995; Inaoki et al., 1995).

Bone lining cells cover bone surfaces particularly in the adult skeleton that are quiescent in terms of bone formation and resorption, which are the active processess determining bone remodeling. They are flattened spindle-shaped cells with oval nuclei and few organelles and communicate with osteocytes via gap junctions. In certain circumstances they appear to proliferate and differentiate into osteoblasts. Lining cells may be involved in the control of mineral homeostasis by controlling ion fluxes between bone and interstitial fluids. In addition as part of the hemopoietic microenvironment they may play a role in regulation of hemopoiesis (Miller et al., 1989; Islam et al., 1990).

The major bone-forming cells lying directly on the layer of unmineralized bone matrix, or osteoid, which they secrete, are the osteoblasts. These cells exhibit variable activities and the morphological appearance of the active cell reflects this activity. The active osteoblast is a cuboidal cell with an eccentrically placed nucleus, which resides away from the cell surface nearest to mineralized bone. The osteoblast contains abundant endoplasmic reticulum and Golgi complex indicative of its intense synthetic capacity and maintains contact by means of cellular extensions with its surface neighbors and entombed osteocytes (Palumbo et al., 1990). Together with the immediate predecessor, the preosteoblast (Young, 1962; Owen, 1963), but unlike other cells in this lineage, the osteoblast characteristically exhibits alkaline phosphatase expression on the plasma membrane (Doty and Schofield, 1976). Behind these layers of cuboidal cells, spindle-shaped fibroblastic cells with oval nuclei probably make up the osteoprogenitors and stem cell layers.

B. Phenotypic Markers

Morphological identification of cells within the osteoblast lineage is supplemented by knowledge of the biochemical characteristics *in vivo* of the cells as they progress down the differentiation pathways (Bianco et al., 1993). However, the lack of specific, identifiable, morphological features within the early progenitor cell populations is paralleled by the sparse knowledge concerning synthesis and expression of particular proteins by these cells. Much more is known about the proteins synthesized particularly by the most differentiated cells following the extensive analyses of bone matrix, and the application of molecular biology techniques over the past few years (Triffitt 1987b; 1996; Robey et al., 1993).

Type-I collagen is the major protein produced by bone cells and smaller amounts of other noncollagenous proteins, and other products are produced by osteoblasts during bone matrix formation; these include osteonectin, osteopontin, osteocalcin or bone gla-protein, matrix gla- protein, bone sialoprotein, bone acidic glycoprotein-75, thrombospondin, decorin, and biglycan (Termine, 1993). Antibodies to these proteins and the corresponding specific mRNAs are useful for identification of cells of the osteoblast lineage in *in vitro* studies, together with their stages of differentiation as they are expressed in a temporal fashion (Ibaraki et al., 1992). However, only osteocalcin is specifically produced by osteoblastic cells and this production is by those cells late in the differentiation pathway when a competent or mineralized bone matrix is present (Owen et al., 1991). In addition the expression of parathyroid hormone receptors and an associated adenylate cyclase activation are characteristic of osteogenic cells (Rodan and Rodan 1984).

The possibilities for the production by monoclonal antibody techniques of new antibody markers for early osteogenic progenitors are being actively pursued by a number of research groups. At the moment, however, there are no defining phenotypic features available for recognition of these stem cells. Fluorescence activated cell sorting (FACS) has been used for partial characterization and purification of mouse marrow osteogenic cells (Van Vlasselaer et al., 1994) using two-color cell sorting, with Sca-1 expression and wheat germ agglutinin binding, together with light scatter characteristics. Those cells with high forward (FSC) and perpendicular (SSC) light scatter contained the osteogenic progenitors as seen in human marrow (Simmons and Torok-Storb 1991). Subsequent cell sorting by using a variety of antibodies, indicated that these progenitors have the phenotype FSC^{high} SSC^{high} Lin^- $Sca\text{-}1^+$ $WGA\text{-}^{bright}$ $KM16^+$ $Sab\text{-}1^+$ $Sab\text{-}2^+$ $Thy1.2^-$ $c\text{-}kit^-$. In separate studies, a monoclonal antibody (STRO-1) characterizes colony-forming units fibroblastic (CFU-F) in adult human bone marrow (Simmons and Torok-Storb 1991). Separated $STRO\text{-}1^+$ cell populations give rise to osteogenic, fibroblastic, adipogenic, and smooth muscle cells (Gronthos et al., 1994) showing that the STRO-1 positive cell populations contain osteoprogenitors. Antibodies (SH2, SH3, and SH4) to human mesenchymal progenitor cells have been produced and recently SB10 antibody which appears to be against even earlier progenitors, has been described (Haynesworth et al., 1992; Bruder et al., 1995; 1997). We have characterized two monoclonal antibodies raised against early human marrow stromal progenitors, HOP-11 and HOP-26 (Joyner et al., 1997). These antibodies react specifically with the earliest progeny of human marrow CFU-F. HOP-26 is reactive with a cell surface epitope, while HOP-11 reacts with an intracellular antigen. These types of reagents are required to further characterize the early stages of differentiation.

C. Interconversion Potentials

Progression down a cell differentiation pathway in a specific lineage normally results in restricted differentiation potential. Nevertheless, some evidence indicates that there is some plasticity in the phenotypes that constitute the mesenchymal lineages. As mentioned previously the inactive, fibroblastic, lining cells that line resting bone surfaces *in vivo* appear capable of proliferation and differentiation into functional osteoblasts and other cells (Islam et al., 1990). *In vitro*, interconversion of phenotypically defined adipocytes appear to be capable of reproliferation and differentiation in an alternative, osteogenic pathway when implanted in diffusion chambers *in vivo* (Bennett et al., 1991). Factors determining the cellular commitment to a particular multipotential pathway and to a restricted phenotype are under intense study currently but how much interconversion between the phenotypes may be possible is unknown at the present time. Recent studies on the modulation of osteogenesis and adipogenesis by human serum, indicate the presence of adipogenic factors in human serum or the absence of a factor permissive for osteogenesis (Oreffo et al., 1997). The ability to possibly manipulate the cell phenotype towards osteogenesis has important implications with respect to disease conditions such as osteoporosis in which there is increased fat deposition in the marrow (Burkhardt et al., 1987). Further consideration of the origins and lineages related to the osteoblast have been published earlier in detailed reviews (Friedenstein, 1976; Owen, 1985; Triffitt, 1987a; 1996; Beresford, 1989; Aubin et al., 1993; Triffitt et al., 1998).

D. Experimental Systems

The study of bone cell biology has been facilitated by the development of a variety of *in vitro* and *in vivo* osteoblast models (Mundy et al., 1991). In particular, cell culture has proved a powerful tool in the elucidation of osteoblast function with the development of cell culture models including primary mesenchymal populations derived from bone marrow or periosteum, transformed and nontransformed cell lines and, recently, selective immortalized cell populations.

The use of cells isolated from calvarial digests of fetal or neonatal rodent calvariae and embryonic chick calvariae was originally developed by Peck, Wong, and Cohen (Peck et al., 1973; Wong and Cohn 1975). Osteoblast-like cells isolated from these tissues are enriched in alkaline phosphatase, synthesise type I collagen, osteocalcin and a variety of other noncollagenous proteins and elaborate an extracellular matrix in a temporal and highly regulated process (Owen et al., 1990, 1991). This cell culture model has proved responsive to osteotropic hormones and mechanical stimuli, and cultured calvarial cell populations form osteogenic tissue when implanted *in vivo* within diffusion chambers (Simmons et al., 1982). Prolonged cell culture (post-confluence) gives rise to bone nodules—foci of cells which express elevated alkaline phosphatase and mineralize in the presence of ascorbic acid and β-glycerophosphate (Ecarot-Charier et al., 1988, Nefussi et al., 1985, Nagata et al., 1991). The bone formed, although avascular, resembles mineralized bone tissue and exhibits some of the characteristics of woven bone. Over the last few years, this system has been used to characterize and unravel the genetic events accompanying bone formation (Stein and Lian 1995). Nevertheless, there may be species-specific as well as distinct differences in cells from fetal calvaria and cells present in the mature or osteoporotic animal.

Recently, osteocyte cultures have been generated by sequentially treating calvaria from newborn rats with collagenase and EDTA. The cells generated expressed essentially no alkaline phosphatase, had well-developed dendritic processes and expressed casin kinase II (Mikuni-Takagaki et al., 1995). The use of collagenase and EDTA treatment was originally developed by Van der Plaas and Nijweide (1992) for the isolation and purification of osteocytes from chick calvaria. These authors achieved over 95% purity in these cultures and numerous interconnecting cell processes similar to the osteocyte network in bone were observed.

Delineation of the osteogenic lineage has been aided by studies using the adherent marrow stromal cells recovered after culture of a single cell suspension of marrow cells onto tissue culture plastic (Friedenstein et al., 1987). The stromal population consists of a heterogenous group of histogenetically distinct cell types (fibroblastic, endothelial, macrophage-monocytic cells) but, more significantly, contains the putative mesenchymal stem cells. Freshly isolated marrow stromal fibroblastic cells, therefore, represent primitive osteoprogenitors and a variety of culture techniques allows examination of their terminal differentiation potentials, activity, and numbers in normal and diseased states (Oreffo et al., 1998).

Much of the early work on the hormonal regulation and phenotype of the osteoblast was performed utilising osteosarcoma cell lines. However, these tumor derived cell lines may not reflect the true phenotype of their nontransformed counterparts. A recent approach to circumvent this problem has been the development of immortalised human osteoblast cell lines, using retroviral transduction with the SV40 large T antigen to produce homogenous populations of human osteoblast cells (Keeting et al., 1992, Harris et al., 1995).

The organ culture model pioneered by Fell and Mellanby in the 1920s has allowed the study of bone tissue as a whole, albeit in the absence of a functional blood supply (reviewed in Fell 1952). This model has been applied extensively in the delineation of the bone resorption process and to study bone mineralisation (Raisz 1965, Mundy et al., 1976, Howard et al., 1982). Results obtained using the organ culture model, in common with all the aforementioned models, is dependent on the culture conditions used.

The complex interplay of factors that are involved in the regulation and modulation of osteoblast activity and bone formation indicate the need for caution in extrapolation from the *in vitro* environment and the requirement for appropriate studies in parallel using *in vivo* models. Widely used experimental systems include i) the ovariectomised rat, ii) segmental bone defect, iii) subcutaneous implantation of demineralized bone matrix, and iv) diffusion chamber implantation.

The relative inexpense and wide availability of the rodent has resulted in its use as the experimental animal of choice in most of these studies. The ovariectomised rat, which demonstrates a dramatic loss in bone mass following estrogen loss, is now used, especially in the pharmaceutical industry, as a surrogate model of osteoporosis (reviewed in Kalu 1991). The segmental bone defect and the subcutaneous implant models provide a robust model of osteogenesis for the assessment of bone inductive agents such as the bone morphogenetic proteins on development of the osteoblast lineage (Horisaka et al., 1991, Johnson et al., 1992, Yasko et al., 1992). Furthermore, the bone formed in the subcutaneous implant is histologically and biochemically identical to normal bone. The diffusion chamber model has been used to study the osteogenic capacity of skeletally derived cells in a variety of species including mice, rat, and rabbit (Ashton et al., 1980, Mardon et al., 1987, Bruder et al., 1990). Using this model, Gundle and colleagues (1995) have recently shown the consistent formation of bone tissue, although avascular, by human marrow stromal and trabecular bone-derived fibroblastic cells grown in the presence of dexamethasone or implanted with porous hydroxyapatite.

E. Biological and Other Factors Affecting Proliferation And Differentiation

Parathyroid hormone (PTH) plays a central role in concert with 1,25$(OH)_2D_3$ in maintaining serum calcium and phosphate levels (for review see Dempster et al., 1993). As indicated earlier, PTH acts directly on osteoprogenitors and osteoblasts via PTH receptors, although its effects on bone formation are complex. PTH can inhibit bone collagen and osteocalcin synthesis, acting at the level of gene transcription (Kream et al., 1990). In contrast, intermittent PTH administration results in bone formation which is mediated, in part, by the production of local growth factors, including insulin-like growth factor-1 (IGF-1) and transforming growth factor beta (TGFβ) (Slovik et al., 1986, Pfeilschifter and Mundy 1987, Canalis et al., 1989). Recently Onyia and co-workers (Onyia et al., 1995) have shown, *in vivo*, that hPTH1-34 can upregulate cell differentiation in trabecular bone cells of young rats via transient stimulation of the early response genes c-*fos*, c-*jun*, c-*myc*, and IL-6 while downregulating cell proliferation

The principal active metabolite of vitamin D_3 ,1,25$(OH)_2D_3$, plays an important role in mineralisation through the control of calcium homeostasis and in the paracrine and autocrine regulation of bone cells (reviewed in Bikle 1994). The effects of 1,25$(OH)_2D_3$ on osteoblasts, which are known to express receptors for the hormone, are dependent on the differentiation state and proliferative capacity of the osteoblast (Narbeitz et al., 1983; Owen et al., 1991). A variety of osteotropic agents including PTH, glucocorticoids, estradiol, and 1,25$(OH)_2D_3$ itself regulate osteoblast 1,25$(OH)_2D_3$ receptor number. The absence of receptors for 1,25$(OH)_2D_3$ on osteoclasts and the observed stimulation of bone resorption by the hormone, has led to the suggestion of 1,25$(OH)_2D_3$ induced osteoclast activating factor by cells of the osteoblast lineage (McSheehy and Chambers 1986). More recently Morrison and co-workers have suggested the major genetic component responsible for bone mass is linked to polymorphism in the gene for the vitamin D receptor, results which have produced a renewed analysis of the role of vitamin D in bone (Morrison et al., 1994).

The observed increase in bone resorption following the menopause or after ovariectomy in premenopausal women indicates the importance of estrogens in bone (Richelson et al., 1984, Turner et al., 1994). Receptors for estrogen are expressed at low levels on both osteoprogenitors and osteoblasts, which has resulted in difficulties in demonstrating a direct effect on bone formation (Eriksen et al.,

1988,). *In vitro* studies indicate estrogen can modulate osteoblast differentiation and proliferation and the stimulation of growth factors (Ernst et al., 1989). Recently, Manolagas and colleagues reported the inhibition of interleukin-6 (IL-6) production, in bone marrow stromal and osteoblastic cell lines, an effect mediated through inhibition of IL-6 gene transcription via the estrogen receptor (Ettinger et al., 1985,Pottratz et al., 1994; Manolagas and Jilka 1995). This inhibition of cytokine secretion in osteoblasts by estrogens is thought to play a key role in estrogen deficiency- related bone loss (Girasole et al., 1992, Manolagas and Jilka 1995).

The reduction of bone density in men, associated with testosterone deficiency, and the observed maintenance of normal bone mass in women with androgen excess and undetectable estrogen levels suggests a role for androgens in skeletal homeostasis. *In vitro*, androgens modulate cell differentiation and increase cell proliferation in human osteoblast-like cells which have been shown to express androgen receptors (Colvard et al., 1989, Kasperk et al., 1989).

Glucocorticoids exert a dramatic effect on cellular differentiation. Dexamethasone, a synthetic glucocorticoid, has been shown to induce the osteoblast phenotype acting to stimulate osteoprogenitor cell differentiation in cultures from a variety of animal species including rat, mouse, and rabbit as well as in human marrow stromal cultures (Benayahu et al., 1989; Bennett et al., 1991; Beresford et al., 1994; Cheng et al., 1994; Locklin et al., 1995). Despite the reported effects of glucocorticoids on osteoblasts, the identification of receptors on osteoblasts and the known effects of pharmacological doses of corticosteroids resulting in glucocorticoid-induced osteoporosis and the inhibition of fracture repair, the precise role of these steroids on osteoblasts and bone metabolism remains ill-defined (Lukert and Raisz 1990).

A number of other systemic hormones including insulin, thyroid hormone, PTH related protein (PTHrP) and growth hormone have profound effects on normal cell growth and differentiation and on bone. The impairment of skeletal growth and mineralisation in individuals with diabetes mellitus indicates the importance of insulin in bone cell metabolism (Levin et al., 1976). *In vitro* studies indicate insulin acts on mature osteoblasts to stimulate matrix synthesis although insulin does not appear to increase the number of mature osteoblasts (Canalis 1980). Thyroid hormone is necessary for normal bone growth, however thyroid hormone acts predominantly to stimulate bone resorption (Mundy et al., 1976). Tri-iodothyronine (T_3), the most active form of thyroid hormone, has been shown to increase cell proliferation and alkaline phosphatase activity at physiological concentrations in primary rat calvarial cells and to inhibit both parameters at higher concentrations (Ernst and Froesch 1987). Ishida and coworkers (1995) using the fetal rat calvarial model showed T_3 inhibited osteoblastic cell differentiation and nodule formation, however, in the presence of dexamethasone low concentrations of T_3 stimulated osteoprogenitor cell differentiation.

PTHrP, originally identified as the cause of hypercalcemia in malignancy, shares the same receptor as PTH, binds with comparable affinity, and displays a range of similar biological activities to PTH (Abou-Samra et al., 1992). Recent studies indicate a mutation in the PTH/PTHrP receptor is the likely cause of Jansen-type metaphyseal chondrodysplasia, a rare form of short-limbed dwarfism associated with severe hypercalcaemia (Schipani et al., 1995). Growth hormone is required for normal skeletal growth. Absence of the hormone in childhood results in pituitary dwarfism and an excess results in gigantism in children and acromegaly in adults.

Among the many proteins sequestered within bone matrix, TGFβ and other related members, which encompasses the bone morphogenetic proteins, have a significant role to play in the modulation of osteoblast activity and bone turnover (reviewed in Centrella et al., 1994). TGFβ, the major component of bone matrix, acts on committed or determined osteogenic precursor cells stimulating proliferation and chemotaxis to create a pool of committed osteoblast cells (Pfeilschifter et al., 1990, Reddi 1995) (Figure 4). TGFβ will inhibit the expression of genes associated with bone formation (type I collagen, alkaline phosphatase, osteopontin, and osteocalcin) in differentiated mineralizing osteoblasts in the fetal rat calvarial model (Harris et al., 1994), providing further evidence for an effect on the precursor osteoblast. Several *in vivo* studies indicate administration of TGFβ results in new bone formation (Noda and Camilliere 1989, Marcelli et al., 1990) although, unlike the bone morphogenetic proteins (BMPs), TGFβ and related isoforms are ineffective in initiating bone formation in extraskeletal sites (Reddi 1992).

Identification of factors that induce bone in skeletal and extraskeletal sites was a significant milestone in skeletal research. Like the TGFβ proteins, the BMPs are found in the bone matrix and exert dramatic effects on cell of the osteogenic lineage (Wozney et al., ,1988; Luyten et al., 1989; Celeste et al., 1990; Wang et al., 1990; Wozney, 1993; Gimble et al., 1995). To date, some 30 members have been identified and the question of possible redundancy in biological activity remains

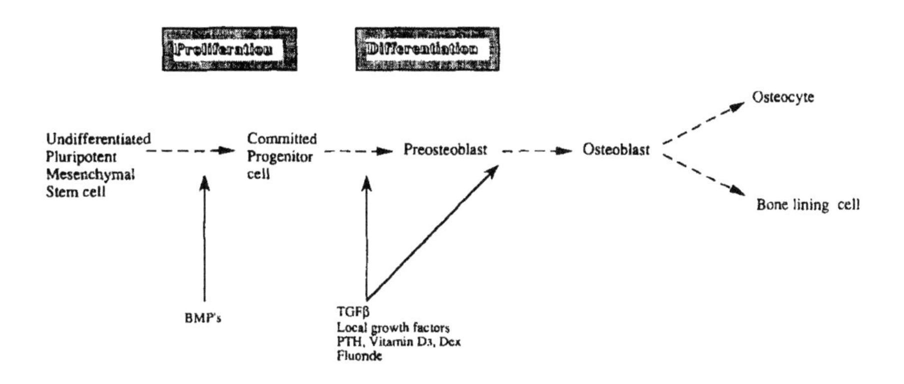

Figure 4. Some biological factors affecting stages of osteogenesis.

obscure. These molecules are believed to be the primordial signaling molecules for initiating bone cell differentiation (Figure 4) from inducible osteogenic precursors or mesenchymal stem cells and thus initiate the first steps in osteogenesis and are unique in their ability to stimulate ectopic bone formation in extraskeletal sites (Hughes et al., 1995, Reddi 1995). Osteoblasts have been shown to express BMPs as part of their differentiation process in long-term cultures of fetal rat calvaria (Harris et al., 1994). In fetal rat osteoblasts, BMP-2 stimulated protein, DNA and collagen synthesis and, unlike TGFβ which inhibited alkaline phosphatase activity, BMP-2 stimulated alkaline phosphatase activity (Chen et al., 1991). *In vivo* studies have demonstrated the ability of BMP-2 to aid healing of nonunions and partial or complete segmental bone defects in human patients (Horisaka et al., 1991, Johnson et al., 1992).

Prostaglandins (PGs), especially PGE_2, are important local factors in bone cell metabolism (Kawaguchi et al., 1995). *In vivo* PGs stimulate bone formation (Mori et al., 1992). *In vitro* studies using fetal rat calvariae, show low concentrations of PGE_2 stimulate cell proliferation, collagen, and noncollagen protein synthesis (Raisz and Fall 1990). The production of PGs and the characteristic response of osteoblasts to PGs has been extensively documented, however there is now emerging evidence that the anabolic actions of PGE_2 on bone may be via the recruitment of osteoprogenitor cells from the nonadherent mesenchymal precursor cells in bone marrow (Scutt and Bertram, 1995). Furthermore, the production of PGE_2 and prostacyclin in osteocytes following loading implicates a role for these molecules in cell signaling within bone (Rawlinson et al., 1993, 1995). Expression of the PG EP_1, EP_2, and EP_3 receptor subtypes in bone tissue has been reported (Kasugai et al., 1995). Furthermore, the observation of the EP_2 receptor in human osteoblast-like cells and in early rat bone marrow cultures suggests a possible role for PGs in bone cell differentiation and development (Oreffo et al., 1991; Kasugai et al., 1995).

A number of other bone-derived growth factors exist within the bone matrix which clearly have important effects on cells of the osteoblast lineage including IGF-1 and IGF-11, Platelet-derived growth factor (PDGF), and the fibroblast growth factors (FGFs) (Hauschka et al., 1986, 1988; Mohan and Baylink 1991). IGF-I and IGF-II act independently to stimulate bone collagen synthesis in osteoblasts as well as osteoblast proliferation (Canalis 1980; McCarthy et al., 1989). The activity of both proteins are regulated by specific binding proteins (IGFBPs) and these IGFBPs are themselves regulated by a variety of osteotropic agents (Schmid et al., 1989; Hayden et al., 1995). *In vitro* studies using fetal rat calvarial bone cells indicate bFGF is a bone cell mitogen, is synthesized by osteoblasts but also decreases the expression of the osteoblast phenotype (McCarthy et al., 1989; Pitaru et al., 1993). However, *in vivo* studies indicate low dose administration of bFGF to growing rats stimulates endosteal and endochondral bone formation (Nagai et al., 1995). PDGF stimulates alkaline phosphatase, collagen synthesis, and cell proliferation in rat calvarial osteoblasts although no effect on cell proliferation in human osteoblasts was observed (Wergedal et al., 1990). *In vivo* studies on normal healing human fractures indicate expression of a PDGFβ chain in osteoblasts during bone formation (Andrew et al., 1995) thus, the lack of response in cell culture may reflect a species-specific difference in PDGF used or possibly a difference in PDGF isoforms expressed.

A number of other cytokines, in particular IL-1 and IL-6, are implicated in the modulation of osteoblast activity and regulation of bone remodeling (Manolagus, 1995; Manolagus et al., 1995). IL-1 has potent effects on both bone resorption and bone formation. In bone organ culture, IL-1 remains the most potent stimulator of bone resorption an effect mediated, in part, by IL-1 induced prostaglandin synthesis (Gowen et al., 1983). In fetal rat calvariae and human bone cells, IL-1 stimulates cell proliferation (Gowen et al., 1985; Canalis 1986). *In vivo* studies show long-term administration of IL-1 over the calvariae of mice increased bone turnover (Boyce et al., 1989). Over the last few years much attention has focused on IL-6 as a regulator of normal osteoblast function. Interest in this cytokine centers around the possible role of IL-6 in the mediation of the bone loss associated with osteopaenia (reviewed in Manolagas and Jilka 1995). IL-6 is produced by osteoblast-like cells and the production of this cytokine is stimulated by the addition of local bone resorption agents such as tumor necrosis factor α and IL-1.

Vitamin A and related molecules—the retinoids—have profound effects on cellular differentiation, growth, and the modulation of bone cells. Vitamin A deficiency and excess have opposing effects on bone metabolism. In hypovitaminosis A, bone thickness is increased in a number of sites while hypervitaminosis A results in increased bone resorption (Fell, 1952).The modulation of collagen synthesis, cell proliferation and differentiation following exposure to retinol and retinoic acid indicate the osteoblast is a target for vitamin A. Retinoic acid has been shown to induce and stimulate alkaline phosphatase activity in preosteoblastic cells (Ng et al., 1989) and, in contrast, to inhibit cell proliferation and alkaline phosphatase activity in differentiated osteoblasts (Oreffo et al., 1985, Zhou et al., 1991). Zhou et al., (1991) have shown the mRNA for RARα,β,γ, in malignant, nontransformed, and immortalized osteoblasts and a recent study by Williams et al., (1995) indicates retinoids modify regulation of endogenous gene expression by $1.25(OH)_2D_3$ and thyroid hormone in rat osteoblast-like osteosarcoma cell lines. The precise role of retinoids and their interactions with other members of the steroid hormone receptor family in the modulation of osteoblast activity is poorly understood.

Vitamin C or ascorbic acid is required for normal osteoblastic cell differentiation. Studies using fetal rat calvariae and the MC-3T3-E1 murine osteoblast cell line indicate the action of the vitamin in the induction of the appearance of the osteoblast phenotype (alkaline phosphatase, osteocalcin) is related to the known actions of ascorbic acid on collagen synthesis (Owen et al., 1990, Franceschi et al., 1994).Further the importance of ascorbic acid in bone metabolism is underscored by observations following Vitamin C deficiency: impaired fracture repair and wound healing in the adult and impaired bone formation and dentition.

Bone is an abundant source of and target tissue for vitamin K, which is essential for the γ-carboxylation of the osteoblast-specific protein osteocalcin, enabling osteocalcin binding to hydroxyapatite (Price 1985, Hauschka and Carr 1982). *In vitro*, Vitamin K_1 and vitamin K_2 (Menatetrenone) enhance osteocalcin production and mineralization in cultured human osteoblasts (Koshihara et al., 1992).

It is now widely accepted that mechanical loading is a key element for the maintenance of bone mass and morphology. Bone resorption and bone growth are strongly influenced by mechanical loading and *in vivo* animal studies show a steady-state adaptation of bone to its mechanical environment (Skerry et al., 1989, Dallas et al., 1993, Dodds et al., 1993). The observations of increased osteocytic activity following mechanical loading and the distribution of osteocytes within bone has led to the proposal that osteocytes are biological sensors within bone able to signal changes in bone remodeling in response to mechanical load (Lanyon, 1993). A normal by-product of functional loading which appears to modulate bone cell activity is the electric field. Enhancement of cell proliferation has been observed in primary osteoblasts and osteosarcoma cell lines exposed to electric fields (reviewed in Rubin et al., 1993). *In vivo*, the inability to apply a functional load following fracture provides the premise for the use of electric fields in treatment of delayed fracture union and, despite the controversy surrounding this area, efficacy has been demonstrated in more than one double-blind clinical trial (Sharrard 1990, Mammi et al., 1993).

A number of studies indicate fluoride will stimulate bone formation by a direct effect on osteoblastic cell proliferation, alkaline phosphatase activity, and collagen synthesis (Farley et al., 1983, 1990; Khoker and Dandona 1990). The molecular mechanisms by which fluoride, which remains one of the most potents agents for the stimulation of new bone formation, exerts its mitogenic activity is unclear. It has been postulated that fluoride enhances protein tyrosine phosphorylation in osteoblasts by enhancing tyrosine kinase activity (Burgener et al., 1995). Questions regarding the mechanical competence of the newly formed bone and the incidence of periarticular bone pain and gastric- intestinal complications make the use of fluoride to increase spinal bone density controversial (Søgaard et al., 1994).

The induction of osteomalacia and a dynamic bone disease as observed following aluminium accumulation in patients with renal failure indicates the influence of aluminium on bone metabolism (reviewed in Goodman et al., 1993). *In vivo*, aluminium inhibits mineralization via inhibition of hydroxyapatite formation and dissolution. *In vitro*, inhibition of mineralization of osteoid nodules in rat calvariae has been observed (Bellows et al., 1995). The effects of aluminium on osteoblasts are less clear, with conflicting reports of both inhibition and stimulation of osteoblast proliferation. Zinc, the most abundant trace metal in bone mineral is required for growth in humans. Hurley et al., (1969) showed zinc stimulated osteoblast proliferation and bone mineralization in weanling rats and, like fluoride, its mechanism of action on bone metabolism is also unclear.

V. DEFECTS AND DISEASES

The pathophysiology of age-related osteoporosis remains poorly understood (reviewed in Silver and Einhorn, 1995). It appears to be due to a decreased ability to form osteoblasts (Manolagas et al., 1995). In Paget's disease, although the primary abnormality appears to reside in the osteoclasts (Basle et al., 1986; Roodman, 1995), the activity of osteoblasts is greatly increased. Extraskeletal deposition of calcium and phosphate resulting in mineral and bone formation is associated with a number of disorders which can be divided into three principal disorders: (i) metastatic calcification as observed in hypercalcemia or hyperphosphatemia occurs when the calcium-phosphate solubility product in extracellular fluid is exceeded. Amorphous calcium phosphate and subsequently hydroxyapatite deposition occurs at sites such as the kidneys and lungs in hypercalcemia and the periarticular subcutaneous tissues in hyperphosphatemia. (ii) Dystrophic calcification, as associated with dermatomyositis and a variety of connective tissue disorders, occurs when mineral is deposited into metabolically impaired or necrotic tissue (Whyte 1993). (iii) Ectopic ossification can be classified into either a) bone acquired from trauma induced myositis ossificans such as follows neurological injury, or inherited as in the extremely disabling condition of fibrodysplasia ossificans progressiva (FOP). In FOP, endochondral ossification appears to be the mechanism of bone formation although how and why bone forms within the skeletal muscles is not known (Smith and Triffitt, 1986). The crucial involvement of aberrant BMP production in extraskeletal sites in the initiation of bone formation in the musculature in this condition appears highly likely (Kaplan et al., 1990).

VI. SUMMARY

The advent of cell culture and molecular biology has seen an explosion in information on the osteoblast lineage, bone cell differentiation, and the regulation and modulation of bone turnover. However, despite the identification of receptors on cells of the osteoblast lineage to virtually every factor implicated in bone formation (including TGFβ, FGF, PDGF, BMPs, PGs, steroid hormones) and their study in *in vitro* systems, their exact functions during proliferation and differentiation of the osteoblast lineage remain indistinct. The relationships of these factors to normal bone formation and remodeling, fracture repair, and to pathological conditions such as osteoporosis, Paget's disease, and ectopic ossification are also obscure. The challenge for the biologist and clinician remains the unraveling of the regulation and modulation of the pathways of osteoblast generation and function, and an understanding of the precise and ordered mechanism of bone formation seen *in vivo*.

ACKNOWLEDGMENTS

J.T.T. is a Permanent Member of the Medical Research Council External Scientific Staff and the authors acknowledge the generous support of the Medical Research Council.

REFERENCES

Abou-Samra, A.B., Juppner, H., Force, T., Freeman, M.W., Kong, X. F., Schipani, E., Urena, P., Richards, J., Bonventre, J. V., Potts, Jr., J. T., Kronenberg, H. M., and Segre, G.V. (1992). Expression cloning of a common receptor for parathyroid hormone and parathyroid hormone-related peptide from rat osteoblast-like cells: A single receptor stimulates intracellular accumulation of both cAMP and inositol triphosphates and increases intracellular free calcium. Proc. Natl. Acad. Sci. U.S.A. 89, 2732-2736.

Andrew, J.G., Hoyland, J.A., Freemont, A.J., and Marsh, D.R. (1995). Platelet-derived growth factor expression in normally healing human fractures. Bone 16, 455-460.

Ashton, B.A., Allen, T.D., Howlett, C.R., Eaglesom, C.C., Hattori, A., and Owen, M. (1980). Formation of bone and cartilage by marrow stromal cells in diffusion chambers in vivo. Clin. Orthop. Rel. Res. 151, 294-307.

Aubin, J.E., Turksen, K., and Heersche, J.N.M. (1993). Osteoblastic cell lineage. In: Cellular and Molecular Biology of Bone. (Noda, M., Ed.), pp. 1-45. Academic Press Inc., San Diego.

Basle, M.F., Fournier, J.G., Rozenblatt, S., Rebel, A., and Bouteille, M. (1986). Measles virus RNA detected in Paget=s disease bone tissue by in situ hybridization. J. Gen. Virol. 67, 907-913.

Bellows, C.G., Aubin, J.E., and Heersche, J.N.M. (1995). Aluminium inhibits both initiation and progression of mineralization of osteoid nodules formed in differentiating rat calvaria cell cultures. J. Bone Miner. Res. 10, 2011-2016.

Benayahu, D., Kletter, Y., Zipori, D., and Weintroub, S. (1989). Bone marrow-derived stromal cell line expressing osteoblastic phenotype in vitro and osteogenic capacity in vivo. J. Cell Sci. 140, 1-7.

Bennett, J.H., Joyner, C.J., Triffitt, J.T., and Owen, M.E. (1991). Adipocytic cells cultured from marrow have osteogenic potential. J. Cell Sci. 99, 131-139.

Beresford, J.N. (1989). Osteogenic stem cells and the stromal system of bone and marrow. Clin. Orthop. Rel. Res. 240, 270-280.

Beresford, J.N., Bennett, J.H., Devlin, C., Leboy, P.S., and Owen, M.E. (1992). Evidence for an inverse relationship between the differentiation of adipocytic and osteogenic cells in rat marrow stromal cell cultures. J. Cell Sci. 102, 341-351.

Beresford, J.N., Joyner, C.J., Devlin, C., and Triffitt, J.T. (1994). The effects of dexamethasone and 1,25-dihydroxyvitamin D_3 on osteogenic differentiation of human marrow stromal cells in vitro. Arch. Oral Biol. 39, 941-947.

Bianco, P., Riminucci, M., Bonnucci, E., Termine, J. D., and Robey, P. G. (1993). Bone sialoprotein (BSP) secretion and osteoblast differentiation: Relationship to bromodeoxyuridine incorporation, alkaline phosphatase and matrix deposition. J. Histochem. Cytochem. 41, 183- 191.

Bikle, D. D. (1994). Role of vitamin D, its metabolites and analogs in the management of osteoporosis. Rheum. Dis. Clin. North Am. 20, 759-775.

Boyce, B. F., Aufdemorte, T. B., Garrett, I. R., Yates, A. J. P., and Mundy, G. R. (1989). Effects of interleukin-1 on bone turnover in normal mice. Endocrinology 125, 1142-1150.

Bruder, S. P., Gazit, D., Passi-Even, L., Bab, I., and Caplan, A. I. (1990). Osteochondral differentiation and the emergence of stage-specific osteogenic cell-surface molecules by bone marrow cells in diffusion chambers. Bone Min. 11, 141-151.

Bruder, S. P., Horowitz, M.C., Mosca, J.D., and Haynesworth, S.E. (1997). Monoclonal antibodies reactive with human osteogenic cell surface antigens. Bone 21, 225-235.

Bruder, S.P., Lawrence, E.G., and Haynesworth, S.E. (1995). Identification and characterization of human osteogenic cell surface differentiation antigens. J. Bone Miner. Res. 10 (Suppl. I), S416.

Burgener, D., Bonjour, J. P., and Caverzasio, J. (1995). Fluoride increases tyrosine kinase activity in osteoblastlike cells: Regulatory role for the stimulation of cell proliferation and Pi transport across the plasma membrane. J. Bone Miner. Res. 10, 164-171.

Burger, E.H., Klein Nulend, J., Van der Plas, A., and Nijweide, P.J. (1995). Function of osteocytes in bone—their role in mechanotransduction. J. Nutr. 125, 2020S-2023S.

Burkhardt, R., Kettner, G., Bohm, W., Schmidmeier, M., Schlao, R., Frisch, B., Mallman, B., Eisenmenger, W., and Gilg, T. (1987). Changes in trabecular bone, haematopoiesis and bone marrow vessels in aplastic anaemia, primary osteoporosis and old age: A comparative histomorphometric study. Bone 8, 157-164.

Canalis, E. (1980). Effect of insulinlike growth factor I on DNA and protein synthesis in cultured rat calvaria. J. Clin. Invest. 66, 709-719.

Canalis, E. (1986). Interleukin-1 has independent effects on deoxyribonucleic acid and collagen synthesis in cultures of rat calvariae. Endocrinology 118, 74.

Canalis, E., Centrella, M., Burch, W., and McCarthy, T.L. (1989). Insulinlike growth factor I mediates selective anabolic effects of parathyroid hormone in bone cultures. J. Clin. Invest. 83, 60-65.

Caplan, A.I. (1995). Osteogenesis imperfecta, rehabilitation medicine, fundamental research and mesenchymal stem cells. Connective Tiss. Res. (In press.)

Celeste, A.J., Ianazzi, J.A., Taylor, R.C., Hewick, R.M., Wang, E.A., and Wozney, J.M. (1990). Identification of transforming growth factor family members present in bone-inductive protein purified from bovine bone. Proc. Natl. Acad. Sci. U.S.A. 87, 9843-9847.

Centrella, M., Horowitz, M.C., Wozney, J.M., and McCarthy, T.L. (1994). Transforming growth factor-β gene family members and bone. Endocrine Reviews 15, 27-39.

Chen, T.L., Bates, R.L., Dudley, A., Glenn Hammonds, Jr., R., and Amento, E.P. (1991). Bone morphogenetic protein-2b stimulation of growth and osteogenic phenotypes in rat osteoblastlike cells: Comparison with TGF-β_1. J. Bone Miner. Res. 6, 1387-1393.

Cheng, S-L., Yang, J.W., Rifns, L., Zhang, S.F., and Avioli, L.V. (1994). Differentiation of human bone marrow osteogenic stromal cells in vitro. Induction of the osteoblast phenotype by dexamethasone. Endocrinology 134, 277-286.

Chrischilles, E., Shireman, T., and Wallace, R. (1994). Costs and health effects of osteoporotic fractures. Bone 15, 377-386.

Colvard, D.S., Eriksen, E. E., Keeting, P.E., Wilson, E.M., Lubahn, D.B., French, F.S., Riggs, B.L., and Spelsburg, T.C. (1989). Identification of androgen receptors in normal human osteoblastlike cells. Proc. Natl. Acad. Sci. U.S.A. 86, 854-857.

Consensus Development Conference. (1991). Prophylaxis and treatment of osteoporosis. Osteoporosis Int. 1, 114-117.

Dallas, S.L., Zaman, G., Pead, M.J., and Lanyon, L.E. (1993). Early strain-related changes in cultured embryonic chick tibiotarsi parallel those associated with adaptive modeling in vivo. J. Bone Miner. Res. 8, 251-259.

Dempster, D.W., Cosman, F., Parisien, M., Shen, V., and Lindsay, R. (1993). Anabolic actions of PTH on bone. Endocrine Reviews 14, 690-709.

Dodds, R.A., Ali, N., Pead, M.J., and Lanyon, L.E. (1993). Early loading-related changes in the activity of glucose 6-phosphate dehydrogenase and alkaline phosphatase in osteocytes and periosteal osteoblasts in rat fibulae in vivo. J. Bone Miner. Res. 8, 261-267.

Doty, S.B. and Schofield, B.H. (1976). Enzyme histochemistry of bone and cartilage cells. Prog. Histochem. Cytochem. 8, 1-38.

Ducy, P., Zhang, R., Geoffroy, V., Ridall, A.L., and Karsenty, G. (1997). Osf2/Cbfa1: A transcriptional activator of osteoblast differention. Cell 89, 747-754.

Ecarot-Charrier, B., Shepard, N., Charette, G., Grynpas, M., and Glorieux, F.H. (1988). Mineralization in osteoblast cultures: A light and electron microscopic study. Bone 9, 147-154.

Eriksen, E.F., Colvard, D.S., Berg, N.J., Graham, M.L., Mann, K.G., Spelsberg, T.C., and Riggs, B.L. (1988). Evidence of estrogen receptors in normal human osteoblastlike cells. Science 241, 84-86.

Ernst, M., Heath, J.K., and Rodan, G.A. (1989). Estradiol effects on proliferation, messenger ribonuclear acid for collagen and insulinlike growth factor I and parathyroid hormone-stimulated adenylate cyclase activity in osteoblastic cells from calvaria and long bones. Endocrinology 125, 825-833.

Ernst, M. and Froesch, E.R. (1987). Triiodothyronine stimulates proliferation of osteoblastlike cells in serum-free culture. FEBS Lett. 220, 163-166.

Ettinger, B., Genant, M. K., and Cann, C.E. (1985). Long-term estrogen replacement therapy prevents bone loss and fractures. Ann. Intern. Med. 102, 319-324.

Farley, J.R., Tarbaux, N., Hell, S., and Baylink, D.J. (1990). Mitogenic action(s) of fluoride on osteoblastlike cells: Determinants of the response in vitro. J. Bone Miner. Res. 5, S107-S113.

Farley, J.R., Wergedal, J.E., and Baylink, D.J. (1983). Fluoride directly stimulates proliferation and alkaline phosphatase activity of bone-forming cells. Science 222, 330-332.

Fell, H.B. (1952). The effect of hypervitaminosis A on embryonic limb-bones cultivated in vitro. J. Physiol. 116, 320-349.

Franceschi, R.T., Iyer, B.S., and Cui, Y. (1994). Effects of ascorbic acid on collagen matrix formation and osteoblast differentiation in murine MC3T3-E1 cells. J. Bone Miner. Res. 9, 843- 854.

Friedenstein, A.J. (1976). Precursor cells of mechanocytes. Int. Rev. Cytol. 47, 327-355.

Friedenstein, A.J. (1973). Determined and inducible osteogenic precursor cells. In: *Hard Tissue Growth, Repair, and Remineralization*. (Elliott, K. and Fitzsimmons, D.W., Eds.) Ciba Foundation Symposium 11, Associated Sci. Pubs., Amsterdam, The Netherlands.

Friedenstein, A.J., Chailalhyan, R.K., and Gerasimov, U.V. (1987). Bone marrow osteogenic stem cells: In vitro cultivation and transplantation in diffusion chambers. Cell Tissue Kinetics 20, 263-272.

Gimble, J.M., Morgan, C., Kelly, K., Wu, X., Dandapani, V., Wang, C-S, and Rosen, V. (1995). Bone morphogenetic proteins inhibit adipocyte differentiation by bone marrow stromal cells. J. Cellular Biochem. 58, 1-10.

Girasole, G., Jilka, R.L., Passeri, G., Boswell, S., Boder, G., Williams, D.C., and Manolagas, S.C. (1992). 17 -Estradiol inhibits interleukin-6 production by bone marrow-derived stromal cells and osteoblasts in vitro: A potential mechanism for the antiosteoporotic effect of estrogens. J. Clin. Invest. 89, 883-891.

Goodman, W.G., Coburn, J.W., Ramirez, J.A., Slatopolsky, E., and Salusky, I.B. (1993). Renal osteodystrophy in adults and children. In: Primer on the Metabolic Bone Diseases and Disorders of Mineral Metabolism. (Favus, M.J., Ed.), 2nd ed., pp. 304-323. Raven Press, New York.

Gowen, M., Wood, D.D., Ihrie, E.J., McGuire, M.K.B., and Russell, R.G.G. (1983). An interleukin 1Blike factor stimulates bone resorption in vitro. Nature 306, 378-380.

Gowen, M., Wood, D.D., and Russell, R.G.G. (1985). Stimulation of the proliferation of human bone cells in vitro by human monocyte products with interleukin-1 activity. J. Clin. Invest. 75, 1223.

Grigoriadis, A.E., Heersche, J.N.M., and Aubin, J.E. (1988). Differentiation of muscle, fat, cartilage, and bone from progenitor cells present in a bone-derived clonal cell population. Effect of dexamethasone. J. Cell Biol. 106, 2139-2151.

Gronthos, S., Graves, S.E., Ohta, S., and Simmons, P.J. (1994). The Stro-1^+ fraction of adult human bone marrow contains the osteogenic precursors. Blood 84, 4164-4173.

Gundle, R., Joyner, C.J., and Triffitt, J.T. (1995). Human bone tissue formation in diffusion chamber culture in vivo by bone-derived cells and marrow stromal fibroblastic cells. Bone 16, 597-601.

Ham, A.W. (1969). Histology. 6th Ed. J.B. Lippincott Co., Philadelphia.

Harris, S.A., Enger, R.J., Riggs, B.L., and Spelsberg, T.C. (1995). Development and characterization of a conditionally immortalized human fetal osteoblastic cell line. J. Bone Miner. Res. 10, 178-186.

Harris, S.E., Bonewald, L.F., Harris, M.A., Sabatini, M., Dallas, S., Feng, J., Ghosh- Choudhury, N., Wozney, J., and Mundy, G.R. (1994). Effects of TGF on bone nodule formation and expression of bone morphogenetic protein-2, osteocalcin, osteopontin, alkaline phosphatase, and type-I collagen mRNA in prolonged cultures of fetal rat calvarial osteoblasts. J. Bone Miner. Res. 9, 855-863.

Hauschka, P.V., Mavrakos, A.E., Iafrati, M.D., Doleman, S.E., and Klagsbrun, M. (1986). Growth factors in bone matrix. J. Biol. Chem. 261, 12665-12674.

Hauschka, P.V., Chen, T.L., and Mavrakos, A.E. (1988). Polypeptide growth factors in bone matrix. In: Cell and Molecular Biology of Vertebrate Hard Tissues. Ciba Foundation Symposium 136.(Evered, D. and Harnett, S., Eds.), pp. 207-230. John Wiley and Sons, Chichester.

Hauschka, P.V., and Carr, S.A. (1982). Calcium-dependent α-helical structure in osteocalcin. Biochemistry 21, 2538-2547.

Hayden, J.M., Mohan, S., and Baylink, D.J. (1995). The insulinlike growth factor system and the coupling of formation to resorption. Bone 17, 93S-98S.

Haynesworth, S.E., Goshima, J., Goldberg, V.M., and Caplan A.I. (1992). Characterization of cells with osteogenic potential from human marrow. Bone 13, 81-88.

Horisaka, Y., Okamoto, Y., Matsumoto, N., Yoshimura, Y., Kawada, J., Yamashita, K., and Tomomichi, T. (1991). Subperiosteal implantation of bone morphogenetic protein adsorbed to hydroxyapatite. Clin. Orthop. 268, 303-312.

Howard, G.A., Carlson, C.A., and Baylink, D.J. (1982). Coupled bone metabolism in vitro: Embryonic chick limbs in organ culture. Prog. Clin. Biol. Res. 101, 259.

Hughes, F.J., Collyer, J., Stanfield, M., and Goodman, S.A. (1995). The effects of bone morphogenetic protein-2, -4, and -6 on differentiation of rat osteoblast cells in vitro. Endocrinology 136, 2671-2677.

Hurley, L.S., Gowan, J., and Milhaud, G. (1969). Calcium metabolism in manganese-deficient and zinc-deficient rats. Proc. Soc. Exp. Biol. Med. 130, 856-860.

Ibaraki, K., Termine, J.D., Whitson, S.W., and Young, M.F. (1992). Bone matrix mRNA expression in differentiating fetal bovine osteoblasts. J. Bone Miner. Res. 7, 743-754.

Inaoki, T., Lean, J.M., Bessho, T., Chow, J.W.M., Mackay, A., Kokubo, T., and Chambers, T.J. (1995). Sequential analyses of gene expression after an osteogenic stimulus: c-fos expression is induced in osteocytes. Biochem. Biophys. Res. Commun. 217, 264-270.

Ishida, H., Bellows, C.G., Aubin, J.E., and Heersche, J.N.M. (1995). Tri-iodothyronine (T_3) and dexamethasone interact to modulate osteoprogenitor cell differentiation in fetal rat calvaria cell cultures. Bone 16, 545-549.

Islam, A., Glomski, C., and Henderson, E.S. (1990). Bone lining (endosteal) cells and hematopoiesis: A light microscopic study of normal and pathologic human bone-marrow in plastic embedded sections. Anat. Rec. 227, 300-306.

Johnson, E.E., Urist, M.R., and Finerman, G.A.M. (1992). Resistant nonunions and partial or complete segmental defects of long bones. Treatment with implants of a composite of human bone morphogenetic protein (BMP) and autolyzed, antigen-extracted, allogeneic (AAA) bone. Clin. Orthop. Rel. Res. 277, 229-237.

Joyner, C.J., Bennett, A., and Triffitt, J.T. (1997). Identification and enrichment of human osteoprogenitor cells by using differentiation state-specific monoclonal antibodies. Bone 21, 1-6.

Kalu, D.N. (1991). The ovariectomized rat model of postmenopausal bone loss. Bone Miner. 15, 175-191.

Kaplan, F.S., Tabas, J.A., and Zasloff, M.A. (1990). Fibrodysplasia Ossificans Progressiva: A clue from the fly? Calcif. Tiss. Int. 47, 117-125.

Kasperk, C.H., Wergedal, J.E., Farley, J.R., Linkhart, T.A., Turner, R.T., and Baylink, D.J. (1989). Androgens directly stimulate proliferation of bone cells in vitro. Endocrinology 124, 1576-1578.

Kasugai, S., Oida, S., Iimura, T., Arai, N., Takeda, K., Ohya, K., and Sasaki, S. (1995). Expression of prostaglandin E receptor subtypes in bone: Expression of EP_2 in bone development. Bone 17, 1-4.

Kawaguchi, H., Pilbeam, C.C., Harrison, J.R., and Raisz, L.G. (1995). The role of prostaglan din's in the regulation of bone modulation. Clin. Orth. Rel. Res. 313, 36-46.

Keeting, P.E., Scott, R.E., Colvard, D.S., Anderson, M.A., Oursler, M.J., Spelsberg, T.C., and Riggs, B.L. (1992). Development and characterization of a rapidly proliferating, well-differentiated cell line derived from normal adult human osteoblastlike cells transfected with SV40 large T-antigen. J. Bone Miner. Res. 7, 127-132.

Khokher, M.A. and Dandona, P. (1990). Fluoride stimulates [3H] thymidine incorporation and alkaline phosphatase production by human osteoblasts. Metabolism 39, 1118-1121.

Komori, T., Yagi, H., Nomura, S., Yamaguchi, A., Sasaki, K., Deguchi, K., Shimizu, Y., Bronson, R.T., Gao, Y.H., Inada, M., Sato, M., Okamoto, R., Kitamura, Y., Yoshiki, S., and Kishimoto, T. (1997). Targeted disruption of Cbfa1 results in a complete lack of bone formation owing to maturational arrest of osteoblasts. Cell 89, 755-764.

Koshihara, Y., Hoshi, K., and Shiraki, M. (1992). Enhancement of mineralization of human osteoblasts by vitamin K_2 (Menaquinone 4). J. Clin. Exp. Med. 161, 439-440 Kream, B.E., Peterson, D.N., and Raisz, L.G. (1990). Parathyroid hormone blocks the stimulatory effect of insulinlike growth factor-I on collagen synthesis in cultured 21-day fetal rat calvariae. Bone 11, 411-415.

Lajtha, L.G. (1982). Cellular kinetics of haemopoiesis. In: Blood and Its Disorders. (Hardisty, R.M. and Weatherall, D.J., Eds), 2nd ed., pp. 57-74. Blackwell Scientific Publications, Oxford, England.

Lanyon, L., Rubin, C., and Baust, G. (1986). Modulation of bone loss during calcium insufficiency by controlled dynamic loading. Calcif. Tissue Int. 38, 209-216.

Lanyon, L.E. (1993). Osteocytes, strain detection, bone modeling, and remodeling. Calcif. Tissue Int. 53 (Suppl.), S102-S107.

Levin, M.E., Boisseau, V.C., and Avioli, L. (1976). Effects of diabetes mellitus on bone mass in juvenile and adult onset diabetes. N. Engl. J. Med. 294, 241-245.

Locklin, R.M., Williamson, M.C., Beresford, J.N., Triffitt, J.T., and Owen, J.T. (1995). In vitro effects of growth factors and dexamethasone on rat marrow stromal cells. Clin. Orthop. Rel. Res. 313, 27-35.

Lukert, B.P. and Raisz, L.G. (1990). Glucocorticoid-induced osteoporosis: Pathogenesis and management. Ann. Intern. Med. 112, 352-364.

Luyten, F.P., Cunningham, N.S., Ma, S., Muthukumaran, N., Hammonds, R.G., Nevins, W.B., Wood, W.I., and Reddi, A.H. (1989). Purification and partial amino-acid sequence of osteogenin, a protein initiating bone differentiation. J. Biol. Chem. 264, 13377-13380.

Mammi, G.I., Rocchi, R., Cadossi, R., Massari, R.L., and Traina, G.C. (1993). The electrical stimulation of tibial osteotomies. Clin. Orthop. Rel. Res. 288, 246-253.

Manolagas, S.C. (1995). Role of cytokines in bone resorption. Bone 17, 635-675.

Manolagas, S.C., Bellido, T., and Jilke, R.L. (1995). New insights into the cellular, biochemical, and molecular basis of postmenopausal and senile osteoporosis: Roles of IL-6 on SP-130. Int. J. Immunopharmacol. 17, 109-116.

Manolagas, S.C., and Jilka, R.L. (1995). Bone marrow, cytokines, and bone remodeling. In: Mechanisms of Disease (Epstein, F.H., Ed.). N. Engl. J. Med. 332, 305-311.

Marcelli, C., Yates, A.J., and Mundy, G.R. (1990). In vivo effects of human recombinant transforming growth factor-β on bone turnover in normal mice. J. Bone Miner. Res. 5, 1087-1096.

Mardon, H.J., Bee, J., von der Mark, K., and Owen, M.E. (1987). Development of osteogenic tissue in diffusion chambers from early precursor cells in bone marrow of adult rats. Cell Tissue Res. 250, 157-165.

McCarthy, T.L., Centrella, M., and Canalis, E. (1989). Regulatory effect of insulinlike growth factors I and II on bone collagen synthesis in rat calvaria cultures. Endocrinology 124, 301-309.

McSheehy, P.M.J. and Chambers, T.J. (1986). 1,25-dihydroxyvitamin D stimulates rat osteoblastic cells to release a soluble factor that increases osteoblastic bone resorption. J. Clin. Invest. 80, 425-429.

Mikuni-Takagaki, Y., Kakai, Y., Satoyoshi, M., Kawano, E., Suzuki, Y., Kawase, T., and Saito, S. (1995). Matrix mineralization and the differentiation of osteocytelike cells in culture. J. Bone Miner. Res. 10, 231-242.

Miller, S.C., De Saint George, L., Bowman, B.M., and Jee, W.S.S. (1989). Bone lining cells: Structure and function. Scanning Microscopy 3, 953-961.

Mohan, S., and Baylink, D.J. (1991). Bone growth factors. Clin. Orthop. Rel. Res. 263, 30-48.

Mori, S., Jee, W.S.S., and Li, X.J.(1992).Production of new trabecular bone in osteopenic ovariectomized rats by prostaglandin E_2. Calcif. Tissue Int. 50, 80-87.

Morrison, N.A., Qi, J.C., Tokita, A., Kelly, P.J., Crofts, L., Nguyen, T.V., Sambrook, P. N., and Eisman, J.A. (1994). Prediction of bone density from vitamin D receptor alleles. Nature 367, 284-287.

Mundlos, S., Otto, F., Mundlos, C., Mulliken, J.B., Aylsworth, A.S., Albrigh, S., Lindhout, D., Cole, W.G., Henn, W., Knoll, J.H.M., Owen, M.J. , Mertelsmann, R., Zabel, B.U., and Olsen, B.R. (1997). Mutations involving the transcription factor CBFA1 cause cleidocranial dysplasia. Cell 89, 773-779.

Mundy, G.R., Roodman, G.D., Bonewald, L.F., Oreffo, R.O.C., and Boyce B. (1991). Assays for bone resorption and bone formation. In: Peptide Growth Factors, Part C. Methods in Enzymology. 198, 502-510.

Mundy, G.R., Shapiro, J.L., Bandelin, J.G., Canalis, E.M., and Raisz, L.G. (1976). Direct stimulation of bone resorption by thyroid hormones. J. Clin. Invest. 58, 529-534.

Nagai, H., Tsukuda, R., and Mayahara, H. (1995). Effects of basic fibroblast growth factor (bFGF) on bone formation in growing rats. Bone 16, 367-373.

Nagata, T., Bellows, C.G., Kasugai, S., Butler, W.T., and Sodek, J. (1991). Biosynthesis of bone proteins [SPP-1 (secreted phosphoprotein-1, osteopontin), BSP (bone sialoprotein), and SPARC (osteonectin)] in association with mineralized-tissue formation by fetal-rat calvarial cells in culture. J. Biochem. 274, 513-520.

Narbeitz, R., Stumpf, W.E., Sar, M., Huang, S., and Deluca, H.F. (1983). Autoradiographic localisation of target cells for 1,25-dihydroxyvitamin D in bones from fetal rats. Calif. Tiss. Int. 35, 177-182.

Nefussi, J.R., Boy-Lefevre, M.R., Boulekbache, H., and Forest, N. (1985). Mineralization in vitro of matrix formed by osteoblasts isolated by collagenase digestion. Differentiation 29, 160-168.

Ng, K.W., Manjii, S.S., Young, M.F., and Findlay, D.M. (1989). Opposing influences of glucocorticoid and retinoic acid on transcriptional control in proosteoblasts. Mol. Endocrinol. 3, 2079-2085.

Noda, M. and Camilliere, J.J. (1989). In vivo stimulation of bone formation by transforming growth factor beta. Endocrinology 124, 2991-2994.

Onyia, J.E., Bidwell, J., Herring, J., Hulman, and Hock, J.M. (1995). In vivo, human parathyroid hormone fragment (hPTH 1-34) transiently stimulates immediate early-response gene expression, but not proliferation, in trabecular bone cells. Bone 17, 479-484.

Oreffo, R.O.C., Bord, S., and Triffitt, J.T. (1998). Skeletal progenitor cells and ageing human populations. Clin. Sci. 94, 549-555.

Oreffo, R.O.C., Francis, M.J.O., and Triffitt, J.T. (1985). Vitamin A effects on UMR 106 cells are not mediated by specific cytosolic receptors. Biochem. J. 232, 599-603.

Oreffo, R.O.C., Virdi, A.S., and Triffitt, J.T. (1997). Modulation of osteogenesis and adipogenesis by human serum in human bone marrow cultures. Eur. J. Cell Biol. 74, 251-261.

Oreffo, R.O.C., Wells, N., and Johnstone, D. (1991). The presence of prostaglandin receptors on osteoclasts and osteoblasts. J. Bone Miner. Res. 6 (Suppl 1), S208.

Otto, F., Thornell, A.P., Crompton, T., Denzel, A., Gilmour, K.C., Rosewell, I.R., Stamp, G.W.H., Beddington, R.S.P., Mundlos, S., Olsen, B.R., Selby, P.B., and Owen, M.J. (1997). *Cbfa1*, a candidate gene for cleidocranial dysplasia syndrome, is essential for osteoblast differentiation and bone development. Cell 89, 765-771.

Owen, M. (1963). Cell population kinetics of an osteogenic tissue. J. Cell Biol. 19, 19-32.

Owen, M. (1985). Lineage of osteogenic cells and their relationship to the stromal system. In: Bone and Mineral Research, 3. (Peck, W.A., Ed.), pp. 1-25. Elsevier Science Publishers.

Owen, M. and Friedenstein, A.J. (1988). Stromal stem cells: Marrow-derived osteogenic precursors. In: Cell and Molecular Biology of Vertebrate Hard Tissues. Ciba Foundation Symposium 136. pp. 42-60. Wiley, Chichester.

Owen, T.A., Aronow, M.S., Barone, L.M., Bettencourt, B., Stein, G.S., and Lian, J.B. (1991). Pleiotropic effects of vitamin D on osteoblast gene expression are related to the proliferative and differentiated state of the bone cell phenotype: Dependency upon basal levels of gene expression, duration of exposure, and bone matrix competency in normal rat osteoblast cultures. Endocrinology 128, 1496-1504.

Owen, T.A., Aronow, M., Shalhoub, V., Barone, L.M., Wilming, L., Tassinari, M.S., Kennedy, M.B., Pockwinse, S., Lian, J.B., and Stein, G.S. (1990). Progressive development of the rat osteoblast phenotype in vitro: Reciprocal relationships in expression of genes associated with osteoblast proliferation and differentiation during formation of the bone extracellular matrix. J. Cell Physiol. 143, 420-430.

Palumbo, C., Palazzini, S., and Marotti, G. (1990). Morphological study of intercellular function during osteocyte differentiation. Bone 11, 401-406.

Peck, W.A., Carpenter, J., Messinger, K., and DeBra, D. (1973). Cyclic 3'5'-adenosine monophosphate in isolated bone cells. Response to low concentrations of parathyroid hormone. Endocrinology 92, 692.

Pfeilschifter, J., and Mundy, G.R. (1987). Modulation of type β transforming growth factor activity in bone cultures by osteotropic hormones. Proc. Natl. Acad. Sci. USA. 84, 2024-2028.

Pfeilschifter, J., Wolf, U., Naumann, A., Minne, H.W., Mundy, G.R., and Ziegler, R. (1990). Chemotactic response of osteoblastlike cells to transforming growth factorβ. J. Bone Miner. Res. 5, 825-830.

Pitaru, S., Kotev-Emeth, S., Noff, D., Kaffuler, S., and Savion, N. (1993). Effect of basic fibroblast growth factor on the growth and differentiation of adult stromal bone marrow cells: enhanced development of mineralized bonelike tissue in culture. J. Bone Miner. Res. 8, 919- 929.

Pottratz, S.T., Bellido, T., Mocharla, H., Crabb, D., and Manolagas, S.C. (1994). 17β-estradiol inhibits expression of human interleukin-6 promoter-reporter constructs by a receptor-dependent mechanism. J. Clin. Invest. 93, 944-950.

Price, P.A. (1985). Vitamin K-dependent formation of bone gla protein (osteocalcin) and its function. Vitam. Horm. 42, 65-108.

Raisz, L.G. (1965). Bone resorption in organ culture. Factors influencing the response to parathyroid hormone. J. Clin. Invest. 44, 103-116.

Raisz, L.G., and Fall, P.M. (1990). Biphasic effects of prostaglandin E_2 on bone formation in cultured fetal rat calvariae: Interaction with cortisol. Endocrinology 126, 1654-1659.

Rawlinson, S.C.F., Mohan, S., Baylink, D.J., and Lanyon, L.E. (1993). Exogenous prostacyclin, but not PGE_2, produces similar responses in both G6PD activity and RNA production as mechanical loading, and increases IGF-II release, in adult cancellous bone in culture. Calcif. Tissue Int. 53, 324-329.

Rawlinson, S.C.F., Mosley, J.R., Suswillo, R.F.L., Pitsillides, A.A., and Lanyon, L.E. (1995). Calvarial and limb bone cells in organ and monolayer culture do not show the same early responses to dynamic mechanical strain. J. Bone Miner. Res. 10, 1225-1232.

Reddi, A.H. (1992). Regulation of cartilage and bone differentiation by bone morphogenetic proteins. Curr. Opinion Cell Biol. 4, 850-855.

Reddi, A.H. (1995). Bone morphogenetic proteins, bone marrow stromal cells, and mesenchymal stem cells. Maureen Owen revisited. Clin. Orthop. Rel. Res. 313, 115-119.

Richelson, L.S., Wahner, H.W., Melton, L.J., and Riggs, B.L. (1984). Relative contributions of aging and estrogen deficiency to postmenopausal bone loss. N. Engl. J. Med. 311, 1273-1275.

Robey, P.G., Fedarko, N.S., Hefferan, T.E., Bianco, P., Vetter, U.K., Grzesik, W., Friedenstein, A., Van der Pluijm, G., Mintz, K.P., Young, M.F., Kerr, J.M., Ibaraki, K., and Heegaard, A.M. (1993). Structure and molecular regulation of bone matrix proteins. J. Bone Min. Res. 8, S483-S487.

Rodan, G.A. and Rodan, S.B. (1984). Expression of the osteoblast phenotype. In: Bone and Mineral Research. Edit.2. (Peck, W.A., Ed.), pp. 244-286. Elsevier, Amsterdam.

Roodman, G.D. (1995). Osteoclast function in Paget=s disease and multiple myeloma. Bone 17, 57S-61S.

Rubin, C.T., Donahue, H.J., Rubin, J.E., and McLeod, K.J. (1993). Optimization of electric field parameters for the control of bone remodeling: Exploitation of an indigenous mechanism for the prevention of osteopenia. J. Bone Miner. Res. 8, S573-S581.

Saito, T., Dennis, J.E., Lennon, D.P., Young, R.G., and Caplan, A.I. (1995). Myogenic expression of mesenchymal stem cells within myotubes of mdx mice in vitro and in vivo. Tissue Engineering, (In press.)

Schipani, E., Kruse, K., and Juppner, H. (1995). A constitutively active mutant PTH/PTHrP receptor in Jansen type metaphyseal chondrodysplasia. Science 268, 98-100.

Schmid, C., Ernst, M., Zapf, J., and Froesch, E.R. (1989). Release of insulinlike growth factor carrier proteins by osteoblasts: Stimulation by estradiol and growth hormone. Biochem. Biophys. Res. Commun. 160, 788-794.

Scutt, A., and Bertram, P. (1995). Bone marrow cells are targets for the anabolic actions of prostaglandin E_2 on bone: Induction of a transition from nonadherent to adherent osteoblast precursors. J. Bone Miner. Res. 10, 474-487.

Sharrard, W.J.W. (1990). A double-blind trial of pulsed electromagnetic fields for delayed union of tibial fractures. J. Bone Joint Surg. 72B, 347-355.

Silver, J.J., and Einhorn, T.A. (1995). Osteoporosis and aging. Clin. Orthop. Rel. Res. 316, 10-20.

Siminovitch, L., McCulloch, E.A., and Till, J.E. (1963). The distribution of colony forming cells among spleen colonies. J. Cell Comp. Physiol. 62, 327-336.

Simmons, P.J., Kent, G.N., Jilka, R.L. Scott, D.M., Fallon, M., and Cohn, D.V. (1982). Formation of bone by isolated, cultured osteoblasts in millipore diffusion chambers. Calcif. Tissue Int. 34, 291-294.

Simmons, P.J. and Torok-Storb, B. (1991). Identification of stromal cell precursors in human bone marrow by a novel monoclonal antibody STRO-1. Blood 78, 55-62

Skerry, T.M., Bitensky, L., Chayen, J., and Lanyon, L.E. (1989). Early strain-related changes in enzyme activity in osteocytes following bone loading in vivo. J. Bone Miner. Res. 4, 783-788.

Slovik, D.M., Rosenthal, D.I., Doppelt, S.H., Potts, Jr, J.T., Daly, M.A., Campbell, J.A., and Neer, R.M. (1986). Restoration of spinal bone in osteoporotic men by treatment with human parathyroid hormone (1-34) and 1,25-dihydroxyvitamin D. J. Bone Miner. Res. 1, 377-381.

Smith, R. and Triffitt, J.T. (1986). Bones in muscles: The problems of soft tissue ossification. Quart. J. Med., (new series) 61, 985-990.

Søgaard, C. H., Mosekilde, Li., Richards, A., and Mosekilde, Le. (1994). Marked decrease in trabeculae bone quality after five years of sodium fluoride therapy—assessed by biomechanical testing of iliac crest bone biopsies in osteoporotic parients. Bone 15, 393-399.

Stein, G.S. and Lian, J.B. (1995). Molecular mechanisms mediated proliferation- differentiation interrelationships during progressive development of the osteoblast phenotype: Update 1995. Endocrine Reviews 4, 290-297.

ten Dijke, P., Yamashita, H., Sampath, T.K., Reddi, A.H., Estevez, M., Riddle, D.L., Icijo, H., Heldin, C.H., and Miyazono, K. (1994). Identification of type-I receptors for osteogenic protein-1 and BMP-4. J. Biol. Chem. 269, 16985-16988.

Termine, J.D. (1993). Bone matrix proteins and the mineralization process. In: Primer on the metabolic bone diseases and disorders of mineral metabolism (Favus, M.J., Ed.), 2nd ed., pp. 21-25. Raven Press, New York.

Triffitt, J.T. (1987a). Initiation and enhancement of bone formation. Acta Orthop. Scand. 58, 673-684.

Triffitt, J.T. (1987). The special proteins of bone tissue. (1987). Clin. Sci. 72, 399-408.

Triffitt, J.T. (1996). The stem cell of the osteoblast. In: *Principles of Bone Cell Biology*, (Bilizekian, J., Raisz, L., and Rodan, G, Eds.), pp. 39-50. Academic Press, San Diego.

Triffitt, J.T., Joyner, C.J., Oreffo, R.O.C., and Virdi, A.S. (1998). Osteogenesis: Bone development from primitive progenitors. Biochem. Soc. Trans. 26, 21-27.

Turner, R.T., Riggs, L., and Spielsberg, T.C. (1994). Skeletal effects of estrogen. Endocrine Reviews 15, 129-154.

Van Der Plas, A. and Nijweide, P.J. (1992). Isolation and purification of osteocytes. J. Bone Miner. Res. 7, 389-396.

Van Vlasselaer, P., Falla, N., Snoek, H., and Mathein, E. (1994). Characterization and purification of osteogenic cells from murine bone marrow by two-color cell sorting using anti-Sca-1-monoclonal antibody and wheat germ agglutinin. Blood 84, 753-763.

Waller, E.K., Olivens, J., Lund Johansen, F., Huang, S., Nguyen, M., Gur, G.R., and Terstappen, L. (1995). 'The common stem cell= hypothesis reevaluated: Human fetal bone marrow contains separate populations of hematopoietic and stromal progenitors. Blood 85, 2422-2435.

Wang, E.A., Rosen, V., D=Alessandro, J.S., Bauduy, M., Cordes, P., Harada, T., Israel, D.I., Hewick, R.M., Kerns, K.M., LaPan, P., Luxenberg, D.P., McQuaid, D., Moutsatos, I., Nove, J., and Wozney, J.M. (1990). Recombinant human bone morphogenetic protein induces bone formation. Proc. Natl. Acad. Sci. U.S.A. 87, 2220-2224.

Wergedal, J.E., Mohan, S., Lundy, M., and Baylink, D.J. (1990). Skeletal growth factor â and other growth factors known to be present in bone matrix stimulate proliferation and protein synthesis in human bone cells. J. Bone. Miner. Res. 5, 179-186.

Whyte, M.P. (1993). Extraskeletal calcification and ossification. In: Primer On the Metabolic Bone Diseases and Disorders of Mineral Metabolism. (Favus, M., Ed.), pp. 386-395. Raven Press, New York.

Williams, G.R., Bland, R., and Sheppard, M.C. (1995). Retinoids modify regulation of endogenous gene expression by vitamin D_3 and thyroid hormone in three osteosarcoma cell lines. Endocrinology 136, 4304-4314.

Wong, G.L., and Cohn, D.V. (1975). Target cells in bone for parathormone and calcitonin are different: Enrichment for each cell type by sequential digestion of mouse calvaria and selective adhesion to polymeric surfaces. Proc. Natl. Acad. Sci. U.S.A. 72, 3167.

Wozney, J.M. (1993). Bone morphogenetic proteins and their gene expression. In: Cellular and Molecular Biology of Bone. (Noda, M., Ed.), pp. 131-167. Academic Press, San Diego.

Wozney, J.M., Rosen, V., Celeste, A.J., Mitsock, L.M., Whitters, M.J., Kriz, R.W., Hewick, R.M., and Wang, E.A. (1988). Novel regulators of bone formation: Molecular clones and activities. Science 242, 1528-1534.

Yamaguchi, A. and Kahn, A.J. (1991). Clonal osteogenic cell lines express myogenic and adipocytic developmental potential. Calcif. Tissue Int. 49, 221-225.

Yasko, A.W., Lane, J.M., Fellinger, E.J., Rosen, V., Wozney, J.M., and Wang, E.A. (1992). The healing of segmental bone defects, induced by recombinant human bone morphogenetic protein (rhBMP-2): A radiographic, histological, and biochemical study in rats. J. Bone Joint Surg. 74-A, 659-670.

Young, R.W. (1962). Regional differences in cell generation time in growing rat tibia. Exp. Cell Res. 26, 562-567.

Zhou, H., Hammonds, R.G., Findlay, D.M., Fuller, D.M., Martin, T.J., and Ng, K.W. (1991). Retinoic acid modulation of mRNA levels in malignant, nontransformed, and immortalized osteoblasts. J. Bone Miner. Res. 6, 767-777.

OSTEOBLAST RECEPTORS

Janet E. Henderson and David Goltzman

I. INTRODUCTION

The osteoblast regulates skeletal growth and homeostasis not only by depositing and mineralizing the extracellular bone matrix but also by transducing signals which lead to the ultimate destruction of that matrix by osteoclasts. To this end, mature osteoblasts retain autocrine control over their own growth and differentiation

Advances in Organ Biology
Volume 5B, pages 499-512.

ISBN: 0-7623-0390-5

as well as regulating the recruitment and activity of osteoclasts. Until quite recently much of the information regarding the regulation of bone growth and turnover had been accumulated in the clinical setting in individuals with deficiency or excess of systemically active hormones such as $1,25(OH)_2D_3$ or parathyroid hormone (PTH). Recent improvements in cell and tissue culture techniques coupled with the explosion in transgenic technology has permitted the development of highly sophisticated models in which to study the molecular mechanisms involved in bone cell biology. With the use of these models it is becoming increasingly evident that the orderly progression of osteoblast development is regulated by the interaction of numerous local and systemic factors through their interaction with specific receptors expressed by the osteoblastic cells. In this chapter we briefly review the current state of knowledge regarding those receptors and the signal transduction pathways which lead to gene regulation following their activation.

II. G PROTEIN-COUPLED RECEPTORS

The importance of PTH in the regulation of skeletal homeostasis has been recognized since the early part of the century and, yet, its mechanism of action in bone is only now being elucidated. An initial breakthrough came with the hypothesis that the cellular target of the action of PTH might be the osteoblast (Rodan and Martin 1981). This hypothesis achieved experimental validity when it was shown that PTH directly stimulated cAMP accumulation in UMR-106 osteosarcoma cells *in vitro* (Partridge et al., 1982). Subsequently, *in vivo* studies using an autoradiographic approach demonstrated the presence of PTH receptors both on mature osteoblasts and preosteoblastic cells initially termed parathyroid hormone target cells or PT cells (Rouleau et al., 1988, 1990). The picture has been additionally complicated by the revelation that a PTH-related protein (PTHrP) binds to the same receptor as PTH on osteoblastic cells and activates the same transduction pathways with equal efficacy (Abou-Samra et al., 1992; Pausova et al. 1994). Amino-terminal PTH and PTHrP can elicit both anabolic and catabolic activity when infused into rats (Thompson et al., 1988; Hock et al., 1989). On the other hand, endogenously produced PTHrP regulates the growth and differentiation of skeletal cells in an autocrine/paracrine manner, a function which does not appear to be sub-served by PTH (Amizuka et al., 1995). Taken together, these observations suggest the existence of a highly complex interrelationship between the systemic hormone (PTH) and the locally produced protein of the same gene family, i.e., PTHrP.

Much of the early work characterizing the structure-function relationships between the PTH/PTHrP receptor and its ligand was accomplished using the rodent UMR and ROS 17/2.8 osteoblastic cell lines (Chorev and Rosenblatt 1994). Cloning of the common PTH/PTHrP receptor from these sources (Abou-Samra et al., 1992; Pausova et al., 1994) has enabled the construction of chimeric and mutant receptors to more precisely define the mechanistic interactions between the multiple

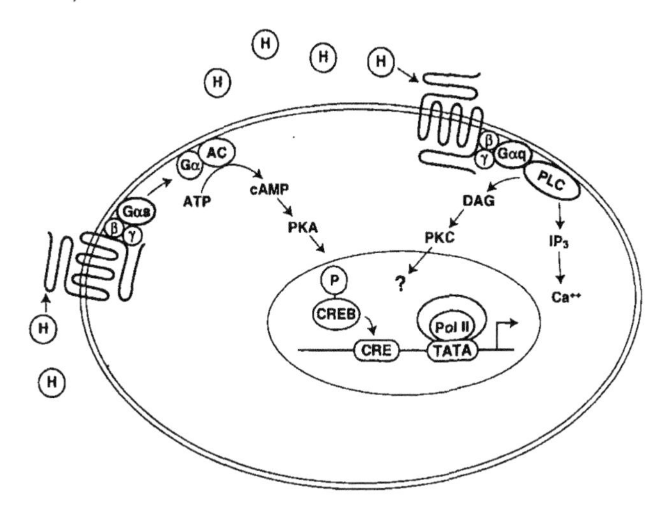

Figure 1. G-protein-coupled receptors. Ligands such as PTH, PTHrP, and PGs interact with sequences in the extracellular domain and in the transmembrane loops of their cognate G protein coupled receptors. This interaction is thought to promote a conformational change allowing for the intracellular interaction of the sub-unit of the heterotrimeric guanyl nucleotide binding protein (G) with either adenylate cyclase (AC) (G_s) or with phospholipase C (PLC) (G_q). In the former case, a cascade of intracellular events is initiated which culminates in actions such as transcriptional activation of genes containing a cAMP response element (CRE). In the latter instance, signaling through either PKC or intracellular calcium (Ca^{2+}) is initiated. Abbreviations: CREB, cAMP response element; DAG, diacyl glycerol; H, hormone; IP_3, inositol triphosphate; PKA, protein kinase A; PLC, phospholipase C; Pol II, DNA polymerase II.

signal transduction pathways linked to this receptor in skeletal cells. In addition, it has permitted identification of the PTH-receptor in bone and bone marrow by *in situ* hybridization (Amizuka et al., 1995), thus verifying results from previous *in vivo* binding studies identifying PTH target cells in bone marrow (Rouleau et al., 1998).

The seven transmembrane-spanning G protein-coupled receptors characteristically interact with their ligand through amino acids in the extracellular domain and in the membrane-spanning regions (Figure 1). They commonly interact with the heterotrimeric G proteins via the intracellular COOH-terminus (Segre, 1994). Diversity among the a subunits of the G proteins confers specificity to the signaling cascade, $G\alpha_s$ being linked to adenylate cyclase (AC) and $G\alpha_q$ to phospholipase C (PLC) (Simon et al., 1991). Binding of amino-terminal fragments of PTH or PTHrP to their cognate receptor results in transduction through both enzymatic pathways in osteoblastic cells (Azarani et al., 1995). The conversion of ATP to cAMP stimulates protein kinase A (PKA) which is responsible for the phosphorylation of cAMP response element binding protein (CREB) allowing it to associate with other proteins (such as ATF) and bind to the cAMP response element (CRE) of target genes (Siddhanti and

Quarles 1994). Alternatively, activation of PLC results in the generation of inositol triphosphates (IP_3) and stimulation of intracellular calcium release as well as eliciting protein kinase C (PKC) activity through diacylglycerol (DAG).

Other G protein-coupled receptor agonists which directly influence osteoblast activity are the prostaglandins (PGs), primarily those of the E series. Like PTHrP, PGs are produced by osteoblasts within the bone microenvironment and have a similar bifunctional role in stimulating both anabolic and catabolic activity. These actions are also mediated, at least in part, by cAMP (Kawaguchi et al., 1995). Their local production in bone appears to be regulated by a variety of growth factors stored in bone such as transforming growth factors (TGFs), platelet-derived growth factor (PDGF), and fibroblast growth factor (FGF) as well as by systemic hormones such as PTH.

III. STEROID HORMONE RECEPTORS

Steroid hormones act by binding to specific, intracellular receptors (Figure 2). The hormone/receptor complex then translocates to the nucleus where it interacts, as either a monomer or dimer, with a recognition element (HRE) in the promoter region of responsive genes. In the case of vitamin D, two classes of response elements

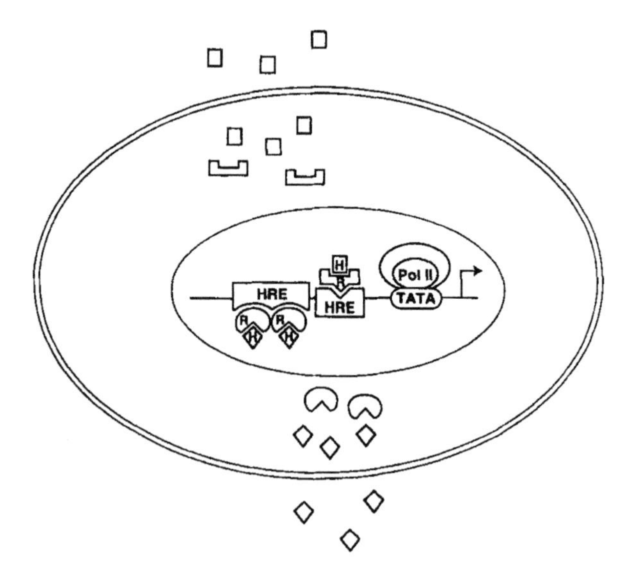

Figure 2. Steroid hormone receptors. Steroid hormones traverse the lipid bilayer of the cell membrane and bind to intracellular receptor elements. The hormone/receptor complex translocates to the nucleus where it exerts its influence on osteoblastic activity by binding, either as a monomer or a dimer, to a recognition element (HRE) in the promoter region of responsive genes. Dimers can be either homodimers or heterodimers. Abbreviations: H, hormone; R, receptor.

(VDRE) have been identified, one of which interacts with the monomeric vitamin D receptor (VDR) and the other with heterodimeric VDR complexed with a retinoid receptor (Truss and Beato, 1993). These VDREs may transduce either positive or negative transcriptional responses (Kremer et al., 1996). Molecular analysis of the receptor proteins has revealed a modular organization of functional domains required for ligand binding, dimerization, nuclear translocation, and DNA binding. In osteoblasts, the presence of receptors for vitamin D, estrogen, retinoic acid, glucocorticoids, and thyroid hormone has been well documented (Canalis, 1993).

In contrast to vitamin D and estrogen, which are generally associated with anabolic effects in bone, glucocorticoids elicit an overall catabolism of bone and their excessive use can result in osteopenia. This effect is mediated directly, by inhibiting replication of osteoblast precursors, through inhibition of expression of type I collagen, osteocalcin, and tissue inhibitor of metalloproteinase (TIMP), as well as through stimulation of interstitial collagenase expression (Delaney et al., 1994). In addition, glucocorticoids can exert an indirect inhibitory effect on bone growth by downregulating expression of insulin-like growth factor (IGF-1) and its binding proteins in osteoblasts (Delaney et al., 1994).

The specific response of an osteoblast to a particular class of hormone can be a function of modulation of receptor numbers, differential expression of HREs, requirements for additional transcriptional transactivators, or interaction with other signaling pathways (Truss and Beato, 1993). It has been suggested that cross-talk may occur, for example, between the PKC and estrogen receptor (ER) signaling pathways (Migliaccio et al., 1993). Steroid hormone receptors are phosphoproteins, and it has been suggested that their phosphorylation status could influence both intracellular trafficking and interactions with other transcriptional regulatory proteins and DNA (Truss and Beato, 1993).

IV. SERINE/THREONINE RECEPTOR KINASES

The osteoinductive actions of the bone morphogenetic protein (BMP)/(TGFβ)/activin super-family of growth factors are mediated, at the local level, through their high affinity receptors on osteoblasts and their precursors. While the BMPs appear to be important in the early induction of osteogenic precursors from mesenchymal stem cells, TGFβ is thought to influence later stages of osteoblast development (Reddi, 1995). However, the observed complexity of stimulatory and inhibitory effects of TGFβ on bone cell function could be related to differentiation-dependent interactions between BMP and TGFβ signal transduction (Centrella et al., 1995). TGFβ is released from osteoblastic cells as an inactive heterodimer which is activated to its disulfide-bonded mature form by proteolytic cleavage of the latency associated (precursor) peptide (Bonewald and Dallas, 1995). Mature TGFβ can then interact with either cell surface receptors or with matrix proteins such as decorin and thrombospondin. TGFβ signaling is

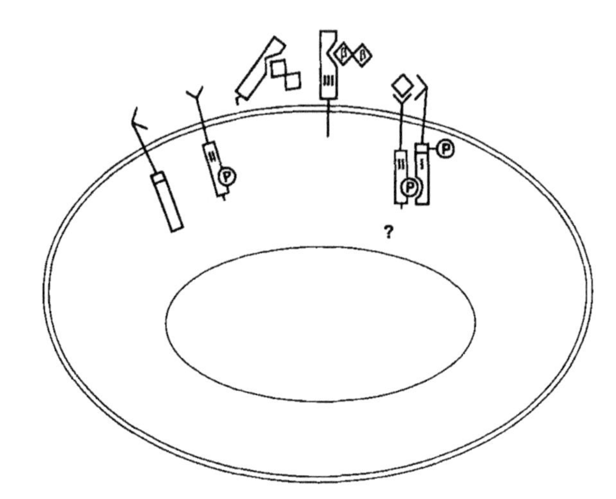

Figure 3. Serine/threonine receptor kinases. The TGFβ super-family of growth factors, including BMPs and activin, exert their pleiotropic effects on cells of the osteoblastic lineage by binding to cell surface receptors with intrinsic serine/threonine kinase activity. In the case of TGFβ, the ligand is thought to bind in the dimeric form to either soluble or membrane-bound type III receptors (betaglycan) which then present the ligand to the membrane-bound type II receptor. This interaction allows for recruitment of the type I receptor into the complex and its subsequent phosphorylation by the constitutively active type II receptor kinase. The intracellular cascades linked to receptor activation are currently under study but the family of signal transducers known as Smad has been implicated in cell signaling by members of the TGFβ super family.

thought to occur through a heteromeric receptor system involving two transmembrane receptors which exhibit serine-threonine kinase activity (receptors I and II) and a membrane-anchored proteoglycan (receptor III or betaglycan) (Figure 3).

The ligand, which is first presented by the type III receptor to the constitutively active type II kinase for binding, is subsequently recognized by the type I receptor which is then recruited into the complex and transphosphorylated by the type II receptor. In addition to its essential role as a membrane-bound molecule which captures and presents TGFβ to the signaling receptors, betaglycan has also been identified as a soluble protein and is thought to act as a receptor antagonist in this capacity (Attisano et al., 1994). Perhaps the ratio of membrane-bound to soluble betaglycan is a determinant of the availability of TGFβ *in vivo*, and hence, of its ultimate biological activity.

An increasing amount of information is accumulating about events that occur between stimulation of these serine/threonine kinase-type receptors and gene regulation induced by members of the TGFβ family of peptide growth factors. A recent report has, however, identified a novel member of the mammalian MAPKKK family, TAK1, which was shown to be activated in MC3T3-E1 osteoblastic cells in response to TGFβ stimulation (Yamaguchi et al., 1995). The observations suggest that this kinase may be involved in the initiation of a cascade culminating in osteoblastic gene regulation. Cloned BMP receptors also appear to have intrinsic serine/threonine kinase activity (Yamaji et al., 1994). Recently a protein family of signal transducers termed Smad (Similar to mothers against decapeptaplegic) has been discovered and implicated in cell signaling and stimulation of gene transcription by members of the TGFβ super family (Massagué et al., 1997).

V. RECEPTOR TYROSINE KINASES

The pleiotropic actions of many bone-derived growth factors are mediated through activation of cell surface receptors which have intrinsic tyrosine kinase activity. These receptor tyrosine kinases (RTKs) are grouped into four classes on the basis of structural similarity (Ullrich and Schlessinger, 1990). Class I (epidermal growth factor EGF/TGFα family) and class II (IGF family receptors), are characterized by cysteine-rich repeats in the extracellular domain, while those of class III (PDGF family) and Class IV (FGF family) contain immunoglobulin-like repeats in the ligand-binding domain. Signal transduction is initiated by ligand-induced receptor oligomerization and autophosphorylation leading to recruitment of intracellular signaling proteins.

Cells of the osteoblast lineage express receptors from each of the RTK subclasses and secrete many of their ligands, which also appear to be stored within the bone microenvironment (Baylink et al., 1993). These observations suggest that the RTK agonists modulate bone cell function in an autocrine/paracrine manner. Of these agonists the IGFs appear to be of major importance. IGF-1 expression by osteoblasts has been shown to be regulated by PTH, estrogen, thyroid hormone, growth hormone, and glucocorticoids and may well mediate some of the anabolic effects of these hormones on skeletal metabolism (Siddhanti and Quarles, 1994). Osteoblasts also produce IGF binding proteins which are thought to modulate the action of IGF in bone by sequestering the ligand in the vicinity of its receptor, thus prolonging its bioactivity (Andress and Birnbaum, 1992).

Although little is known regarding the actions of FGFs on mature osteoblasts, class IV receptors for the FGF ligands have been demonstrated on osteoblastic cells (Siddhanti and Quarles, 1994). Various mutations in different FGF receptor subtypes have been shown to have profound consequences for skeletal development (Muenke and Schell, 1995). FGFs comprise a family of nine ligands which bind with high affinity to four different receptor subclasses. The prototypical ligand, basic (β) FGF is secreted by mature osteoblasts and has also been shown to be stored in bone tissue (Baylink et al., 1993). Like other RTKs, ligand binding induces receptor oligomerization, resulting in transphosphorylation of one cytoplasmic domain by the other (Figure 4). It appears that binding of βFGF to its high affinity receptor is dependent on the presence of cell surface heparin sulfate proteoglycan (HSPG), which functions as a low affinity receptor for this growth factor (Ornitz and Leder, 1992).

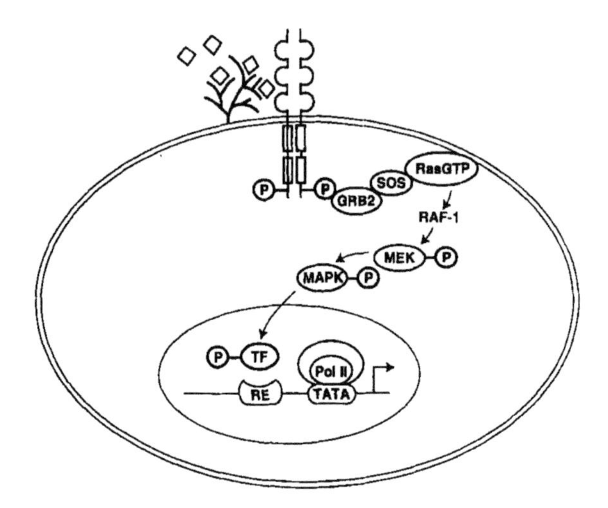

Figure 4. Receptor tyrosine kinases. Receptors with intrinsic tyrosine kinase activity mediate the effects of many bone derived growth factors such as insulin-like growth factor (IGF), platelet-derived growth factor (PDGF), epidermal growth factor (EGF), and fibroblast growth factor (FGF). In the case of basic FGF (βFGF), ligand binding to the high affinity receptor is facilitated by interaction with cell-surface heparin sulfate proteoglycan (HSPG). Ligand binding results in receptor dimerization and tyrosine phosphorylation of adaptor proteins such as GRB2. The subsequent activation of a kinase cascade culminates in phosphorylation of transcription factors (TF) by mitogen activated protein kinase (MAPK), also known as extracellular signal regulated kinase (ERK), allowing them to interact with response elements (RE) in the promoter proximal region of target genes. Abbreviations: GRB2, growth factor receptor binding protein; SOS, son of sevenless; MEK, MAPK kinase or ERK kinase.

It has been proposed that the interaction between βFGF and HSPG is a prerequisite for binding to the high affinity RTK, perhaps by facilitating presentation of the ligand to the RTK (Klagsbrun and Baird, 1991). Autophosphorylation of the RTK leads to recruitment of the growth factor receptor binding protein (GRB2) with the mammalian homologue of a complex of the *Drosophila* son of sevenless gene product (SOS), which then forms a link to the membrane-bound protein, Ras (Marx, 1993). Subsequent GDP/GTP exchange on Ras renders the protein competent to activate the MAP kinase cascade, culminating in phosphorylation of transcriptional proteins and gene regulation.

VI. CYTOKINE RECEPTORS

The osteoblast derives from a pluripotent, mesenchymal stem cell present in bone marrow which, when isolated and cultured under appropriate conditions, gives rise to cells which form calcified nodules of bone (Grigoriadis et al., 1988). These cells have been shown to express cytokines such as those of the interleukin (IL-6) family (Malaval et al., 1995) which are thought to act in an autocrine manner in the regulation of osteoblast growth and differentiation (Franchimont and Canalis, 1995; Romas et al., 1995) in addition to stimulating osteoclastogenesis (Manolagas and Jilka, 1995). Furthermore, expression of these cytokines by osteoblastic cells has been reported to be regulated by calciotropic hormones such as PTH/PTHrP and $1,25(OH)_2D_3$ (Romas et al., 1995), by gonadal steroids (Manolagas and Jilka, 1995), and by the soluble form of the IL-6 receptor (sIL-6r) (Franchimont and Canalis, 1995). Knowledge of the mechanisms regulating interactions between IL-6-related cytokines and their receptors on osteoblastic cells is, therefore, of obvious importance to our overall understanding of skeletal growth and metabolism.

Like other cytokines, those belonging to the IL-6 family are pleiotropic and exhibit functional overlap. They bind to hetero-oligomeric cell surface receptors which share a common signal transduction β subunit, gp130, that associates with different ligand binding a subunits (Figure 5). IL-6 target cells can express both the a subunit and gp130 on the surface or only gp130. In the latter case, a soluble form of the a subunit binds the ligand prior to complexing with gp130 (Hibi et al., 1996).

Figure 5. Cytokine receptors. Receptors which transduce the signals of cytokines such as IL-6 are not thought to have intrinsic kinase activity but appear to facilitate the phosphorylation of associated proteins including the Janus kinase (JAK). IL-6 binds to either the soluble or membrane-bound form of the α subunit of the receptor which then forms a hetero-oligomeric complex with the signal transducing β subunit, or gp130. Activation of gene transcription via the JAK/STAT pathway is thought to occur through a series of phosphorylation events resulting in dimerization of phosphorylated STAT monomers which bind to response elements of target genes. Abbreviations C, cytokine; RE, response element.

Following ligand binding it has been proposed that activation of the Janus kinase (JAK) signal transducers of transcription (STAT) pathway is initiated through tyrosine phosphorylation of a JAK which associates constitutively with gp130 (Lutticken et al., 1993). Phosphorylation of selected tyrosine residues in the intracellular domain of gp130 by the activated JAK generates recognition motifs for the amino acid region known as the src-homology 2 (SH2) domain of the adaptor protein STAT3 (Stahl et al., 1995). Upon binding, STAT3 is activated after being phosphorylated by JAK. Subsequent association of the activated STATs forms the transcriptionally active dimer which binds to responsive genes (Wen et al., 1995). The mitogen activated protein kinase (MAPK) pathway is an alternative route which has been implicated in signal transduction of ligands binding to the IL-6 receptor (Hibi et al., 1996).

VII. RECEPTORS INVOLVED IN OSTEOBLAST/MATRIX INTERACTIONS

A. Matrix Attachment

In addition to the receptors required for transduction of signals from cytokines, growth factors, and hormones, the surface of osteoblasts is equipped with receptors which allow it to interact with its extracellular environment. These include integrins, which serve to anchor the cell through their interaction with matrix components, as well as receptors for plasminogen activators (PAs) which are involved in the proteolytic degradation of surrounding tissue by the serine protease, plasmin.

Integrins represent a family of transmembrane receptors which appear as heterodimers of α and β subunits on the surface of cells. These receptors mediate attachment of the cell to R-G-D recognition motifs in components of the extracellular matrix, such as collagen and vitronectin, which represent the ligands for integrin receptors (Parsons, 1996). Ligand binding results in receptor clustering and coupling of the cytoplasmic domain to cytoskeletal proteins. This forms the basis for complexing of additional cytoskeletal elements to form a focal adhesion. It is believed that the β subunits initiate coupling to the cytoskeleton while the α subunits confer receptor specifictiy (Clark and Brugge, 1995).

Transduction through integrin receptors may occur via pathways that are also linked to growth factor receptors (Ras-GTP and MAPK) as well as G protein coupled receptors (PLC and PKC) (Clark and Brugge, 1995). Recruitment of the nonreceptor protein tyrosine kinase, or focal adhesion kinase (FAK), appears to play a pivotal role in focal adhesion formation and may link integrin receptor activation to common downstream events (Parsons, 1996)

In osteoblastic cells, binding of type I collagen to the $\alpha_2\beta_1$ integrin heterodimer was shown to regulate cellular differentiation by decreasing TGFβ binding to its receptor (Takeuchi et al., 1996). It was suggested that collagen, synthesized and deposited under the stimulatory influence of TGFβ, binds to the $\beta_2\beta_1$ integrin receptor resulting in inhibition of TGFβ binding by an as yet unidentified mechanism. In view of the known inhibitory influence of TGFβ on osteoblast differentiation, this could represent a mechanism whereby the osteoblast is permitted to proceed through its program of differentiation.

B. Matrix Degradation

Like many other cells, osteoblasts have been shown to express components of the plasminogen activator/plasmin proteolytic pathway in a well-regulated manner. Thus, urokinase-type plasminogen activator (uPA) (Allan et al., 1991), the uPA receptor (uPAR) (Rabbani et al., 1991), and plasminogen activator inhibitor 1 (PAI-1) (Allan et al., 1991) are all expressed by osteoblast-like cells. In addition to localizing proteolytic activity to the cell surface by binding uPA, activation of the uPAR by the amino-terminal fragment of uPA results in mitogenic activity in osteoblastic cells (Rabbani et al., 1992). The existence of this nonproteolytic role for uPA, as well as evidence for reciprocal changes in the expression of uPA and PAI-1, suggests the presence of a self-limiting system of proteolysis coupled to a mechanism for matrix replacement (Allan et al., 1991). This may indeed have important implications for the coupling of bone resorption to bone formation which is necessary for the maintenance of skeletal integrity.

VIII. CONCLUSIONS

Osteoblasts express many different types of receptors which permit them to interact with factors present in the prevailing microenvironment and, thus, to respond in an appropriate manner. Systemic factors, such as PTH, steroidal and thyroid hormones, can influence osteoblastic activity to conform to the needs of the organism as a whole. These agents function by initiating signaling cascades via either plasma membrane or nuclear receptors. Osteoblasts also possess plasma membrane receptors for a wide variety of locally produced factors such as PTHrP, PGs, BMPs, TGFs, and cytokines. Through autocrine/paracrine loops these agents are capable of profoundly influencing the growth and differentiated function of osteoblasts and their precursors. Additionally, specific plasma membrane receptors allow osteoblasts to interact with matrix proteins and also permit the cell to reshape the matrix by facilitating site-directed, focal proteolysis. The entire repertoire of osteoblast receptors, whether at the plasma membrane or in the nucleus, as well as the availability of their cognate ligands, may vary with the degree of cellular maturity as well as with the developmental stage of the organism as a whole. Finally, post-receptor signal transduction mechanisms may mediate discrete responses to a given ligand but cross-talk between signaling pathways also offers the opportunity for redundancy in controlling certain downstream events. This complex scheme involving interplay between genetically programmed and environmentally influenced events enables the osteoblast to fulfill its pivotal role of coordinating both the anabolic and catabolic activity necessary for skeletal growth and remodeling.

REFERENCES

Abou-Samra, A.B., Jüppner, H., Force, T., Freeman, M.W., Kong, X.F., Schipani, E., Urena, P., Richards, J., Bonventre, J.V., Potts, Jr., J.T., Kronenberg, H.M., and Segre, G. V. (1992). Expression cloning of a common receptor for parathyroid hormone and parathyroid hormone-related peptide from rat osteoblastlike cells: A single receptor stimulates intracellular accumulation of both cAMP and inositol trisphosphates and increases intracellular free calcium. Proc. Natl. Acad. Sci. USA, 89, 2732-2736.

Allan, E.H., Zeheb, R., Gelehrter, T.D., Heaton J.H., Fukumoto, S., Yee, J.A., and Martin, T.J. (1991). Transforming growth factor β inhibits plasminogen activator (PA) activity and stimulates production of urokinase-type PA, PA inhibitor-1 mRNA and protein in rat osteoblastlike cells. J. Cell Physiol., 149, 34-43.

Amizuka, N., Karaplis, A.C., Henderson, J.E., Warshawsky, H., Lipman, M.L., Matsuki, Y., Ejiri, S., Tanaka, M., Izumi, N., Ozawa, H., and Goltzman, D. (1995). Haploinsufficiency of parathyroid hormone-related peptide (PTHrP) results in abnormal postnatal bone development. Dev. Biol., 175, 166-176.

Andress, D.L. and Birnbaum, R.S. (1992). Human osteoblast-derived insulinlike growth factor (IGF) binding protein-5 stimulates osteoblast mitogenesis and potentiates IGF action. J. Biol. Chem. 267, 22467-22472.

Attisano, L., Wrana, J.L., Lopez-Casillas, F., and Massague, J. (1994). TGF-β receptors and actions. Biochim Biophys Acta 1222, 71-80.

Azarani, A., Orlowski, J., and Goltzman, D. (1995). Parathyroid hormone and parathyroid hormone-related peptide activate the Na^+/H^+ exchanger NHE-1 isoform in osteoblastic cells (UMR-106) via a cAMP-dependent pathway. J. Biol. Chem. 270, 23166-23172.

Baylink, D.J., Finkelman, R.D., and Mohan, S. (1993). Growth factors to stimulate bone formation. J. Bone Miner. Res. 8, S565-572.

Bonewald, L.F. and Dallas, S.L. (1994). Role of active and latent transforming growth factor β in bone formation. J. Cell Biochem. 55, 350-357.

Canalis, E., (1993). Regulation of bone remodelling. In: Primer on the Metabolic Bone Diseases and Disorders of Mineral Metabolism. (Favus, M.J., Ed.), pp. 33-37. Raven Press, New York.

Centrella, M., Casinghino, S., Kim, J., Pham, T., Rosen, V., Wozney, J., and McCarthy, T.L. (1995). Independent changes in type-I and type-II receptors for transforming growth factor β induced by bone morphogenetic protein 2 parallel expression of the osteoblast phenotype. Mol. Cell Biol. 15, 3273-3281.

Chorev, M. and Rosenblatt, M. (1994). Structure-function analysis of parathyroid hormone and parathyroid hormone-related protein. In: The parathyroids: Basic and clinical aspects (Bilezekian, J.P., Marcus, R., and Levine, M.A., Eds.) pp. 139-156. Raven Press, New York.

Clark, E.A. and Brugge, J.S. (1995). Integrins and signal transduction pathways: The road taken. Science 268, 233-239.

Delaney, A.M., Dong, Y., and Canalis, E. (1994). Mechanism of glucocorticoid action in bone cells. J. Cell Biochem. 56, 295-302.

Franchimont, N. and Canalis, E. (1995). Soluble interleukin-6 receptor (sIL-6r) enhances the transcriptional autoregulation of interleukin-6 (IL-6) in osteoblasts. J. Bone Miner. Res. 10, S159.

Grigoriadis, A.E., Heersche, Y.N.M., and Aubin, J.E. (1988). Differentiation of muscle, fat, cartilage, and bone from progenitor cells present in a bone-derived clonal cell population: Effect of dexamethasone. J. Cell Biol. 106, 2139-2151.

Hibi, M., Nakajima, K., and Hirano, T. (1996). IL-6 cytokine family and signal transduction: A model of the cytokine system. J. Mol. Med. 74, 1-12.

Hock, J.M., Fonesca, J., Gunness-Hey, M., Kemp, B.E. and Martin, T.J. (1989). Comparison of the anabolic effects of synthetic parathyroid hormone-related protein (PTHrP) 1-34 and PTH 1-34 on bone in rats. Endocrinology 125, 2022-2027.

Kawaguchi, H., Pilbeam, C.C., Harrison, J.R. and Raisz, L.G. (1995). The role of prostaglandins in the regulation of bone metabolism. Clin. Ortho. Rel. Res. 313, 36-46.

Klagsbrun, M. and Baird, A. (1991). A dual receptor system is required for basic fibroblast growth factor activity. Cell 67, 229-231.

Kremer, R., Sebag, M., Champigny, C., Meerovitch, K., Hendy, G.N., White, J., and Goltzman, D. (1996). Identification and characterization of 1,25-dihydroxyvitamin D3-responsive repressor sequences in the rat parathyroid hormone-related peptide gene. J. Biol. Chem., 271, 16310-16316.

Lutticken, C., Wegenka, U.M., Yuan, J., Buschmann, J., Schindler, C., Ziemiecki, A., Harpur, A.G., Wilks, A.F., Yasukawa, K., Taga, T., Kishimoto, T., Barbieri, G., Pellegrini, S., Sendtner, M., Heinrich, P.C., and Horn, F. (1993). Association of transcription factor APRF and protein kinase Jak 1 with the interleukin-6 signal transducer gp 130. Science 263, 89-92.

Malaval, L., Gupta, A.K., and Aubin, J.E. (1995). Leukemia inhibitory factor inhibits osteogenic differentiation in rat calvaria cell cultures. Endocrinology 136, 1411-1418.

Manolagas, S.C. and Jilka, R.J. (1995). Bone marrow cytokines and bone remodeling. N. Engl. J. Med. 332, 305-311.

Marx, J. (1993). Forging a path to the nucleus. Science, 260, 1588-1590.

Massagué, J., Hata, A., and Liu, F. (1997). TGF-β signaling through the Smad pathway. Trends Cell Biol. 7, 187-192.

Migliaccio, S., Wetsel, W.C., Fox, W.M., Washburn, T.F., and Korach, K.S. (1993). Endogenous protein kinase-C activation in osteoblastlike cells modulates responsiveness to estrogen and estrogen receptor levels. Mol. Endocrinol. 7, 1133-1143.

Muenke, M. and Schell, U. (1995). Fibroblast-growth-factor receptor mutations in human skeletal disorders. Trends Genet. 11, 308-313.

Ornitz, D.M., and Leder, P. (1992). Ligand specificity and heparin dependence of fibroblast growth factors 1 and 3. J. Biol. Chem. 267, 16305-16311.

Parsons, J.T. (1996). Integrin-mediated signaling: Regulation by protein tyrosine kinases and small GTP-binding proteins. Curr. Opin. Cell Biol. 8, 146-152.

Partridge, N.C., Kemp, B.E., Livesey, S.A., and Martin, T.J. (1982). Activity ratio measurements reflect intracellular activation of adenosine 3',5'-monophosphate-dependent protein kinase in osteoblasts. Endocrinology 111, 178-183.

Pausova, Z., Bourdon, J., Clayton, D., Mattei, M.G., Seldin, M.F., Janicic, N., Riviere, M., Szpirer, J., Levan, G., Szpirer, C., Goltzman, D., and Hendy, G.N. (1994). Cloning of a parathyroid hormone/parathyroid hormone-related peptide (PTHrP) receptor cDNA from a rat osteosarcoma (UMR 106) cell line. Genomics 20, 20-26.

Rabbani, S.A., Desjardins, J., Bell, A.W., Banville, D., and Goltzman, D. (1991). An amino-terminal fragment of urokinase isolated from a prostatic cancer cell line (PC-3) is mitogenic for osteoblastlike cells. Biochem. Biophys. Res. Comm. 173, 1058-1064.

Rabbani, S.A., Mazar, A.P., Bernier, S.M., Haq, M., Bolivar, I., Henkin, J., and Goltzman, D. (1992). Structural requirements for the growth factor activity of the amino-terminal domain of urokinase. J. Biol. Chem. 267, 14151-14156.

Reddi, A.H. (1995). Bone morphogenetic proteins, bone marrow stromal cells, and mesenchymal stem cells. Clin. Orthop. Rel. Res. 313, 115-119.

Rodan, G.A. and Martin, T.J. (1981). Role of osteoblasts in hormonal control of bone resorption—a hypothesis. Calcified Tiss. Int. 33, 349-351.

Romas, E., Udagawa, N., Hilton, D.J., Martin, T.J., and Ng, K.W. (1995). Osteotropic factors regulate interleukin-11 production by osteoblasts. J. Bone Miner. Res. 10, S142.

Rouleau, M.F., Mitchell, J., and Goltzman, D. (1988). In vivo distribution of parathyroid hormone receptors in bone: Evidence that a predominant target osseous cell is not the mature osteoblast. Endocrinology 123, 187-191.

Rouleau, M.F., Mitchell, J., and Goltzman, D. (1990). Characterization of the major parathyroid hormone target cell in the endosteal metaphysis of rat long bones. J. Bone Miner. Res. 5, 1043-1053.

Segre, G.V. (1994). Receptors for parathyroid hormone and parathyroid hormone-related protein. In: *The Parathyroids: Basic and Clinical Aspects.* (Bilezekian, J.P., Marcus, R., and Levine, M.A., Eds.), pp. 213-218. Raven Press, New York.

Siddhanti, S.R. and Quarles, L.D. (1994). Molecular to pharmacologic control of osteoblast proliferation and differentiation. J. Cell Biochem. 55, 310-320.

Simon, M.I., Strathmann, M.P., and Gautam, N. (1991). Diversity of G proteins in signal transduction. Science 252, 802-808.

Stahl, N., Farrugella, T.J., Boulton, T.G., Zhong, Z., Darnell, J.E., and Yancopoulos, G. (1995). Choice of STATS and other substrates specified by modular tyrosine-based motifs in cytokine receptors. Science 267, 1349-1353.

Takeuchi, Y., Nakayama, K., and Matsumoto, T. (1996). Differentiation and cell surface expression of transforming growth factor-β receptors are regulated by interaction with matrix collagen in murine osteoblastic cells. J. Biol. Chem. 271, 3938-3944.

Thompson, D.D., Seedor, J.G., Fisher, J.E., Rosenblatt, M., and Rodan, G.A. (1988). Direct action of the parathyroid hormonelike human hypercalcemic factor on bone. Proc. Natl. Acad. Sci. USA 85, 5673-5677.

Truss, M. and Beato, M. (1993). Steroid hormone receptors: Interaction with deoxyribonucleic acid and transcription factors. Endocrine Rev. 14, 459-475.

Ullrich, A. and Schlessinger, J. (1990). Signal transduction by receptors with tyrosine kinase activity. Cell 61, 203-212.

Wen, Z., Zhong, Z., and Darnell, J.E. (1995). Maximal activation of transcription by Stat1 and Stat3 requires both tyrosine and serine phosphorylation. Cell 82, 241-250.

Yamaguchi, K., Shirakabe, K., Shibuya, H., Irie, K., Oishi, I., Ueno, N., Taniguchi, T., Nishida, E., and Matsumoto, K. (1995). Identification of a member of the MAPKKK family as a potential mediator of TGF-β signal transduction. Science 270, 2008-2011.

Yamaji, N., Celeste, A.J., Thies, R.S., Song, J.J., Bernier, S.M., Goltzman, D., Lyons, K.M., Nove, J., Rosen, V., and Wozney, J.M. (1994). A mammalian serine/threonine kinase receptor specifically binds BMP-2 and BMP-4. Biochem. Biophys. Res. Comm. 205, 1944-1951.

COLLAGENASE AND OTHER OSTEOBLAST ENZYMES

Anthony Vernillo and Barry Rifkin

I. INTRODUCTION

Bone remodeling depends on the precise regulation of bone resorption and formation. Thus, an imbalance between these processes may lead consequently to pathologic bone loss. Osteoblasts synthesize a collagen-containing matrix which mineralizes to form mature bone. Collagen is the major component of the bone matrix; it represents at least 90% of the organic matrix of bone and is mostly type I collagen (Wright and Leblond, 1981). Destruction or excessive resorption of this

Advances in Organ Biology
Volume 5B, pages 513-528.

ISBN: 0-7623-0390-5

connective tissue matrix is likely a most critical step in the pathogenesis of the lytic bone diseases, including rheumatoid arthritis, skeletal malignancies, and periodontitis. Interstitial collagenase (matrix metalloproteinase-1, MMP-1) activity, as well as the activity of other MMPs (e.g., gelatinases [MMP-2 and MMP-9] and stromelysin [MMP-3]), may be important in both the normal and pathologic remodeling or resorption of the bone collagen extracellular matrix. The activity of the bone cell MMPs is regulated *in vivo* by the tissue inhibitors of MMPs (TIMPs) (Birkedal-Hansen et al., 1993). Bone resorbing hormones (e.g., parathyroid hormone, PTH) act through receptors found on osteoblasts (primary bone-forming cells) that synthesize bone matrix proteins and initiate mineralization (Aubin et al., 1982). Multinucleated osteoclasts are responsible for the resorption of the mineralized matrix (Delaisse and Vaes, 1992).

The interaction of PTH with receptors on osteoblasts initiates events that activate osteoclasts (Mundy, 1992); however, the exact nature of the signals that mediate such cell-to-cell communication is not well understood. Thus, PTH-stimulated resorption apparently involves the interaction of at least two cell types: the osteoblast and the osteoclast. Furthermore, it has been shown that several classes of MMPs (see below) are also secreted by both osteoblastic and osteoclastic cells, suggesting several potential roles for these enzymes in the bone remodeling process (Vaes, 1972).

This chapter will focus largely on osteoblast collagenase and related MMPs and their putative roles in skeletal remodeling as degradative enzymes; related discussion of osteoclast MMPs must also be included. To a lesser extent, a discussion of the plasmin and plasminogen activator system and its role in bone remodeling will be presented in the latter part of this chapter. Finally, the role of alkaline phosphatase, a major osteoblast marker enzyme, in bone metabolism will not be presented because this chapter will focus on degradative enzymes in bone remodeling rather than on the action of enzymes in the mineralization process itself.

II. MATRIX METALLOPROTEINASES AND THEIR INHIBITORS

The MMPs are a large family (MMPs 1–13) containing, in part, the secreted metal-dependent proteases that collectively degrade virtually all components of the extracellular matrix (ECM). These enzymes are secreted in a latent or proform (i.e., proMMP) that must be activated in the extracellular milieu to degrade the ECM (Alexander and Werb, 1989; Matrisian, 1992; Birkedal-Hansen et al., 1993). At least three subclasses of matrix degrading metalloproteinases have been identified and classified by their substrate specificities: the collagenases, which are the only enzymes capable of degrading fibrillar collagens; the gelatinases, which primarily degrade nonfibrillar and denatured collagens; and the stomelysins, which have a broad specificity and degrade matrix proteoglycans, glycoproteins, and some native and denatured collagens. More recently, the MT (membrane-type) -MMP subclass has been identified and is composed of at least four members mainly characterized by the occurrence of a putative transmembrane domain and whose proposed role is the proteolytic activation of other MMPs like gelatinase A and MMP-13 (Knauper et al., 1996). It was also shown that MT1-MMP (MMP-14) is highly expressed in rabbit osteoclasts (Sato et al., 1997).

MMPs are inhibited by the TIMPs (Birkedal-Hansen et al., 1993). TIMP-1, a 28.5 kDa glycoprotein, is produced by many cell types. This protein recognizes the active forms of all of the MMPs, binds them in a 1:1 complex, and inhibits their enzyme activity. A second inhibitor, TIMP-2, has been isolated in a complex with gelatinase A (72kDa gelatinase, MMP-2) and, like TIMP-1, recognizes the active forms of all of the MMPs. TIMPs-1 and 2 are also synthesized by osteoblastic cells (Cook et al., 1994). TIMP-3, identified in chickens and humans, is the newest member of the family of MMP inhibitors and is tightly associated with the ECM (Leco et al., 1992). A balance between levels of the activated metalloproteinases and their natural inhibitors, in addition to their spatial localization, controls the net MMP activity within tissues.

III. THE ROLE OF COLLAGENASE IN BONE REMODELING

The discovery of a latent, trypsin-activated form of collagenase (i.e., procollagenase) in tadpole tissue cultures by Harper and Gross (1972) gave impetus to subsequent studies on the regulation of collagenase. Sakamoto et al. (1975) first reported a correlation between bone resorption in organ culture and the release of collagenase by the addition of PTH extract. Puzas and Brand (1979) first showed collagenolytic activity from enzymatically isolated bone cells (an osteoblast-enriched preparation). Luben and Cohen (1976) also had isolated populations of osteoblasts and provided evidence for a direct hormonal action on cells, presumably through PTH receptors.

That osteoblasts appeared to be the target cells for bone resorbing hormones (e.g., PTH) thus led to the hypothesis, as proposed by Sakamoto and Sakamoto (1982), that osteoblasts played a pivotal role in bone resorption. Osteoblasts were apparently the collagenase-synthesizing cells as demonstrated by immunocytochemistry (i.e., anti-mouse bone collagenase antibody), whereas osteoclasts, the principal resorbing cells of bone, were not reactive for MMP-1 (Sakamoto and Sakamoto, 1984). Therefore, under the influence of bone resorption-stimulating hormones, it was the osteoblast interstitial collagenase (MMP-1) which would degrade the surface type I collagen (osteoid) and initiate resorption (Sakamoto et al., 1979). Livesey et al. (1982) showed that the action of PTH upon osteoblasts was predominantly through activation of adenylate cyclase and activation of cyclic AMP-dependent protein kinase. A number of specific postreceptor events follow such activation, including stimulation of plasminogen activator (PA) (Allan et al., 1986) and the production of collagenase and TIMP (Sakamato et al., 1975; Heath et al., 1984; Partridge et al., 1987). A role for the osteoblast-derived collagenase in bone resorption was further supported by the observations of Chambers and Fuller

(1985) who showed that isolated osteoclasts cultured on calvarial explants did not resorb the mineralized matrix unless the osteoid layer was first removed, either by pretreatment with osteoblasts or collagenase. Thus, collagenase in particular could catalyze a rate-limiting step in bone resorption if this thin layer of osteoid (unmineralized collagen) next to osteoblasts presents a barrier to osteoclastic activity.

In addition to their role as target cells for calciotropic hormones such as PTH, osteoblasts interact with osteoclasts to mediate the resorption of bone. For example, McSheehy and Chambers (1986) had shown that PTH was without effect on isolated and cultured osteoclasts, but if osteoblasts and osteoclasts were cultured together, then PTH enhanced osteoclastic bone resorption. Indeed, the studies of Chambers on the model of disaggregated osteoclasts had been particularly important in establishing such an osteoblast-mediated hormonal stimulation of osteoclastic activity (Chambers, 1982; Chambers et al., 1984). Similarly, additional bone resorption inducers such as 1,25-dihydroxyvitamin D_3 (McSheehy and Chambers, 1987), interleukin-1 (IL-1) (Thomson et al., 1986), or tumor necrosis factor-(TNF) α and TNF-β (Thomson et al., 1987), which had no effect on disaggregated osteoclasts alone, caused a significant increase in osteoclastic resorption when cocultured with osteoblasts. Finally, Rodan and Martin (1981) had also suggested an initiator role (from a morphological point of view) for the osteoblast in bone resorption; in response to PTH, contracted osteoblasts might then yield a greater bone surface for the subsequent attachment of osteoclasts.

Information on bone metabolism rapidly accrued with the availability of cloned rodent osteoblastic cell lines (e.g., UMR 106, ROS 17/2). Agents that had been known to stimulate bone resorption (e.g., PTH, 1,25-dihydroxyvitamin D_3, prostaglandin) in osteoblast-osteoclast cocultures also upregulated the production of collagenase from isolated clones of osteoblastic cells (Partridge et al., 1987). Partridge et al. (1983) had initially characterized the ultrastructural and biochemical properties of four clonal osteogenic osteosarcoma lines (transformed cells); these osteoblastic clones (UMR cells) exhibited a stable phenotype through many passages in culture (e.g., high alkaline phosphatase activity and PTH activation of adenylate cyclase) and are still widely used to study osteoblast structure and function. Otsuka et al. (1984) also assayed collagenase and TIMP from ROS 17/2 cells in culture. We have also shown differences in the proportion of gelatinase species between unstimulated UMR 106-01 and ROS 17/2.8 osteoblastic clones; however, the significance of these differences is not yet well understood. For example, our laboratory demonstrated in preliminary studies that UMR 106-01 cells expressed 72 kDa gelatinase A predominantly, whereas unstimulated ROS 17/2.8 cells had approximately equal proportions of gelatinase A (72 kDa) and B (92 kDa) as analyzed by gelatin zymography (Vernillo et al., unpublished observations). Furthermore, PTH-stimulated UMR cells secreted predominantly gelatinase B (Vernillo et al., unpublished observations).

The mechanism of PTH stimulated collagenase synthesis and the complex regulation of MMP activity by TIMPs from osteoblastic clones was further examined in

UMR 106-01 cells. Partridge et al. (1987) assayed MMP-1 (interstitial collagenase) activity from culture media after limited trypsinization (to activate the latent extracellular collagenase or procollagenase); in response to PTH, the cells produced not only significant amounts of enzyme (12–48 hours) but also TIMP (72–96 hours). The levels of collagenase were markedly curtailed after the appearance of TIMP. These data suggested a complex pattern in the regulation of collagenase and its inhibitor. Furthermore, UMR 106-01 osteoblastic cells synthesized two inhibitors, a 20 kDa TIMP-2 and a 30 kDa TIMP-1 (Roswit et al., 1992). More recently, it was also shown that PTH stimulated a twofold increase in rat TIMP-2 transcription from UMR 106-01 cells; this TIMP shared a high degree of homology with human TIMP-2. However, this stimulation was not inhibited by cycloheximide, suggesting a primary effect of the hormone (Cook et al., 1994). This effect on TIMP-2 was in contradistinction to PTH regulation of MMP-1 in these same cells; the regulation of the latter was protein synthesis-dependent (Clohisy et al., 1994). PTH in its stimulation of TIMP-2 mRNA also appeared to act through a signal transduction pathway involving protein kinase A (PKA) (Cook et al., 1994).

Another subtle mechanism (apart from the action of TIMPs) exists to control collagenase activity. UMR 106-01 osteoblastic cells have a receptor specific for rat collagenase. This receptor functions to eliminate extracellular collagenase by internalizing the enzyme into the cells; such a cell-mediated binding mechanism was rapid and saturable (Omura et al., 1994). Finally, human osteoblasts were also shown to synthesize the MMPs, collagenase, gelatinase B (92 kDa gelatinase, MMP-9), and stromelysin when treated with PTH (Meikle et al., 1992); this particular finding gives the putative role of collagenase additional relevance in the human bone diseases.

Work from Partridge and collaborators (Partridge et al., 1983, 1987; Hamilton et al., 1985; Scott et al., 1992) had expanded the model of Chambers and Fuller (1985) in which the osteoblast played a key role in the *initiation* of bone resorption. Thus, the osteoblast is initially activated by a calciotropic hormone (PTH) that upregulates the synthesis and secretion of procollagenase. Procollagenase may, in turn, be activated through the PA/plasmin pathway (Hamilton et al., 1985) (see more detailed discussion below). Furthermore, our laboratory first showed that a reactive oxygen species, sodium hypochlorous acid (NaOCl), also activated extracellular osteoblastic procollagenase *in vitro* (Ramamurthy et al., 1993) (Table 1).

The generation of reactive oxygen species from inflammatory cells (i.e., neutrophils) had been shown to activate latent neutrophil collagenase and, thus, may also be critical in the pathogenesis of connective tissue degradation (Weiss et al., 1985). Consequently, the activation of the latent collagenase initiates the degradation of osteoid, followed by migration of osteoclasts into the osteoid-free areas to resorb the mineralized matrix. These models and data strongly supported the concept of an osteoclast-mediated degradation of the calcified bone matrix with the osteoblast as the initiator of resorption.

Osteoblasts can contribute to the resorption process and it has long been recognized that the processes of resorption and formation are coupled. The destruction of

Table 1. Effect of Doxycycline on NaOCl Activation of UMR-106-01 Procollagenase[a]

Treatment Groups	*% Collagen Lysis*
Enzyme (E) alone	7.2 ± 0.8
E + 1.2 mM APMA	48.5 ± 4.0
E + 5 mM NaOCl	42.5 ± 3.5[b]
E + 5 mM NaOCl + 400 mM doxycycline	5.2 ± 2.0[c]
E + 5 mM NaOCl + 200 mM doxycycline	12.1 ± 2.0[d]
E + 5 mM NaOCl + 100 mM doxycycline	28.7 ± 3.5[e]
E + 5 mM NaOCl + 50 mM doxycycline	22.0 ± 3.0[f]
E + 5 mM NaOCl + 25 mM doxycycline	36.0 ± 4.0[g]
E + 5 mM NaOCl + 12.5 mM doxycycline	39.0 ± 7.0
E + 5 mM NaOCl + 6.0 mM doxycycline	48.7 ± 6.0

Notes: [a]Results from a collagenase assay for each group represent the mean ± standard deviation (SD) and are compared for statistical significance (*p* value) by analysis of variance (ANOVA).
[b]NaOCl (5 mM) + enzyme compared to 400 mM doxycycline, $p < 0.01$.
[c]Doxycycline, 400 mM compared to 200 mM, $p < 0.01$.
[d]Doxycycline, 200 mM compared to 100 mM, $p < 0.01$.
[e]Doxycycline, 100 mM compared to 50 mM, $p < 0.05$.
[f]Doxycycline, 50 mM compared to 25 mM, $p < 0.01$.
[g]Doxycycline (25–6.0 mM), not statistically significant.
From Ramamurthy, N.S. et al., Reactive Oxygen Species Activate and Tetracyclines Inhibit Rat Osteoblast Collagenase. *Journal of Bone and Mineral Research* Volume 8, Number 10, 1993, pp. 1247-1253. Reprinted by permission of Blackwell Science, Inc.

Table 2. Osteoblast Enzymes and Their Putative Roles in Bone Resorption

Enzyme	*Putative Role*
Collagenase (MMP-1)	Degradation of fibrillar collagen
Gelatinase A (MMP-2; 72-kDa gelatinase; Type IV collagenase)	Degradation of native and denatured fibrillar collagen; degradation of nonfibrillar collagen
Gelatinase B (MMP-9; 92-kDa gelatinase; Type IV collagenase)	Degradation of denatured fibrillar collagen; degradation of nonfibrillar collagen
Stromelysin	Degradation of matrix proteoglycans, glycoproteins, and some native and denatured collagens
Plasmin/PlasminogenActivator	Activation of latent matrix metalloproteinases (proMMPs)

connective tissue (i.e., pathologic resorption) is an essential step in the pathogenesis of osseous diseases such as periodontitis, rheumatoid and osteoarthritis, and the lytic bone diseases associated with malignancy (Table 2).

The earlier work of Chambers and Fuller (1985) supported the role of osteoblast collagenase in bone resorption because isolated osteoclasts cultured on calvarial explants did not resorb mineralized matrix unless the osteoid layer was first removed, either by pretreatment with osteoblasts or collagenase. Indeed, studies with disaggregated rat osteoclasts were particularly important in establishing osteoblast-mediated hormonal stimulation of osteoclastic activity (Chambers et al., 1985; McSheehy and Chambers, 1986). However, unlike the study of osteoblast metabolism, investigative work in osteoclast biology had still lagged because it was difficult to isolate large numbers of purified osteoclasts. Nonetheless, the central role of the osteoclast in resorption was never in doubt, based on classical studies, and this view has been confirmed partly due to recent isolation techniques that facilitated study of its function (Collin-Osdoby et al., 1991; Oursler et al., 1991).

The degradation of bone occurs in an extracellular compartment (the subosteoclastic resorption zone) acidified by proton transport at the osteoclast ruffled border membrane (Baron et al., 1985). The resulting low pH (4–5) permits the dissolution of the mineral phase, exposes the organic phase, and denatures the helical structure of the collagen molecule. Thus, this low pH likely optimizes the degradative action of lysosomal acid proteinases (i.e., cathepsins) (Vaes, 1988; Baron, 1989). However, the precise nature (temporal and spatial) of the interactions between major osteoblastic and osteoclastic enzyme degrading systems (e.g., matrix metalloproteinases and cysteine proteinases) during bone resorption is not yet completely understood. Although cooperation between these systems likely exists (Everts et al., 1992), our understanding of the interactions remains incomplete possibly because the localization of collagenase to osteoclasts in particular is still equivocal. Furthermore, the exact nature of the signal(s) that allow osteoblasts and osteoclasts to communicate (i.e., cell-to-cell interactions) in bone remodeling is poorly understood.

That cysteine proteinases (i.e., lysosomal cathepsins) likely play an earlier role than collagenase in the complex cascade of bone degradation was supported from several investigations. Conceivably, the collagenolytic action of the cysteine proteinases, optimal at pH 4–5, could be exerted preferentially in the most acid portion of the bone-resorption lacuna and in the immediate vicinity of the ruffled border. In contrast, neutral collagenase could be predominantly active deeper in the lacuna at the interface between demineralized and mineralized matrix where the pH is likely more neutral due to the buffering capacity of the dissolved salts. It is even likely that the concerted, sequential action of both enzymes may render collagen degradation much more efficient compared to the isolated action of each separately (Delaisse and Vaes, 1992). Collagenase could also degrade the fringes of yet undegraded, but already demineralized, collagen, likely remaining at the base of the resorption lacuna when the osteoclast detaches (Delaisse and Vaes, 1992) and, thereby, allowing a sudden neutralization of the pH at that site. This already demineralized collagen could also be denatured at an acidic pH and at physiologic temperature. Moreover, denatured collagen at this fringe could be degraded by osteoclast gelatinase (see below) or, possibly, osteoblast gelatinase. If the pH were suddenly neutralized, then the acid-requiring cysteine proteinases would be inactive, creating a permissive environment for the action of collagenase (and possibly neutral gelatinase) and the completion of the resorbing process.

Studies have suggested that cathepsins are not only critical in osteoclastic resorption but that gelatinase may also act in cooperation with cathepsins to remodel bone. Isolated avian osteoclasts resorbed both the organic and inorganic components of bone (Blair et al., 1986) and an acid collagenase resembling mammalian cathepsin B was isolated and characterized from these cells (Blair et al., 1993). Such an observation implied that cysteine proteinases, not collagenase, might be sufficient to degrade bone matrix; that implication received further confirmation from the work of Delaisse et al. (1987) whereby the addition of TIMP (an inhibitor of collagenase) did not inhibit bone resorption *in vitro*. Furthermore, Fuller and Chambers (1995) found no evidence of expression of mRNA for collagenase in rat osteoclasts by *in situ* hybridization, indicating that these cells did not synthesize collagenase. However, Reponen et al. (1992, 1994a) had shown by *in situ* hybridization that a related MMP (gelatinase B, or 92 kDa type IV collagenase) was widely distributed in murine mesenchymal tissues during development. Therefore, it was proposed that this enzyme might be utilized for the turnover of bone matrix, possibly as a gelatinase required for the removal of denatured collagen fragments (i.e., gelatin) generated by other proteases (Reponen et al., 1994b); such a proposal was consistent with their finding that gelatinase was localized to cells of the osteoclast lineage located at the bone surface (Reponen et al., 1994b). Northern blot analysis detected this enzyme (gelatinase B) in related rodent (i.e., rabbit) osteoclasts (Tezuka et al., 1994). Furthermore, studies have shown predominant expression of gelatinase B (Wucherpfennig et al., 1994) in human osteoclasts. Our laboratory assayed cathepsin L and B activities from avian osteoclasts and further demonstrated inhibition of bone resorption by disaggregated rat osteoclasts *in vitro* with the addition of a selective inhibitor of cathepsin L (Rifkin et al., 1991). These studies collectively implied that the substrate for gelatinase (i.e., denatured collagen fragments) arises from the action of cathepsin. Thus, cooperation between gelatinase and cathepsin(s) may, in turn, be facilitated during resorption because gelatinase is also predominantly expressed in osteoclasts (Tezuka et al., 1994; Wucherpfennig et al., 1994). Finally, Aimes and Quigley (1995) reported that 72 kDa gelatinase (gelatinase A, MMP-2) from chicken fibroblasts degraded type I native collagen fibrils at a slower rate than MMP-1. In studies from our laboratory (Rifkin et al., 1994), the MMPs from avian osteoclast preparations degraded native collagen into α^A and α^B breakdown products much slower than from neutrophils or from diabetic rat skin extract. In view of the above studies (Fuller and Chambers, 1995), the collagenolytic activity in osteoclast media reported previously from our laboratory (Rifkin et al., 1994) was probably due to MMP-9 (gelatinase B) (Tezuka et al., 1994) rather than interstitial collagenase (MMP-1). Although osteoclast MMP-9 has not yet been shown to degrade intact and denatured collagen, like gelatinase A, it nonetheless has the potential to play a key, singular role in the degradation of both forms of collagen in the remodeling of bone.

However, Delaisse et al. (1993) have shown specific localizations of collagenase to the underlying bone resorbing compartment of osteoclasts from mice, rats, and rabbits using an anti-mouse collagenase antiserum and an affinity-purified IgG fraction that specifically immunoblotted and immunoprecipitated procollagenase. Okamura et al. (1993) had also detected collagenase mRNA in

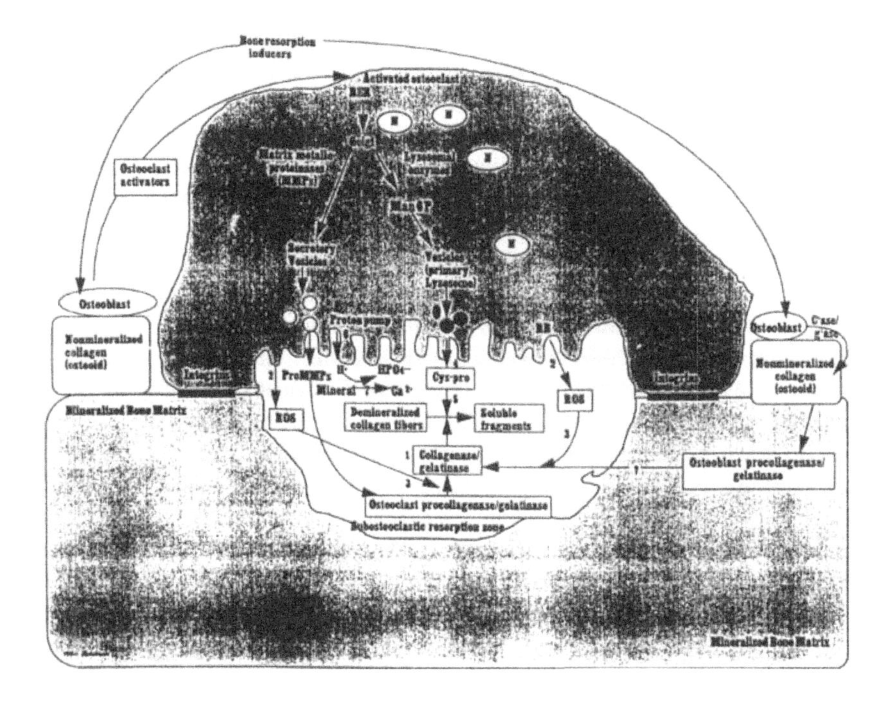

Figure 1. Potential sites for tetracycline (TC) or chemically modified tetracycline (CMT) molecule action on bone cell metabolism. TCs and CMTs have both anti-matrix metalloproteinase (anti-MMP) and anti-osteoclast properties. TCs and their chemically modified analogues may act (1) to inhibit directly active extracellular collagenase/gelatinase (C'ase/G'ase); (2) to reduce the available concentration of osteoclast-generated superoxide radicals (reactive oxygen species [ROS]), and thereby (3) inhibit the superoxide radical conversion of extracellular procollagenase/progelatinase (proMMPs) to active enzyme (the existence of osteoclast procollagenase/collagenase is still controversial); (4) to reduce the secretion of the lysosomal osteoclast cysteine proteinases (Cys-pro) such as cathepsin L, and thereby (5) reduce resorption of bone collagen; (6) to reduce the secretion of acid (protons, H^+) from the osteoclast ruffled border (RB), and thereby alter the pH optima (pH 4–5) for lysosomal enzymes; and (7) to inhibit acid solubilization of bone mineral. Abbreviations: N, nucleus; RER, rough endoplasmic reticulum; Man 6 P, mannose-6-phosphate. (Reprinted by permission of The New York Academy of Sciences from Inhibition of Matrix Metalloproteinases: Therapeutic Potential Volume 732 of the *Annals of the New York Academy of Sciences*, September 6, 1994.)

odontoclasts of bovine roots, indicating that these actively resorbing cells were also synthesizing collagenase. Furthermore, stimulated bone resorption *in vitro* was inhibited with TIMPs-1 and 2 (Hill et al., 1993). Work from our laboratory with tetracyclines and their chemically modified analogues (CMTs) also suggested that collagenase activity and bone resorption were coupled processes (Rifkin et al., 1992). We have shown that tetracyclines (TCs) and CMTs (tetracycline analogues that have been modified to lose their antimicrobial activity) are both potent inhibitors of bone cell matrix metalloproteinases, a potential therapeutic property independent of antimicrobial action (Figure 1) (Rifkin et al., 1993, 1994; Vernillo et al., 1994;1997).

Tetracyclines and CMTs inhibited not only extracellular collagenase activity from the culture media of UMR 106-01 osteoblastic cells (functional assay) but also bone resorption as assayed in two separate systems using fetal rat long bones and disaggregated rat osteoclasts (resorption pit assay) (Rifkin et al., 1992). At 50 mM concentrations, doxycycline and CMT-1 also inhibited osteoblastic extracellular gelatinase activity from clones of ROS 17/2 and UMR 106 cells (Vernillo et al., 1993). Furthermore, tetracyclines and CMTs affected intracellular MMP pathways. Doxycycline (antimicrobial TC) and CMT-1 (nonantimicrobial TC analogue) inhibited PTH-stimulated collagenase synthesis in UMR 106-01 cultures 63% and 78%, respectively, as shown by ELISA against rat collagenase (Vernillo et al., unpublished observations). Minocycline (antimicrobial TC) and CMT-1 also inhibited PTH-stimulated collagenase synthesis from tetracycline-treated UMR 106-01 cells as suggested by a significant inhibition of collagenolytic activity in media (64% and 90%, respectively) and analyzed by functional assay (Ramamurthy et al., 1990).

Two independent studies showed the role of matrix metalloproteinases as not only significant but also, perhaps, novel and distinct in resorption (Blavier and Delaisse, 1995; Witty et al., 1996). PTH-induced resorption in fetal rat limb bones was associated with the production of the MMPs, collagenase and gelatinase B, and inhibited with recombinant TIMP-1 (Witty et al., 1996). Furthermore, MMPs appeared obligatory for the migration of preosteoclasts to the developing marrow cavity of fetal mouse metatarsals in culture (Blavier and Delaisse, 1995). That finding was supported by the fact that inhibitors of MMPs prevented migration of preosteoclasts, whereas a cysteine proteinase inhibitor had no effect (Blavier and Delaisse, 1995). Thus, it has been proposed that MMPs in preosteoclasts (gelatinase B) and interstitial collagenase in hypertrophic chondrocytes (i.e., MMP-13) may have a new and distinct role apart from the one that MMPs may play in the subosteoclastic resorption compartment (Delaisse and Vaes, 1992; Everts et al., 1992; Delaisse et al., 1993), namely as a major component of a mechanism that determines where and when the osteoclasts will attack bone (Blavier and Delaisse, 1995). Therefore, the role of collagenase in bone resorption may still be very critical even though its localization to and synthesis by osteoclasts have remained controversial and its interaction(s) with cysteine proteinases are yet to be defined.

That the role of collagenase in tissue remodeling is likely very critical is supported further by additional studies (Krane and Jaenisch, 1992; Krane, 1995; Liu et al., 1995). Krane and Jaenisch (1992) through the genetic approach of site-directed mutagenesis had initially altered genes encoding critical amino acid sequences in the collagen substrate, to assess its effect on susceptibility to collagenase. Subsequently, Krane (1995) altered the amino acid sequences around the collagenase cleavage site by site-directed mutagenesis of the murine Colla-I gene. The mutation was introduced into the endogenous Colla-I gene by homologous recombination in embryonic stem cells to determine the role of collagenase *in vivo*. Liu et al. (1995) showed that such a targeted mutation at the known collagenase cleavage site in mouse type I collagen (i.e., Gly 775 and Ile 776 of the alpha 1 (I) chain) impaired tissue remodeling.

IV. THE PLASMINOGEN ACTIVATOR/PLASMIN PATHWAY

Osteoblasts play a central and complex role in bone metabolism both through the formation of bone (the traditionally described role of the osteoblast) and its interaction or coupling with the bone-resorbing osteoclast. Presumably and in part through the action of collagenase, an extremely potent and specific, neutral metalloproteinase, the osteoblast can exert profound effects on tissue remodeling. For example, Walker implicated collagenase in bone remodeling three decades ago when he had observed the apparent reversal of osteopetrosis in grey-lethal (gl) mice by PTH administration; this reversal was correlated with the presence of collagenolytic activity in the cell-free extracts taken from the PTH-treated homozygotes (gl/gl) (Walker, 1966). This activity is most likely achieved physiologically through the action of plasmin on secreted, inactive collagenase (procollagenase) (Eeckhout and Vaes, 1977; Thomson et al., 1989).

Another neutral proteinase, PA, is regulated in osteoblasts. Several bone-resorbing hormones have been shown to promote PA activity in osteoblast-like cells (Hamilton et al., 1985); others (Partridge et al., 1987; Thomson et al., 1989) subsequently supported the hypothesis that the significance of PA formation was to activate procollagenase and consequently facilitate osteoclastic resorption. Osteoblasts also synthesize transforming growth factor (TGFβ) which is stored in the bone matrix as latent TGFβ (Carrington et al., 1988) and in turn can be activated by plasmin (Allan et al., 1991). Thus, plasmin generated from PA may activate TGFβ. Such an activation mechanism also illustrates that the activities of bone forming osteoblasts and bone resorbing osteoclasts are coupled and regulated in physiological bone metabolism by several endocrine and paracrine factors, including peptide and steroid hormones as well as a variety of cytokines and growth factors (Vaes, 1988). Finally, TGFβ limits the extent of its own activation (and consequently restricts the rate of bone formation by osteoblasts) by directly inhibiting PA activity through the synthesis of the specific PA inhibitor-1, (PAI-1) (Pfeilschifter et al., 1990).

V. SUMMARY

The osteoblast plays a central but complex role in bone metabolism. The precise nature of the mechanisms by which skeletal remodeling occurs is not yet entirely understood. However, osteoblast neutral proteinases such as the potent MMPs (collagenase and gelatinase), as well as PA, apparently exert profound effects on the remodeling of the ECM partly by facilitating the coupling of osteoblast bone formation to bone resorption. PTH stimulation of osteoblast collagenase may facilitate subsequent osteoclastic resorption and recent studies using site-directed mutagenesis at the collagenase cleavage site for type I collagen strongly suggest that collagenase may be critical for tissue remodeling. Furthermore, interactions with osteoblast and osteoclast gelatinases as well as the osteoclast cathepsins (lysosomal cysteine proteinases) may be significant for the degradation of the ECM. A greater understanding of bone remodeling mechanisms should provide additional insights into the pathogenesis of the human lytic bone diseases.

ACKNOWLEDGMENTS

Research was supported, in part, by the National Institute of Dental Research Grants RO1 DE-09576 and R37 DE-03987.

REFERENCES

Aimes, R.T. and Quigley, J. (1995). Matrix metalloproteinase-2 is an interstitial collagenase, J. Biol. Chem. 270, 5872-5876.

Alexander, C.M. and Werb, Z. (1989). Proteinases and extracellular matrix remodeling. Curr. Opin. Cell Biol. 1, 974-982.

Allan, E.H., Hamilton, J.A., Medcalf, R.L., Kubota, M., and Martin, T.J. (1986). Cyclic AMP-dependent and -independent effects on tissue type plasminogen activator in osteogenic sarcoma cells: Evidence from phosphodiesterase inhibition and parathyroid hormone antagonists. Biochim. Biophys. Acta 888, 199-207.

Allan, E.H., Zeheb, R., Gelehrter, T.D., Heaton, J.H., Fukumoto, S., Yee, J.A., and Martin, T.J. (1991). Transforming growth factor β stimulates production of urokinase-type plasminogen activator mRNA and plasminogen activator inhibitor-1 mRNA and protein in rat osteoblastlike cells. J. Cell Physiol. 149, 34-43.

Aubin, J.E., Heersche, J.N.M., and Merrilees, M.J. (1982). Isolation of bone cell clones with differences in growth, hormone responses, and extracellular matrix production. J. Cell Biol. 92, 452-461.

Baron, R. (1989). Molecular mechanisms of bone resorption by the osteoclast. Anat. Rec. 224, 317-324.

Baron, R., Neff, L., Louvard, D., and Courtoy, P.J. (1985). Cell-mediated extracellular acidification and bone resorption: Evidence for a low pH in resorbing lacunae and localization of a 100-kD lysosomal membrane protein at the osteoclast ruffled border. J. Cell Biol. 101, 2210-2222.

Birkedal-Hansen, H., Moore, W.G.I., Bodden, M.K., and Windsor, L. (1993). Matrix metalloproteinases: A review. Crit. Rev. Oral Biol. Med. 4, 455-463.

Blair, H.C., Kahn, A.J., Crouch, E.C., Jeffrey, J.J., and Teitelbaum, S.L. (1986). Isolated osteoclasts resorb the organic and inorganic components of bone. J. Cell Biol. 102, 1164-1172.

Blair, H.C., Teitelbaum, S.L., Grosso, L.E., Lacey, D.L., Tan, H.-L., McCouri, D.W., and Jeffrey, J.J. (1993). Extracellular matrix degradation at acid pH: Avian osteoclast acid collagenase isolation and characterization. Biochem. J. 294, 873-884.

Blavier, L. and Delaisse, J.M. (1995). Matrix metalloproteinases are obligatory for the migration of preosteoclasts to the developing bone marrow cavity of primitive long bones. J. Cell Sci. 108, 3649-3659.

Carrington, J.L., Roberts, A.B., Flanders, K.C., Roche, N.S., and Reddi, A.H. (1988). Accumulation, localization, and compartmentation of transforming growth factor-β during endochondral bone development. J. Cell Biol. 107, 1969-1975.

Chambers, T.J. (1982). Osteoblasts release osteoclasts from calcitonin-induced quiescence. J. Cell Sci. 57, 247-252.

Chambers, T.J., Athanasou, N.A., and Fuller, K. (1984). Effect of parathyroid hormone and calcitonin on the cytoplasmic spreading of isolated osteoclasts. J. Endocrinol. 102, 281-286.

Chambers, T.J. and Fuller, K. (1985). Bone cells predispose endosteal surfaces to resorption by exposure of bone mineral to osteoclastic contact. J. Cell Sci. 76, 155-163.

Chambers, T.J., McSheehy, P.M.J., Thomson, B.M., and Fuller, K. (1985). The effect of calcium-regulating hormones and prostaglandins on bone resorption by osteoclasts disaggregated from neonatal rabbit bones. Endocrinology 116, 234-239.

Clohisy, J.C., Connolly, T.J., Bergman, K.D., Quinn, C.O., and Partridge, N.C. (1994). Prostanoid-induced expression of matrix metalloproteinase-1 messenger ribonucleic acid in rat osteosarcoma cells. Endocrinology 135, 1447-1454.

Collin-Osdoby, P., Oursler, M.J., Webber, D., and Osdoby, P. (1991). Osteoclast-specific monoclonal antibodies coupled to magnetic beads provide a rapid and efficient method of purifying avian osteoclasts. J. Bone Miner. Res. 6, 1353-1356.

Cook, T.F., Burke, J.S., Bergman, K.D., Quinn, C.O., Jeffrey, J.J., and Partridge, N.C. (1994). Cloning and regulation of rat tissue inhibitor of metalloproteinases-2 in osteoblastic cells. Arch. Biochem. Biophys. 311, 313-320.

Delaisse, J.M., Boyde, A., Maconnachie, E., Ali, N.N., Sear, C.H.J., Eeckhout, Y., Vaes, G., and Jones, S.J. (1987). The effects of inhibitors of cysteine-proteinases and collagenase on the resorptive activity of isolated osteoclasts. Bone 8, 305-313.

Delaisse, J.M., Eeckhout, Y., Neff, L., Francois-Gillet, C., Henriet, P., Su, Y., Vaes, G., and Baron, R. (1993). (Pro)collagenase (matrix metalloproteinase-1) is present in rodent osteoclasts and in the underlying bone-resorbing compartment. J. Cell Sci. 106, 1071-1082.

Delaisse, J.M. and Vaes, G. (1992). Mechanism of mineral solubilization and matrix degradation in osteoclastic bone resorption. In: *The Biology and Physiology of the Osteoclast.* (Rifkin, B. R. and Gay, C. V., Eds.), pp. 289-314, CRC Press, Boca Raton, FL.

Eeckhout, Y. and Vaes, G. (1977). Further studies on the activation of procollagenase, the latent precursor of bone collagenase. Effects of lysosomal cathepsin B, plasmin and kallikrein, and spontaneous activation. Biochem. J. 166, 21-31.

Everts, V., Delaisse, J.M., Korper, W., Hiehof, A., Vaes, G., and Beertsen, W. (1992). Degradation of collagen in the bone resorbing compartment underlying the osteoclast involves both cysteine-proteinases and matrix metalloproteinases. J. Cell Physiol. 150, 221-231.

Fuller, K. and Chambers, T.J. (1995). Localization of mRNA for collagenase in osteocytic, bone surface and chondrocytic cells but not osteoclasts. J. Cell Sci. 108, 2221-2230.

Hamilton, J.A., Lingelbach, S.R., Partridge, N.C., and Martin, T.J. (1985). Regulation of plasminogen activator production by bone resorbing hormones in normal and malignant osteoblasts. Endocrinology 116, 2186-2191.

Harper, E. and Gross, J. (1972). Collagenase, procollagenase, and activator relationships in tadpole tissue cultures. Biochem. Biophys. Res. Commun. 48, 1147-1152.

Heath, J.K., Atkinson, S.J., Meikle, M.C., and Reynolds, J.J. (1984). Mouse osteoblasts synthesize collagenase in response to bone resorbing agents. Biochim. Biophys. Acta 802, 151-154.

Hill, P.A., Reynolds, J.J., and Meikle, M.C. (1993). Inhibition of stimulated bone resorption in vitro by TIMP-1 and TIMP-2. Biochim. Biophys. Acta 1177, 71-74.

Khaüper, V., Will, H., López-Otin, C., Smith, B., Atkinson, S.J., Stanton, H., Hembey, R.M., and Murphy, G. (1996). Cellular mechanisms for human procollagenase-3 (MMP13) activation. J. Biol. Chem. 271, 17124-17131.

Krane, S.M. (1995). Is collagenase (matrix metalloproteinase-1) necessary for bone and other connective tissue remodeling? Clin. Orthop. Rel. Res. 313, 47-53.

Krane, S.M. and Jaenisch, R. (1992). Site-directed mutagenesis of type-I collagen: Effect on susceptibility to collagenase. Matrix (Suppl.) 1, 64-67.

Leco, K.J., Khokha, R., Pavloff, N., Hawkes, S.P., and Edwards, D.R. (1992). Tissue inhibitor of metalloproteinases-3 (TIMP-3) is an extracellular matrix-associated protein with a distinctive pattern of expression in mouse cells and tissues. J. Biol. Chem. 267, 17321-17326.

Liu, X., Wu, H., Byrne, M., Jeffrey, J., Krane, S., and Jaenisch, R. (1995). A targeted mutation at the known collagenase cleavage site in mouse type-I collagen impairs tissue remodeling. J. Cell Biol. 130, 227-237.

Livesey, S.A., Kemp, B.E., Re, C.A., Partridge, N.C. and Martin, T.J. (1982). Selective hormonal activation of cyclic AMP-dependent protein kinase isoenzymes in normal and malignant osteoblasts. J. Biol. Chem. 257, 14989-14998.

Luben, R.A. and Cohen, D.V. (1976). Effects of parathormone and calcitonin on citrate and hyaluronate metabolism in cultured bone. Endocrinology 98, 413-419.

Matrisian, L.M. (1992). The matrix-degrading metalloproteinases. BioEssays 14, 455-463.

McSheehy, P.M.J. and Chambers, T.J. (1986). Osteoblastic cells mediate osteoclastic responsiveness to parathyroid hormone. Endocrinology 118, 824-828.

McSheehy, P.M.J. and Chambers, T.J. (1987). 1,25-Dihydroxyvitamin D, stimulates rat osteoblastic cells to release a soluble factor that increases osteoclastic bone resorption. J. Clin. Invest. 80, 425-431.

Meikle, M.C., Bord, S., Hembry, R.M., Compston, J., Croucher, P.I., and Reynolds, J.J. (1992). Human osteoblasts in culture synthesize collagenase and other metalloproteinases in response to osteotropic hormones and cytokines. J. Cell Sci. 103, 1093-1099.

Mundy, G.R. (1992). Local factors regulating osteoclast function. In: *The Biology and Physiology of the Osteoclast.* (Rifkin, B.R. and Gay, C.V., Eds.), pp. 171-185, CRC Press, Boca Raton, FL.

Okamura, T., Shimokawa, H., Takagi, Y., Ono, H., and Sasaki, S. (1993). Detection of collagenase mRNA in odontoclasts of bovine root-resorbing tissue by in situ hybridization. Calcif. Tissue Int. 52, 325-330.

Omura, T.H., Noguchi, A., Johanns, C.A., Jeffrey, J.J., and Partridge, N.C. (1994). Identification of a specific receptor for interstitial collagenase on osteoblastic cells. J. Biol. Chem. 269, 24994-24998.

Otsuka, K., Sodek, J., and Limeback, H. (1984). Synthesis of collagenase and collagenase inhibitors by osteoblastlike cells in culture. Eur. J. Biochem. 145, 123-129.

Oursler, M.J., Collin-Osdoby, P., Anderson, F., Li, L., Webber, D., and Osdoby, P. (1991). Isolation of avian osteoclasts: Improved techniques to preferentially purify viable cells. J. Bone Miner. Res. 4, 375-385.

Partridge, N.C., Alcorn, D., Michelangeli, V.P., Ryan, G., and Martin, T.J. (1983). Morphological and biochemical characterization of four clonal osteogenic sarcoma cell lines of rat origin. Cancer Res. 43, 4388-4394.

Partridge, N.C., Jeffrey, J.J., Ehlich, L.S., Teitelbaum, S.L., Fliszar, C., Welgus, H.G., and Kahn, A.J. (1987). Hormonal regulation of the production of collagenase and a collagenase inhibitor activity by rat osteogenic sarcoma cells. Endocrinology 120, 1956-1962.

Pfeilschifter, J., Erdmann, J., Schmidt, W., Naumann, A., Minne, H.W., and Ziegler, R. (1990). Differential regulation of plasminogen activator and plasminogen activator inhibitor by osteotropic factors in primary cultures of mature osteoblasts and osteoblast precursors. Endocrinology 126, 703-711.

Puzas, J.E., and Brand, J.S. (1979). Parathyroid hormone stimulation of collagenase secretion by isolated bone cells. Endocrinology 104, 559-569.

Ramamurthy, N.S., Vernillo, A.T., Greenwald, R.A., Lee, H.-M., Sorsa, T., Golub, L.M., and Rifkin, B.R. (1993). Reactive oxygen species activate and tetracyclines inhibit rat osteoblast collagenase. J. Bone Miner. Res. 8, 1247-1253.

Ramamurthy, N.S., Vernillo, A.T., Lee, H-M., Golub, L.M., and Rifkin, B.R. (1990). The effect of tetracyclines on collagenase activity in UMR 106-01 rat osteoblastic osteosarcoma cells. Res. Commun. Chem. Pathol. Pharmacol. 70, 323-335.

Reponen, P., Sahlberg, C., Huhtala, P., Hurskainen, T., Thesleff, I., and Tryggvason, K. (1992). Molecular cloning of murine 72-kDa type-IV collagenase and its expression during mouse development. J. Biol. Chem. 267, 7856-7862.

Reponen, P., Sahlberg, C., Munaut, C., Thesleff, I., and Tryggvason, I. (1994a). High expression of 92-kDa type-IV collagenase (gelatinase B) in the osteoclast lineage during mouse development. J. Cell Biol. 124, 1091-1102.

Reponen, P., Sahlberg, C., Munaut, C., Thesleff, I., and Tryggvason, K. (1994b). High expression of 92-kDa type IV collagenase (Gelatinase) in the osteoclast lineage during mouse development. In: *Inhibition of Matrix Metalloproteinases: Therapeutic Potential.* (Greenwald, R.A., and Golub, L.M., Eds.), pp. 472-475, The New York Academy of Sciences, New York.

Rifkin, B.R., Golub, L.M., Sanavi, F., Vernillo, A.T., Kleckner, A.P., McNamara, T.F., Auszmann, J.M., and Ramamurthy, N.S. (1992). Effects of tetracyclines on rat osteoblast collagenase activity and bone resorption in vitro. In: *The Biological Mechanisms of Tooth Movement and Craniofacial Adaptation*. (Davidovitch, Z., Ed.), pp. 85-90, EBSCO Media, Birmingham, AL.

Rifkin, B.R., Golub, L.M., and Vernillo, A.T. (1993). Blocking periodontal disease progression by inhibiting tissue-destructive enzymes: A potential therapeutic role for tetracyclines and their chemically-modified analogs. J. Periodontol. 64, 819-827.

Rifkin, B.R., Vernillo, A.T., Golub, L.M., and Ramamurthy, N.S. (1994). In: *Inhibition of Matrix Metalloproteinases: Therapeutic Potential.* (Greenwald, R.A., and Golub, L.M., Eds.), pp. 165-180, The New York Academy of Sciences, New York.

Rifkin, B.R., Vernillo, A.T., Kleckner, A.P, Auszmann, J.M., Rosenberg, L.R., and Zimmerman, M. (1991). Cathepsin B and L activities in isolated osteoclasts. Biochem. Biophys. Res. Commun. 179, 63-69.

Rodan, G.A., and Martin, T.J. (1981). Role of osteoblasts in hormonal control of bone resorption—a hypothesis. Calcif. Tissue Int. 33, 349-351.

Roswit, W.T., McCourt, D.W., Partridge, N.C., and Jeffrey, J.J. (1992). Purification and sequence analysis of two rat tissue inhibitors of metalloproteinases. Arch. Biochem. Biophys. 292, 402-410.

Sakamoto, S. and Sakamoto, M. (1982). Biochemical and immunohistochemical studies on collagenase in resorbing bone in tissue culture. J. Periodont. Res. 17, 523-526.

Sakamoto, M., and Sakamoto, S. (1984). Immunocytochemical localization of collagenase in isolated mouse bone cells in culture. Biomed. Res. 5, 29-38.

Sakamoto, S., Sakamoto, M., Goldhaber, P., and Glimcher M.J. (1975). Collagenase and bone resorption: Isolation of collagenase from culture medium containing serum after stimulation of bone resorption by addition of parathyroid extract. Biochem. Biophys. Res. Commun. 63, 172-178.

Sakamoto, S., Sakamoto, M., Goldhaber, P., and Glimcher, M.J. (1979). Collagenase activity and morphological and chemical bone resorption induced by prostaglandin E_2 in tissue culture. Proc. Soc. Exp. Biol. Med. 161, 99-105.

Sato, T., del Carmen Ovejero, M., Peng, H., Heegaard, A-M., Kumegawa, M., Foged, N.T., and Delaisse, J-M. (1997). Identification of the membrane-type matrix metalloproteinase MT1-MMP in osteoclasts. J. Cell Sci. 110, 589-596.

Scott, D.K., Brakenhoff, K.D., Clohisy, J.C., Quinn, C.O., and Partridge, N.C. (1992). Parathyroid hormone induces transcription of collagenase in rat osteoblastic cells by a mechanism using cyclic adenosine 3',5'-monophosphate and requiring protein synthesis. Mol. Endocrinol. 6, 2153-2159.

Tezuka, K., Nemoto, K., Tezuka, Y., Sato, T., Ikeda, Y., Kobori, M., Kawashima, H., Eguchi, H., Hakeda, Y., and Kumegawa, M. (1994). Identification of matrix metalloproteinase 9 in rabbit osteoclasts. J. Biol. Chem. 269, 15006-15009.

Thomson, B.M., Atkinson, S.J., McGarrity, A.M., Hembry, R.M., Reynolds, J.J., and Meikle, M.C. (1989). Type-I collagen degradation by mouse calvarial osteoblasts stimulated with 1,25-dihydroxyvitamin D_3: Evidence for a plasminogen-plasmin-metalloproteinase activation cascade. Biochim. Biophys. Acta 1014, 125-132.

Thomson, B.M., Mundy, G.R., and Chambers, T.J. (1987). Tumour necrosis factors α and β induce osteoblastic cells to stimulate osteoclastic bone resorption. J. Immunol. 138, 775-779.

Thomson, B.M., Saklatvala, J., and Chambers, T.J. (1986). Osteoblasts mediate interleukin 1 responsiveness of bone resorption by rat osteoclasts. J. Exp. Med. 164, 104-112.

Vaes, G. (1972). The release of collagenase as an inactive proenzyme by bone explants in culture. Biochem. J. 126, 275-289.

Vaes, G. (1988). Cellular biology and biochemical mechanism of bone resorption. A review of recent developments on the formation, activation, and mode of action of osteoclasts. Clin. Orthop. Rel. Res. 231, 239-271.

Vernillo, A.T., Ramamurthy, N.S., Golub, L.M., and Rifkin, B.R. (1994). The nonantimicrobial properties of tetracycline for the treatment of periodontal disease. Curr. Opin. Periodontol. 2, 111-118.

Vernillo, A.T., Ramamurthy, N. S., Golub, L.M., Greenwald, R.A., and Rifkin, B.R. (1997). Tetracyclines as inhibitors of bone loss in vivo. In: *Studies in Stomatology and Craniofacial Biology on the Threshold of the 21st century.* (Baum, B.J. and Cohen, M.M., Eds.), pp. 499-522. IOS Press, Amsterdam.

Vernillo, A.T., Ramamurthy, N.S., Lee, H.M., Mallya, S., Auszmann, J., Golub, L.M., and Rifkin, B.R. (1993). ROS and UMR osteoblast gelatinases: Tetracycline inhibition. J. Dent. Res. 72, 367a.

Walker, D.G. (1966). Elevated bone collagenolytic activity and hyperplasia of parafollicular light cells of the thyroid in parathormone-treated grey-lethal mice. Z. Zellforsch. 72, 100-124.

Weiss, S.J., Peppin, G., Ortiz, X., Ragsdale, C., and Test, S.T. (1985). Oxidative autoactivation of latent collagenase by human neutrophils. Science 277, 747-749.

Witty, J.P., Foster, S.A., Stricklin, G.P., Matrisian, L.M., and Stern, P.H. (1996). Parathyroid hormone-induced resorption in fetal rat limb bones is associated with production of the metalloproteinases collagenase and gelatinase B. J. Bone Miner. Res. 11, 72-78.

Wright, G.M. and Leblond, C.P. (1981). Immunohistochemical localization of procollagens. III. Type-I procollagen antigenicity in osteoblasts and prebone (osteoid). J. Histochem. Cytochem. 29, 791-804.

Wucherpfennig, A.L., Li, Y., Stetler-Stevenson, W.G., Rosenberg, A.E., and Stashenko, P. (1994). Expression of 92-kD type-IV collagenase/gelatinase B in human osteoclasts. J. Bone Miner. Res. 9, 549-556.

BIOLOGY OF OSTEOCYTES

P.J. Nijweide, N.E. Ajubi, E.M. Aarden, and

A. Van der Plas

I. INTRODUCTION

Osteocytes are the most abundant cells of mature bone. At active sites of bone formation some of the osteoblasts lining the bone surface are incorporated in the bone matrix and differentiate into osteocytes. During this process, the cells diminish in size and lose part of their cytoplasmic organelles. Still, even mature osteocytes are capable of (limited) matrix protein production (collagen, osteocalcin, fibronectin,

Advances in Organ Biology
Volume 5B, pages 529-542.

ISBN: 0-7623-0390-5

osteopontin). Within each osteon all osteocytes are connected to one another via long, slender, branched cytoplasmic processes and gap junctions. This morphology is not merely enforced on the cells by the surrounding matrix. Osteocytes isolated from the tissue with an immunodissection method and osteocyte-specific antibodies assume a similar configuration in the absence of matrix *in vitro*.

The osteocytic network is the key to the understanding of osteocyte function. Osteocytes are probably the mechanosensor cells of bone. Loading of bone results in strain forces in the matrix which evoke cellular responses either directly or via fluid shear stress produced by increased fluid flow in the lacuno-canalicular system. Osteocytes possess a large number of cell-matrix adhesion sites, both integrins and others (CD44). These probably serve as foci for the transfer of extracellular strain signals to the intracellular compartment, the cytoskeleton, and related signal transduction systems. Osteocytes may regulate their response by adjusting the production of extracellular matrix proteins and expression of adhesion receptors. Upon loading osteocytes have been found to secrete several factors, notably prostaglandins, nitric oxide (NO), and insulin-like growth factor (IGF). These and other yet unknown factors may activate the bone remodeling system of osteoblasts and osteoclasts upon loading.

II. BONE AS A MECHANOSENSITIVE TISSUE

The skeleton has the ability to compensate for changes in the mechanical environment by altering its mass and structure (Wolff, 1882). This ability to adapt to functional demands involves the activities of bone cells capable of bone remodeling, i.e., osteoclasts and osteoblasts. These cells are, however, present on the surfaces of bone, not in the bone matrix itself, and occupy only a minor part of these surfaces in the adult. It is therefore unlikely that these effector cells of bone remodeling are directly involved in the response of bone to mechanical stimuli. It is much more likely that they are governed in their activities by chemical stimuli produced by sensors in the bone matrix. The location and configuration of osteocytes are ideally suited for the sensing of changes in strain in the bone matrix as a result of changes in mechanical loading. Osteocytes are regularly spaced throughout the bone matrix. The morphology of cell bodies connected with each other via long, thin cell processes ensure an enormous cell-matrix surface area (Figure 1).

Especially since the original proposal for a function of osteocytes in blood calcium homeostasis through the process of osteocytic osteolysis (Bélanger, 1969) was largely refuted (Boyde, 1980; Marotti et al., 1990), osteocytes have not received serious attention. The mineralized matrix around the cells has for a long time hampered in-depth studies into the properties and abilities of osteocytes. The recent increased interest in the mechanoregulation of bone as one of the major determinants of bone morphology and structure has, however, reemphasized their potential importance. Several *in vivo* and *in vitro* studies have produced direct evidence for a

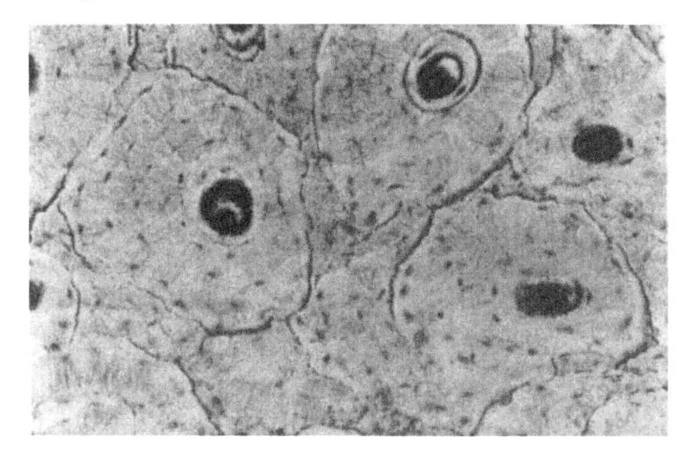

Figure 1. Osteons in mature human bone. Osteocytes are arranged in concentric circles around the central haversian channel containing a blood vessel. Note the numerous canaliculi radiating from the osteocyte lacunae. Magnification: 130x.

role as mechanosensor cells. *In vivo* experiments, using functionally isolated turkey ulnae or rat caudal vertebrae, have shown that intermittent loading increased the number of osteocytes expressing glucose-6-phosphate dehydrogenase (G6PD) activity (Pead et al., 1988; Skerry et al., 1989), as well as collagen type I production (Sun et al., 1995) and IGF-I mRNA (Lean et al., 1995). These experiments demonstrated that three of the most important cellular metabolic processes are influenced by mechanical stress: energy production, matrix synthesis, and intercellular signal transduction. *In vitro* studies have elaborated on the production of signal molecules by osteocytes upon mechanical stimulation. Prostaglandins (PGE_2, PGI_2) and NO were found to be early responders to mechanical loading (Klein-Nulend et al., 1995a,b; Pitsillides et al., 1996).

At this time, little is known about how mechanical loading induced strain in the bone matrix is transferred to the osteocytes, which intracellular signal transduction pathways are activated in the osteocytes, and which intercellular signals are produced that regulate the remodeling response of bone. It has, however, become evident that the unique network of osteocytes in bone plays a decisive role in the mechanoregulation of the skeleton.

III. FORMATION OF OSTEOCYTES

Osteocytes represent the ultimate differentiation stage of the osteoblast lineage which starts with the multipotential mesenchymal or stromal stem cell and progresses via the intermediate stages of osteoprogenitor cell, preosteoblast and osteoblast to finally reach the osteocyte stage (Nijweide et al., 1986; Figure 2). The

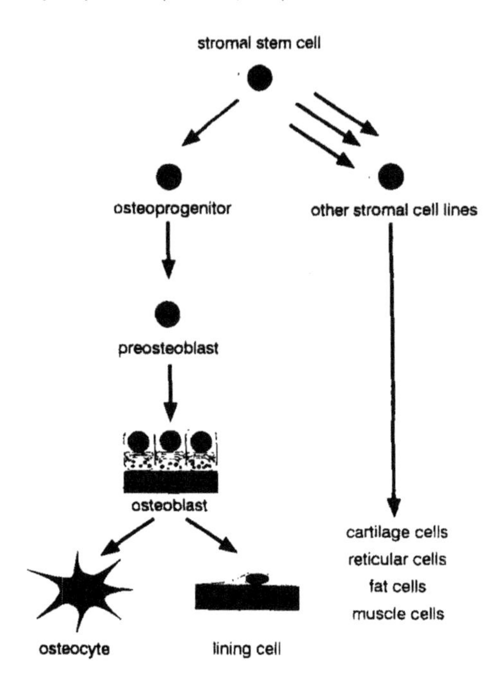

Figure 2. Osteogenic differentiation. Reprinted from O.L.M. Bijvoet et al. (eds), Bisphosphonates on bones, 1995, p. 32, with kind permission from Elsevier Science-NL, Sara Burgerhartstraat 25, 1055 KV Amsterdam, The Netherlands.

immediate precursor cells of the osteocytes, the osteoblasts, are responsible for formation of bone matrix. Active osteoblasts are found on the advancing surfaces of growing bone. They are arranged as a monolayer of polarized cells. Bone matrix formation chiefly takes place on the bone side of the monolayer, while on the vascular side a layer of preosteoblasts stands ready to fill in gaps in the osteoblast layer. During the process of matrix formation some of the osteoblasts become incorporated and stay behind while the bone matrix front and the other osteoblasts advance. The cells remaining behind change in morphology from plumb, cuboidal osteoblasts into spindle shaped osteocytes. The mechanism that decides which osteoblasts become osteocytes is not yet elucidated. It is tempting to assume that the signal that activates the osteoblast-osteocyte differentiation is produced by earlier embedded osteocytes (Palumbo et al., 1990a). As the number of osteocytes per unit of surface area is much smaller than that of osteoblasts, the degree of interconnection between osteoblasts and osteocytes may differ between osteoblasts. A high degree of osteocyte-osteoblast coupling may be a decisive factor in the number of signals reaching a particular osteoblast and therefore in the determination of the destiny of that osteoblast (Marotti et al., 1992). Although the nature of the signal is completely unknown, it is thought to inhibit local matrix formation. As the linear appositional production rate of the surrounding cells remains unchanged, the involved osteoblast will be incorporated in the matrix (Palumbo et al., 1990a,b; Nefussi et al., 1991).

Osteoblasts are connected to one another by short processes and gap junctions (Jeansonne et al., 1979). During the incorporation in the matrix and differentiation into osteocytes the processes between the cells become necessarily elongated because the distances between neighboring cells increase. Nevertheless, the stellate shape of osteocytes is not merely imposed upon them by the intermediate matrix. Isolated osteocytes regain this typical configuration as soon as they are seeded on a support in the absence of matrix (Van der Plas and Nijweide, 1992) (see Figure 3).

Although the change in morphology during the transition of osteoblast to preosteocyte (early osteoid osteocyte, early osteocytic osteoblast) immediately underneath the osteoblast layer is quite dramatic, this first stage of osteocytic differentiation still has a large endoplasmic reticulum and Golgi area, and relatively many mitochondria (Nijweide et al., 1981). Deeper into the bone the cells lose more and more cell organelles and become smaller. Some of the osteocytes may ultimately die or will be destroyed during bone resorption (Elmardi et al., 1990). Others may live on for many years. For example, in human auditory ossicles that show no or very little bone remodeling, the osteocyte life-span may go up to 80 years (Marotti et al., 1990).

IV. THE OSTEOCYTE NETWORK

At first, the matrix around the newly embedded cells is not calcified (osteoid osteocytes). Later, the mineralization front, lagging somewhat behind the matrix formation front, reaches and passes the cells, resulting in new bone enclosing now mature osteocytes. It is possible that the osteocytes themselves are involved in the regulation of matrix mineralization either by secreting calcification-regulating noncollagenous proteins or by adapting their surrounding matrix by the action of exo-enzymes (Mikuni-Takagaki et al., 1995). Of particular interest is that matrix calcification stops at some distance from the cell body and its cellular processes. Again secretion of calcification-regulating noncollagenous proteins by the osteocytes may be involved.

The above described processes of bone matrix formation, osteocyte differentiation, and matrix mineralization result in the creation in bone of a complex and continuous network of cavities (lacunae) in which the cell bodies are situated connected to one another by minute channels (canaliculi) containing the cellular processes. These processes possess at mutual adhesion sites gap junctions (Doty, 1981). Thus, bone possesses two networks, one intracellular, open for the passage of ions (electrical coupling) and small molecules capable of penetrating the gap junctional pores, and the other extracellular.

The extracellular network provides avenues for the diffusion of nutrients and waste products in and out of the bone, and appears to be penetrable for medium sized molecules such as horseradish peroxidase (Doty and Schofield, 1972; Dillaman et al., 1991). At first, the function of the extracellular network was considered to be primarily that of a transport system for nutrients and waste products, necessary to keep the deeply embedded osteocytes alive. However, it seems, unlikely that such an intricate system of osteocytes in their lacunae and canaliculi would be positioned just to maintain the system. The proposed function of the osteocyte—or rather of the complete functional syncytium of osteocytes including the lacuno-canalicular system—as a mechanosensory system is much more intriguing. Osteocytes have been shown to respond to mechanical loading with the production of paracrine factors (Klein-Nulend et al., 1995a,b; Lean et al., 1995; Pitsillides et al., 1996). The lacuno-canalicular system offers a transport system for these factors to the effector cells, osteoblasts and osteoclasts. Even more fascinating is the idea of Weinbaum et al. (1994) that the lacuno-canalicular system itself provides the mechanism by which the strain signal in bone, as the result of mechanical loading, is transferred to the osteocytes. According to these authors, mechanical loading of bone causes compression of the bone matrix and subsequently results in the squeezing out of interstitial fluid from the lacuno-canalicular system. Intermittent loading then would cause a repeated influx and outflow of fluid, and this flow would induce fluid shear stress along the osteocytic processes. Osteoblasts (Reich et al., 1990) and osteocytes (Klein-Nulend et al., 1995a) are sensitive to fluid flow.

V. OSTEOCYTE ISOLATION

As was mentioned earlier, the encapsulation of osteocytes in bone matrix has severely hampered studies into the metabolism and functions of osteocytes. Therefore, the development of monoclonal antibodies that specifically recognize antigens on the cell surface of osteocytes (Nijweide and Mulder, 1986; Bruder and Caplan, 1990) and the use of these antibodies for osteocyte isolation (Van der Plas and Nijweide, 1992) were major strides forward. In the isolation procedure, magnetic beads to which rabbit-antimouse IgG antibodies are covalently coupled, are coated with the osteocyte-specific antibody MAb OB7.3. These beads are then incubated together with a mixed bone cell preparation enzymatically isolated from calvariae of 18-day-old fetal chickens. Fetal chicken have the advantage that the osteoid zone covering the calcified matrix of the calvariae is wider than the osteoid zones in fetal mouse or rat calvariae (Nijweide et al., 1981). A few successive collagenase treatments alternated with mild EDTA decalcifications releases a reasonable number of osteocytes from chick calvariae. Osteocytes attach via MAb OB7.3 to the beads and are removed together with the beads from the cell suspension with a magnet. Finally, the beads may be removed from the osteocytes by incubation with an excess of antibody.

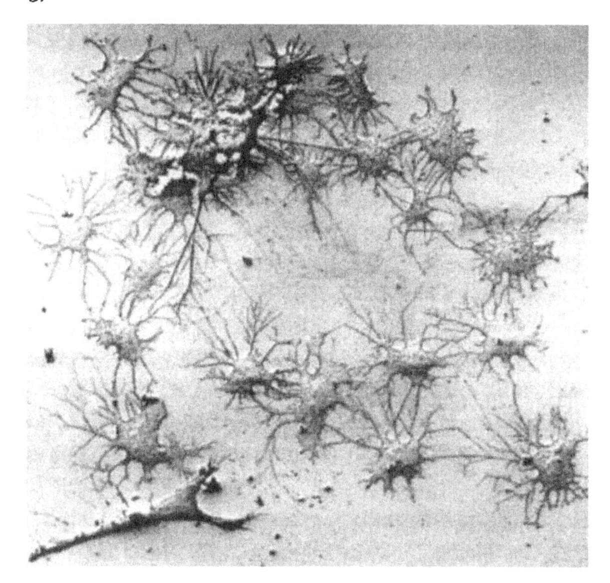

Figure 3. Isolated chicken osteocytes cultured for 24 hours. Osteocytes in culture form a network of interconnected cell processes similar to that of osteocytes *in vivo*. Note the fibroblast-like cell at the bottom possessing a completely different phenotype. Magnification: 630x.

Suspended osteocytes have a globular morphology. However, as soon as they are seeded on a substratum and attach, osteocytes form fingerlike extrusions in all directions. Somewhat later, the extrusions in the plane of the support elongate to long, slender, often branched processes, while the extrusions perpendicular on the support surface disappear. After 24 hours of culture, the osteocytes have formed multiple, smooth connections between cell processes resulting in the formation of a network very similar to the osteocyte network in bone, albeit two-dimensional (Figure 3; Van der Plas and Nijweide, 1992).

Isolated osteocytes were found to be very variable in alkaline phosphatase activity. Their mean activity was lower than that of isolated osteoblasts. They appeared not to be able to proliferate. When seeded on dentine slices, osteocytes did not show to any observable extent resorption activity, even in the presence of parathyroid hormone (PTH). Nevertheless, osteocytes do possess PTH receptors and respond to PTH with an increased cAMP production (Van der Plas et al., 1994).

VI. MATRIX PRODUCTION AND CELL ADHESION

During differentiation from osteoblasts, osteocytes lose cellular body volume up to 70% (Palumbo, 1986). In the most mature stage they have retained only few cell organelles. Osteocytes therefore do not have the appearance of strongly secretory cells. Their position inside the bone matrix probably does not require them to produce proteins to any large extent. Nevertheless, osteocytes *in situ* and after isolation have been found to be able to produce a number of matrix proteins. In immunocytochemical and *in situ* hybridization studies on bone sections, osteocytes were found positive for osteocalcin protein (Bronckers et al., 1985; Vermeulen et al., 1989; Boivin et al., 1990) and osteocalcin mRNA (Ikeda et al., 1992). Osteopontin protein (Mark et al., 1988; Chen et al., 1991, 1993a), osteopontin mRNA (Arai et al., 1993; Chen et al., 1993b), osteonectin protein (Chen et al., 1991), and osteonectin mRNA (Metsäranta et al., 1989) were also demonstrated in osteocytes *in situ. In vitro* isolated osteocytes express collagen type I, fibronectin, osteocalcin, osteopontin, and osteonectin proteins (Aarden et al., 1994). Of particular interest is the concentrated presence of osteopontin and bone sialoprotein in the perilacunar matrix of some osteocytes (McKee et al., 1992; Ingram et al., 1993).

Besides collagenous and noncollagenous proteins, bone matrix contains proteoglycans. Electron microscopical studies (Jande, 1971) had already shown that the osteocyte body and its processes are surrounded by a thin layer of uncalcified material containing collagen and proteoglycans. These findings were recently confirmed and extended by Sauren et al. (1992). These authors used the cationic dye cuprolinic blue to show the presence of proteoglycans in the bone matrix. Although proteoglycans appeared to be present throughout the matrix, the pericellular matrix both around the cell body and processes was decorated with proteoglycans. As the size of the cuprolinic blue positive rods around the osteocytes was larger than of those in the intercellular matrix, the authors were convinced that the proteoglycans were secreted by the osteocytes themselves.

What is the likely function of the (limited) protein/proteoglycan production of osteocytes? First, osteocytes may be involved in promoting calcification of the intercellular (osteoid) matrix while on the other hand hampering the calcification of the pericellular sheet immediately around the osteocytes. Second, through the production of matrix proteins and proteoglycans, osteocytes may be able to regulate their attachment to the bone matrix. If fluid flow is the driving force behind the mechanosensing mechanism, attachment of the osteocytes to the bone matrix is of crucial importance. The cell-matrix attachment sites will be the foci where the mechanosignal is conveyed from the extracellular to the intracellular compartment. Several types of adhesion receptors have been demonstrated on osteocytes. Isolated osteocytes have been shown to adhere to fibronectin and vitronectin via RGD peptide dependent integrins (Aarden et al., 1996). For adhesion to other matrix proteins, RGD-independent, anti-β_1 integrin subunit antibody (Neff et al., 1982) blockable integrins were shown to be involved (Aarden et al., 1996). In addition also the nonintegrin-adhesion receptor CD44 was shown to be highly expressed on osteocytes (Hughes et al., 1994; Nakamura et al., 1995).

VII. INTRA- AND INTERCELLULAR SIGNAL TRANSDUCTION

The presence of cellular responses to mechanical stress factors in organisms as diverse as microbes, plants, and animals reflects the fundamental importance of stress. In animals and humans, muscle and endothelium are probably the best studied tissues for their responses to mechanical stress. In the muscle, stretch is clearly the important stress factor, while for endothelium fluid shear stress resulting from the blood flow determines the mechanical stimulus. As mentioned earlier, in bone, osteocytes probably function as mechanosensors. They may respond to strain caused by cellular deformation as the result of matrix deformation or to an indirect effect of matrix deformation which is the increased interstitial fluid flow through the lacuno-canalicular system (Weinbaum et al., 1994). In both instances, twisting of osteocyte-matrix attachment sites may be the pathway of transfer of the strain signal from matrix to cell (Ingber, 1991; Wang et al., 1993).

Transmembrane attachment receptors, such as integrins and CD44, present in osteocytes, are linked to the cytoskeleton of microfilaments. Especially, the osteocytic processes are packed with microfilaments (King and Holtrop, 1975). Disruption of the actin cytoskeleton has been shown to abolish calcium responses that arise due to mechanical loading in fibroblasts (Glogauer et al., 1995) and endothelial cells (Diamond et al., 1994). This suggests that cytoskeletal elements might be directly linked to ion channels and that intracellular Ca^{2+} may play an important role as second messenger in strain related intracellular events.

Several *in vitro* studies have shown that mechanical loading causes a rapid increase of free intracellular Ca^{2+} concentrations in osteoblasts (Jones et al., 1991; Williams et al., 1994) and fibroblasts (Glogauer et al., 1995). These very rapid, transient increases were also observed in several studies involving endothelial cells subjected to various regimes of mechanical loads. The calcium response can be elicited by directly activating Ca^{2+}-ion channels or by activating phospholipase C (PLC) which stimulates the formation of diacylglycerol (DAG) and inositoltrisphophate (IP_3), from phospholipids. IP_3 causes the release of calcium from intracellular stores (Chen et al., 1992; Diamond et al., 1994). Other ions, like Na^+ and K^+, may also play a significant role in the mechanotransduction. Mechanical loading activates K^+ currents in vascular endothelial cells leading to hyperpolarization (Olesen et al., 1988). The K^+ currents may be a direct effect of mechanical loading (membranal perturbation of ion channels) or can be activated by Ca^{2+} fluxes. These K^+ fluxes may modulate the responses of cells to mechanical loading (Cooke et al., 1991).

Little is known whether these various mechanisms also play a role in the mechanoactivation of osteocytes. It has been demonstrated, however, that osteocytes do respond to mechanical loading with an increased release of PGE_2, PGI_2, and NO (Jubi et al., 1996; Klein-Nulend et al., 1995a,b; Pitsillides et al., 1996). The production of PG involves the activation of various enzymes and requires calcium. PGs are synthesized from arachidonic acid by the action of cyclo-oxygenase (PGHsynthase) which can be activated by NO (Hajjar et al., 1995). The substrate, arachidonic acid, is liberated from phospholipids by the action of phospholipase A_2, an enzyme that has been implicated in osteoblast-like cells to be involved in mechanosignal transduction (Binderman et al., 1988).

PGs play a pivotal role in the response of osteocytes to mechanical loading. Virtually all effects that have been described in osteocytes are PG-dependent, i.e., can be blocked by indomethacin, an inhibitor of PG synthesis (Pead and Lanyon, 1989; Chow and Chambers, 1994). Thus PGs appear to have paracrine effects in the osteocyte network. Whether PGs also transduce the mechanosignal to the bone surface cells and act as activators of the remodeling system of osteoblasts and osteoclasts remains to be investigated. IGF (Lean et al., 1995) is, in any case, a good candidate for such a function.

VIII. CONCLUSIONS

Considering that the osteocyte is by far the most abundant cell type in bone, osteocytes should be expected to play a decisive role in bone metabolism. On the other hand, especially mature osteocytes possess relatively few organelles, which limits their possible secretory function. The most striking property of osteocytes is their ability to form and maintain a network of intercommunicating cells. Osteocytes may be involved in osteoid calcification (Mikuni-Takagaki et al., 1995) or fine regulation of blood calcium homeostasis by regulating calcium exchange (Bonucci, 1990). Their most important function, we believe, is that of mechanosensor cells. The functional osteocyte syncytium, the configuration of the lacuno-canalicular network, the (although limited) ability of osteocytes to produce collagenous and noncollagenous matrix proteins and the abundant presence of matrix attachment receptors (integrins, CD44) on osteocytes, are all properties that make osteocytes perfectly suited for the sensing and processing of mechanosignals. Up until now hard evidence for such a role is still missing. Recently, however, the first, apparently direct interaction between osteocytes and osteoclasts has been described (Tanaka et al., 1995). Osteocytes were found to secrete (a) factor(s) that increased osteoclast formation and activity. Whether mechanical loading may influence this process has still to be evaluated.

REFERENCES

Aarden, E.M., Wassenaar, A.M., Alblas, M.J., and Nijweide, P.J. (1996). Immunocytochemical demonstration of extracellular matrix proteins in isolated osteocytes. Histochem. Cell Biol. 106, 495-501.

Aarden, E.M., Nijweide, P.J., Van der Plas, A., Alblas, M.J., Mackie, E.J., Horton, M.A. and Helfrich, M.H. (1996). Adhesive properties of isolated chick osteocytes in vitro. Bone 18, 305-313.

Ajubi, N.E., Klein-Nulend, J., Nijweide, P.J., Vrijheid-Lammers, T., Alblas, M.J., and Burger, E.H. (1996). Pulsating fluid flow increases prostaglandin production by cultured chicken osteocytes—a cytoskeleton-dependent process. Biochem. Biophys. Res. Commun. 225, 62-68.

Arai, N., Ohya, K., and Ogura, H. (1993). Osteopontin mRNA expression during bone resorption: An in situ hybridization study of induced ectopic bone in the rat. Bone Miner. 22, 129-145.

Bélanger, L.F. (1969). Osteocytic osteolysis. Calcif. Tissue Res. 4, 1-12.

Binderman, I., Zor, K., Kaye, A.M., Shimshomi, Z., Harrell, A. and Somjen, D. (1988). The transduction of mechanical force into biochemical events in bone cells may involve activation of phospholipase A_2. Calcif. Tissue Int. 42, 261-266.

Boivin, G., Morel, G., Lian, J.B., Anthoine-Terrier, C., Dubois, P.M. and Meunier, P.J. (1990). Localization of endogenous osteocalcin in neonatal rat bone and its absence in articular cartilage: Effect of warfarin treatment. Virchows Archiv. A. Pathol. Anat. 417, 505-512.

Bonucci, E. (1990). The ultrastructure of the osteocyte. In: Ultrastructure of Skeletal Tissues (Bonucci, E. and Motta, P.M., Eds.), pp. 223-237. Kluwer Academic Publ., London.

Boyde, A. (1980). Evidence against "osteocytic osteolysis". Metab. Bone Dis. Rel. Res. 2 (Suppl.), 239-255.

Bronckers, A.L.J.J., Gay, S., Dimuzio, M.T., and Butler, W.T. (1985). Immunolocalization of γ-carboxyglutamic acid containing proteins in developing rat bones. Collagen Rel. Res. 5, 273-281.

Bruder, S.P. and Caplan, A.I. (1990). Terminal differentiation of osteogenic cells in the embryonic chick tibia is revealed by a monoclonal antibody against osteocytes. Bone 11, 189-198.

Chen, J., Zhang, Q., McCulloch, C.A.G., and Sodek, J. (1991). Immunohistochemical localization of bone sialoprotein in fetal porcine bone tissue: Comparisons with secreted phosphoprotein 1(SPP-1, osteopontin) and SPARC (osteonectin). Histochem. J. 23, 281-289.

Chen, J., Luscinskas, F.W., Connoly, A., Dewey, C.F. and Gimbone, M.A. (1992). Fluid shear stress modulates cytosolic-free calcium in vascular endothelial cells. Am. J. Physiol. 262, C384-C390.

Chen, J., McCulloch, C.A.G., and Sodek, J. (1993a). Bone sialoprotein in developing porcine dental tissues: Cellular expression and comparison of tissue localization with osteopontin and osteonectin. Archs. Oral Biol. 38, 241-249.

Chen, J., Singh, K., Mukherjee, B.B., and Sodek, J. (1993b). Developmental expression of osteopontin (OPN) mRNA in rat tissues: Evidence for a role for OPN in bone formation and resorption. Matrix 13, 113-123.

Chow, J.W. and Chambers, T.J. (1994). Indomethacin has distinct early and late actions on bone formation induced by mechanical stimulation. Am. J. Physiol. 267, E287-E292.

Cooke, J.P., Rossitch, E., Andon, N.A., Loscalzo, J., and Dzau, V.J. (1991). Flow activates an endothelial potassium channel to release an endogenous nitrovasodilator. J. Clin. Invest. 88, 1663-1671.

Diamond, S.L., Sachs, F., and Sigurdson, W.J. (1994). Mechanically induced calcium mobilization in cultured endothelial cells is dependent on actin and phospholipase. Arterioscler. Thromb. 14, 2000-2006.

Dillaman, R.M., Roer, R.D., and Gay, D.M. (1991). Fluid movement in bone: Theoretical and empirical. J. Biomech. 24S, 163-177.

Doty, S.B. and Schofield, B.M. (1972). Metabolic and structural change within osteocytes of rat bone. In: Calcium, parathyroid hormone, and the calcitonins (Talmage, B.V. and Munson, P.L., Eds.), pp. 353-365, Excerpta Medica, Amsterdam.

Doty, S.B. (1981). Morphological evidence of gap junctions between bone cells. Calcif. Tissue Int. 33, 509-512.

Elmardi, A.S., Katchburian, M.V., and Katchburian, E. (1990). Electron microscopy of developing calvaria reveals images that suggest that osteoclasts engulf and destroy osteocytes during bone resorption. Calcif. Tissue Int. 46, 239-245.

Glogauer, M., Ferrier, J., and McCulloch, C.A.G. (1995). Magnetic fields applied to collagen-coated ferricoxide beads induce stretch-activated Ca^{2+} flux in fibroblasts. Am. J. Physiol. 269, C1093-C1104.

Hajjar, D.P., Lander, H.M., Pierce, S.F.A., Upmacis, R.K., and Pomerantz, K.B. (1995). Nitric oxide enhances prostaglandin-H-synthase-1 activity by a heme-independent mechanism: Evidence implicating nitrosothiols. J. Am. Chem. Soc. 117, 3340-3346.

Hughes, D.E., Salter, D.M., and Simpson, R. (1994). CD44 expression in human bone: A novel marker of osteocytic differentiation. J. Bone Miner Res. 9, 39-44.

Ikeda, T., Nomura, S., Yamaguchi, A., Suda, T., and Yoshiki, S. (1992). In situ hybridization of bone matrix proteins in undecalcified adult rat bone sections. J. Histochem. Cytochem. 40, 1079-1088.

Ingber, D. (1991). Integrins as mechanochemical transducers. Curr. Opin. Cell Biol. 3, 841-848.

Ingram, R.T., Clarke, B.L., Fisher, L.W., and Fitzpatrick, L.A. (1993). Distribution of noncollagenous proteins in the matrix of adult human bone: Evidence of anatomic and functional heterogeneity. J. Bone Miner. Res. 8, 1019-1029.

Jande, S.S. (1971). Fine structural study of osteocytes and their surrounding bone matrix with respect to their age in young chicks. J. Ultrastr. Res. 37, 279-300.

Jeansonne, B.G., Feagin, F.F., McMinn, R.W., Shoemaker, R.L., and Rehen, W.S. (1979). Cell-to-cell communication of osteoblasts. J. Dent. Res. 58, 1415-1423.

Jones, D.B., Nolte, H., Scholübbers, J.-G., Turner, E., and Veltel, D. (1991). Biomechanical signal transduction of mechanical strain in osteoblastlike cells. Biomaterials 12, 101-110.

King, G.J. and Holtrop, M.E. (1975). Actinlike filaments in bone cells of cultured mouse calvaria as demonstrated by binding to heavy meromyosin. J. Cell Biol. 66, 445-451.

Klein-Nulend, J., Van der Plas, A., Semeins, C.M., Ajubi, N.E., Frangos, J.A., Nijweide, P.J., and Burger, E.H. (1995a). Sensitivity of osteocytes to biomechanical stress in vitro. FASEB J. 9, 441-445.

Klein-Nulend, J., Semeins, C.M., Ajubi, N.E., Nijweide, P.J., and Burger, E.H. (1995b). Pulsating fluid flow increases nitric oxide (NO) synthesis by osteocytes but not periosteal fibroblasts—correlation with prostaglandin upregulation. Biochem. Biophys. Res. Commun. 217, 640-648.

Lean, J.M., Jagger, C.J., Chambers, T.J., and Chow, J.W. (1995). Increased insulinlike growth factor I mRNA expression in rat osteocytes in response to mechanical stimulation. Am. J. Physiol. 268, E318-E327.

Mark, M.P., Butler, W.T., Prince, C.W., Finkelman, R.D., and Ruch, J.V. (1988). Developmental expression of 44-kDa bone phosphoprotein (osteopontin) and bone γ-carboxyglutamic acid (Gla)-containing protein (osteocalcin) in calcifying tissues of rat. Differentiation 37, 123-136.

Marotti, G., Cane, V., Palazzini, S., and Palumbo, C. (1990). StructureBfunction relationships in the osteocyte. Ital. J. Min. Electrolyte Metab. 4, 93-106.

Marotti, G., Ferretti, M., Muglia, M.A., Palumbo, C., and Palazzini, S. (1992). A quantitative evaluation of osteoblastBosteocyte relationships on growing endosteal surface of rabbit tibiae. Bone 13, 363-368.

McKee, M.D., Glimcher, M.J., and Nanci, A. (1992). High-resolution immunolocalization of osteopontin and osteocalcin in bone and cartilage during endochondral ossification in the chicken tibia. Anat. Rec. 234, 479-492.

Metsäranta, M., Young, M.F., Sandberg, M., Termine, J., and Vuorio, E. (1989). Localization of osteonectin expression in human skeletal tissues by in situ hybridization. Calcif. Tissue Int. 45, 146-152.

Mikuni-Takagaki, Y., Kakai, Y., Satoyoshi, M., Kawano, E., Suzuki, Y., Kawaze, T., and Saito, S. (1995). Matrix mineralization and differentiation of osteocytelike cells in culture. J. Bone Miner. Res. 10, 231-242.

Nakamura, H., Kenmotsu, S., Sakai, H., and Ozawa, H. (1995). Localization of CD44, the hyaluronate receptor, on the plasma membrane of osteocytes and osteoclasts in rat tibiae. Cell Tissue Res. 280, 225-233.

Neff, N.T., Lowrey, C., Decker, C., Tovar, A., Damsky, C., Buck, C., and Horwitz, A.F. (1982). A monoclonal antibody detaches embryonic skeletal muscle from extracellular matrices. J. Cell Biol. 95, 654-666.

Nefussi, J.R., Sautier, J.M., Nicolas, V., and Forest, N. (1991). How osteoblasts become osteocytes: A decreasing matrix-forming process. J. Biol. Buccale 19, 75-82.

Nijweide, P.J., Van der Plas, A., and Scherft, J.P. (1981). Biochemical and histological studies on various bone cell preparations. Calcif. Tissue Int. 33, 529-540.

Nijweide, P.J. and Mulder, R.J.P. (1986). Identification of osteocytes in osteoblastlike cultures using a monoclonal antibody specifically directed against osteocytes. Histochem. 84, 343-350.

Nijweide, P.J., Burger, E.H., and Feyen, H.H.M. (1986). Cells of bone: Proliferation, differentiation, and hormonal regulation. Phys. Rev. 66, 855-886.

Olesen, S.P., Chapman, D.E., and Davies, P.F. (1988). Hemodynamic shear stress activates a K^+ current in vascular endothelial cells. Nature 331, 168-170.

Palumbo, C. (1986). A three-dimensional ultrastructural study of osteoid-osteocytes in the tibia of chick embryos. Cell Tissue Res. 246, 125-131.

Palumbo, C., Palazzini, S., and Marotti, G. (1990a). Morphological study of intercellular junctions during osteocyte differentiation. Bone 11, 401-406.

Palumbo, C., Palazzini, S., Zaffe, D., and Marotti, G. (1990b). Osteocyte differentiation in the tibia of newborn rabbit: An ultrastructural study of the formation of cytoplasmic processes. Acta Anat. 137, 350-358.

Pead, M.J., Suswillo, R.F.L., Skerry, T.M., Vedi, S., and Lanyon, L.E. (1988). Increased ^{3}H-uridine levels in osteocytes following a single short period of dynamic loading in vivo. Calcif. Tissue Int. 43, 92-96.

Pead, M.J. and Lanyon, L.E. (1989). Indomethacin modulation of load-related stimulation of new bone formation in vivo. Calcif. Tissue Int. 45, 34-40.

Pitsillides, A.A., Rawlinson, S.C.F., Suswillo, R.F.L., Bourrin, S., Zaman, G., and Lanyon, L.E. (1995). Mechanical strain-induced NO production by bone cells: A possible role in adaptive bone (re)modeling? FASEB J. 9, 1614-1622.

Reich, K.M., Gay, C.V., and Frangos, J.A. (1990). Fluid shear stress as a mediator of osteoblast cyclic adenosine monophosphate production. J. Cell Physiol. 143, 100-104.

Sauren, Y.M.H.F., Mieremet, R.H.P., Groot, C.G., and Scherft, J.P. (1992). An electron microscopic study on the presence of proteoglycans in the mineralized matrix of rat and human compact lamellar bone. Anat. Rec. 232, 36-44.

Skerry, T.M., Bitensky, L., Chayen, J., and Lanyon, L.E. (1989). Early strain-related changes in enzyme activity in osteocytes following bone loading in vivo. J. Bone Miner. Res. 4, 783-788.

Sun, Y.Q., McLeod, K.J., and Rubin, C.T. (1995). Mechanically induced periosteal bone formation is paralleled by the upregulation of collagen type-1 mRNA in osteocytes as measured by in situ reversed transcript polymerase chain reaction. Calcif. Tissue Int. 57, 456-462.

Tanaka, K., Yamaguchi, Y., and Hakeda, Y. (1995). Isolated chick osteocytes stimulate formation and bone-resorbing activity of osteoclastlike cells. J. Bone Miner. Metab. 13, 61-70.

Van der Plas, A. and Nijweide, P.J. (1992). Isolation and purification of osteocytes. J. Bone Miner. Res. 7, 389-396.

Van der Plas, A., Aarden, E.M., Feyen, J.H.M., de Boer, A.H., Wiltink, A., Alblas, M.J., de Ley, L., and Nijweide, P.J. (1994). Characteristics and properties of osteocytes in culture. J. Bone Miner. Res. 9, 1697-1704.

Vermeulen, A.H.M., Vermeer, C., and Bosman, F.T. (1989). Histochemical detection of osteocalcin in normal and pathological bone. J. Histochem. Cytochem. 37, 1503-1508.

Wang, N., Butler, J.P., and Ingber, D.E. (1993). Mechanotransduction across the cell surface and through the cytoskeleton. Science 260, 1124-1127.

Weinbaum, S., Cowin, S.C., and Zeng, Y. (1994). A model for the excitation of osteocytes by mechanical loading-induced bone fluid shear stresses. J. Biomech. 27, 339-360.

Williams, J.L., Ianotti, J.P., Ham, A., Bleuit, J., and Chen, J.H. (1994). Effect of fluid shear stress on bone cells. Biorheology 31, 163-170.

Wolff, J.D. (1882). Das Gesetz der Transformation der Knochen. A. Hirschwald, Berlin.

CELL-CELL COMMUNICATION IN BONE

Roberto Civitelli

I. INTRODUCTION

Bone remodeling is a dynamic process consisting of repeated sequences of bone resorptive and formative cycles, which in turn require a well-coordinated cellular ac-

Advances in Organ Biology
Volume 5B, pages 543-564.

ISBN: 0-7623-0390-5

tivity among osteoblasts, osteoclasts, and osteocytes (Raisz and Rodan, 1990). Bone is remodeled in specific areas of the skeleton where repair of an injury or replacement of aging bone is necessary. In these areas, new bone is deposited in packets, functionally defined as basic multicellular units (BMU) (Frost, 1986; Parfitt, 1988), in which osteoclasts first remove the old or damaged bone, and bone forming cells are subsequently recruited to refill the resorbed space. The morphology of the osteoblastic cell layer which covers the surface undergoing active bone formation is reminiscent of an epithelium (Raisz and Rodan, 1990). Like in epithelial tissues, adjacent osteoblasts come in direct contact with one another via junctional structures (Doty, 1981, 1988; Palumbo et al. 1990).

Cell-cell communication may presumably allow osteoblasts to establish and maintain a functional bone-forming "syncitium." If an effective intercellular network is necessary for osteoblasts to produce bone within each BMU, junctional structures that provide the nodes of this network may be critical regulators of osteoblast function (Doty, 1981). Gap junctions are the best known mechanism by which cells directly exchange signals with each other, and play an essential physiologic role in highly cooperative organs, such as the heart, the liver, and the uterus (Sheridan and Atkinson, 1985; Ramon and Rivera, 1986). In the past few years, a number of studies have led to the definition of the molecular structure and function of gap junctions in bone cells and their regulation by hormones and local factors. From these studies, direct cell-cell communication is beginning to take shape as a fundamental biologic mechanism that controls bone modeling and remodeling.

II. GAP JUNCTIONS AS INTERCELLULAR CHANNELS

Gap junctions are essentially transcellular channels formed by the juxtaposition of two hemichannels facing each other on the plasma membrane of two adjacent cells (Revel and Karnovsky, 1967). Each hemichannel, also called a connexon, is composed of six subunits, or connexins. When two connexons on opposing membranes are in contact, a channel is formed which allows effective continuity between two cells. The functionality of these transcellular channels is commonly assessed by monitoring the cell-to-cell diffusion of fluorescent molecules that permeate the junctions (Stewart, 1978). The size and charge of fluorescent dyes can be exploited to probe for the molecular permeability of the channels. Because ions can also flow through gap junctions, electrophysiologic methods have been developed to allow a precise analysis of transjunctional currents and unitary channel conductances. These methods are based on a double-cell configuration of the patch-clamp technique (Veenstra and DeHaan, 1986; Veenstra and Brink, 1992). As explained later, dye coupling and electric conductance may not go hand in hand in all circumstances. Different domains of the connexin molecule may be involved in the regulation of size or charge permeability and ion gating.

Connexins are a family of gap junction proteins whose structure somewhat resembles that of the α subunit of ion channels. They are membrane proteins, with four transmembrane spanning domains, two extracellular loops, and one intracellular loop. Both carboxyl and amino-termini reside inside the cell. Unlike membrane receptors, the extracellular domains of connexin molecules are highly conserved, which allows binding to the same connexin on the opposing membrane (Beyer et al., 1990). Conversely, both the intracellular loop and the long carboxyl-terminal intracellular tail are widely different among the various connexins, both in sequence and in length (Beyer et al., 1990). Following the purification and cloning of connexin32 (Cx32) in liver (Paul, 1989) and connexin43 (Cx43) in heart tissue (Beyer et al., 1987), about a dozen different connexins have been identified in other tissues and species, all products of different genes (Goodenough et al., 1996; White et al., 1995). In the last few years, considerable progress has been made in our understanding of the molecular structure and function of each connexin and of the complex mechanisms that control the formation and permeability of the intercellular channels. It is now known that the pores of the transcellular channel formed by the different connexins differ in size and electric charge, which could provide the connexon with specificity for certain molecules or ions (Goodenough et al., 1996; White et al., 1995). Studies of animals lacking specific connexin genes have demonstrated the critical physiological importance of intercellular transfer of messenger molecules for the normal function of many tissues (Paul 1995; Nicholson and Bruzzone, 1997). As detailed subsequently in this review, we have demonstrated that different connexins impart different molecular permeabilities to the gap junctions they form, and that different gap junctional permeabilities translate into specific gene regulatory mechanisms in osteoblasts.

III. FUNCTIONAL GAP JUNCTIONS IN THE SKELETAL TISSUE

Structures resembling gap junctions were first identified in bone by electron microscopy in the early 1970s (Doty and Schofield, 1972; Stanka, 1975). A decade later, Doty and co-workers, using the lanthanum nitrate method for staining electron micrographic slides, confirmed the nature of these structures as gap junctions (Doty, 1981). In these studies, gap junctions were consistently observed between adjacent osteoblasts, osteoblasts and osteocytes, and osteoblasts and periosteal fibroblasts, but not in osteoclasts (Doty, 1981). Such a diffuse distribution of gap junctions among cells of the osteoblastic lineage and fibroblasts, with the exclusion of osteoclasts, was later confirmed by a number of ultrastructural studies in histologic sections of bone (Shapiro, 1988; Miller et al., 1989; Palumbo et al., 1990; Jones et al., 1993), as well as *in vitro* cultures of bone cells (Bhargava et al., 1988).

The functionality of gap junctions was demonstrated by monitoring the cell-to-cell diffusion of microinjected fluorescein dyes in rat calvaria osteoblasts (Jeasonne et al., 1979), and in explants of canine odontoblasts (Ushiyama, 1989). With the identification of connexins, the molecular nature of gap junctions in bone has been defined in a variety of cell systems, and correlated with the degree of cell coupling. Expression of Cx43, but not Cx32 or Cx26, was reported at sites of cell-cell contact in cultures of rat calvaria cells (Schirrmacher et al., 1992) and osteogenic sarcoma cells (Schiller et al., 1992). Expression of Cx43 protein was associated with dye and electric coupling between rat bone cells, thus pointing to Cx43 as the major gap junction protein that mediates intercellular communication between osteoblastic cells.

Our group has demonstrated that cell coupling and expression of connexins is widely heterogenous among different osteoblastic cell systems. For example, a high degree of cell coupling can be observed in monolayer cultures of human normal trabecular bone cells and osteoprogenitor marrow stromal cells, whereas the human osteogenic sarcoma cell line SaOS-2 is poorly coupled (Civitelli et al., 1993). In agreement with the previous studies (Schiller et al., 1992; Schirrmacher et al., 1992), we have found that both rat calvaria osteoblastic cells and the osteogenic sarcoma cells ROS 17/2.8 are very well coupled, whereas UMR 106-01 cells pass dye poorly. All osteoblastic cells express Cx43, and expression of Cx43 mRNA and protein is correlated with the degree of coupling (Civitelli et al., 1992, 1993; Steinberg et al., 1994). In general, all cells that couple the relative abundance of phosphorylated and nonphosphorylated forms of Cx43 are similar, indicating that the expression of Cx43 protein is probably more important than its phosphorylation state for a cell's ability to transfer dye (Civitelli et al., 1993; Steinberg et al., 1994). However, in the poorly coupled sarcomatous SaOS-2 cells, primarily nonphosphorylated forms of Cx43 have been detected (Civitelli et al., 1993c; Donahue et al., 1995c), suggesting that in this particular cell line the phosphorylation profiles of Cx43 may determine the degree of dye coupling. These observations suggest that a number of regulatory steps may be involved in determining gap junctional permeability in osteoblastic cell systems.

A direct link between Cx43 expression and dye coupling was established by Yamaguchi et al. (1994b) who were able to abolish cell-to-cell diffusion of fluorescent dyes between mouse MC3T-E1 osteoblast-like cells by intracellular injection of an anti-Cx43 antibody. Like some of their rat and human counterparts, these mouse cells express abundant Cx43 and are highly coupled (Yamaguchi et al., 1994b). A recent ultrastructural study has confirmed the presence of gap junctions formed by Cx43 between osteoblasts, osteocytes, and chondrocytes in rat calvaria (Jones et al., 1993). In that study, Cx43 labeling was also observed between osteoclasts and overlying mononuclear cells at sites of active bone resorption (Jones et al., 1993), thus raising the hypothesis that intercellular communication via gap junctions formed by Cx43 may be involved in osteoclast development. Notably, however, no evidence of heterotypic gap junctional structures between osteoblastic and osteoclastic cells was found (Jones et al., 1993).

Further studies have demonstrated that Cx43 is not the only connexin expressed by bone cells. All human and rat osteoblastic cells (except the ROS 17/2.8) also express Cx45, albeit in lower abundance compared to Cx43 (Civitelli et al., 1993; Steinberg et al., 1994). However, the abundance of Cx45 mRNA and protein expression does not seem to be related to the degree of dye coupling. In fact, Cx45 is expressed in higher abundance by cells that exhibit a very low level of dye coupling, i.e., the rat UMR 106 and the human SaOS-2 osteosarcoma cells. Nonetheless, Cx45 is invariably found at the borders between cells, a situation compatible with functional gap junctions (Civitelli et al., 1993; Steinberg et al., 1994). This observation raises the possibility that in osteoblastic cells gap junctions formed by Cx45 may possess a different permeability than those formed by Cx43, perhaps reflecting a different size and/or charge selectivity.

Furthermore, as discussed later in this chapter, experiments with small-sized dyes and electrophysiologic measurements of gap junctional conductance have proven that cells that prevalently express Cx45, such as UMR 106-01 and SaOS-2, are electrically coupled although they diffuse dyes poorly (Steinberg et al., 1994). In addition to Cx45, sizable amounts of Cx46 can also be detected in rat osteoblastic cells. However, Cx46 is exclusively localized within intracellular compartments and does not oligomerize (Koval et al., 1997). Thus, the function of this connexin in osteoblastic cells remains uncertain. To date, there is no evidence that other known connexins such as Cx26, Cx32, Cx40, or Cx47 are present in osteoblasts.

Cartilage may seem an unlikely tissue for intercellular junctions, simply because in normal mature cartilage chondrocytes are isolated cells without direct contacts with one another. However during cartilage development, the mesenchymal chondrogenic precursors must condense before chondrogenesis can occur. During this phase, cells become tightly compacted and gap junctions appear between adjacent cells (Langille, 1994; Minkoff et al., 1994). Therefore, direct intercellular communication seems to be important for these early steps of cartilage development when recruitment, proliferation, and differentiation of precursors occurs. Donahue et al. (1995a) have demonstrated that Cx43 is expressed by adult bovine articular chondrocytes in culture, and that these cells are functionally coupled. Based on these observations, it seems possible that gap junctional communication is reestablished in mature cartilage in conditions that lead to cell proliferation or tissue repair, as it occurs for example in osteoarthritis (Hamerman, 1989).

IV. DIRECT CELL-CELL CONTACT VIA CELL ADHESION MOLECULES

There is growing evidence suggesting that the formation of functional gap junction channels may be regulated by a higher hierarchical system, represented by cell recognition and adhesion via cell adhesion molecules. At least four structurally distinct classes of cell adhesion molecules exist (Albelda and Buck, 1990), and a functional correlation has been observed between gap junctions and members of the immunoglobulin and the cadherin superfamilies. A communication deficient mouse sarcoma cell line could be induced to form epithelial sheets with abundant cell-cell contacts and functional gap junctions after transfection with either liver cell adhesion molecule (L-CAM i.e., E-cadherin) or N-cadherin (Mege et al., 1988; Matsuzaki et al., 1990). Similarly, transfection of E-cadherin into communication deficient cell lines restores gap junctional communication and changes the pattern of Cx43 phosphorylation (Musil et al., 1990; Jongen et al., 1991), and inhibition of cell-cell adhesion by anti-cadherin antibodies not only suppresses the formation of gap junctions but also the differentiation of neural ectodermal tissue (Keane et al., 1988). Thus, cell adhesion molecules may be involved in cell signaling by two mechanisms, direct activation of intracellular signaling pathways, and indirect modulation of metabolic and electric coupling through gap junctions.

Work performed during the past few years has demonstrated that cell adhesion molecules of the immunoglobulin and cadherin superfamilies are expressed in the skeletal tissue and that they may be important for bone remodeling. In developing chick bone, neural cell adhesion molecule (N-CAM) is expressed transiently during osteoblast differentiation. Expression of this molecule does not start until osteogenic condensation and it is subsequently lost when osteoblasts become terminally differentiated into osteocytes (Lee and Chuong, 1992). Expression of cadherin-like molecules by rat osteoblasts had been postulated by two groups in preliminary reports using reverse transcription-polymerase chain reactioon techniques (Kunth et al., 1993; Stueckle et al., 1993). Sequence analysis indicates that one of these molecules isolated from rodent osteoblasts is closely related to N-cadherin, probably representing an alternatively spliced variant (Stueckle et al., 1993; Suva et al., 1994). Other investigators have also reported the cloning of a cadherin from rat and human osteogenic sarcoma cells (Okazaki et al., 1994). This molecule was tentatively called OB-cadherin, to indicate its origin from osteoblastic cells. However, analysis of the full DNA sequence clearly suggests that OB-cadherin is the mouse homologue of human cadherin-11 (Okazaki et al., 1994; Tanihara et al., 1994). Expression of cadherin-11 by osteoblasts is very consistent with recent data indicating that in embryos this cadherin appears during formation of limb buds and somites, where it may play a role in cell condensation and segregation (Kimura et al., 1995). Thus cadherin-11, a cell adhesion molecule present primarily in cells of mesenchymal origin (Hoffmann and Balling, 1995; Kimura et al. 1995), is expressed at sites of embryonic bone formation.

Work from our group has confirmed the presence of cadherin-11 in several human osteoblastic cells models, although osteoblasts, like many other cell systems, express multiple cadherins, including cadherin-4 (the human homologue of chicken R-cadherin), and N-cadherin-like molecules (Cheng et al., 1998). Importantly, osteoblast cadherins are regulated by corticosteroids and parathyroid hormone (PTH) both *in vitro* and *in vivo* (Cheng et al., 1994; Suva et al., 1994), and the pattern of cadherin expression changes with different stages of osteoblast differentiation (Cheng et al., 1998). The potential importance of cadherin-mediated cell-cell adhesion in the skeletal tissue is further stressed by the observation that N-cadherin is expressed in a specific spatio-temporal manner in developing limb buds during chondrogenesis *in vivo,* and that antibodies against N-cadherin could inhibit the differentiation of cultured mesenchymal cells from limb buds into chondrocytes (Oberlender and Tuan, 1994). Further support to the role of cadherins in bone formation comes from our recent studies demonstrating that inhibition of cadherin-mediated cell–cell adhesion by synthetic peptides dramatically reduces the osteogenic potential of human bone marrow stromal cells under stimulation with bone morphogenetic protein-2 (Cheng et al., 1998). Cadherin-mediated adhesion is also essential for fusion of mononuclear precursors into multinucleated osteoclasts, a critical step in bone resorption (Mbalaviele et al., 1995). E-cadherin, and not N-cadherin, has been implicated as the molecule that mediates homotypic adhesion between osteoclast precursors in this regard. Studies using inhibitory peptides or neutralizing antibodies showed that *in vitro* bone resorption can be abolished by inhibiting cadherin-mediated homophilic binding, thus demonstrating on a functional basis the critical role of cadherins in osteoclast development (Mbalaviele et al., 1995).

The present data clearly underscores the critical role of cell-cell adhesion in the development of skeletal tissue and for the remodeling of adult bone. It is foreseeable that the progress on cell adhesion will help understand some of the fundamental mechanisms by which cell-cell communication via gap junctions is regulated in bone cell networks. Ultimately, cell-cell adhesion and communication may be seen as two aspects of the same general biologic phenomenon.

V. HETEROTYPIC CONTACT AND COMMUNICATION

Another role in which direct cell-cell contact and communication may prove critical in adult bone is the regulation of bone resorption. In theory, such a mechanism may provide the means by which signals can be exchanged between cells of the osteoblastic and osteoclastic lineages. In some systems, physical contact between stromal cells or osteoblasts and osteoclast precursors is necessary for osteoclast development (Takahashi et al., 1988; Udagawa et al., 1989). As pointed out above, E-cadherin is involved in mediating the fusion of mononuclear precursors to form multinucleated bone resorbing osteoclasts (Mbalaviele et al., 1995). Although there is no clear evidence that E-cadherin is expressed by osteoblasts, it is possible that other cell adhesion molecules that may be expressed by cells of the osteoclastic lineage may mediate direct contact between stromal cells and osteoclast precursors (Mbalaviele et al., 1995). Interaction between vascular cell adhesion molecule (VCAM)–containing integrins has been proposed as a possible mechanism in this regard (Duong et al., 1994), and VCAM expression has been described in osteoblasts (Tanaka et al., 1995). More recent data shows that heterotypic interactions between hemopoietic osteoclastic and stromal bone marrow cells can occur via cadherin-6 isoforms, and inhibition of cadherin-6 expression or function inhibits the stromal cell ability to support osteoclast differentiation (Mbalaviele et al., 1998). Furthermore, cells of the osteoclastic and osteoblastic lineage express protocadherin-2 (Cheng et al., 1994; Sakai et al., 1995). Although not much is known about these molecules, protocadherins do impart adhesive properties to cells (Sano et al., 1993). Therefore, it is conceivable that a temporally and spatially regulated expression of different types of cell adhesion molecules may coordinate osteoblasts/osteoclasts interactions in the bone microenvironment.

More recently, members of the immunoglobulin superfamily, including intercellular adhesion molecule (ICAM-1), VCAM-1, and leukocyte function-associated antigen (LFA-3) have been identified in human osteoblasts (Tanaka et al., 1995). Importantly, not only do ICAM-1 and VCAM-1 allow adhesion of T cells to osteoblasts, but they also provide the mechanism by which T lymphocytes stimulate cytokine production by osteoblasts (Tanaka et al., 1995). These studies offer an excellent example of how heterotypic cell-cell contact through immunoglobulin-like cell adhesion molecules can be relevant to bone remodeling.

On the other hand, there is no evidence that functional gap junctions exist between cells of the osteoblast and osteoclast lineages. In mouse stromal/spleen cells cocultures, an established system of osteoclast development (Udagawa et al., 1989), we have failed to see direct diffusion of dye molecules between the mononuclear stromal cells and multinucleated osteoclastic cells (unpublished observations). Although direct interactions between precursors of the two cell lineages are theoretically possible, the available data seem to discount the hypothesis that differentiated osteoclasts and osteoblasts directly interact through gap junctional communication. Nonetheless, heterotypic coupling is possible between other types of cells resident in bone. A preliminary report suggests that osteoblastic cells can exchange dye molecules with endothelial cells (Melchiore et al., 1994), although the physiologic relevance of this type of interaction remains to be determined.

VI. REGULATION OF GAP JUNCTIONAL COMMUNICATION AND CONNEXIN EXPRESSION IN OSTEOBLASTS

Early work from other members of our group had demonstrated that prostaglandin E_2 (PGE_2) induced a rapid appearance of gap junctional structures in fetal calvaria bone cultures in association with a shape change from flat to stellate morphology (Shen et al., 1986). Following that initial observation, we have observed that PGE_2 increases transfer of lucifer yellow in both the highly coupled ROS 17/2.8 and the poorly coupled UMR 106-01 cells (Civitelli et al., 1998) (Figure 1). Also consistent with the previous report, the stimulatory effect of PGE_2 on dye coupling was rapid and persisted for at least four hours of incubation. PGE_2 had no effect on Cx43 protein expression, and its stimulatory action on dye coupling was insensitive to cycloheximide, suggesting that the increase in dye coupling induced by PGE_2 is independent of Cx43 gene expression.

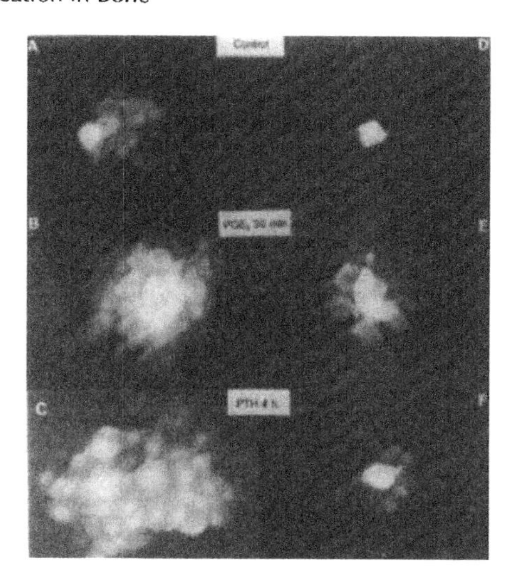

Figure 1. Effect of PTH and PGE_2 on dye coupling in rat osteogenic sarcoma cells. Confluent cultures of ROS 17/2.8 (left) or UMR 106-01 (right) cells were incubated with either 10^{-6} M PGE_2 for 30 minutes or with 10^{-7} M PTH for four hours. One cell in each microscopic field was microinjected with lucifer yellow. Snapshots were taken 3–5 minutes after the injection. The fluorescent dye diffuses form the microinjected cells to neighboring cells, and the number of cells that take the fluorescent dye from the microinjected cell serves as a measure of gap junctional communication (dye coupling).

Because Cx43 abundance at cell-cell contact sites was increased after PGE_2 incubation, it is likely that the prostanoid may affect the posttranslational processing of Cx43, leading to an increased assembly of preformed connexins into hemichannels.

Other factors active on bone remodeling can modulate gap junctional communication in osteoblasts. The most studied so far is PTH which stimulates gap junctional communication in the UMR 106-01, as well as in other human osteogenic sarcoma and rat calvaria cells (Schiller et al., 1992; Donahue et al., 1995c). The hormonal effect on cell coupling is paralleled by a time- and dose-dependent increase of steady-state Cx43 mRNA levels, thus implying an increased protein synthesis (Schiller et al., 1992). We have been able to confirm this hypothesis in both the poorly coupled UMR 106-01 and the well coupled ROS 17/2.8 cells (Figure 1). The stimulatory effect of PTH on cell coupling is biphasic, with a rapid increase of dye transfer independently of new protein synthesis, and a sustained effect that is associated with increased Cx43 mRNA (Figure 2) and protein expression. This prolonged stimulation of gap junctional communication by PTH requires new protein synthesis and is probably the result of transcriptional upregulation of the Cx43 promoter (Civitelli et al., 1998). A similar stimulatory action is induced by prostaglandin E_2 (PGE_2), an important regulator of bone remodeling (Civitelli et al., 1998). These effects of PTH and PGE_2 appear to be mediated by cAMP production be-

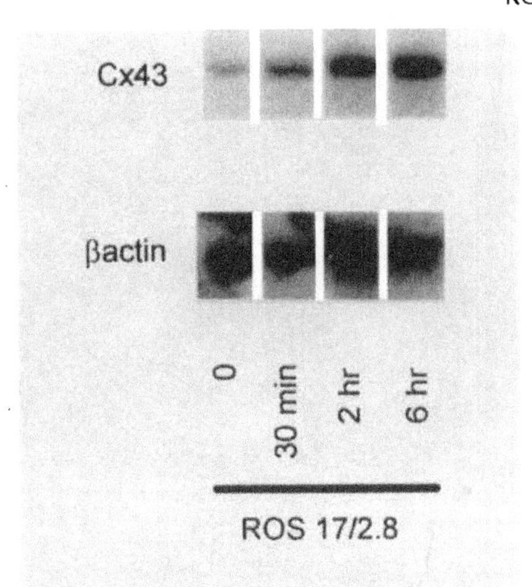

Figure 2. Effect of PTH on Cx43 mRNA. ROS 17/2.8 cultures were exposed to 10^{-7} M PTH for different times, and total RNA was extracted using standard procedures. Membranes were hybridized with a ^{32}P-labeled cDNA probe for Cx43, then washed and rehybridized with a human β-actin probe.

cause they can be reproduced by cAMP analogs (Schiller et al., 1992) and prevented by PTH antagonists (Donahue et al., 1995c). Thus, both PTH and PGE_2, potent stimulators of bone resorption, increase gap junctional communication between osteoblasts by modulating Cx43 expression or function by different mechanisms. The physiologic correlates of these effects are still hypothetical. The resorptive action of these two factors requires an intervening cell which can transfer the hormonal signal to the osteoclast and/or its precursors (McSheehy and Chambers, 1986a,b), and direct cell-cell communication may provide a convenient means of signal exchanges between cells of the two lineages (see above). In addition, an increased gap junctional communication may play a role in the "anabolic" action of PTH, by enhancing the metabolic coupling within the osteoblastic network and thus facilitating the diffusion of locally generated signals, such as mechanical strain.

Both retinoic acid and transforming growth factor-β (TGF-β) have been reported to increase gap junctional communication and Cx43 mRNA and protein expression in a human osteoblastic cell line whereas $1{,}25(OH)_2D_3$ had no effect (Chiba et al., 1994). On the contrary, other investigators found that both TGFβ and BMP-2 actually decreased cell coupling in to mouse MC3T3-E1 cells (Rudkin et al., 1996). Aside from the differences in cell models, these results are intriguing since both retinoic acid and TGFβ may inhibit osteoblast differentiated function, although their action on bone-forming cells is complex (Canalis et al., 1991). Nonetheless, because of the importance of some members of the TGF-β superfamily for osteogenesis and osteoblast differentiation (Wang et al., 1990; Rosen et al., 1994), regulatory effects of these factors on intercellular communication are expected to occur. Whether this regulation is part of the osteoblast differentiation program remains unclear.

Regulation of intercellular communication by pH (Spray et al., 1982) may be particularly important in bone remodeling, because of the pH dependency of mineral deposition and dissolution (Cuervo et al., 1971; Baron et al., 1985). Yamaguchi et al. (1995) have carefully analyzed this regulatory aspect of gap junctional communication in the mouse MC3T3-E1 cell line. As it occurs in other cell systems, extracellular acidification and the attendant cytoplasmic acidification decrease cell coupling, whereas alkalinization increases gap junctional communication. The mechanisms by which pH regulates cell coupling are complex, and involve rapid effects on channel permeability, as well as decreased Cx43 protein synthesis for prolonged exposures to a low ambient pH (Yamaguchi et al., 1995).

Mechanical and physical factors can also modulate gap junctional communication. The number of gap junctions in bone declines in weightlessness conditions (Doty and Morey-Holton, 1982), and application of cyclical stretch, an anabolic stimulus for osteoblasts (Harter et al., 1995), increases intercellular communication (Ziambaras et al., 1998). In our hands, this rapid effect is mediated by interference with Cx43 protein turnover and phosphorylation (Ziambaras et al., 1998). In vivo, Cx43 protein expression is increased in periodontal ligament after experimental tooth movement, and in osteocytes after tooth extraction (Su et al., 1997). Furthermore, reports that have remained preliminary suggest a regulatory action of magnetic fields on gap junction function and connexin expression (Donahue et al., 1995b; Yamaguchi et al., 1994). These observations and the key role of mechanical stimuli in the regulation of bone remodeling underscore the physiologic importance of physical factors in modulating intercellular signals via gap junctional communication.

VII. HETEROGENEITY OF CONNEXINS IN OSTEOBLASTS: FUNCTIONAL SPECIFICITY OR REDUNDANCY?

As discussed above, osteoblastic cells express multiple connexins in various degrees of abundance. Although several lines of evidence have established that gap junctions formed by Cx43 provide the transcellular channels that allow dye coupling between osteoblasts, it is also clear that Cx45 as well can form functional gap junctions between osteoblasts. Such a diversity of connexins may reflect either specific functional roles of each protein, or simply represent the expression of a redundant system. Our group has provided some initial insights giving support to the former hypothesis. When electric coupling was examined in UMR 106-01 cell pairs, sizable currents were consistently found, despite the fact that UMR 106-01, which expresses abundant Cx45, is not chemically coupled (Steinberg et al. 1994). However, total transjunctional conductance was lower in UMR 106-01 (1–2 or

8–11 nS) than in ROS 17/2.8 cell pairs (4–31 nS), which express Cx43 but not Cx45. Transjunctional conductance was also strongly voltage-dependent in many UMR 106-01 cell pairs, in contrast to the voltage-independent transcellular currents detected in ROS 17/2.8 cells. The voltage-dependency and a lower total transjunctional conductance in UMR 106-01 cells is compatible with the electrophysiologic characteristics of gap junctions formed by Cx45 (Veenstra et al. 1992). Thus, UMR 106-01 cells express gap junctions that are permeable to ions, although they are not permeable to molecules of the size of lucifer yellow (mw = 623 Da).

Experiments in which UMR 106-01 were stably transfected with either Cx43 or Cx45 confirmed that Cx43 confers the ability to diffuse small molecules (dye coupling). UMR 106-01 transfected with Cx43 exhibited a high degree of dye coupling, whereas UMR 106-01 overexpressing Cx45 was not different than the parent cells (Steinberg et al., 1994). These results demonstrate that gap junction channels formed by Cx45 in osteoblasts have a different molecular permeability than those formed by Cx43, and point to specific roles of different connexins in mediating intercellular communication between bone cells.

Further studies demonstrated that, when expressed in the same cells, Cx45 and Cx43 are able to interact in forming gap junctions, and that the gating properties of Cx45 prevail in the resulting channels. This type of interaction was directly assessed by overexpressing Cx45 in stable ROS 17/2.8 cell transfectants. Diffusion of lucifer yellow and transjunctional conductance were reduced by more than 50% in these clones, compared to the parent ROS 17/2.8 cells (Koval et al., 1995a) (Figure 3). However, transfer of smaller fluorescent molecules, such as hydroxycoumarin (mw = 206 Da), was reduced by only 35% in the ROS 17/2.8 overexpressing Cx45. Furthermore, transfection of a truncated form of Cx45, lacking the last 37 amino acids at the intracellular carboxyl-terminus did not alter dye coupling but it reduced electric conductance (Figure 3), demonstrating that the cytoplasmic domain of Cx45 may be involved in determining the molecular permeability of the channel, or the type of interaction between connexins (Koval et al., 1995a). Therefore, expression of multiple connexins may enable cells to achieve different modalities of cell-cell communication. This in turn may determine the type of molecules or ions that can diffuse through the intercellular channels, thus providing a mechanism for modulating signal exchange among cells in a network.

Initial studies indicate that diffusion of intercellular signals can in fact be different in cells expressing different connexins. As reported for other non-osseous cell systems, a gentle touch of one osteoblast generates a sudden and transient increase in cytosolic calcium $[Ca^{2+}]_i$, which spreads rapidly to neighboring cells with a variable time lag (Xia and Ferrier, 1992). These mechanically-induced $[Ca^{2+}]_i$ waves can be generated in many osteoblastic cell models, and they require functional gap junctions, since they can be abolished by inhibitors of gap junctional communication (Xia and Ferrier, 1992). However, not all $[Ca^{2+}]_i$ waves require gap junctions.

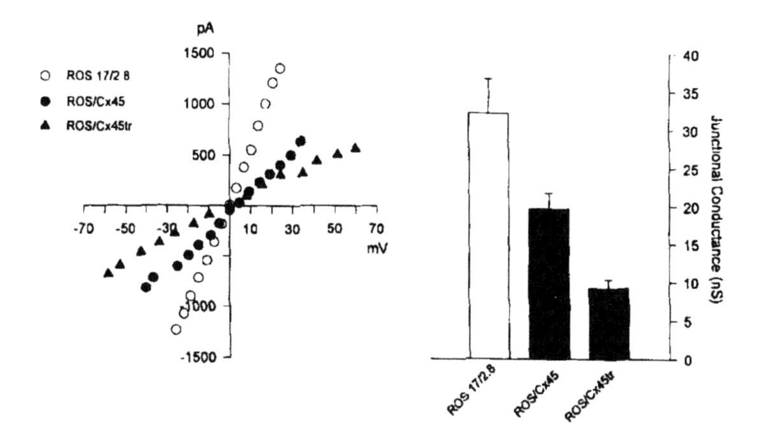

Figure 3. Gap junctional conductance in ROS 17/2.8 transfectants. Pairs of ROS/Cx45, ROS/Cx45tr, or their parent clone were analyzed with the double whole-cell patch voltage clamp technique. Intensity/voltage (I/V) curves for each cell line are plotted on the left, and the average total junctional conductances (g_j) are illustrated on the right. The I/V relationships are linear in all transfectants within -60 and +60 mV range, indicating no voltage dependency of g_j. The average g_j is significantly lower in ROS/Cx45 and ROS/Cx45tr versus ROS 17/2.8 (p<0.01, t test for unpaired sample).

We have found that propagation of $[Ca^{2+}]_i$ waves among osteoblastic cell lines and human bone marrow stromal cells is mediated by two different mechanisms: by autocrine activation of P2Y (nucleotide) receptors through released ATP and by gap junctional communication. In the former model, ATP released from the mechanically stimulated cell activates P2Y receptors in neighboring cells with consequent release of intercellular calcium stores and propagation of a fast $[Ca^{2+}]_i$ wave to many cells in the population. Gap junction–dependent $[Ca^{2+}]_i$ waves propagate slowly and do not require release of intracellular calcium stores by inositol trisphosphate (Jørgenson et al., 1997). It is possible that these different modalities of intercellular calcium signaling may serve specific functions.

Based on the studies mentioned above, one would expect that molecules such as cAMP (mw ~350 Da) or inositol trisphosphate (mw = 420 Da) should permeate through Cx43 channels, but that their permeability to Cx45 channels would be lower. As predicted, using the cAMP probe, FlCRhR (Adams et al., 1991) we have observed direct cell-to-cell diffusion of cAMP in human osteoblast cultures, which express primarily Cx43. Upon microinjection of cAMP in FlCRhR-loaded cells, cAMP-dependent fluorescent signal rapidly spread throughout the cytoplasm of the cell injected, and after a latency of a few seconds it diffused into the neighboring cells. These experiments provide a direct demonstration that physiologically relevant molecules, such as second messengers, can be rapidly but selectively propagated from cell to cell via gap junctions.

VIII. ROLE OF DIRECT CELL-CELL COMMUNICATION IN THE SKELETAL TISSUE

The role of gap junctional communication in osteoblast differentiation and function is still the object of investigation. Some observational studies have described an association between gap junctions and bone cell function, but the evidence is indirect and far from conclusive. While the expression of Cx43 increased in a SV40-immortalized human osteoblast cell line with the acquisition of a differentiated phenotype (Chiba et al., 1993), loss of gap junctions was observed to precede calcification in rat calvaria cells by electron microscopy (Yamazaki et al., 1989). A recent study on gap junction proteins during bone formation in chick embryos may help shed some light on the physiologic role of connexins in bone development. Immunohistochemical localization using specific anti-connexin antibodies revealed that Cx43 was more concentrated on mesenchymal cells during the earliest stages of osteogenic differentiation, preceding the appearance of osteogenic cells (Minkoff et al., 1994). Although Cx43 was present throughout mandibular bone formation, its expression was not appreciably regulated with further differentiation. Conversely, Cx45 distribution was more restricted to areas of active bone formation, gradually increasing in abundance at successive stages of development (Minkoff et al., 1994). Although a specific role of each connexin cannot be established with these studies, the close spatial and temporal association between gap junction proteins and osteogenesis is consistent with the hypothesis that connexin expression may be instrumental for initiation of bone formation and the development of the osteoblastic phenotype.

In the chick embryo, Cx43 expression occurred in the interval between commitment of mesenchyme to osteogenesis and the appearance of the osteogenic phenotype (Minkoff et al., 1994), pointing to a role of Cx43 in cell condensation before osteogenesis begins. Such conclusion is consistent with observations in developing chick limb buds and micromass cultures of mammalian mesenchyme (Coelho and Kosher, 1991; Langille, 1994). Gap junctional communication is present during condensation of embryonic mesenchymal cells, immediately before the onset of chondrogenesis (Coelho and Kosher, 1991). This phase is very brief, and it is rapidly followed by chondroblast differentiation and matrix production. Cartilage cells become separated by abundant matrix and direct cell-cell contacts are lost (Langille, 1994). The recent observation of Cx43 expression by cultured articular chondrocytes (Donahue et al., 1995a) makes it very likely that Cx43 is involved in chondroblast condensation. Thus, the early stages of osteogenesis and chondrogenesis seem to follow very similar molecular programs. The two programs diverge in the later phase of differentiation, reflecting the strikingly different features of mature bone and cartilage. Whereas osteoblasts form a tightly packed layer above the bone surface and remain connected through anchoring and communicating junctions, chondrocytes are isolated cells within the mature cartilage. Such anatomical differences underline the different requirements for tissue turnover and cell recruitment between adult bone and cartilage.

Recent advances have underscored the critical physiologic role of gap junctional communication for bone-forming cells. Using different ostoblastic cell lines that express different connexins, we have found that altering gap junctional communication by manipulating the relative expression of Cx43 and Cx45 affects the osteoblast phenotype (Lecanda et al., 1998). Transfection of Cx45 in cells that express primarily Cx43 (ROS 17/2.8 and MC3T3-E1) decreased both dye transfer and expression of osteocalcin (OC) and bone sialoprotein (BSP), genes pivotal to bone matrix formation and calcification. Conversely, transfection of Cx43 into cells that express predominantly Cx45 (UMR 106-01) increased both cell coupling and expression of OC and BSP. Using OC and BSP promoter–luciferase constructs, we demonstrated that these effects were the consequence of transcriptional downregulation or upregulation by Cx45 or Cx43, respectively (Lecanda et al., 1998). Thus, the type of gap junctional communication provided by Cx43 is permissive for a full elaboration of the mature osteoblastic phenotype. This important role of Cx43 gap junctions for osteoblast function also emerges from studies of Van der Molen and colleagues (1996), showing that responsiveness to PTH is decreased in ROS 17/2.8 cells rendered communication-deficient by transfection with an antisense Cx43 cDNA construct. The type of signals that diffuse from cell to cell through junctional channels and regulate gene expression remain elusive. As reviewed above, permeability to cyclic nucleotides or inositol phosphates changes with expression of different connexins, and spontaneous oscillations in intracellular ionic concentration or in membrane polarity may be sensitive to intercellular communication. Discovering the intercellular signals that diffuse through gap junctions and modulate gene expression represent an exciting challenge for future research in this area.

IX. DERANGED CELL-CELL COMMUNICATION AS A POTENTIAL MECHANISM OF DISEASE

Experience in other cell systems suggests that cell recognition and adhesion precede and control cell-cell communication via gap junctions (Keane et al., 1988; Mege et al., 1988; Matsuzaki et al., 1990; Musil et al., 1990; Jongen et al., 1991). Based on this premise, one can hypothesize that during normal remodeling of adult bone, commitment of undifferentiated bone marrow stromal cells to osteogenic precursors is associated with the expression of specific cell adhesion molecules. Contact with neighboring cells and the matrix may define the differentiation pathways for osteoblast precursors, and provide the molecular signals for homing to the bone surface in areas of active bone formation. As differentiation progresses and cells become adherent to bone, fully structured anchoring and gap junctions will form. Mature tight and adherens junctions or desmosomes provide anchorage to the surrounding cells and matrix through the cytoskeleton, and gap junctions allow direct exchange of ions and small molecules with neighboring cells (Civitelli, 1995). Thus, junctional structures may be thought as the "nodes" of the network linking bone forming cells within each

BMU. Coordinated activity among different BMUs may thus provide the positional information that drives the deposition of bone mineral in specific areas of the bone, where replacement of old tissue or repair of an injury is required.

In theory, the consequences of a deranged intercellular communication on the skeletal tissue can be multiple. An impaired transcellular signaling may lead to more difficult integration of input signals with consequent slower system response to chemical or mechanical stimuli. In addition, inefficient signal exchanges among cells within BMUs may cause an asynchronous activity within each BMU, with consequent disruption of bone remodeling. Furthermore, age- or menopause-related abnormalities in the expression of cell adhesion molecules and/or connexins may also impair the ability of osteoblast precursors to differentiate, with a consequent decrease in osteoblast number. A decreased population of bone forming cells has been quite consistently observed in animal models of aging (Wronski et al., 1989; Egrise et al., 1992; Rohol et al., 1994). Although it is still uncertain whether a decreased responsiveness of aging cells or changes in the bone microenvironment or both are responsible for this defect, some recent studies seem to suggest that a reduced capacity of osteoprogenitor cells to differentiate into mature osteoblasts occurs with aging (Egrise et al., 1992; Rohol et al., 1994). If cell-cell recognition and adhesion is important for osteoblast differentiation, deregulation of cell-cell contact and signal exchanges within the bone microenvironment may lead to decreased bone formation, thus providing a possible pathogenetic mechanism of the decreased mean wall thickness of trabecular packets that characterizes the histologic appearance of postmenopausal and aging osteoporotic bone (Eriksen, 1986; Parfitt, 1988). Additionally, since cell-cell contact and communication may be required for osteoclast development (Suda et al., 1992), abnormal expression of cell adhesion molecules may also lead to decreased osteoclast activity and translate into reduced activation frequency of the remodeling cycle at the tissue level. Since abnormalities in bone remodeling constitute the physiopathologic basis of metabolic bone diseases, such as osteoporosis, an impaired intercellular signaling may contribute to the pathogenetic mechanism of these conditions.

X. SUMMARY

Direct cell-to-cell interactions via anchoring and communicating or "gap" junctions are general biologic phenomena characteristic of highly cooperative tissues and organs, such as the heart, the uterus, and epithelia. Gap junctions are aqueous transcellular channels that allow exchange of small molecules and ions between adjacent cells. Each hemichannel on facing plasma membranes is an homohexameric structure formed by gap junction proteins, connexins. Functional gap junctions exist between osteoblasts, osteocytes, and periosteal fibroblasts. In bone, gap junctions are formed primarily by connexin43 (Cx43), which mediates diffusion of ions and small molecules from cell to cell, but also by connexin45 (Cx45), which imparts different molecular permeability to the junctions it forms. Gap junctional communication and Cx43 expression can be modulated by factors that regulate bone remodeling, such as PTH, PGE_2, TGFβ, and ambient pH. Cx43 and Cx45 are present in embryonic bone and cartilage, where their appearance coincides with mesenchymal cell condensation and precedes chondrocyte and osteoblast differentiation. Contact between bone cells also occurs through cell adhesion molecules of the cadherin and immunoglobulin superfamilies. Not only are cadherins expressed by osteoblasts, where they may represent the master switches for the osteoblast differentiation program, but they are also critical for osteoclast maturation. The physiologic role of cell-cell adhesion and communication in bone is currently being detailed. It is foreseeable that derangement of cell-cell interactions that may occur with aging or as a consequence of hormonal imbalance may directly affect bone remodeling.

ACKNOWLEDGMENTS

The author wishes to acknowledge the contribution of Thomas Steinberg, MD and Eric Beyer, MD PhD, Washington University, as co-investigators in most of the work reviewed herein. Gratitude also goes to Su-Li Cheng, PhD, for the work on osteoblast cadherins; Alexandra Kemendy, PhD, for the electrophysiology experiments; Pamela Warlow, Tracy Nelson, and Shu-Fang Zhang, for technical assistance in many aspects of these studies; Linda Halstead and Marilyn Roberts for the human tissue cultures; and to Brian Bacskai, PhD and Roger Tsien, PhD, University of California at San Diego, for the experiments with FlCRhR. Most of the work described in this review has been supported by NIH grants AR41255 and DK46686.

REFERENCES

Adams, S.R., Harootunian, A.T., Buechler, Y.J., Taylor, S.S., and Tsien, R.Y. (1991). Fluorescence ratio imaging of cyclic AMP in single cells. Nature 349, 694-697.

Albelda, S.M. and Buck, C.A. (1990). Integrins and other cell adhesion molecules. FASEB J. 4, 2868-2880.

Baron, R., Neff, L., Luovard, D., and Courtoy, P.J. (1985). Cell-mediated extracellular acidification and bone resorption: Evidence for a low pH in resorbing lacunae and localization of a 100-kDa lysosomal membrane protein on the osteoclast ruffled border. J. Cell Biol. 101, 2210-2222.

Beyer, E.C., Paul, D.L., and Goodenough, D.A. (1987). Connexin43: A protein from rat heart homologous to a gap junction protein from liver. J. Cell Biol. 105, 2621-2629.

Beyer, E.C., Paul, D.L., and Goodenough, D.A. (1990). Connexin family of the gap junction proteins. J. Membr. Biol. 116, 187-194.

Bhargava, U., Bar-Lev, M., Bellows, C.G., and Aubin, J.E. (1988). Ultrastructural analysis of bone nodules formed in vitro by isolated fetal rat calvaria cells. Bone 9, 155-163.

Canalis, E., McCarthy, T.L., and Centrella, M. (1991). Growth factors and cytokines in bone cell metabolism. Annu. Rev. Med. 42, 17-24.

Cheng, S.-L., Lecanda, F., Davidson, M., Warlow, P.M., Zhang, S.-F., Zhang, L., Suzuki, S., St.John, T., and Civitelli, R. (1998). Human osteoblasts express a repertoire of cadherins, which are critical for BMP-2-induced osteogenic differentiation. J. Bone Miner. Res. 13, 633-644.

Cheng, S., Zhang, S., Warlow, P.M., Davidson, M., Suzuki, S., Avioli, L.V., and Civitelli, R. (1994). Developmental expression of integrins and cadherins by human bone marrow stromal osteoprogenitor cells and osteoblasts. J. Bone Miner. Res. 9, S151. (Abstract.)

Chiba, H., Sawada, N., Oyamada, M., Kojima, T., Nomura, S., Ishii, S., and Mori, M. (1993). Relationship between the expression of the gap junction protein and osteoblast phenotype in a human cell line during cell proliferation. Cell Struct. Funct. 18, 419-426.

Chiba, H., Sawada, N., Oyamada, M., Kojima, T., Iba, K., Ishii, S., and Mori, M. (1994). Hormonal regulation of connexin 43 expression and gap junctional communication in human osteoblastic cells. Cell Struct. Funct. 19, 173-177.

Civitelli, R., Warlow, P.M., Robertson, A.J., Avioli, L.V., Beyer, E.C., and Steinberg, T.H. (1992). Intercellular communication correlates with expression of connexin43 in osteoblastic cells. In: Calcium Regulating Hormones and Bone Metabolism. (Cohn, D.V., Gennari, C., and Tashijan, Jr., A.H., Eds.), pp. 206-211, Elsevier Science Publishers B.V. Amsterdam, The Netherlands.

Civitelli, R., Beyer, E.C., Warlow, P.M., Robertson, A.J., Geist, S.T., and Steinberg, T.H. (1993). Connexin43 mediates direct intercellular communication in human osteoblastic cell networks. J. Clin. Invest. 91, 1888-1896.

Civitelli, R. (1995). Cell-cell communication in bone. Calcif. Tissue Int. 56, S29-S31.

Civitelli, R., Ziambaras, K., Warlow, P.M., Lecanda, F., Nelson, Harley, J., Atal, N., Beyer, E.C., and Steinberg, T.H. (1998). Regulation of connexin43 expression and function by prostaglandin E2 (PGE2) and parathyroid hormone (PTH) in osteoblastic cells. J. Cell. Biochem. 68, 8-21.

Coelho, C.N.D. and Kosher, R.A. (1991). Gap junctional communication during limb cartilage differentiation. Dev. Biol. 144, 47-53.

Cuervo, L.A., Pita, J.C., and Howell, D.S. (1971). Ultramicroanalysis os pH, pCO_2, and carbonic anhydrase activity at calcifying sites in cartilage. Calcif. Tissue Res. 7, 220-231.

Donahue, H.J., Guilak, F., Vander Molen, M.A., McLeod, K.J., Rubin, C.T., Grande, D.A., and Brink, P.R. (1995a). Chondrocytes isolated from mature articular cartilage retain the capacity to form functional gap junctions. J. Bone Miner. Res. 10, 1359-1364.

Donahue, H.J., Li, Z., Zhou, Z., and Simon, B. (1995b). Pulsed electromagnetic fields affect steady-state levels of Cx43 mRNA in osteoblastic cells. J. Bone Miner. Res. 10, S207. (Abstract.)

Donahue, H.J., McLeod, K.J., Rubin, C.T., Andersen, J., Grine, E.A., Hertzberg, E.L., and Brink, P.R. (1995c). Cell-to-cell communication in osteoblastic networks: Cell line-dependent hormonal regulation of gap junction function. J. Bone Miner. Res. 10, 881-889.

Doty, S.B. (1981) Morphological evidence of gap junctions between bone cells. Calcif. Tissue Int. 33, 509-511.

Doty, S.B. (1988). Cell-to-cell communication in bone tissue. In: The Biological Mechanisms of Tooth Eruption and Root Resorption. (Z. Davidovitch, Ed.), pp. 61-69, EBSCO Media, Birmingham, AL.

Doty, S.B. and Morey-Holton, E.R. (1982). Changes in osteoblastic activity due to simulated weightless conditions. The Physiologist 25, S141-S142.

Doty, S.B. and Schofield, B.H. (1972). Metabolic and structural changes within osteocytes of rat bone. In: Calcium, Parathyroid Hormone, and the Calcitonins. (R.V. Talmage and P.L. Munson, Eds.), pp. 353-364, Excerpta Medica, Amsterdam, The Netherlands.

Duong, L.T., Tanaka, H., and Rodan, G.A. (1994). VCAM-1 involvement in osteoblast-osteoclast interaction during osteoclast differentiation. J. Bone Miner. Res. 9, S131. (Abstract.)

Egrise, M.D., Vienne, A., Neve, P., and Schoutens, A. (1992). The number of fibroblastic colonies formed from bone marrow is decreased and the in vitro proliferation rate of trabecular bone cells increased in aged rats. Bone 13, 355-361.

Eriksen, E.F. (1986). Normal and pathological remodeling of human trabecular bone: Three-dimensional reconstruction of the remodeling sequence in normal and in metabolic bone disease. Endocr. Rev. 7, 739-409.

Frost, H.M. (1986). Intermediary organization of the skeleton. C.R.C. Press, Boca Raton, FL.

Goodenough, D.A., Goliger, J.A., and Paul, D.L. (1996). Connexins, connexons, and intercellular communication. (Review). Annu. Rev. Biochem. 65, 475-502.

Hamerman, D. (1989). The biology of osteoarthritis. N. Engl. J. Med. 320, 1322-1330.

Harter, L.V., Hruska, K.A., and Duncan, R.L. (1995). Human osteoblastlike cells respond to mechanical strain with increased bone matrix protein production independent of hormonal regulation. Endocrinology 136, 528-535.

Hoffmann, I. and Balling, R. (1995). Cloning and expression analysis of a novel mesodermally expressed cadherin. Dev. Biol. 169, 337-346.

Jeasonne, B.G., Faegin, F.A., McMinn, R.W., Shoemaker, R.L., and Rehm, W.S. (1979). Cell-to-cell communication of osteoblasts. J. Dent. Res. 58, 1415-1419.

Jones, S.J., Gray, C., Sakamaki, H., Arora, M., Boyde, A., Gourdie, R., and Green, C. (1993). The incidence and size of gap junctions between the bone cells in rat calvaria. Anat. Embryol. 187, 343-352.

Jongen, W.M.F., Fitzgerald, D.J., Asamoto, M., Piccoli, C., Slaga, T.J., Gros, D., Takeichi, M., and Yamasaki, H. (1991). Regulation of connexin43-mediated gap junctional intercellular communication by Ca^{2+} in mouse epidermal cells is controlled by cadherin. J. Cell Biol. 114, 545-555.

Jørgensen, N.R., Geist, S.T., Civitelli, R., and Steinberg, T.H. (1997). ATP- and gap junction–dependent intercellular calcium signaling in osteoblastic cells. J. Cell Biol. 139, 497-506.

Keane, R.W., Mehta, P.P., Rose, B., Honig, L.S., Lowenstein, W.R., and Rutishauser, U. (1988). Neural differentiation, NCAM-mediated adhesion, and gap junctional communication in neuroectoderm. A study in vitro. J. Cell Biol. 106, 1307-1319.

Kemler, R. (1992). Classical cadherins. Sem. Cell Biol. 3, 149-155.

Kimura, Y., Matsunami, H., Inoue, T., Shimamura, K., Uchida, N., Ueno, T., Miyazaki, T., and Takeichi, M. (1995). Cadherin-11 expressed in association with mesenchymal morphogenesis in the head, somite, and limb bud of early mouse embryos. Dev. Biol. 169, 347-358.

Koval, M., Geist, S.T., Westphale, E.M., Kemendy, A.E., Civitelli, R., Beyer, E.C., and Steinberg, T.H. (1995a). Transfected connexin45 alters gap junction permeability in cells expressing endogenous connexin43. J. Cell Biol. 130, 987-995.

Koval, M., Harley, J.E., Hick, E., and Steinberg, T.H. (1997). Connexin46 is retained as monomers in a trans-Golgi compartment of osteoblastic cells. J. Cell Biol. 137, 847-857.

Kunth, K., Becker, K.F., and Atkinson, M.J. (1993). CD44 and cadherin adhesion molecules are expressed in osteoblasts. Calcif. Tissue Int. 52, S41. (Abstract.)

Langille, R.M. (1994). Chondrogenic differentiation in cultures of embryonic rat mesenchyme. Microsc. Res. Tech. 28, 455-469.

Lecanda, F., Towler, D.A., Ziambaras, K., Cheng, S.-L., Koval, M., Steinberg, T.H., and Civitelli, R. (1998). Gap junctional communication modulates gene expression in osteoblastic cells. Mol. Biol. Cell. 9, 2249-2258.

Lee, Y. and Chuong, C. (1992). Adhesion molecules in skeletogenesis: I. Transient expression of neural cell adhesion molecules (NCAM) in osteoblasts during endochondral and intramembranous ossification. J. Bone Miner. Res. 7, 1435-1446.

Matsuzaki, F., Mege, R.M., Jaffe, S.H., Friedlander, D.R., Gallin, W.J., Goldberg, J.I., Cunningham, B.A., and Edelman, G.M. (1990). cDNAs of cell adhesion molecules of different specificity induce changes in cell shape and border formation in cultured S180 cells. J. Cell Biol. 110, 1239-1252.

Mbalaviele, G., Chen, H., Boyce, B.F., Mundy, G.R. and Yoneda, T. (1995). The role of cadherin in the generation of multinucleated osteoclasts from mononuclear precursors in murine marrow. J. Clin. Invest. 95, 2757-2765.

Mbalaviele, G., Nishimura, R., Myoi, A., Niewolna, M., Reddy, S.V., Chen, D. Feng, J., Roodman, G.D., Mundy, G.R., and Yoneda, T. (1998). Cadherin-6 mediates the heterotypic interactions between the hemopoietic osteoblast cell lineage and stromal cells in a murine model of osteoclast differentiation. J. Cell Biol. 141, 1467-1476.

McSheehy, P.M.J. and Chambers, T.J. (1986a). Osteoblastlike cells in the presence of parathyroid hormone release soluble factor that stimulates osteoclastic bone resorption. Endocrinology 119, 1654-1659.

McSheehy, P.M.J. and Chambers, T.J. (1986b). Osteoblastic cells mediate osteoclastic responsiveness to parathyroid hormone. Endocrinology 118, 824-828.

Mege, R.M., Matsuzaki, F., Gallin, W.J., Goldberg, J.I., Cunningham, B.A., and Edelman, G.M. (1988). Construction of epithelioid sheets by transfection of mouse sarcoma cells with cDNAs for chicken cell adhesion molecules. Proc. Natl. Acad. Sci. USA 85, 7274-7278.

Melchiore, S., Huang, J., Ma, D., Brandi, M.L., and Yamaguchi, D.T. (1994). Heterotypic gap junctional intercellular communication between bovine bone endothelial cells and MC3T3-E1 osteoblastlike cells. J. Bone Miner. Res. 9, S237. (Abstract.)

Miller, S.C., de Saint-Georges, L., Bowman, B.M., and Jee, W.S. (1989). Bone lining cells: structure and function. [Review]. Scanning Microsc. 3, 953-960.

Minkoff, R., Rundus, V.R., Parker, S.B., Hertzberg, E.L., Laing, J.G., and Beyer, E.C. (1994). Gap junction proteins exhibit early and specific expression during intramembranous bone formation in the developing chick mandible. Anat. Embryol. 190, 231-241.

Musil, L.S., Cunningham, B.A., Edelman, G.M., and Goodenough, D.A. (1990). Differential phosphorylation of the gap junction protein connexin43 in junctional communication-competent and -deficient cell lines. J. Cell Biol. 111, 2077-2088.

Nicholson, S.M., and Bruzzone, R. (1997). Gap junctions: Getting the message through (Review). Current Biology 7, R340-R344.

Oberlender, S.A. and Tuan, R.S. (1994). Expression and functional involvement of N-cadherin in embryonic limb chondrogenesis. Development 120, 177-187.

Okazaki, M., Takeshita, S., Kawai, S., Kikuno, R., Tsujimura, A., Kudo, A., and Amann, E. (1994). Molecular cloning and characterization of OB-cadherin, a new member of cadherin family expressed in osteoblasts. J. Biol. Chem. 269, 12092-12098.

Palumbo, C., Palazzini, S., and Marotti, G. (1990). Morphological study of intercellular junctions during osteocyte differentiation. Bone 11, 401-406.

Parfitt, A.M. (1988). Bone remodeling: Relationship to the amount and structure of bone, and the pathogenesis and prevention of fractures. In: Osteoporosis: Etiology, Diagnosis, and Management. (Riggs, B.L. and Melton, III, L.J., Eds.), pp. 45-93, Raven Press, New York.

Paul, D.L. (1989). Molecular cloning of cDNA for rat liver gap junction protein. J. Cell Biol. 103, 123-134.

Paul, D.L., (1995). New functions for gap junctions. (Review). Curr. Op. Cell Biol. 7, 665-672.

Raisz, L.G. and Rodan, G.A. (1990). Cellular basis of bone turnover. In: Metabolic Bone Diseases and Clinically Related Disorders. (Avioli, L.V. and Krane, S.M., Eds.), 2nd ed., pp. 1-41, WB Saunders Co., Philadelphia.

Ramon, F. and Rivera, A. (1986). Gap junction channel-a physiological viewpoint. Prog. Biophys. Mol Biol. 48, 127-153.

Revel, J.P. and Karnovsky, M.J. (1967). Hexagonal array of subunits in intercellular junctions of the mouse heart and liver. J. Cell Biol. 33, C7-C12.

Rohol, P.J.M., Blauw, E., Zurcher, C., Dormans, J.A.M.A., and Theuns, H.M. (1994). Evidence for a dimisished maturation of preosteoblast into osteoblasts during aging in rats, an ultrastructural analysis. J. Bone Miner. Res. 9, 355-366.

Rosen, V., Nove, J., Song, J.J., Thies, S., Cox, K., and Wozney, J.M. (1994). Responsiveness of clonal limb bud cell lines to bone morphogenetic protein 2 reveals a sequential relationship between cartilage and bone cell phenotypes. J. Bone Miner. Res. 9, 1759-1768.

Rudkin, G.H., Yamaguchi, D.T., Ishida, K., Peterson, W.J., Bahadosingh, F., Thye, D., and Miller, T.A. (1996). Transforming growth factor-β, osteogenin, and bone morphogenetic protein-2 inhibit intercellular communication and alter cell proliferation in MC3T3-E1 cells. J. Cell Physiol. 168, 433-441.

Sakai, D., Tong, H., and Minkin, C. (1995). Osteoclast molecular phnotyping by random cDNA sequencing. Bone 17, 111-119.

Sano, K., Tanihara, H., Heimark, R.L., Obata, S., Davidson, M., St. John, T., Taketani, S., and Suzuki, S. (1993). Protocadherins: A large family of cadherin-related molecules in central nervous system. EMBO J. 12, 2249-2256.

Schiller, P.C., Mehta, P.P., Roos, B.A., and Howard, G.A. (1992). Hormonal regulation of intercellular communication: Parathyroid hormone increases connexin43 gene expression and gap-junctional communication in osteoblastic cells. Mol. Endocrinol. 6, 1433-1440.

Schirrmacher, K., Schmitz, I., Winterhager, E., Traub, O., Brummer, F., Jones, D., and Bingmann, D. (1992). Characterization of gap junctions between osteoblastlike cells in culture. Calcif. Tissue Int. 51, 285-290.

Shapiro, F. (1988). Cortical bone repair. The relationship of the lacunarBcanalicular system and intercellular gap junctions to the repair process. J. Bone Joint Surg. 70, 1067-1081.

Shen, V., Rifas, L., Kohler, G., and Peck, W.A. (1986). Prostaglandins change cell shape and increase intercellular gap junctions in osteoblasts cultured from rat fetal calvaria. J. Bone Miner. Res. 1, 243-249.

Sheridan, J.D. and Atkinson, M.M. (1985). Physiological roles of permeable junctions: Some possibilities. Annu. Rev. Physiol. 47, 337-353.

Spray, D.C., Stern, J.H., Harris, A.L., and Bennett, M.V.L. (1982). Gap junctional conductance: Comparison of sensitivities to H^+ and Ca^{2+} ions. Proc. Natl. Acad. Sci. USA 79, 441-445.

Stagg, R.B. and Fletcher, W.H. (1990). The hormone-induced regulation of contact-dependent cellBcell communication by phosphorylation. Endocr. Rev. 11, 302-325.

Stanka, P. (1975). Occurrence of cell junctions and microfilaments in osteoblasts. Cell Tissue Res. 159, 413-422.

Steinberg, T.H., Civitelli, R., Geist, S.T., Robertson, A.J., Hick, E., Veenstra, R.D., Wang, H., Warlow, P.M., Westphale, E.M., Laing, J.G., and Beyer, E.C. (1994). Connexin43 and connexin45 form gap junctions with different molecular permeabilities in osteoblastic cells. EMBO J. 13, 744-750.

Stewart, W.W. (1978). Functional connections between cells as revealed by dye-coupling with a highly fluorescent naphthalimide tracer. Cell 14, 741-759.

Stueckle, S.M., Towler, D.A., Rosenblatt, M., and Suva, L.J. (1993). Isolation and characterization of novel adhesion molecules: Evidence for cadherins in osteoblasts. J. Bone Miner. Res. 8, S126. (Abstract.)

Su, M., Borke, J.L., Donahue, H.J., Li, Z., Warshawsky, N.M., Russell, C.M., and Lewis, J.E. (1997). Expression of connexin 43 in rat mandibular bone and periodontal ligament (PDL) cells during experimental tooth movement. J. Dent. Res. 76, 1357-1366.

Suda, T., Takahashi, N., and Martin, T.J. (1992). Modulation of osteoclast differentiation. Endocr. Rev. 13, 66-80.

Suva, L.J., Towler, D.A., Harada, S., LaFage, M., Stueckle, S.M., and Rosenblatt, M. (1994). Hormonal regulation of osteoblast-derived cadherin expression in vitro and in vivo. J. Bone Miner. Res. 9, S123. (Abstract.)

Takahashi, N., Akatsu, T., Udagawa, N., Sasaki, T., Yamaguchi, A., Mosley, J.M., Martin, T.J., and Suda, T. (1988). Osteoblastic cells are involved in osteoclast formation. Endocrinology 123, 2600-2603.

Tanaka, Y., Morimoto, I., Nakano, Y., Okada, Y., Hirota, S., Nomura, S., Nakamura, T., and Eto, S. (1995). Osteoblasts are regulated by the cellular adhesion through ICAM-1 and VCAM-1. J. Bone Miner. Res. 10, 1462-1469.

Tanihara, H., Sano, K., Heimark, R.L., St. John, T., and Suzuki, S. (1994). Cloning of five human cadherins clarifies characteristic features of cadherin extracellular domain and provides further evidence for two structurally different types of cadherin. Cell Adhes. Commun. 2, 15-26.

Udagawa, N., Takahashi, N., Akatsu, T., Sasaki, T., Yamaguchi, A., Kodama, H., Martin, T.J., and Suda, T. (199). The bone marrow-derived stromal cell lines MC3T3-G2/PA6 and ST2 support osteoclastlike cell differentiation in cocultures with mouse spleen cells. Endocrinology 125, 1805-1813.

Ushiyama, J. (1989). Gap junctions between odontoblasts revealed by transjunctional flux of fluorescent tracers. Cell Tissue Res. 258, 611-616.

Van der Molen, M.A., Rubin, C.T., McLeod, K.J., McCauley, L.K., and Donahue, H.J. (1996). Gap junctional intercellular communication contributes to hormonal responsiveness in osteoblastic networks. J. Biol. Chem. l271, 12165-12171.

Veenstra, R.D. and Brink, P.R. (1992). Patch-clamp analysis of gap junctional currents. In: Cell-Cell Interactions. (Stevenson, B.R., Gallin W.J., and Paul, D.L., Eds.), pp. 167-201, IRL Press, Oxford.

Veenstra, R.D. and DeHaan, R.L. (1986). Measurement of single channel currents from cardiac gap junctions. Science 233, 972-974.

Wang, E.A., Rosen, V., D'Alessandro, J.S., Bauduy, M., Cordes, P., Harada, T., Israel, D.I., Hewick, R.M., Kerns, K.M., and LaPan, P. (1990). Recombinant human bone morphogenetic protein induces bone formation. Proc. Natl. Acad. Sci. USA 87, 2220-2224.

White, T.W., Bruzzone, R., and Paul, D.L. (1995). The connexin family of intercellular channel forming proteins. (Review). Kidney Int. 48, 1148-1157.

Wronski, T.J., Dann, L.M., Scott, K.S., and Cintron, N. (1989). Long-term effects of ovariectomy and aging on the rat skeleton. Calcif. Tissue Int. 45, 360-366.

Xia, S. and Ferrier, J. (1992). Propagation of a calcium pulse between osteoblastic cells. Biochem. Biophys. Res. Commun. 186, 1212-1219.

Yamaguchi, D.T., Huang, J., Ma, D., and Wang, P. (1994a). Inhibition of gap junction intercellular communication by 60 Hz magnetic fields in MC3T3-E1 cells. J. Bone Miner. Res. 9, S237. (Abstract.)

Yamaguchi, D.T., Ma, D., Lee, A., Huang, J., and Gruber, H.E. (1994b). Isolation and characterization of gap junctions in the osteoblastic MC3T3-E1 cell line. J. Bone Miner. Res. 9, 791-803.

Yamaguchi, D.T., Huang, J.T., and Ma, D. (1995). Regulation of gap junction intercellular communication by pH in MC3T3-E1 osteoblastic cells. J. Bone Miner. Res. 10, 1891-1899.

Yamazaki, K., Ichimura, S., Allen, T.D., and Nakagawa, T. (1989). Cell-mediated calcification in collagen gel cultures of fetal rat calvaria cells. Loss of gap junctions precedes calcification. J. Bone Min. Metab. 7, 136-147.

Ziambaras, K., Lecanda, F., Steinberg, T.H., and Civitelli, R. (1998). Cyclic stretch enhances gap junctional communication between osteoblastic cells. J. Bone Miner. Res. 13, 218-228.

THE COLLAGENOUS AND NONCOLLAGENOUS PROTEINS OF CELLS IN THE OSTEOBLASTIC LINEAGE

Pamela Gehron Robey and Paolo Bianco

Advances in Organ Biology
Volume 5B, pages 565-589.

ISBN: 0-7623-0390-5

I. INTRODUCTION

The field of bone matrix biochemistry has witnessed major advances within the last decade due to the seminal work of Termine and co-workers, who developed techniques whereby the organic components of mineralized matrix could be isolated by the use of dissociating and demineralizing solvents (Termine et al., 1980a,b). Purification of chemical amounts of protein lead to the development of antibodies that have been utilized to look at when and where a particular protein is produced during development. While complete identification of the pattern of bone matrix proteins as a function of osteoblastic maturation is far from complete, the information that is available has been essential for the development and testing of hypotheses for a particular protein's function, and its potential utility as a biochemical marker for measuring skeletal status.

II. ORIGIN OF BONE MATRIX PROTEINS

While the major structural proteins are primarily the products of cells in the osteoblastic lineage, a significant percentage of the organic phase is synthesized elsewhere in the body and deposited within bone due to affinity for hydroxyapatite. These exogenous proteins originate primarily in the liver and arrive in bone via the bloodstream, although other cell types associated with bone, such as cells in the blood vessel wall, marrow stromal and hematopoietic cells, and even osteoclasts, may contribute components to bone matrix. In addition, it is now apparent that many proteins, in particular those with growth factor activity, are synthesized exogenously and endogenously by cells in the osteoblastic lineage. The reason for the dual origin of such factors is not clear, and it is possible that the origin of a particular growth factor may influence the response by cells in the osteoblastic lineage.

Development of cell culture model systems that recapitulate bone formation has been another essential tool facilitating the identification of proteins synthesized endogenously, of the pathways by which their genes are activated and transcribed, and of subsequent events including posttranslational modification, intracellular traffic, secretion, and deposition. Cell cultures have provided much valuable information on the major structural elements that are produced by cells in the osteoblastic lineage, as well as on the identity of less abundant elements. A lengthy list of connective tissue proteins, growth factors and cytokines has been accumulated using these systems; however, there are some potential caveats in dealing with this mountain of data. The development of polymerase chain reaction (PCR) amplification whereby a single copy of an mRNA can be amplified using an appropriate primer set has contributed significantly to the list. It is now known that gene transcription is not operated by an "on or off" switch, but rather by a "dimmer" switch. Consequently, extremely low levels of mRNA, which can now be identified by reverse transcription- (RT) PCR, may not be physiologically relevant to the cell. Identification of an mRNA species also does not indicate that it is actually translated into protein. In addition, many proteins are induced *in vitro* due to the prevailing culture conditions that are not present *in vivo*. Consequently, functionality must also be demonstrated to implicate low abundancy proteins as being involved in bone metabolism. However, with these tools, it is possible to ask the question: what is the composition of the supramolecular complex that is calcifiable, and how do cells in the osteoblastic lineage control this process?

III. PATTERNS OF BONE FORMATION

The formation of mineralized matrix is not a one-step process whereby a certain mixture of proteins is deposited, seeded with calcium and phosphate ions, and hydroxyapatite appears. Rather, it is a process that is accomplished by a series of cells, cells in the osteoblastic lineage, that are at different stages of maturation (Owen et al., 1990; Strauss et al., 1990; Robey et al., 1992). These cells must establish the appropriate extracellular milieu in which deposition of hydroxyapatite can be initiated and subsequently propagated. Accordingly, cell maturation culminates with initiation of matrix mineralization (i.e., these two processes do not occur in a parallel fashion, rather one is ending as the other begins). Matrix protein profiles change during osteoblast differentiation, but they also change during deposition of bone proper. "Early" and "late" events in bone formation and protein profiles have been defined with respect to time in culture and corresponding phases of cell growth and *in vitro* mineralization, respectively. However, this does not allow the appropriate time frame for understanding what is "early" and "late" during bone formation. Mineralization is not a late event, as one is led to believe based on cell culture models in which mineralization is temporally "late." In embryonic bone formation, initial deposition of bone matrix and it's mineralization are simultaneous events.

There are several different patterns of bone formation, including 1) intramembranous bone which forms in the absence of an initial cartilage rudiment, 2) endochondral bone where the cartilage rudiment undergoes hypertrophy and after vascular invasion is replaced by bone, and 3) bone turnover in the adult skeleton where new bone is formed on a preexisting bone template. In light of the different environments that bone formation can occur in, there is the possibility of site-specific differences in the way in which cells in the osteoblastic lineage are regulated due to differences in the surrounding environment, and in the biosynthetic activity of the cells themselves in that they may perform different mRNA splicing and posttranslational modifications in order to produce a protein that is tailor-made for the environment in which it is to function. However, there is a basic scheme of bone formation clearly identifiable irrespective of the site, and a qualitatively consistent pattern of bone matrix protein expression.

IV. COMPONENTS OF BONE MATRIX

While it was initially thought that the components of bone matrix would be composed of a set of unique proteins that would magically initiate the deposition of hydroxyapatite, it quickly became apparent that the proteins synthesized by cells in the osteoblastic lineage are not dissimilar to the proteins synthesized by other soft connective tissues such as skin and tendon. However, there are differences in the biosynthetic patterns between these tissues with respect to the percentage of collagenous and noncollagenous proteins, not only synthesized by the cell, but in the amount that is actually deposited and maintained within the mineralized matrix.

Bone is composed of 90% mineral and 10% organic matrix such that the overall amount of protein present in bone is far less than that in other tissues. Of the 10% organic matrix, collagen accounts for 90% (therefore 9% of the total mass) with the remaining 10% contributed by the noncollagenous proteins. While the types of proteins are similar to other connective tissues, the relative amounts compared to collagen are greatly reduced. In addition, bone matrix proteins are chemically different from their soft connective tissue counterparts due to differential splicing of mRNA and different posttranslational modifications such as glycosylation, phosphorylation and sulfation to proteins have a unique profile. The differences between bone and soft connective tissue proteins have not been completely documented, but it is clear that they can have a significant impact on the function of a protein. It is possible that these differences could also be exploited by the development of assays where bone versus nonbone protein could be separated out and used to determine changes in the skeleton as compared to changes in soft connective tissue. Taking advantage of these chemical differences would require the generation of reagents, such as antibodies or molecular probes, that could recognize an epitope or a splice variant, followed by correlation with histomorphometric parameters of different phases of bone metabolism and known disease states.

A. Collagen

Collagenous proteins are the major structural proteins of all connective tissues (Vuorio and de Crombrugghe 1990; Burgeson and Nimni 1992; Mayne and Brewton 1993). A collagen molecule consists of three polypeptide chains called α chains that are characterized by a repeating Gly-X-Y triplet sequence where X is often hydroxyproline and Y is proline. Collagen is also unique in that it contains hydroxylysine, which participates in crosslinking of the triple helices, and can be glycosylated to form galactosyl-hydroxylysine and glucosyl-galactosyl-hydroxylysine. Subseqently, three chains fold around each other to form a highly stable triple-helical molecule. Head to tail association of molecules, along with lateral associations that are slightly staggered, results in the formation of a collagen fibril that contains a gap or hole region where mineral is often first associated.

There are at least 16 types of collagen, each characterized by their component α-chains. The molecules can be either homotrimeric, composed of three identical α chains, or heterotrimeric with two or three different types of chains. Collagen molecules can form many different macromolecular structures ranging from the ropelike banded structures of the interstitial collagens (types I, II, III, V, XI), networks (types IV, VIII, X), cords (VI, VII), and FACITs (fibril associated collagens with interrupted triple helices) (IX, XII, XIV). The nature of these structures is dictated to some extent by the length of the triple helical region, and whether noncollagenous (nontriple helical) regions are interspersed within the helix (which conveys flexibility to the otherwise rigid structure) and/or at the amino- and carboxy-termini. Consequently, the physical and metabolic properties of the particular collagen type are thereby adapted to suit a particular need within a given connective tissue.

Although there is a great deal of overlap, expression of the different collagen types is in a somewhat tissue/structure specific pattern. Type I collagen is found in almost all connective tissues, predominantly in skin, bone and tendon, and type III, V, XI, and XII are often closely associated and may regulate the diameter growth of type I collagen fibrils. Type II collagen is found almost exclusively in cartilage where its fibrils are coated by type IX collagen, and in vitreous humor and vertebral disk. Type X is virtually exclusive to hypertrophic cartilage. Type IV is found in basement membranes and type VII in fibrils that extend from cell surfaces to underlying basement membranes. Type VIII is found in a specialized basement membrane, Descemet's membrane, and is related to type X.

B. Noncollagenous Proteins

Bone matrix contains representatives of many different classes of proteins. A significant proportion is contributed by proteoglycans. This class is characterized by the presence of a protein core to which glycosaminoglycans (GAGs) are covalently attached (Hardingham and Fosang, 1992). GAGs are composed of repeating disaccharide units that are assembled by the sequential addition of UDP-monosaccharides to a lipid (dolichol) carrier, which ultimately transfers the chain to a Ser or Thr residue on the core protein. GAGs include chondroitin sulfate, dermatan sulfate, heparan sulfate, and keratan sulfate, all of which contain variable amounts of sulfate substitution. Within the proteoglycan family, there are subfamilies, loosely defined by the chemical nature of their protein cores and the types of GAGs that are attached. Bone also contains γ-carboxy glutamic acid containing proteins (Hauschka et al., 1989). Most γ-carboxylated proteins are functionally related to the blood coagulation cascade. However, several proteins found in bone matrix share the amino acid sequence that is the substrate for γ-carboxylase and are consequently γ-carboxylated. The remaining proteins are a heterogeneous collection of glycoproteins that are often phosphorylated and/or sulfated. Many of these glycoproteins contain cell attachment sequences (RGD and related sequences) that allow these proteins to bind to the integrin class of cell surface receptors (Ginsberg et al., 1995).

The structural and functional features of all of the bone matrix proteins have been extensively reviewed (Robey et al., 1992; Robey and Boskey, 1996). What follows below is a description of the major structural proteins that are synthesized by cells in the osteoblastic lineage as they pass through the different stages of maturation that are currently recognized.

V. EXPRESSION OF BONE MATRIX PROTEINS AS A FUNCTION OF OSTEOBLASTIC MATURATION

A. The Preosteogenic (and Nonmineralizing) Matrix

During early stages of bone formation as exemplified by the intramembranous pattern, mesenchymal stem cells—which are a population of cells that contain osteogenic precursors (Caplan, 1994)—reside in a loose interstitial type of connective tissue that must ultimately be destroyed at a later stage in order to make room for what will become bone matrix proper. That this matrix is ultimately destroyed is evidenced by the fact that these components are not detected in mineralized matrix proper. Consequently, it is apparent that this matrix also does not serve as a template upon which mineralization is formed. As a corollary to this statement, it may be that cells at different stages of maturation do not necessarily sequentially add another set of matrix proteins to a preformed matrix, building it up, so to speak, for the final event of matrix mineralization.

While mesenchymal stem cells are the osteogenic precursors during bone development, the source of osteogenic precursors is most likely marrow stromal fibroblasts during bone remodeling and turnover (Friedenstein, 1995). Surprisingly little is known about the matrix surrounding these normally quiescent cells in the bone microenvironment. This population of cells has received a great deal of attention of late. When they are explanted *in vitro*, they begin to differentiate into a number of different lineages (depending on the culture conditions) with time in culture. Consequently, the pattern of expression of bone matrix proteins expressed in culture, which closely resembles the pattern expressed by osteoblastic cells, may not relate to what marrow stromal fibroblasts are expressing *in situ*.

In spite of the incomplete characterization of preosteogenic matrices, produced after committment of either mesenchymal or stromal cells, a number of components have been identified including collagen, versican, hyaluronan, and matrix gla protein. (Table 1).

Collagens

The preosteogenic matrix contains a mixture of collagen types. As in all connective tissues, type I collagen predominates. However, unlike bone matrix proper, type III and type V collagen are also present (Gehron Robey, unpublished results).

Table 1. Proteins Present in the Preosteogenic Matrix

Component	Gene	Structure	Function
Type I collagen	COL1A1 17q21.3-22 18 kb, 51 exons 7.2 and 5.9 mRNA COL1A2 7q21.3-22 35 kb, 52 exons 6.5 and 5.5 kb mRNA	$[\alpha 1(I)_2 \alpha 2(I)]$ $[\alpha 1(I)]_3$ triple helical molecule composed of Gly-X-Y repeating sequence, contains Hyp, Hyl, Gal-Hyl, Glu-Gal-Hyl	Serves as a scaffolding upon which nucleators of hydroxyapatite deposition are oriented, the major structural protein of bone
Type III collagen	COL3A1 2q24.3-q31	$[\alpha 1(III)]_3$	May coat type I collagen fibrils and regulate diameter
Type V collagen	COL5A1 COL5A2 2q24.3-q31 COL5A3	$[\alpha 1(V)_2 \alpha 2(V)]$ $[\alpha 1(I)\alpha 2(V)\alpha 3(V)]$	Present in a pericellular pattern, may regulate fibril size
Versican PG-M	5q12-14 90 kb, 15 exons 10, 9, 8 kb mRNAs	360 kDa core protein, 12 CS chains of 45 kDa, N and C terminal globular domains, EGF, CRP-like sequences	May associate with hyaluronan, demarcates area that will become bone, is destroyed prior to mineralization
Hyaluronan	Multigene complex	Long-chain repeating disaccharides, unsulfated	May influence proliferation, differentiation, cell migration
Matrix gla protein	12p 3.9 kb, 4 exons, 0.7 kb mRNA	~15 kDa, five gla residues, one disulfide bridge, phosphoserine residues	May function in cartilage metabolism

It is thought that the ratio of type I to type III collagen determines the final fibril size. Type V collagen is also thought to participate in the regulation of fibril diameter, and is often localized in a pericellular environment. In addition to these "interstitial" collagen types, FACIT collagens, types XII and XIV, have been found to associate with type I collagen fibrils and may do so in this matrix as well. This matrix may be removed by the action of collagenases produced by cells in the osteogenic tissue; however, its degradation has not been well documented.

Versican (PG-M)

Versican is a member of the aggrecan-like subclass of proteoglycans, and is so named due to it's rather ubiquitous pattern of expression in a large variety of connective tissues. It is composed of a core protein of 360 KDa with ~12 chondroitin sulfate side chains of 45 KDa. The core contains globular domains at both the amino- and carboxy-terminus. The G1 globular domain is similar to the hyaluronic acid binding protein, link protein, while the G3 domain contains homologies to epidermal growth factor (EGF), selectins, and C-reactive protein (CRP) (Zimmermann and Ruoslahti, 1989).

During early stages of bone formation, versican or a highly related molecule is produced at high levels by osteogenic precursors (Fisher, 1985). PG-M, a splice variant of versican (Shinomura et al., 1993), has been identified in condensing limb bud mesenchyme, and it is proposed that these molecules demarcate areas that will ultimately become skeletal tissue. It has been reported that removal of proteoglycans must occur prior to matrix mineralization as determined by studies using radiolabeled sulfate as a marker (Klein-Nulend et al., 1987). It is most likely versican that is being degraded as osteogenesis occurs. Interestingly, it has been reported that EGF stimulates proliferation of osteogenic cells, and it is possible that the EGF-like sequence that is released as versican is being degraded could stimulate subsequent bone formation. PG-100, a low abundance proteoglycan has also been found in undifferentiated fibroblasts and in preosteoblasts, but it is not known precisely when and where it is produced (Bosse et al., 1994).

Hyaluronan

Although hyaluronan is a GAG, it is not sulfated or covalently attached to a core protein, or assembled by the same biosynthetic route as the other GAGs. In contrast, it is assembled by an enzyme(s) on the outer surface of the cell membrane (reviewed in Robey and Boskey, 1996). Large amounts of hyaluronan are present in the early stages of bone formation (Knudson and Toole, 1985), and high levels are produced by preosteogenic cells *in vitro* (Fedarko et al., 1990). Hyaluronan is most likely contributing to the microenvironment created by marrow stromal fibroblasts, in particular in those that are actively supporting hematopoiesis (Minguell, 1993).

In cartilage, it is known that hyaluronan associates noncovalently with the large, cartilage specific proteoglycan, aggrecan, an interaction that is stabilized by link protein. These interactions lead to the formation of a bottle-brush-like structure consisting of numerous aggrecan molecules that are associated with a single hyaluronan molecule, each aggrecan-hyaluronan interaction stabilized by link protein (Hardingham et al., 1994). It is thought that this structure conveys resilience to cartilaginous tissues. In soft connective tissues, versican has been found to associate with hyaluronan (Bignami et al., 1993); however, this interaction has not been demonstrated using versican isolated from developing bone. By analogy to its functions in other tissues, hyaluronan may play a role in cell migration and adhesion, proliferation and differentiation, and in maintenance of a hydrated state. As osteogenesis proceeds into later stages, hyaluronan is removed, the process of which has not been described in bone.

Matrix Gla Protein (MGP)

One of the γ-carboxylated proteins in bone, MGP, has a molecular weight of ~15 KDa, although it migrates considerably heavier by standard polyacrylamide gel electrophoresis. It contains 5 gla residues, as well as phosphoserine. It is noted for its great insolubility in standard aqueous solutions (Price et al., 1983, 1994; Cancela et al., 1990).

MGP expression begins at early stages of development, not only in bone, but also in cartilage, lung, kidney, and heart. In bone, its level remains high in subsequent stages as well (Otawara and Price, 1986). Other studies indicate that it may be a potential marker for chondrogenesis (Luo et al., 1995). Clear information on the pattern of MGP expression by cells in the osteoblastic lineage through either immunohistochemistry or *in situ* hybridization are not available, although time of the appearance would suggest that it is in the preosteogenic matrix.

The function of MGP in bone metabolism may be mediated indirectly through its action in cartilage. In experiments where animals were treated with warfarin to inhibit γ-carboxylation resulting in reduced tissue content of gla-proteins, there was premature closure of the growth plate (Price, 1989). This implies that MGP, along with its relative, osteocalcin (discussed below), may play a role in bone remodeling. A transgenic animal null for MGP expression is currently under development (Karsenty, personal communication) and characterization of the phenotype expressed by this animal will undoubtedly shed some light on MGP's function.

B. Matrix of the Presumptive Periosteum

Preosteogenic cells receive a number of signals that induce them to progress towards maturation. This next step in maturation is marked by a number of events, including a limited number of cell divisions (as indicated by tritiated thymidine incorporation and labeling by bromo-deoxy-uridine) and the induction of alkaline phosphatase (Bianco et al., 1993a). Although alkaline phosphatase is found in many tissues and made by cells that are intimately associated with bone formation, such as vascular endothelial cells and marrow stromal fibroblasts, it is clear that the initiation of expression of alkaline phosphatase during development is the hallmark of impending osteogenesis. Preosteoblasts are the first identifiable cells in bone development to express this enzymatic activity, and in many cases are closely associated with incoming capillaries.

The matrix surrounding these preosteoblastic cells is not totally dissimilar to that of the preosteogenic matrix in that it contains a collagen scaffolding consisting of types I and III collagen, at least. The major difference appears to be the loss of versican, and the induction of two other small proteoglycans, decorin and biglycan (Bianco et al., 1990) (Table 2). While it is apparent that proteoglycans are degraded prior to matrix mineralization, it is most likely versican which contains a large number of sulfated GAG chains that is being lost, compared to decorin and biglycan, which contain only a few. Although versican is replaced by equivalent amounts of decorin and biglycan on a molar basis, the differences in their GAG content account for the apparent decrease in sulfate content prior to matrix mineralization. Consequently, this net decrease in sulfate content does not necessarily indicate that proteoglycans are inhibitory to the mineralization process.

Table 2. Components of the Presumptive Periosteum*

Protein	Gene	Structure	Function
Type- XII	COL12A1	$[\alpha1(XII)]_3$	May associate with type I collagen fibrils
Alkaline phosphatase	1 50 kb, 12 exons, alternative promoter and one RFLP 2.5, 4.1, 4.7 kb mRNA	Disulfide bonded dimer of identical subunits (80 kDa), bone specific post translational modifications	The hallmark of the osteoblastic lineage, may bind to calcium and transport it, may hydrolyze inhibitors
Decorin	12q21-23 45 kDa, 9 exons alternative promoters, 1.6 and 1.9 kb mRNA	~38–45 kDa core protein (10 leucine rich repeats), one 40 kDa CS chain	Binds to collagen fibrils and may regulate diameter, binds to TGFβ, fibronectin, thrombospondin
Biglycan	Xq27 7 kb, 8 exons 2.1 and 2.6 kb mRNAs	~38–45 kDa core protein (12 leucine rich repeats), two 40 kDa CS chains	Binds to collagen and TGFβ, involved in morphogenetic events

Note: Proteins that are found in mesenchyme as well as in the presumptive perrosteam—matrix gla protein, type-I collagen, type-III collagen, type-V collagen—are in Table 1.

Decorin and Biglycan (PG-II and PG-I)

These two small proteoglycans are representatives of a class whose core protein is characterized by a 24 amino acid leucine-rich repeat sequence. In decorin, this sequence is repeated 10 times while in biglycan there are 12 repeats (Fisher et al., 1989). The three-dimensional (3D) structure of an RNAse inhibitor, another protein with the leucine-rich repeat, has been recently reported by X-ray crystallography (Kobe and Deisenhofer, 1993). Because of the pattern of leucine residues, the protein forms a series of loops, creating a highly active binding surface. The decorin and biglycan cores also contain amino- and carboxy-terminal globular domains stabilized by disulfide bonding, resulting in core proteins with molecular weights of ~38 KDa for both proteoglycans. Attachment sites for chondroitin sulfate (bone) or dermatan sulfate chains (soft connective tissues) of ~40 KDa are close to the amino-terminus, one site for decorin and two sites for biglycan. The fully glycosaminoglycated and glycosylated proteoglycans have apparent molecular weights of ~130 KDa and ~270 KDa, respectively.

In addition to decorin and biglycan, there are other proteins with the leucine-rich repeat sequence in bone matrix, some of which may also be proteoglycans. Osteoglycin, previously thought to be a bone morphogenetic protein (Ujita et al., 1995), fibromodullin, a keratan sulfate proteoglycan found in cartilage and in bone (Oldberg et al., 1989), and a newly described protein, osteoadherin (Wendel and Heinegaard, personal communication), all have the leucine-rich repeat sequence in their primary structure. However, it is not yet known at what point of osteoblastic maturation and matrix mineralization these proteins are formed.

During subperiosteal bone formation, both decorin and biglycan first appear in cells that can be identified as preosteoblasts. In a broader survey, decorin, named for its ability to "decorate" type I collagen fibers, matches the expression of type I collagen. Consequently, it is somewhat ubiquitous and uniformly distributed. Biglycan, on the other hand, is expressed in unique sites undergoing morphogenetic events that demarcate the boundaries of cartilage and bone. This pattern of expression is very reminiscent of the pattern observed for several of the bone morphogenetic proteins (BMPs) such as BMP-5 and BMP-7. As osteogenesis proceeds, both decorin and biglycan are highly expressed in the osteoblastic layer, but only biglycan is maintained by cells that become buried in mineralized matrix (osteocytes) (Bianco et al., 1990).

In vitro studies indicate that decorin is a regulator of collagen fibril growth. Biglycan also binds to type I collagen fibrils (reviewed in Kresse et al., 1994). Both proteins bind to transforming growth factor β (TGFβ), and it is likely that they regulate this growth factor's activity and availability to cells (Ruoslahti et al., 1991). This is particularly intriguing with respect to biglycan since BMPs are members of the TGFβ superfamily. Both proteoglycans also bind to fibronectin, and inhibit attachment of cells to this protein *in vitro* (Robey and Grzesik, 1995). Although decorin and biglycan are similar in many respects, the different patterns of expression indicate distinct functions within bone metabolism. Decorin has a low affinity for Ca^{2+} and hydroxyapatite whereas biglycan at low concentrations does have an affinity for hydroxyapatite. Prevailing evidence suggests that decorin functions in matrix organization while biglycan functions in tissue morphogenesis and development (reviewed in Robey and Boskey, 1996).

C. Cement

At the beginning of an individual bone formation phase at a particular locale, osteogenic cells begin to produce at least two proteins that have been implicated in the nucleation of hydroxyapatite deposition. These cells would be histologically defined as newly differentiated, mature osteoblasts due to their location within developing bone, their large size and intracellular organization with a large number of mitochondria, expanded rough endoplasmic reticulum and extensive Golgi and post-Golgi system. In bone development *in vivo*, mineralization is a "prime" event, and is associated with the expression and deposition of certain noncollagenous proteins. Expression of these proteins leads to the deposition of a non-collagenous structural phase identified by electron microscopy where the first recognizable deposition of mineral occurs. This phase includes the cement lines marking the onset of a formative phase, and "spots" of material with identical immunoreactivity interspersed among collagen fibers in bone matrix that was added later. For this reason, it has been called "cement" (Riminucci et al., 1995) (Table 3). In bone turnover,

Table 3. Components of Cement

Protein	Gene	Structure	Function
Osteopontin	4q13-21 8.2 kb, 7 exons, multiple alleles, one splice variant 1.6 kb mRNA	~44–75 kDa protein that is differentially glycosylated and phosphorylated, polyasp stretch, RGD in mid-region	Binds to hydroxyapatite, cell attachment factor, inhibits NOS, inhibitor of viral infection
Bone Sialoprotein	4q13-21 15 kb, 7 exons 2.0 mRNA	~46–75 kDa, 50% carbohydrate, tyrosine sulfated, polyglu stretches, RGD at C terminus	Nucleates hydroxyapatite precipitation, cell attachment factor

cement lines are clearly identifiable indicating where bone resorption has taken place and where bone formation has filled in the gap. In developing bone (McKee and Nanci, 1995), such lines are less distinct, but can be observed with certain immunohistochemical techniques.

Osteopontin (BSP-1, SPP, sp66)

Osteopontin, independently discovered and described in several tissues, has a protein backbone of ~44 KDa daltons that can be differentially posttranslationally modified by glycosylation and phosphorylation, depending on the cell origin. A differential splice variant has been identified that removes a potential phosphorylation site from the product made by cells in the osteoblastic lineage, but the consequences of this change on protein function are not known (Kerr et al., 1991). It has a number of unique chemical features including a polyaspartyl sequence at the amino-terminus and an RGD sequence in the mid region of the molecule (Denhardt and Guo, 1993).

Osteopontin is synthesized by a large variety of epithelial cells including those of the kidney, mammary, and salivary glands, in addition to smooth muscle cells, certain neuronal cells, and in teeth. In bone, it can be found in ablated marrow at a point where there is a high degree of proliferation (Suva et al., 1995) which may correspond to the fact that this protein is dramatically upregulated upon transformation of normal cells, a process that leads to increased proliferation. At later stages of bone formation, it is produced by newly formed osteoblasts, and accumulated in mineralized matrix (Mark et al., 1988). It is also found in a particular mononuclear population in bone marrow. Interestingly, osteoclasts have also been found to contain osteopontin mRNA and protein (Merry et al., 1993), similar to what has been reported for bone sialoprotein (BSP) (Bianco et al., 1991). Consequently, in the adult skeleton where bone turnover has occurred, cement lines may also be the product of osteoclasts as well as of osteoblasts.

The functions for osteopontin are extremely varied, ranging from a regulator of proliferation, a mediator of cell adherence in osteoblasts and osteoclasts via $\alpha_v\beta_3$ integrin, and a stimulator of intracellular signaling pathways (in osteoclasts) (Butler, 1995). Osteopontin inhibits the production of nitric oxide synthase which may modulate osteoblastic and osteoclastic metabolism (Hwang et al., 1994), and it is identical to a protein that regulates viral resistance (Patarca et al., 1993). Osteopontin binds to hydroxyapatite with high affinity although it blocks formation of hydroxyapatite in an *in vitro* assay. These data suggest that osteopontin may not be a nucleator of hydroxyapatite crystal formation, but may control the growth of these crystals by binding to one of the crystal faces (reviewed in Robey and Boskey, 1996).

Bone Sialoprotein (BSP-II)

The other protein that has been intimately associated with matrix mineralization is BSP first identified in a fragmented form by Herring (1977), and subsequently by Fisher et al. (1983). The protein backbone has a molecular weight of ~34 KDa, but it is highly modified by the glycosylation (50% of the final mass is composed of carbohydrate, 12% being sialic acid), phosphorylation, and sulfation. There are several stretches of polyglutamic acid at the amino-terminus and an RGD sequence near the carboxy-terminus. Tyrosine-rich regions that are variably sulfated, depending on factors yet to be elucidated, are present at the amino-terminus and flanking the RGD region (Fisher et al., 1990; Midura et al., 1990).

BSP has a much more limited pattern of expression than osteopontin. In the adult organism, it is expressed primarily by osteoblasts, odontoblasts, and osteoclasts. However, during development, it is produced by trophoblasts of the placenta and hypertrophic chondrocytes (Bianco et al., 1991). It is also found in particular pathologies such as in metastatic breast cancer (Bellahcene et al., 1994), and most likely in prostate cancer and atherosclerotic plaques as is osteopontin (Brown et al., 1994). Interestingly, bone sialoprotein exhibits a different pattern of secretion from the other bone matrix proteins. Immunolocalization at the light and electron microscopic level indicate a distinct accumulation of BSP in the Golgi and post-Golgi structures, and it appears to be organized into discrete packets of protein that become immediately mineralized upon secretion by the cell (Bianco et al., 1993b).

Based on the presence of the RGD sequence, it has been postulated that BSP functions as a cell attachment factor for both osteoblasts and osteoclasts via the $\alpha_v\beta_3$ integrin (Mintz et al., 1993; Ross et al., 1993; Grzesik and Robey, 1994). However, BSP also has a high affinity for hydroxyapatite, and *in vitro* analysis indicates that it is a nucleator of hydroxyapatite nucleation. It is not clear how these two potential functions are reconciled within one protein. In addition, the expression of BSP by cells that are in the process of invading surrounding tissue (breast and prostate cancer, trophoblasts, osteoclasts) and some of which home to bone (metastatic breast and prostate cancer), suggests other intriguing functions for this protein.

D. Osteoid

After newly formed osteoblasts have initiated matrix mineralization by the deposition of "cement", the fully mature osteoblast begins to produce massive amounts of additional proteins that compose osteoid. Osteoid is unmineralized bone matrix that is deposited in a somewhat vectorial fashion towards the mineralization front (established by the deposition of cement) away from the layer of osteoblasts. While these cells do not appear to maintain synthesis of osteopontin or BSP, they upregulate the levels of biglycan and decorin and, in addition, begin to produce high levels of other RGD-containing proteins (thrombospondin and fibronectin), another major glycoprotein of bone (osteonectin), and type I collagen, exclusive of other collagen types (see Table 4).

Thrombospondin

This multifunctional glycoprotein is composed of three identical subunits that have a dumbbell shape (globular domains at the amino- and carboxy-termini connected by a stalk) held together in the stalk region by disulfide bonds. Each subunit contains sequence homologies to fibrinogen at the amino-terminus, properdin-like and EGF-like repeats, and homologies to collagen, von Willebrand factor and a parasite (*P. falciparium*) protein in the stalk region, calmodulin-like repeats and an RGD sequence in the carboxy-terminus. In humans, there are at least three different genes that code for thrombospondin; each differs from one another by the number of repeats that are found in the different regions. It is not known which gene(s) code for the form that is produced by osteoblasts (reviewed in Bornstein, 1992).

First isolated from platelet α granules, thrombospondin has now been shown to be expressed in many connective tissues throughout the body in a developmentally regulated pattern. For example, it is expressed at boundaries such as at the dermal-epidermal junction in the skin, surrounding muscle fibers, and separating glandular epithelium. Low levels are first found in preosteoblasts; however, osteoid and the osteoblastic layer are much more intensely stained for thrombospondin by immunohistochemical techniques. It is also maintained within the mineralized matrix (Robey et al., 1989; Grzesik and Robey 1994).

Thrombospondin contains a large number of binding sites. The amino-terminal globular domain binds to heparin, platelets, and to cells via an RGD-independent fashion. The stalk region binds to types I and V collagen, thrombin, fibrinogen, laminin, plasminogen activator, and its inhibitor. The carboxy-terminus binds to the histidine rich glycoprotein of serum, platelets, cells and has the ability to bind to large quantities of Ca^{2+}. In fact, the availability of the RGD region to bind to cell surface receptors has been found to be influenced by the prevailing Ca^{2+} concentration. *In vitro* analysis indicates that osteoblastic cells adhere to, but do not spread on, thrombospondin in an RGD independent fashion, indicating that a sequence other than RGD is active. The exact cell surface receptor has not been identified, but could be the $\alpha_v\beta_3$ integrin. Thrombospondin also binds to decorin, osteonectin and TGFβ. Based on these findings, thrombospondin is most likely involved in the organization of matrix and a mediator of events between the extracellular environment and the cell (reviewed in Bornstein, 1992; Robey and Boskey, 1996).

Table 4. Components of Osteoid*

Protein	*Gene*	*Structure*	*Function*
Thrombospondins	15q15 6q27 1q21-24 >16 kb, 22 exons 6.1 kb mRNA	~450 kDa with three identical subunits, homologies to fibrinogen, properdin, EGF, collagen, von Willebrand, RGD in C terminal domain	Cell attachment (but not spreading), binds to many matrix proteins and to TGFβ, may function as a matrix organizer
Fibronectin	2p14-16, 1q34-36 50 kb in chicken, 50 exons, multiple splice variants, 6 RFLPs 7.5 kb mRNA	~400 kb dimer with nonidentical subunits, composed of type I, II, and III repeats, RGD near C terminus	Cell attachment protein, binds to many matrix molecules
Osteonectin	5q3-33 20 kb, 10 exons one RFLP 2.2, 3.0 kb mRNA	~35-45 kDa, disulfide bonded, low and high affinity calcium binding sites, phosphorylated	Binds to collagen, calcium, ydroxyapatite, thrombospondin, may influence cell growth, shape and cell-matrix interactions
Type I collagen	Regulated by promoter elements residing between 2.3 and 1.7 kb upstream from the start of transcription	Procollagen is phosphorylated (24 KDa), predominantly Gal-Hyl, hydroxyallysine crosslinking pathway, hydropyridinoline crosslink, thick fibrils	Conveys mechanical strength to bone as fibrils are laid down in a plywoodlike fashion

Note: * Other proteins that are synthesized by pre-osteoblasts as well as osteoblasts: matrix gla protein (see Table 1); decorin and biglycan (see Table 2).

Fibronectin

Fibronectin is perhaps the prototypical RGD-containing glycoprotein. It is composed of two disulfide bonded, nonidentical subunits of ~200,000 daltons, derived from the same gene via differential splicing. Each subunit consists of repeating sequences that have been designated types I, II, and III. Types I and II are between 45–50 amino acids in length with a disulfide loop, whereas type III units are twice as long with no loop. There are several different domains defined by varying combinations of these repeats. Domains I and VIII bind to fibrin, I to heparin and certain bacteria, II to gelatin and collagen, IV to DNA, VI to cell surfaces via the RGD sequence, and VII which also binds to heparin. The RGD region has been characterized by X-ray crystallography and nuclear magnetic resonance and found to be in a loop region that is stabilized by disulfide bonds. There are a large number of potential splicing variants and several have been described to date. The splice variant that is found in bone has not yet been identified (reviewed in Potts and Campbell, 1994).

Fibronectin expression is extremely widespread throughout all connective tissues in the body. In addition, it is one of the most abundant proteins present in serum. Consequently, fibronectin found in mineralized matrix could be synthesized endogenously and/or adsorbed from the circulation. *In vitro* studies clearly show that it is an osteoblastic product. It had been previously reported that fibronectin was expressed during early stages of bone formation. Histological examination indicates that it is highly upregulated in the osteoblastic layer, as its specific receptor, $\alpha_4\beta_1$. It is also maintained at high levels in the mineralized matrix (Grzesik and Robey, 1994). The list of potential functions of fibronectin is lengthy; it has been reported to be involved in virtually every cell process imaginable due to the different types of molecules to which it binds. It acts as an opsonic protein, in keeping with its high levels in serum, and as a cell attachment factor for a large variety of cell types. Interestingly, osteoblastic cells attach to fibronectin in an RGD independent fashion, indicating another sequence is mediating interaction with the cell surface receptor. It's dramatic upregulation in the osteoblastic layer is intriguing in that decorin and biglycan are also highly expressed in this layer as well, both of which have been reported to inhibit cell attachment to fibronectin (Robey and Grzesik, 1995). Consequently, fibronectin may not be functioning in a typical cell-matrix interaction in this particular instance.

Osteonectin (SPARC, Culture Shock Protein, BM-40)

One of the first noncollagenous proteins to be isolated in its intact form is the phosphorylated glycoprotein, osteonectin, so named for its ability to bind to collagen, calcium, and hydroxyapatite (Termine et al., 1981). It has a molecular weight of ~35 KDa and contains a number of intracellular disulfide bonds. The amino-terminus is largely acidic, and assumes an α-helical conformation such that there are a large number of low affinity Ca^{2+} binding sites. It also contains two EF hand structures which bind Ca^{2+} with high affinity, structures that are not usually found in secreted proteins. There is also an ovomucoid-like sequence in the mid region (Bolander et al., 1988).

During development, osteonectin is expressed by many tissues such as in skin, cartilage, teeth, whiskers, and the heart. However, postnatally, it is constitutively expressed by tissues actively involved in ion transport such as salivary, mammary, and renal epithelial cells, and in cells associated with the skeleton (bone and cartilage) (Holland et al., 1987). During bone formation, there is a marked increase in osteonectin expression noted between the preosteoblastic and osteoblastic cells. It is only accumulated in bone matrix and in the α-granules of platelets (Stenner et al., 1986). It is not clear to date why platelets (and megakaryocytes) contain this and other bone-related proteins such as osteopontin and osteocalcin (Bianco et al., 1988).

Based on studies using peptides from different regions of the osteonectin molecule, it has been reported to function in regulation of the cell cycle, cell migration, cell shape, and cell-matrix interactions. In addition, it binds platelet derived growth factor and plasminogen and as such may function in wound regeneration (reviewed in Lane and Sage, 1994). Many of these studies have been performed on endothelial cells, which produce large amounts of osteonectin under certain culture conditions; these activities have not been demonstrated in osteogenic cells to date. However, based on the potential function to disengage cells from a matrix in which they are attached, it is possible that the upregulation of osteonectin in the osteoblastic layer may be associated with the dramatic shape change that is noted between preosteoblasts and osteoblasts. Based on its ability to bind to collagen and calcium, it may also be involved in matrix mineralization, although it does not appear to nucleate hydroxyapatite formation in *in vitro* assays. Consequently, it may serve to transport Ca^{2+} from the intracellular to the extracellular environment. Interestingly, it also binds to thrombospondin, another glycoprotein that is abundant in osteoid. Studies utilizing antibodies to neutralize osteonectin during development indicate that it is involved in neurulation and myotome development (Purcell et al., 1993), and its overexpression in *Caenorhabitis elegans* caused developmental abnormalities (Schwarzbauer and Spencer, 1993).

Collagen(s)

While collagen is present in the preosteogenic matrix, what is synthesized by fully mature osteoblasts and deposited in osteoid is different in both composition and biochemical character. In addition, transcription of "bone collagen" is regulated by a different part of the collagen promoter (D'Souza et al., 1993). Osteoblasts secrete type I collagen at the virtual exclusion of other types such as type III and V, although it can not be ruled out that FACIT collagens are not present. Collagen fibrils in bone are somewhat thicker than what is found in soft connective tissues which may be a result of the lack of these other collagen types that are thought to regulate collagen fibril diameter. In addition, type I collagen in bone is posttranslationally modified in a different fashion. The amino-terminal precursor peptide is phosphorylated and it is not known if this post-translational modification of procollagen occurs in soft connective tissues. Once cleaved, the phosphorylated amino propeptide remains, at least in part, in the mineralized matrix (Fisher et al., 1987) where it was initially identified as the 24 KDa phosphoprotein of bone. Bone type I collagen contains mainly galactosyl-hydroxylysine as opposed to glucosyl-galactosyl-hydroxylysine, and the pattern of crosslinking and maturation of these crosslinks is also different. This latter modification is now the basis for a urine biochemical assay that can measure the degradation of type I collagen from bone as opposed to soft connective tissue on the basis of the different crosslinks that are released by the degradative process (Eyre, 1995).

Based on analysis that predicts a β pleated sheet structure for a nucleator, it would appear that type I collagen does not have the appropriate 3D conformation to nucleate hydroxyapatite deposition (Addadi and Weiner, 1985). However, it is clear that without type I collagen, bone does not form or mineralize normally as shown in patients with Osteogenesis Imperfecta where a mutation in either chain of type I collagen is the underlying cause. Studies have indicated that matrix mineralization is initiated in a relatively collagen-free zone by proteins such as osteopontin and BSP (Riminucci et al., 1995). Once it is initiated, however, mineralization may proceed somewhat spontaneously by crystal growth and propagation. As collagen is deposited in the osteoid, the crystals grow into the gap region in the collagen molecule. This hypothesis is highly speculative, and is not totally in keeping with data that has been accumulated from studies on mineralized turkey tendon. But it must be pointed out that those studies can not preclude association of noncollagenous proteins with the type I collagen fibrils and that mineralization in this case is occurring in a relatively cell free environment, which is not the case in bone.

Table 5. Components of Bone Matrix Proper*

Protein	*Gene*	*Structure*	*Function*
Osteocalcin	1 3.9 kb, 4 exons 0.6 kb mRNA	~5 kDa, 3-5 gla residues, one disulfide bridge	May regulate the recruitment and activity of osteoclasts
Albumin	4q11-22 17 kb, 15 exons	69 kDa, non-glycosylated, one disulfide bond	May regulate hydroxyapatite crystal growth.
α2-HS glycoprotein	3 two RFLP 1.5 kb mRNA	Human analogue of fetuin, ala-ala and pro-pro repeats, cystatin-like domains	May control cell proliferation
Other serum proteins	IgGs, transferrin		
Enzymes and inhibitors	Collagenase and TIMPs, plasminogen activator and inhibitor, matrix phosphoprotein kinase		
Morphogenetic proteins	TGFβ, bone morphogenetic proteins		
Growth fators	Insulin-like growth factors and binding proteins		

Note: * Other proteins produced at earlier stages of differentiation, but maintained in bone matrix proper: matrix gla protein (see Table 1); decorin and biglycan (see Table 2); osteopontin and BSP (see Table 3); thrombospondin, fibronectin, osteonectin, and type I collagen (see Table 4). Also found in bone matrix proper are TIMPs, tissue inhibitors of matrix metalloproteinases.

E. Mineralized Matrix Proper

As matrix apposition occurs, and the mineralization front encroaches upon the osteoblastic layer, certain cells dissociate from the osteoblastic layer by mechanisms yet to be determined. These cells remain in contact with the osteoblastic layer through cell processes that bear gap junctions. The matrix mineralizes around these cell processes to form canaliculae, and the cell body encapsulated within a lacunae becomes the last stage of the lineage, the osteocyte. Other cells in the osteoblastic layer have been postulated to die by apoptosis, or to become inactive to form lining cells on the endosteal surface. The composition of the thin layer of unmineralized matrix underneath lining cells has not been well characterized, but is thought to be similar to osteoid. While osteocytes do not maintain synthesis of many of the structural bone matrix proteins, several are still expressed such as biglycan, fibronectin, and to some extent BSP. The composition of mineralized matrix proper is similar to that of osteoid. However, additions are made by osteocytes (osteocalcin) and other components are brought in by the vasculature (serum proteins), keeping in mind that as bone remodels from woven bone to lamellar bone, every Haversian system that is formed contains a blood vessel and a nerve (see Table 5).

Osteocalcin

The major gla-containing protein in bone, osteocalcin, is a polypeptide of ~5,000 kDa and was the first noncollagenous protein isolated from bone matrix. It contains 3–5 gla residues and one disulfide bond. The proximity of the gla residues on an α-helical region of the molecule conveys a conformation so that the carboxy groups are ideally situated to bind to hydroxyapatite (Hauschka et al., 1989). Although it has been long thought that osteocalcin is exclusively a bone and dentin specific protein, however recent studies in the mouse indicate that there are three genes, one of which is expressed outside of the skeleton (Desbois and Karsenty, 1995). In addition, it has been reported that osteocalcin mRNA is found in platelets, although it is not clear that protein is actually translated from this mRNA (Thiede et al., 1994). Low levels are found in hypertrophic chondrocytes. In human bone, antibodies against osteocalcin intensely stain the mineralization front, but the osteoid is unstained. When an antibody against the propeptide of osteocalcin is utilized, osteocytes are clearly stained, indicating that they are the primary source of osteocalcin in human bone (Kasai et al., 1994). In rodent bone, however, osteoblastic cells on the endosteal surface appear to be the major source of osteocalcin. The significance of this species difference is not clear. However, these data suggest that osteocalcin is secreted as a postmineralization event. In light of studies suggesting a role for osteocalcin in the recruitment of osteoclastic precursors and their ultimate activity as osteoclasts (Glowacki et al., 1991), osteocytes acting as mechanoreceptors may be induced to secrete osteocalcin to signal a bone resorptive event, and osteocalcin secretion by osteocytes may be the turning point between bone formation and bone resorption.

Other Components

With the use of cell cultures and antibodies against specific proteins, cDNA libraries prepared from osteoblastic cells, isolated cDNA probes for *in situ* hybridization, and other molecular approaches, the list of proteins that are present in bone matrix is ever increasing. While these proteins are most likely not functioning in a structural fashion, they do play critical roles in bone metabolism. These proteins include the members of the TGFβ super family (TGFβs and the bone morphogenetic proteins), growth factors such as the insulin-like growth factors and their binding proteins, enzymes and their inhibitors (collagenases and tissue inhibitors of matrix metalloproteinases, plasminogen activator and its inhibitor, matrix phosphoprotein kinases and possibly phosphatases). In addition, proteolipids are formed by cells and are part of the matrix where they may participate in matrix mineralization. Other glycoproteins include bone acid glycoprotein-75 (BAG-75), which has properties similar to osteopontin and BSP (Gorski, 1992), tetranectin which may be involved in matrix mineralization (Wewer et al., 1994), and three additional RGD-containing proteins, vitronectin (Preissner, 1991), fibrillin (Keene et al., 1991), and tenascin. However, there is little information on when and where these proteins are produced during osteoblastic maturation, matrix deposition and mineralization.

Serum Proteins

A large number of serum proteins including IgGs, transferrin, protein S, albumin, and α2-HS glycoprotein have been identified in bone matrix by two-dimensional electrophoresis and are presumably present due to their affinity for hydroxyapatite (Delmas et al., 1984). The adsorption of many of these proteins could be passive; this does not mean that they do not have a function in bone metabolism. Many of these serum derived proteins have growth factor activity, and may serve to stimulate bone formation when liberated by osteoclastic activity. Two of these serum proteins, albumin and α2-HS glycoprotein, are actually concentrated at 50–100 times the serum levels. Albumin has been found to bind to one or more of the hydroxyapatite crystal faces and subsequently influence the growth and shape of crystals (Garnett and Dieppe, 1990). α2-HS glycoprotein is the human analogue of fetuin, a protein found to stimulate cell proliferation of many cell types. It has been reported that α2-HS glycoprotein is expressed in cartilage and in bone; however, this has not been verified by *in vivo* techniques (Ohnishi et al., 1993).

VI. SUMMARY

The structural elements of bone matrix are representative of a wide variety of protein types, including collagens (primarily type I), proteoglycans (versican, decorin, biglycan, and other related proteins), and glycoproteins such as osteonectin and the

RGD-containing proteins (thrombospondin, fibronectin, osteopontin, and BSP). These proteins have a wide variety of potential functions including the regulation of cell proliferation and maturation, matrix organization and mineralization, and possibly in controlling bone resorption. While a significant amount of information has been gained in the last decade on the identity of the proteins produced endogenously and deposited in bone matrix, much more information is still needed to determine at what stage of osteoblastic maturation a protein is produced, how its production is regulated, and its potential function in matrix mineralization.

REFERENCES.

Addadi, L. and Weiner, S. (1985). Interaction between acidic proteins and crystals: Stereochemical requirements in biomineralization. Proc. Natl. Acad. Sci. USA 82, 4110-4114.

Bellahcene, A., Merville, M.P., and Castronovo, V. (1994). Expression of bone sialoprotein, a bone matrix protein, in human breast cancer. Cancer Res., 54, 2823-2826.

Bianco, P., Fisher, L.W., Young, M.F., Termine, J.D., and Robey, P.G. (1990). Expression and localization of the two small proteoglycans biglycan and decorin in developing human skeletal and nonskeletal tissues. J. Histochem. Cytochem. 38, 1549-1563.

Bianco, P., Fisher, L.W., Young, M.F., Termine, J.D., and Robey, P.G. (1991). Expression of bone sialoprotein (BSP) in developing human tissues. Calcif. Tissue Int. 49, 421-426.

Bianco, P., Riminucci, M., Bonucci, E., Termine, J.D., and Robey, P.G. (1993a). Bone sialoprotein (BSP) secretion and osteoblast differentiation: Relationship to bromodeoxyuridine incorporation, alkaline phosphatase, and matrix deposition. J. Histochem. Cytochem. 41, 183-191.

Bianco, P., Riminucci, M., Silvestrini, G., Bonucci, E., Termine, J.D., Fisher, L.W., and Robey, P.G. (1993b). Localization of bone sialoprotein (BSP) to Golgi and post-Golgi secretory structures in osteoblasts and to discrete sites in early bone matrix. J. Histochem. Cytochem. 41, 193-203.

Bianco, P., Silvestrini, G., Termine, J.D., and Bonucci, E. (1988). Immunohistochemical localization of osteonectin in developing human and calf bone using monoclonal antibodies. Calcif. Tissue. Int., 43, 155-61.

Bignami, A., Perides, G., and Rahemtulla, F. (1993). Versican, a hyaluronate-binding proteoglycan of embryonal precartilaginous mesenchyma, is mainly expressed postnatally in rat brain. J. Neurosci. Res., 34, 97-106.

Bolander, M.E., Young, M.F., Fisher, L.W., Yamada, Y., and Termine, J.D. (1988). Osteonectin cDNA sequence reveals potential binding regions for calcium and hydroxyapatite and shows homologies with both a basement membrane protein (SPARC) and a serine proteinase inhibitor (ovomucoid). Proc. Natl. Acad. Sci. USA, 85, 2919-2923.

Bornstein, P. (1992). Thrombospondins. FASEB J. 6, 3290-3299.

Bosse, A., Kresse, H., Schwarz, K., and Muller, K.M. (1994). Immunohistochemical characterization of the small proteoglycans decorin and proteoglycan-100 in heterotopic ossification. Calcif. Tissue Int. 54, 119-124.

Brown, L.F., Papadopoulos-Sergiou, A., Berse, B., Manseau, E.J., Tognazzi, K., Perruzzi, C.A., Dvorak, H.F., and Senger, D.R. (1994). Osteopontin expression and distribution in human carcinomas. Am. J. Pathol. 145, 610-623.

Burgeson, R.E. and Nimni, M.E. (1992). Collagen types. Molecular structure and tissue distribution. Clin. Orthop. 282, 250-272.

Butler, W.T. (1995). Structural and functional domains of osteopontin. Ann. NY Acad. Sci. 760, 6-11.

Cancela, L., Hsieh, C.L., Francke, U., and Price, P.A. (1990). Molecular structure, chromosome assignment, and promoter organization of the human matrix Gla protein gene. J. Biol. Chem. 265, 15040-15048.

Caplan, A.I. (1994). The mesengenic process. Clin. Plast. Surg. 21, 429-435.

D'Souza, R.N., Niederreither, K., and de Crombrugghe, B. (1993). Osteoblast-specific expression of the α 2(I) collagen promoter in transgenic mice: Correlation with the distribution of TGF-β1. J. Bone. Miner. Res. 8, 1127-1136.

Delmas, P.D., Tracy, R.P., Riggs, B.L., and Mann, K.G. (1984). Identification of the noncollagenous proteins of bovine bone by two-dimensional gel electrophoresis. Calcif. Tissue Int. 36, 308-316.

Denhardt, D.T. and Guo, X. (1993). Osteopontin: A protein with diverse functions. FASEB J. 7, 1475-1482.

Desbois, C. and Karsenty, G. (1995). Osteocalcin cluster: Implications for functional studies. J. Cell Biochem. 57, 379-383.

Eyre, D.R. (1995). The specificity of collagen cross-links as markers of bone and connective tissue degradation. Acta Orthop. Scand. Suppl. 266, 166-170.

Fedarko, N.S., Termine, J.D., and Robey, P.G. (1990). High-performance liquid chromatographic separation of hyaluronan and four proteoglycans produced by human bone cell cultures. Anal. Biochem. 188, 398-407.

Fisher, L.W. (1985). The nature of the proteoglycans of bone, EBSCO Media. Birmingham, AL.

Fisher, L.W., McBride, O.W., Termine, J.D., and Young, M.F. (1990). Human bone sialoprotein. Deduced protein sequence and chromosomal localization. J. Biol. Chem. 265, 2347-2351.

Fisher, L.W., Robey, P.G., Tuross, N., Otsuka, A.S., Tepen, D.A., Esch, F.S., Shimasaki, S., and Termine, J.D. (1987). The Mr 24,000 phosphoprotein from developing bone is the NH2-terminal propeptide of the α1 chain of type-I collagen. J. Biol. Chem. 262, 13457-13463.

Fisher, L.W., Termine, J.D., and Young, M.F. (1989). Deduced protein sequence of bone small proteoglycan I (biglycan) shows homology with proteoglycan II (decorin) and several nonconnective tissue proteins in a variety of species. J. Biol. Chem. 264, 4571-4576.

Fisher, L.W., Whitson, S.W., Avioli, L.V., and Termine, J.D. (1983). Matrix sialoprotein of developing bone. J. Biol. Chem. 258 (20), 12723-12727.

Friedenstein, A.J. (1995). Marrow stromal fibroblasts. Calcif. Tissue Int. 56 (Suppl. 1), S17.

Garnett, J. and Dieppe, P. (1990). The effects of serum and human albumin on calcium hydroxyapatite crystal growth [see comments]. Biochem J. 266, 863-868.

Ginsberg, M.H., Froese, S., Shephard, E., Adams, S., Robson, S., and Kirsch, R. (1995). Integrins: Dynamic regulation of ligand-binding integrins, selectins and CAMs—the "glue of life". Biochem. Soc. Trans. 23, 439-446.

Glowacki, J., Rey, C., Glimcher, M.J., Cox, K.A., and Lian, J. (1991). A role for osteocalcin in osteoclast differentiation. J. Cell Biochem. 45, 292-302.

Gorski, J.P. (1992). Acidic phosphoproteins from bone matrix: A structural rationalization of their role in biomineralization. Calcif. Tissue Int. 50, 391-396.

Grzesik, W.J. and Robey, P.G. (1994). Bone matrix RGD glycoproteins: Immunolocalization and interaction with human primary osteoblastic bone cells in vitro. J. Bone Miner. Res. 9, 487-496.

Hardingham, T.E. and Fosang, A.J. (1992). Proteoglycans: Many forms and many functions. FASEB J. 6, 861-870.

Hardingham, T.E., Fosang, A.J., and Dudhia, J. (1994). The structure, function, and turnover of aggrecan, the large aggregating proteoglycan from cartilage. Eur. J. Clin. Chem. Clin. Biochem. 32, 249-257.

Hauschka, P.V., Lian, J.B., Cole, D.E., and Gundberg, C.M. (1989). Osteocalcin and matrix Gla protein: Vitamin K-dependent proteins in bone. Physiol. Rev. 69, 990-1047.

Herring, G.M. (1977). Methods for the study of the glycoproteins and proteoglycans of bone using bacterial collagenase. Determination of bone sialoprotein and chondroitin sulphate. Calcif. Tissue Res. 24, 29-36.

Holland, P.W.H., Harper, S.J., and McVey, J.H. (1987). In vivo expression of mRNA for the Ca++-binding protein SPARC (osteonectin) revealed by in situ hybridization. J. Cell Biol. 105, 473-482.

Hwang, S.M., Lopez, C.A., Heck, D.E., Gardner, C.R., Laskin, D.L., Laskin, J.D., and Denhardt, D.T. (1994). Osteopontin inhibits induction of nitric oxide synthase gene expression by inflammatory mediators in mouse kidney epithelial cells. J. Biol. Chem., 269, 711-715.

Kasai, R., Bianco, P., Robey, P.G., and Kahn, A.J. (1994). Production and characterization of an antibody against the human bone GLA protein (BGP/osteocalcin) propeptide and its use in immunocytochemistry of bone cells. Bone Miner. 25, 167-182.

Keene, D.R., Sakai, L.Y., and Burgeson, R.E. (1991). Human bone contains type-III collagen, type-VI collagen, and fibrillin: Type-III collagen is present on specific fibers that may mediate attachment of tendons, ligaments, and periosteum to calcified bone cortex. J. Histochem. Cytochem. 39, 59-69.

Kerr, J.M., Fisher, L.W., Termine, J.D., and Young, M.F. (1991). The cDNA cloning and RNA distribution of bovine osteopontin. Gene, 108, 237-243.

Klein-Nulend, J., Veldhuijzen, P., van de Stadt, R.J., Van Kampen, G., Kuijer, R., and Burger, E.H. (1987). Influence of intermittent compressive force on proteoglycan content in calcifying growth plate cartilage in vitro. J. Biol. Chem. 262, 15490-15495.

Knudson, C. and Toole, B. (1985). Changes in the pericellular matrix during differentiation of limb bud mesoderm. Dev. Biol. 112, 308-318.

Kobe, B. and Deisenhofer, J. (1993). Crystal structure of porcine ribonuclease inhibitor, a protein with leucine-rich repeats. Nature 366, 751-756.

Kresse, H., Hausser, H., and Schonherr, E. (1994). Small proteoglycans. Exs. 70, 73-100.

Lane, T.F. and Sage, E.H. (1994). The biology of SPARC, a protein that modulates cell-matrix interactions. FASEB J. 8, 163-173.

Luo, G., D'Souza, R., Hogue, D., and Karsenty, G. (1995). The matrix Gla protein gene is a marker of the chondrogenesis cell lineage during mouse development. J. Bone Miner. Res. 10, 325-334.

Mark, M.P., Butler, W.T., Prince, C.W., Finkelman, R.D., and Ruch, J.V. (1988). Developmental expression of 44-kDa bone phosphoprotein (osteopontin) and bone γ-carboxyglutamic acid (Gla)-containing protein (osteocalcin) in calcifying tissues of rat. Differentiation 37, 123-136.

Mayne, R. and Brewton, R.G. (1993). New members of the collagen superfamily. Curr. Opin. Cell Biol. 5, 883-890.

McKee, M.D. and Nanci, A. (1995). Osteopontin and the bone remodeling sequence. Colloidal-gold immunocytochemistry of an interfacial extracellular matrix protein. Ann. NY Acad. Sci. 760, 177-189.

Merry, K., Dodds, R., Littlewood, A., and Gowen, M. (1993). Expression of osteopontin mRNA by osteoclasts and osteoblasts in modelling adult human bone. J. Cell Sci. 104, 1013-1020.

Midura, R.J., McQuillan, D.J., Benham, K.J., Fisher, L.W., and Hascall, V.C. (1990). A rat osteogenic cell line (UMR 106-01) synthesizes a highly sulfated form of bone sialoprotein. J. Biol. Chem. 265, 5285-5291.

Minguell, J.J. (1993). Is hyaluronic acid the organizer of the extracellular matrix in marrow stroma? [editorial; comment]. Exp. Hematol. 21, 7-8.

Mintz, K.P., Grzesik, W.J., Midura, R.J., Robey, P.G., Termine, J.D., and Fisher, L.W. (1993). Purification and fragmentation of nondenatured bone sialoprotein: Evidence for a cryptic, RGD-resistant cell attachment domain. J. Bone Miner. Res. 8, 985-995.

Ohnishi, T., Nakamura, O., Ozawa, M., Arakaki, N., Muramatsu, T., and Daikuhara, Y. (1993). Molecular cloning and sequence analysis of cDNA for a 59-kD bone sialoprotein of the rat: Demonstration that it is a counterpart of human α 2-HS glycoprotein and bovine fetuin. J. Bone Miner. Res. 8, 367-377.

Oldberg, A., Antonsson, P., Lindblom, K., and Heinegard, D. (1989). A collagen-binding 59-kd protein (fibromodulin) is structurally related to the small interstitial proteoglycans PG-S1 and PG-S2 (decorin). EMBO J., 8, 2601-2604.

Otawara, Y. and Price, P.A. (1986). Developmental appearance of matrix GLA protein during calcification in the rat. J. Biol. Chem. 261, 10828-10832.

Owen, T.A., Aronow, M., Shalhoub, V., Barone, L.M., Wilming, L., Tassinari, M.S., Kennedy, M.B., Pockwinse, S., Lian, J.B., and Stein, G.S. (1990). Progressive development of the rat osteoblast phenotype in vitro: Reciprocal relationships in expression of genes associated with osteoblast proliferation and differentiation during formation of the bone extracellular matrix. J. Cell. Physiol. 143, 420-430.

Patarca, R., Saavedra, R.A., and Cantor, H. (1993). Molecular and cellular basis of genetic resistance to bacterial infection: The role of the early T-lymphocyte activation-1/osteopontin gene. Crit. Rev. Immunol. 13, 225-246.

Potts, J.R. and Campbell, I.D. (1994). Fibronectin structure and assembly. Curr. Opin. Cell Biol. 6, 648-655.

Preissner, K.T. (1991). Structure and biological role of vitronectin. Annu. Rev. Cell Biol. 7, 275-310.

Price, P.A. (1989). Gla-containing proteins of bone. Connect Tissue Res 21, 51-57, (Discussion, 57-60.)

Price, P.A., Rice, J.S., and Williamson, M.K. (1994). Conserved phosphorylation of serines in the Ser-X-Glu/Ser(P) sequences of the vitamin K-dependent matrix Gla protein from shark, lamb, rat, cow, and human. Protein Sci. 3, 822-830.

Price, P.A., Urist, M.R., and Otawara, Y. (1983). Matrix Gla protein, a new γ-carboxyglutamic acid-containing protein, which is associated with the organic matrix of bone. Biochem. Biophys. Res. Commun. 117, 765-771.

Purcell, L., Gruia-Gray, J., Scanga, S., and Ringuette, M. (1993). Developmental anomalies of *Xenopus* embryos following microinjection of SPARC antiboies. J. Exp. Zool. 265, 153-164.

Riminucci, M., Silvestrini, S., Bonucci, E., Fisher, L.W., Robey, P.G., and Bianco, P. (1995). The anatomy of bone sialoprotein immunoreactive sites in bone as revealed by combined ultrastructural histochemistry and immunohistochemsitry. Calcif. Tiss. Int. 57, 277-284.

Robey, P.G., Bianco, P., and Termine, J.D. (1992). The cell biology and molecular biochemistry of bone formation. Disorders of Mineral Metabolism. (Favus, M.J. and Coe, F.L., Eds.), 241-263. Raven Press, New York.

Robey, P.G. and Boskey, A.L. (1996). The biochemistry of bone. Osteoporosis. (Marcus, R. Feldman, D., Bilizekian, J.P., and Kelsey, J., Eds.), 95-183. Academic Press, New York.

Robey, P.G. and Grzesik, W.J. (1995). The biochemistry of bone-forming cells: Cell-matrix interactions. Biological Mechanisms of Tooth Eruption, Resorption, and Replacement by Implants. (Davidovitch, Z., Ed.), pp. 167-172. EBSCO Media, Birmingham, AL.

Robey, P.G., Young, M.F., Fisher, L.W., and McClain, T.D. (1989). Thrombospondin is an osteoblast-derived component of mineralized extracellular matrix. J. Cell Biol. 108, 719-727.

Ross, F.P., Chappel, J., Alvarez, J.I., Sander, D., Butler, W.T., Farach-Carson, M.C., Mintz, K.A., Robey, P.G., Teitelbaum, S.L., and Cheresh, D.A. (1993). Interactions between the bone matrix proteins osteopontin and bone sialoprotein and the osteoclast integrin α vs. β 3 potentiate bone resorption. J. Biol. Chem. 268, 9901-9907.

Ruoslahti, E., Yamaguchi, Y., Mann, D.M., and Ruoslahti, E. (1991). Proteoglycans as modulators of growth factor activities: Negative regulation of transforming growth factor-β by the proteoglycan decorin. Cell 64, 867-869.

Schwarzbauer, J.E. and Spencer, C.S. (1993). The *Caenorhabditis elegans* homologue of the extracellular calcium-binding protein SPARC/osteonectin affects nematode body morphology and mobility. Mol. Biol. Cell 4, 941-952.

Shinomura, T., Nishida, Y., Ito, K., and Kimata, K. (1993). cDNA cloning of PG-M, a large chondroitin sulfate proteoglycan expressed during chondrogenesis in chick limb buds. Alternative spliced multiforms of PG-M and their relationships to versican. J. Biol. Chem. 268, 14461-14469.

Stenner, D.D., Tracy, R.P., Riggs, B.L., and Mann, K.G. (1986). Human platelets contain and secrete osteonectin, a major protein of mineralized bone. Proc. Natl. Acad. Sci. USA 83 (18), 6892-6896.

Strauss, P.G., Closs, E.I., Schmidt, J., and Erfle, V. (1990). Gene expression during osteogenic differentiation in mandibular condyles in vitro. J. Cell Biol. 110, 1369-1378.

Suva, L.J., Seedor, J.B., Endo, N., Quartuccio, H.A., Thompson, D.D., Bab, I., and Rodan, G.A. (1995). Pattern of gene expression following rat tibial marrow ablation. J. Bone Miner. Res. 8, 123-129.

Termine, J.D., Belcourt, A.B., Christner, P.J., Conn, K.M., and Nylen, M.U. (1980a). Properties of dissociatively extracted fetal tooth matrix proteins. I. Principal molecular species in developing bovine enamel. J. Biol. Chem. 255, 9760-9768.

Termine, J.D., Belcourt, A.B., Miyamoto, M.S., and Conn, K.M. (1980b). Properties of dissociatively extracted fetal tooth matrix proteins. II. Separation and purification of fetal bovine dentin phosphoprotein. J. Biol. Chem. 255, 9769-9772.

Termine, J.D., Kleinman, H.K., Whitson, S.W., Conn, K.M., McGarvey, M.L., and Martin, G.R. (1981). Osteonectin, a bone-specific protein linking mineral to collagen. Cell 26, 99-105.

Thiede, M.A., Smock, S.L., Petersen, D.N., Grasser, W.A., Thompson, D.D., and Nishimoto, S.K. (1994). Presence of messenger ribonucleic acid encoding osteocalcin, a marker of bone turnover, in bone marrow megakaryocytes and peripheral blood platelets. Endocrinology 135, 929-937.

Ujita, M., Shinomura, T., and Kimata, K. (1995). Molecular cloning of the mouse osteoglycin-encoding gene. Gene 158, 237-240.

Vuorio, E. and de Crombrugghe, B. (1990). The family of collagen genes. Annu. Rev. Biochem. 59, 837-872.

Wewer, U.M., Ibaraki, K., Schjorring, P., Durkin, M.E., Young, M.F., and Albrechtsen, R. (1994). A potential role for tetranectin in mineralization during osteogenesis. J. Cell Biol. 127, 1767-1775.

Zimmermann, D.R. and Ruoslahti, E. (1989). Multiple domains of the large fibroblast proteoglycan, versican. EMBO J. 8, 2975-2981.

THE ROLE OF GROWTH FACTORS IN BONE FORMATION

Lynda F. Bonewald and Sarah L. Dallas

Advances in Organ Biology
Volume 5B, pages 591-613.

ISBN: 0-7623-0390-5

I. INTRODUCTION

Bone is a storage site for growth factors necessary for its growth, repair, and maintenance. The growth factors, known as insulin-like growth factors, (IGFs), transforming growth factor beta (TGFβ), fibroblast growth factors (FGFs), and platelet-derived growth factor (PDGFs), have all been purified from bone (for review see Hauschka, 1990), some in relatively large amounts (see Table 1). These growth factors have been shown to act synergistically to induce new bone formation (Pfeilschifter et al., 1990). The bone morphogenetic proteins (BMPs) were difficult to purify from bone due to their low abundance. The sequences of a series of these proteins were obtained using a novel approach. Highly purified osteogenic material from bone was enzymatically digested to yield peptides which could be sequenced and then the resulting proteins were cloned (Wozney et al., 1988).

Growth factor content of bone declines with age and more significantly after the menopause. Considerable interest has been generated concerning the use of growth factors to treat bone disease such as osteoporosis, for healing of non-union fractures, for treatment of bone loss due to periodontal disease, and many other conditions requiring repair or new bone formation. Attention has also focused on agents that stimulate bone formation through the stimulation of growth factors. For example $1,25(OH)_2D_3$ and intermittent treatment with parathyroid hormone will increase the concentration of IGF and TGFβ in bone (Finkelman et al., 1991; Pfeilschifter et al., 1995).

Table 1. Concentration of Growth Factors in Bone Matrix*

Factor	*ng/gm Bone*
Insulin-like growth factor II	1,500
Transforming growth factor β	450
Insulin-like growth factor I	100
Platelet-derived growth factor	80
Basic fibroblast growth factor	50
Acidic fibroblast growth factor	10
Bone morphogenetic protein	1–2

Note: From Hauschka, 1990; and Wozney, 1992.

Regulation of growth factor actions through their receptors and signaling mechanisms has also been the focus of considerable research. The signaling mechanism for PDGFs, FGFs, and IGFs are through tyrosine kinases, whereas receptors for TGFβs and BMPs appear to cell signal through serine/threonine kinase activity. The signaling pathways from cell membrane to nuclear gene activation are being determined and are very complex. The presence of nonsignaling as well as signaling receptors for the same growth factor is a common feature. Whereas the mechanism of action of high-affinity signaling receptors can clearly be determined, this is not the case for receptors with low affinity that do not appear to signal. It has been hypothesized that the nonsignaling, low-affinity receptors bind with greater efficiency to ligand and therefore accelerate the arrival of ligand to the high-affinity receptors due to "reduced dimensionality," i.e., ligand present only on the cell surface and not in solution (Schlessinger and Lemmon, 1995).

A whole new realm of information concerning the function of growth factors has been generated using transgenic mice which either overexpress the gene of interest or have the functional gene of interest deleted ("knockout" mice). These experiments have usually yielded unexpected information. Very rarely has the phenotype of a knockout mouse been what was expected by the researcher. For example, it was expected that deletion of the TGFβ1 gene would be lethal since TGFβ1 is highly conserved among species and has dramatic effects on many cell types. However, these mice were rescued by maternal transfer of TGFβ1 through the placenta and milk, but died of massive inflammatory reaction, two weeks after weaning (Shull et al., 1992: Kulharni et al., 1993; Letterio et al., 1994). This was the first example of a gene knockout that was not a protein knockout. These experiments emphasized the importance of TGFβ in immune suppression. Therefore, these experiments like many others using transgenic animals have yielded unexpected but useful information (see other examples below).

II. INSULINLIKE GROWTH FACTORS

Difficulties in demonstrating *in vitro* effects of growth hormone—while it was known that growth hormone would stimulate skeletal growth *in vivo*—led to the "somatomedin" hypothesis. This hypothesis stated that growth hormone indirectly stimulated growth through the production of serum factors termed somatomedins. This hypothesis was verified with the discovery of the IGFs; factors produced by the liver upon stimulation with growth hormone (see review by Isaksson et al., 1991).

The IGFs were discovered by their ability to stimulate cartilage sulfation and were originally termed "nonsuppressible insulin-like activity" because although possessing similar activity to insulin (stimulation of DNA synthesis, proteoglycan synthesis, glycosaminoglycan synthesis, and protein synthesis), their activity could not be blocked by the antibody to insulin. When a purified protein of 7.5 kDa was found to possess 48% homology to insulin, it was termed insulin-like growth factor

1. A second purified factor was termed insulin-like growth factor 2 based on its homology to IGF-1. These names replaced the somatomedin term (for review see Jones and Clemmons, 1995).

A. IGF Receptors

Two receptors, the IGF-I and IGF-II receptors, are known to bind specifically to IGFs and will only bind insulin with low affinity. The IGF-I receptor has high homology with the insulin receptor and definitely is responsible for IGF-mediated signaling. The IGF-II/mannose 6-phosphate receptor is identical to the cation-independent mannose 6-phosphate receptor which is normally an intracellular lysosomal binding protein (BP) for lysosomal enzymes. It has no known IGF signaling function.

The IGF-I receptor mediates most of the effects of the IGFs with a K_D of 0.2–1 nM for IGF-1 and a 100–1,000-fold lower affinity for insulin. This receptor is composed of two extracellular α-chains which bind to ligand and two β-chains which span the plasma membrane and possess tyrosine kinase activity. IGF binding to this receptor causes tyrosine phosphorylation of an 185 kDa protein called the insulin receptor substrate (IRS-1) which is also the substrate for the insulin receptor. The multisite phosphorylation of IRS-1 results in the initiation of at least two signaling cascades, one through the MAP kinases and one through the PI-3 kinase pathway. It is not clear how both insulin and the IGFs can utilize the same substrates and have different effects but this may be due to tissue-specific differences in transduction and transcription factors.

The IGF-II/mannose-6-phosphate receptor is a monomeric protein which binds to the mannose 6-phosphate residues on lysosomal enzymes in the trans-Golgi network to translocate the newly synthesized lysosomal enzymes to endosomes. When these receptors are on the cell surface, they bind mannose 6-phosphate containing proteins and endocytose them into endosomes. This receptor can also be proteolytically cleaved releasing soluble receptor into the medium. The physiological significance of this is unknown. It has been postulated that this receptor acts to remove IGF from the extracellular environment.

B. IGF Binding Proteins

The actions of the IGFs appear to be tightly regulated by the IGF BPs. Six have been cloned and sequenced; all are structurally similar and all bind with greater affinity to the IGFs than to insulin. Nearly all IGFs in the circulation are bound to these BPs which can either enhance or inhibit IGF activity. IGF-BPs 1, 2, and 3 appear to potentiate the actions of IGF while IGF-BP4 appears to inhibit IGF activity. The BPs also extend the half-life of IGF in serum. During stress or starvation, proteases are induced which are capable of cleaving the BPs thereby releasing the IGFs. When IGF-BP3 is given simultaneously with IGF-1, the effect of IGF-1 on bone formation is enhanced in ovariectomized rats (Narusawa et al., 1995). IGF-BP5 adheres tightly to the extracellular matrix and potentiates IGF activity at the same time. Interestingly, the IGF BPs may have direct cellular effects independent of their interactions with IGF. These effects may be mediated through cell surface receptors.

C. Use of Transgenic Animals to Determine the Function of IGFs

Overexpressed growth hormone in transgenic mice leads to a dramatic postnatal growth and transgenic animals become twice the size of normals (Palmiter et al., 1982). Induction of IGF-1 expression appears to be responsible for this growth increase. Overexpression of IGF-1 in transgenic mice resulted in an increase of only 30% in body weight over controls in contrast to the overexpressed growth hormone mice (Mathews et al., 1988). This was probably due to suppression of growth hormone and corresponding suppression of IGF levels. Also in contrast to the growth hormone mice, overall length and liver weight were not increased whereas the brains showed a dramatic 50% increase. Mice that were deficient in growth hormone (created by putting the expression of diphtheria toxin under the control of the growth hormone promoter, thereby destroying any cell that produces growth hormone) were crossed with transgenic mice overexpressing IGF-1 (Behringer et al., 1990). In these double transgenic mice, the IGF-1 compensated for the lack of growth hormone showing that the large majority of the effects of growth hormone are mediated through IGF-1.

The importance of IGF-1, IGF-2, and IGF-I receptor was illustrated by creation of mice lacking these genes (DeChiara et al., 1990; Baker et al., 1993; Liu et al., 1993). The transgenic knockout mice lacking IGF-1 were similar to those lacking IGF-2 in that the birth weight of each was 60% of normal mice. IGF-1-deficient neonates had a marked increase in death rate in contrast to the IGF-2 mice that had normal survival rates. Skeletal development is delayed as is ossification in surviving mice. Mice lacking the IGF-I receptor were within 45% of normal birth weight but died within minutes of birth due to an inability to breathe. Therefore in all of these mice morphogenesis was normal but the animals appeared as normally proportioned dwarfs. Mice lacking the IGF-II receptor died in utero.

D. Role of the IGFs in Bone Remodeling

IGF may mediate the coupling process between resorption and formation in bone remodeling. This is based on the observation that IGFs stimulate bone cell growth and that bone resorbing cytokines have been shown to cause the release of IGF from bone organ cultures (for review see Baylink et al., 1993; Hayden et al., 1995). IGF-2 is produced in greater amounts by human bone cells in contrast to rodent bone cells in which IGF-1 is greater. The IGFs appear to be bound in the bone matrix to the IGF-BP5 (Bautista et al., 1991). The amount of IGF in bone appears to decrease with age (Nicolas et al., 1994) with a concomitant increase in IGF-BP4 with age (Rosen et al., 1992). Underproduction of stimulatory components and overproduction of inhibitory components of the IGF system probably leads to a decrease in bone cell proliferation and therefore a reduction in bone formation (Hayden et al., 1995).

IGF may play an important role in the increase in bone formation that occurs upon mechanical loading. IGF-1 expression is increased in osteocytes in response to mechanical stimulation (Lean et al., 1994). Whereas a combination of IGF-1/IGF-BP3 will stimulate increased bone formation in normal and ovariectomized rats, this effect is considerably less in paralyzed limbs from these animals (Narusawa et al., 1995), suggesting that the effects of IGF-1 on bone cells are influenced by loading. This is in contrast to TGFβ which can prevent bone loss due to unloading (Machwate et al., 1995).

E. Clinical Studies

An age-related decline in both growth hormone and IGF occurs which is more dramatically increased after menopause. IGF is produced by bone cells and is influenced by estrogens and other agents known to affect bone metabolism. Attempts have therefore been made to treat osteoporosis in humans with growth hormone and with the IGFs. Treatment with growth hormone has not given encouraging results, while treatment with IGF has (for review see Johansson et al., 1993). In studies in which growth hormone was given with calcitonin or with fluoride, no significant changes were observed in the former group and, in the latter group, an increased number of both osteoblasts and osteoclasts was observed. In a more recent study, administration of growth hormone alone led to an increase in serum osteocalcin, type-1 collagen propeptide and telopeptide, serum levels of IGF-1, insulin, and tri-iodothyronine (Brixen et al., 1995). No effect was observed in serum $1,25(OH)_2D_3$ or parathyroid hormone levels. Whereas this study furnished information on the effects of short-term administration of growth hormone on bone markers, no information was obtained on direct effects on bone formation. In fact significant side effects were observed such as an increase in fasting serum insulin and an increase in body weight associated with edema. Another study compared growth hormone administration to IGF-1 administration on bone markers in humans (Ghiron et al., 1995). The same effects were observed as previous studies using growth hormone and these investigators also observed an increase in bone markers with high dose IGF-1. They did find that low dose IGF-1 appears capable of stimulating bone formation with little or no preliminary resorption phase and therefore may be more useful for treatment of osteoporosis. These studies warrant further clinical long-term investigation.

III. THE TRANSFORMING GROWTH FACTOR β SUPERFAMILY

The TGFβs, and the BMPs are part of a family called the TGFβ superfamily which also includes the activins and inhibins and Müllerian inhibitory substance. More recently the family has expanded to include the growth and differentiation factors (for review see Burt and Law, 1994). These proteins are highly conserved in evolution and appear to be key mediators of growth and development. It has recently emerged that their actions are mediated through a novel family of transmembrane serine-threonine kinase receptors.

New insight into the functions of the TGFβ superfamily has been gained by targeted gene disruption or knockout of individual family members and by analysis of naturally occurring mutations. Such experiments have emphasized the importance of the TGFβ superfamily in development, reproduction, and oncogenesis (for review see Matzuk, 1995). For example, mice lacking BMP-5 are viable but have skeletal and cartilage abnormalities (Kingsley et al., 1992). BMP-7 (also known as Osteogenic Protein-1; OP-1) knockout results in polydactyly in the hindlimbs and abnormalities in the skull, ribs, and kidney (Luo et al., 1995; Lyons et al., 1995). These mice die 24 hours after birth. Growth and differentiation factor 5- (GDF5) deficient mice are viable but have bony defects which are restricted to the limbs (Storm et al., 1994). Activin βA deficiency leads to prenatal lethality, associated with craniofacial defects, and activin βB deficient mice are viable but with eyelid defects (Matzuk, 1995). Lack of inhibin A leads to gonadal and adrenal tumors (Matzuk, 1995). Future gene knockout studies will undoubtedly further our understanding of the precise roles of individual family members in development and in the mature animal.

TGFβ superfamily members signal through a novel family of transmembrane serine-threonine kinase receptors. The cDNAs for TGFβ type I and type II receptors have been cloned (Lin et al., 1992; Franzen et al., 1993; for review see Bonewald, 1996). The receptor signalling process for TGFβ may serve as a model for the rest of the TGFβ superfamily. It is thought that heterodimeric complexes of TGFβ type I and type II serine/threonine kinase receptors must form for signalling to occur (Chen et al., 1993; Wrana et al., 1994), although homodimeric complexes may also form. A type III TGFβ receptor, also known as betaglycan, has also been cloned (Wang et al., 1991). This transmembrane glycoprotein has a short intracellular domain and does not appear to be involved in signalling. It has been proposed that this receptor may function to concentrate ligand before presentation to the type I and II signalling receptors.

Most attention has focused on TGFβ types I and II receptors. However, the TGFβ signaling pathway has eluded scientists for many years. It appears that TGFβ binds to the Type-II receptor followed by recruitment of the Type-I receptor to the complex. The constitutively active Type-II receptor then phosphorylates the Type-I receptor, which in turn initiates the signaling cascade through phosphorylation of transcription factors termed Smad 2, 3, and 4 (For review, see Heldin et al., 1997). The direct substrates for the Type-I receptor appears to be Smad-2 and Smad-3, whereas negative regulators of this interaction include Smad-6 and Smad-7 (For review, see Hu et al., 1998). Smad-4 also known as DPC-4, appears to bring the cytoplasmic Smad-2 and Smad-3 into the nucleus where together they can regulate transcription of target genes. Smad-4 was found to be homologous to a gene deleted in pancreatic carcinomas called 'Deleted in Pancreatic CAncer-4, or DPC-4 (Hahn et al., 1996). The Smads share structural homology through two domains called MH-1 and MH-2. These two regions are highly evolutionarily conserved. The MH-2 domain possesses the constitute transcription activity and is repressed by the MH-1 domain.

Other substrates for the TGFβ type-I receptor have been reported including FKBP12, WD40, and farnesyl transferase (Heldin et al., 1997). However, the signaling function of these pathways is less clear than the Smad pathway.

Receptors for some of the BMPs have been identified. A type I BMP receptor which binds BMPs 2 and 4 has been identified (Koenig et al., 1994) and type I receptors for BMP-7 (OP-1) and BMP-4 have also been cloned (ten Dijke et al., 1994). A type II receptor which binds BMP-7 and, less efficiently, BMP-4 has also been isolated (Rosenzweig et al., 1995). This promiscuity in the binding of the BMP ligands to their receptors may explain in part some of the overlap in function of these proteins. Whereas Smads 2 and 3 are signal transducers for the TGFβ type-I receptor, Smad-1 is the major signal transducer for the BMPs (Massague et al., 1997). Smad 4 can also translocate Smad-1, therefore Smad-4 is a shared and obligate partner of the Smads of both the BMP and TGFβ pathways.

A kinase called TGFβ-activating kinase or TAK-1 has been identified which may mediate both TGFβ and BMP-2 cell signaling (Yamaguchi et al., 1995). This kinase is a member of the MAPKK family. At this time the immediate substrates for the receptors are unknown, as are the downstream substrates for TAK-1.

IV. TRANSFORMING GROWTH FACTOR β

The TGFβs are potent multifunctional cytokines, whose major effects in the body appear to be as regulators of cell growth and differentiation, stimulators of matrix production, and inhibitors of the immune system. The TGFβ family consists of four distinct proteins, TGFβ1, 2, 3, and 5. TGFβ4, originally cloned in the chicken, is actually homologous to mammalian TGFβ1 (Burt and Jakowlew, 1992). TGFβ 1, 2, and 3 are differentially expressed in mammalian tissues. Each binds with different affinities to TGFβ receptors and has slightly different effects, although these differences are often in potency rather than in specificity of the biological effect (for review see Centrella et al., 1994).

A. Multiple Roles for TGFβs in Bone Remodeling

In adult animals, most bone formation occurs at sites of prior bone resorption in a process known as bone remodeling. The sequential cellular events involved in this process include cessation of continued osteoclast activity; recruitment and proliferation of osteoblast precursors in the resorption defect; differentiation of precursors into mature osteoblasts, which secrete osteoid; and, finally, mineralization of the newly formed osteoid. The end result of this process is that the resorption defect is filled with new bone. *In vitro* and *in vivo* studies have shown that TGFβs have potent effects at virtually all stages of this remodelling cycle, as summarized below.

TGFβs generally have inhibitory effects on osteoclastic bone resorption by inhibiting both the formation and activation of osteoclasts (for review see Bonewald and Mundy, 1990). Recently it has been shown that TGFβ1, 2, and 3 may also induce apoptosis (or programmed cell death) of mature osteoclasts, which may account for some of the inhibitory effects (Bursch et al., 1992; Hughes et al., 1995). Large amounts of TGFβ are stored in bone matrix in a latent form. Actively resorbing osteoclasts can both release latent TGFβ from the bone matrix and dissociate it to produce the biologically active form (for review see Bonewald, 1996). This release of matrix-bound TGFβ could provide a negative feedback loop which limits further osteoclastic resorption. It may also initiate the bone formation phase of remodeling (see below), thus serving as a coupling factor which links resorption to subsequent formation.

TGFβ is a potent chemotactic factor that recruits a number of different cell types to sites of repair and inflammation and has been shown to be chemotactic for osteoblast precursors (Pfeilschifter et al., 1990). Thus TGFβ may be important in the recruitment of osteoblast precursors into sites of previous resorption. The chemotactic epitope of TGFβ has been identified as residues 368–374 (Postlethwaite and Seyer, 1995). A synthetic peptide spanning this region induced chemotactic migration of neutrophils, monocytes, and fibroblasts but has yet to be tested in osteoblasts.

In some *in vitro* models TGFβ has been shown to stimulate proliferation of osteoblasts while in others it has been shown to inhibit (for review see Bonewald and Mundy, 1990). These apparently conflicting results are probably due to the stage of differentiation of the osteoblast populations and the different experimental conditions used by different investigators. However, it has been shown by thymidine uptake and autoradiography that in bone organ cultures TGFβ stimulates proliferation of osteoblast precursors, while the already mature osteoblasts are stimulated to produce matrix (Hock et al., 1990). TGFβ is a potent stimulator of matrix production in many cell types, including osteoblasts. Not only does TGFβ stimulate production of matrix proteins, but it further promotes matrix accumulation by inhibiting production of matrix-degrading proteases and stimulating the production of protease inhibitors (for review see Roberts and Sporn, 1990). Thus, once a population of mature osteoblasts has been established in the resorption site, TGFβ may regulate the production of osteoid matrix by these cells through its combined effects on matrix accumulation.

The final stage of the remodeling cycle is the mineralization of newly formed osteoid. Interestingly, this is the one stage where the continued presence of TGFβ may actually be inhibitory (for review see Bonewald and Dallas, 1994). TGFβ inhibits mineralization in primary cultures of rabbit chondrocytes and fetal rat osteoblasts and inhibits expression of osteocalcin in fetal rat osteoblasts and rat osteosarcoma cells. In *in vivo* models, where TGFβ injections over murine calvaria have been shown to stimulate bone formation, it was only after the injections had ceased that the osteoid became mineralized. Thus it appears that once TGFβ has initiated new bone formation, it must be removed before mineralization can occur.

B. Effects of the TGFβs in Bone *In Vivo*

It has emerged from several studies that the TGFβs have potent effects on both osteoblasts and osteoclasts and may be major regulators of bone remodeling (for reviews see Bonewald and Dallas, 1994; Centrella et al., 1994). *In vivo* studies have demonstrated that the TGFβs are potent stimulators of bone formation. Injection of TGFβs 1 or 2 over the calvaria of mice resulted in stimulation of bone formation (see review by Bonewald, 1996) and a single injection of TGFβ1 was shown to induce bone closure in a nonhealing skull defect (Kibblewhite et al., 1993). TGFβ has also been shown to initiate chondrogenesis and osteogenesis when applied to the rat femur (Joyce et al., 1990). Systemic administration of TGFβ stimulated cancellous bone formation in both juvenile and adult rats (Rosen et al., 1994), in the nonload-bearing rat (Machwate et al., 1995) and in the aging mouse model (Gazit et al., 1995). However, Kalu and co-workers (1993) did not find significant effects of TGFβ2 on the loss of cancellous bone after ovariectomy. As TGFβ2 was added only during the stage of accelerated bone resorption following ovariectomy, these studies should be repeated at a later stage when enhanced resorption has subsided.

C. Latent Forms of TGFβ and the Latent TGFβ Binding Protein (LTBP)

TGFβ is produced by virtually all cells as a latent complex, which must be dissociated to release the biologically active peptide. A number of different latent TGFβ complexes have been described; some of which contain the latent TGFβ BP (LTBP) and some of which do not. The 100 kDa precursor latent complex lacking LTBP was originally described in Chinese hamster ovary cells as a recombinant protein. This complex consists of the mature 25 kDa TGFβ homodimer noncovalently associated with a 75 kDa portion of the precursor peptide also known as the latency associated peptide as it confers latency to the complex. Bone cells are the only cells known to produce this precursor complex in high amounts as a naturally occurring form (Bonewald et al., 1991; Dallas et al., 1994). Bone cells also produce the 290 kDa latent TGFβ complex containing the 190 kDa LTBP as originally described in fibroblasts. This protein is not a BP in the same sense as the IGF BPs, as it does not interact with mature TGFβ itself, but is covalently linked to one of the precursor chains.

LTBP is not required for latency, but appears to play a role in the storage of latent TGFβ in the bone extracellular matrix and acts as a cleavage substrate for proteases, such as plasmin, to facilitate release of latent TGFβ from the matrix (Dallas et al., 1995). Immunohistochemistry has shown that LTBP localizes to a network of large fibrillar structures in the extracellular matrix of bone cells, suggesting that LTBP may also have its own independent function as a structural matrix protein. Furthermore, antibodies and antisense oligonucleotides against

LTBP inhibit *in vitro* bone formation in primary osteoblast cultures, indicating that the LTBP fibrillar network may be important for bone formation (Dallas et al., 1995). Therefore, LTBP has two independent functions. One function is to modulate the actions of TGFβ by directing its storage and release from bone. Secondly, and independent of its association with TGFβ, it is an extracellular matrix protein important for new bone formation.

V. BONE MORPHOGENETIC PROTEINS

Founding members of the BMP family were originally defined by their ability to stimulate ectopic bone formation when injected into muscle (Urist, 1965). This is in contrast to the TGFβs, which can only induce bone formation when injected in close proximity to bone. The family of bone morphogenetic proteins now includes BMPs 1 through 13 (Kingsley, 1994; Celeste et al., 1995; Dube and Celeste, 1995), although BMPs 12 and 13 appear to be human homologues of the murine GDFs 7 and 6 (Hattersley et al., 1995; Wolfman et al., 1995). Sequence analysis of BMP-1, which was copurified with other BMPs, reveals that it may actually be a proteolytic enzyme, astacin metalloendopeptidase, rather than a true BMP; however, it may play a role in activation of latent BMPs (Wozney et al., 1988; Fukagawa et al., 1994). The BMPs are more closely related to proteins involved in differentiation during embryogenesis, such as the *Drosophila* decapentaplegic complex, the *Xenopus* Vg-1, and the recently described GDF family than they are to the TGFβs themselves.

A. BMP Expression in Vitro and in Vivo

Studies on the expression patterns of the BMPs provide clues as to their role in bone formation. *In vitro* studies have shown that mRNAs for BMPs 2, 3, 4, and 6 are expressed by osteoblasts as they differentiate (Harris et al., 1994; Chen et al., 1995). Maximal expression of BMP-2, 4, and 6 mRNA correlates with the appearance of mineralized bonelike nodules in these cultures. *In vivo* studies have shown a dramatic increase in expression of BMPs 2 and 4 in animal models of fracture healing, indicating a role in fracture repair (Nakase et al., 1994; Bostrom et al., 1995). Although BMPs 2 through 7 show differential patterns of expression during development, expression is generally seen in bone, cartilage, and/or dental cell-types at various stages of differentiation, i.e., in sites that are consistent with regulatory roles for the BMPs in skeletal and tooth development (Heikinheimo, 1994; Houston et al., 1994; Vukicevic et al., 1994; Lyons et al., 1995).

B. Effects of BMP on Bone in Vitro

In vitro experiments have shown that BMP-2 stimulates the formation and mineralization of bonelike nodules in primary osteoblast cultures (McCuaig et al.,

1995; for review see Ghosh-Choudhury et al., 1994). BMP-2 has also been shown to stimulate markers of bone cell differentiation in osteoblast-like cell lines (Takuwa et al., 1991), and to promote the development of an osteoblast-like phenotype in a number of pluripotent mesenchymal stem cell lines (Yamaguchi et al., 1991; Ahrens et al., 1993; Katagiri et al., 1994) while inhibiting adipocyte differentiation (Gimble et al., 1995). Ghosh-Choudhury and colleagues (1994) have shown that osteoblasts which do not express BMP-2 will not differentiate to form bone spontaneously *in vitro*. However, when these cells are transfected with BMP-2, spontaneous differentiation occurs, suggesting that BMP-2 may be an autocrine factor for bone cell differentiation. Recombinant BMPs 4 through 7 generally have effects similar to BMP-2 in primary osteoblast cultures (Hughes et al., 1995; McCuaig et al., 1995), osteoblast-like cell lines (Maliakal et al., 1994), and on pluripotent mesenchymal cell lines (Ahrens et al., 1993).

C. Role of BMPs in Bone Formation in Vivo

Although it is well established that BMPs 2 through 7 stimulate ectopic bone formation *in vivo,* at present little is known about the role of these BMPs in normal bone formation. Recombinant BMP-2 is perhaps the most well characterized. In animal models it has been shown to promote the bony healing of craniotomies and segmental bone defects (Yasko et al., 1992; Kenley et al., 1994; Lee et al., 1994) and to promote regeneration of alveolar bone (for review see Wozney, 1995). BMP-2 has also been used successfully in humans for the treatment of resistant nonunions and segmental defects of long bones (Johnson et al., 1992).

Data on the other BMPs are less complete. *In vivo* studies have shown that, like BMP-2, BMPs 3 and 7 are useful in promoting the healing of bony defects in animal models (Ripamonti et al., 1992; Cook et al., 1994). At present, BMPs 10 through 13 do not appear to play an important role in bone formation. They do not stimulate ectopic bone formation and may form a subgroup of BMPs that is more intimately involved with morphogenesis of cartilage and other supportive connective tissues such as tendon and ligament (Hattersley et al., 1995; Wolfman et al., 1995).

D. Interactions of BMP with Other Bone Growth Factors

Growth factors do not act alone in the bone microenvironment but in concert and in series with numerous other factors. Studies have shown that both TGFβ1 and activin will enhance BMP-induced ectopic bone formation (Bentz et al., 1991; Ogawa et al, 1992). These growth factors have distinct and overlapping activities. For example, OP-1, also known as BMP-7, has similar chemotactic activities as TGFβ but does not have TGFβ's fibrogenic properties (Postlethwaite et al., 1994). As discussed below, FGF is a potent bone-inducing agent but BMP will actually inhibit limb bud outgrowth induced by FGF (Niswander and Martin, 1993). Limb growth is probably modulated by the combination of the FGF and BMP signaling mechanisms suggesting that limb growth is regulated by a combination of stimulating and inhibiting signals.

VI. FIBROBLAST GROWTH FACTOR

The FGFs were originally named for their ability to stimulate the growth of 3T3 fibroblasts and were originally purified from brain and pituitary extracts (Gospodarowicz, 1974). Two proteins called acidic and basic FGF were responsible for this activity. Acidic FGF is sonamed because it has an acidic isoelectric point of 5.6 and basic FGF has an isoelectric point of 9.6. Basic FGF has 55% homology with acidic FGF. Currently there are nine members of the FGF family, all structurally related. FGF 1, 2, and 9 lack the classical leader piece necessary for secretion but yet are secreted constitutively, probably through an unknown nonclassical pathway (Jackson et al., 1992). The FGFs have numerous biologic activities including the ability to stimulate cell migration, cell proliferation, and cell differentiation (for review see Gospodarowicz et al., 1987). Basic FGF is a well-known angiogenic factor which may prepare basement membrane for migrating endothelial cells. This may be accomplished by the stimulation of plasminogen activator and collagenase by βFGF (Saksela et al., 1987). FGF and TGFβ have been shown to have opposing effects in this system. The FGFs have striking effects on limb formation during embryogenesis. Beads containing FGF 1, 2, or 4 placed in the flank of chick embryos will induce formation of ectopic limb buds which will develop into complete limbs (Cohn et al., 1995). Purified FGF 8 will substitute for the apical ectodermal ridge in limb growth induction. Therefore FGFs are responsible for activation of genes necessary for complete limb development.

A. Effects of FGF on Bone in Vitro

FGFs stimulate bone cell replication in cultures of calvariae and in cultures of osteoblast-like cells (Canalis et al., 1988; McCarthy et al., 1989). Basic FGF appears to be more potent than acidic FGF. Basic FGF enhances the synthesis of TGFβ by osteosarcoma cells and conversely TGFβ can stimulate the synthesis of βFGF in osteoblast-like cells (Hurley et al., 1994). Interestingly, basic FGF causes the activation of latent TGFβ in endothelial cells (Flaumenhaft et al., 1992). It is not known if some of the effects of FGF are mediated through TGFβ. FGF will increase collagen and noncollagen protein synthesis which can be enhanced in the presence of heparin (for review see Mohan and Baylink, 1991). The FGFs do not appear to have any effects on bone resorption or on matrix degradation. Bovine bone cells make basic FGF and store it in their matrix (Globus et al., 1989). Acidic FGF is present at a five- to 10-fold lower concentration than basic FGF in bone matrix (See Table 1).

B. Effects of FGF on Bone Formation in Vivo

The FGFs clearly stimulate new bone formation and enhance bone regeneration. Autografts of irradiated mandibular bone resection sites which contained FGF healed and reestablished their contour in 50% of treated animals whereas autografts without FGF showed 100% lack of healing (Eppley et al., 1991). Intravenous administration of FGF for two weeks into young and aged rats caused a profound increase in bone formation (Mayahara et al., 1993) and FGF enhanced wound repair and endosteal bone formation (Nakamura et al., 1995). Injection of FGF over the calvaria of mice will cause an increase in bone formation similar if not greater in magnitude to that seen with TGFβ (Dunstan et al., 1993) and systemic administration of FGF will prevent bone loss, increase new bone formation, and restore trabecular microarchitecture in ovariectomized rats (Dunstan et al., 1995). One of the major problems with using FGF systemically is its acute hypotensive effects. Development of a mutant FGF which is not hypotensive yet maintains ability to stimulate new bone formation is a desirable research goal.

C. FGF Receptors

At least four receptors for the FGFs are known (Givol and Yayon, 1992). All are high affinity receptors which interact not only with the nine isoforms of FGF but also three heparin sulfate proteoglycans (Klagsbrun and Baird, 1991; Mason, 1994). Heparin proteoglycans are necessary for high-affinity receptor binding. Therefore the diversity for cell signaling by these receptors and their ligands is quite large. These receptors have several immunoglobulin-like domains, a transmembrane domain, and an intracellular tyrosine kinase domain. Little is known as far as FGF signaling in bone cells in concerned.

Defects in FGF receptors can be responsible for inherited disorders of skeletal development. Achondroplasia, a disorder of limb development resulting in short-limbed dwarfism and macrocephaly, is due to a single point mutation in the transmembrane domain of the FGF receptor 3 (Rousseau et al., 1994; Shiang et al., 1994). Hypochondroplasia, a milder form of achondroplasia, is also due to mutations in the FGF receptor 3 (Bellus et al., 1995). A rare syndrome of craniosynostosis known as Crouzon syndrome is due to mutations in the extracellular domain of the FGF receptor 2 (Reardon et al., 1994). Two other forms of craniosynostoses accompanied with other defects such as abnormalities in limb bud development leading to broad thumbs and big toes, known as Pfeiffer syndrome, or with foot anomalies, known as Jackson-Weiss syndrome are also due to mutations in FGF receptor 1 and FGF receptor 2, respectively (Jabs et al., 1994; Muenke et al., 1994). Discovering the underlying defect in these syndromes emphasizes the important functions of the FGFs in

VII. PLATELET-DERIVED GROWTH FACTOR

PDGF was obviously named after its initial purification from platelets but was subsequently found to be made by a variety of normal and transformed cells including bone cells. The factor is a mitogen for all cells of mesenchymal origin, including bone cells (Ross et al., 1986; Kasperk et al., 1990). PDGF is composed of two polypeptide chains, A and B, and can be a homodimer AA, BB, or heterodimer known as PDGF AB. These two chains are 56% identical and their genes are independently regulated. Platelets and serum contain PDGF AB and BB, while skeletal tissues contain primarily PDGF AA.

A. Receptors for PDGF

There are two different types of high affinity receptors for PDGF, one called the α receptor which binds PDGF-AA, BB, and AB and a second, the β receptor which only binds PDGF-BB and AB. Therefore a cellular response depends not only on the isoform of PDGF that is present but also on the type of PDGF receptor present on the cell surface.

B. Effects of PDGF on Bone

PDGF stimulates proliferation of bone cells derived from embryonic chick calvaria and newborn mouse calvaria and stimulates cell replication and collagen and noncollagen protein synthesis in rat calvaria organ cultures (Canalis et al., 1991; for review see Mohan and Baylink, 1991). Unlike FGF, PDGF also stimulates bone resorption and bone collagen degradation possibly through the stimulation of prostaglandin synthesis (Tashjian et al., 1982). As PDGF enhances tissue repair and wound healing, it is likely that PDGF released from platelets at a fracture site also plays a role in fracture repair.

C. In vivo Effects of PDGF on Bone

PDGF injected near periosteum increases bone mass by increasing mesenchymal cell proliferation (Bolander, 1992). A considerable amount of research has been performed concerning the effects of PDGF on periodontal regeneration (for review see Graves and Cochran, 1994). PDGF will increase new bone and cementum formation in dental surgery performed in dogs and may be an important adjuvant to periodontal surgery. Combinations of PDGF with IGF-1 stimulate regeneration of periodontal attachment in monkeys and will increase new bone in contact with implants. More clinical trials utilizing combinations of growth factors are necessary before any conclusions can be drawn concerning their practicality.

VIII. CONCLUSIONS

With the discovery of growth factors came the hope that one day these could be used for the treatment of disease. However, few to date have made it to the clinic and few are prescribed on a routine basis. This is due to the fact that growth factors have multiple activities, some desirable and some not. The activity of some growth factors is difficult to control due to the presence of BPs, short half-life in the circulation, and numerous other mechanisms whereby these factors are naturally controlled. The optimal mode of delivery and targeting of growth factors remains to be determined. With the advent of recombinant technology and protein structure/function studies perhaps new growth factors can be designed and generated with the desired properties suitable for clinical application.

IX. SUMMARY

Several growth factors have been identified which have important critical functions in embryogenesis, development, growth, and repair. The major growth factors that have been identified in bone include IGFs, TGFβ, BMPs, FGF, and PDGF. These are not single factors but families of growth factors. Some of these factors mainly function as regulators of development, others as stimulators of proliferation, inducers of differentiation and others as inducers of matrix formation. The regulation of these growth factors is complicated and intricate. Receptor expression, BPs, latency, matrix storage, secretion, and regulation by cytokines all play important roles in the regulation of these factors. Loss of equilibrium in these intricate control systems can lead to deformity, disease and/or death.

ACKNOWLEDGMENTS

We would like to acknowledge the excellent secretarial assistance of Thelma Barrios.

REFERENCES

Ahrens, M., Ankenbauer, T., Schroder, D., Hollnagel, A., Mayer, H., and Gross, G. (1993). Expression of human bone morphogenetic proteins-2 and -4 in murine mesenchymal progenitor C3H10T1/2 cells induces differentiation into distinct mesenchymal cell lineages. DNA Cell Biol. 12, 871-880.

Baker, J., Liu, J.-P., Robertson, E.J., and Efstratiadis, A. (1993). Role of insulinlike growth factors in embryonic and postnatal growth. Cell 75, 73-82.

Bautista, C.M., Baylink, D.J., and Mohan, S. (1991). Isolation of a novel insulinlike growth factor (IGF) binding protein from human bone: A potential candidate for fixing IGF-II in human bone. Biochem. Biophys. Res. Commun. 176, 756-763.

Baylink, D.J., Finkelman, R.D., and Mohan, S. (1993). Growth factors to stimulate bone formation. J. Bone Miner. Res. 8 (Suppl. 2), S565-S572.

Behringer, R.R., Lewin, T.M., Quaife, C.J., Plamiter, R.D., Brinster, R.L., and D'Ercole, A.J. (1990). Expression of insulinlike growth factor I stimulates normal somatic growth in growth hormone-deficient transgenic mice. Endocrinology 127, 1033-1040.

Bellus, G.A., Hefferon, T.W., Ortiz de Luna, R.I., Hecht, J.T., Horton, W.A., Machado, M., Kaitila, I., McIntoshi, I., and Francomano, CA. (1995). Achondroplasia is defined by recurrent G380R mutations of FGFR3. Am.J. Hum. Genet. 56, 368-373.

Bentz, H., Thompson, A.Y., Armstrong, R., Chang, R.-J., Piez, K.A., and Rosen, D.M. (1991). Transforming growth factor β2 enhances the osteoinductive activity of a bovine bone-derived fraction containing bone morphogenetic protein-2 and 3. Matrix 11, 269-275.

Bolander, M.D. (1992). Regulation of fracture repair by growth factors. Proc. Soc. Exp. Biol. Med. 200, 176-170.

Bonewald, L.F. and Mundy, G.R. (1990). Role of transforming growth factor β in bone remodeling. Clin. Ortho. Related Res. 250, 261-276.

Bonewald L.F., Wakefield L, Oreffo R.O.C., Escobedo A, Twardzik D.R., and Mundy G.R. (1991). Latent forms of transforming growth factor β (TGFβ) derived from bone cultures. Identification of a naturally occurring 100-kDa complex with similarity to recombinant latent TGFβ. Mol. Endo. 5, 741-751.

Bonewald, L.F. Dallas, S.L. (1994). Role of active and latent transforming growth factor β in bone formation. J. Cell. Biochem. 55, 350-357.

Bonewald, L.F. (1996). Transforming growth factor β. In: Principles of Bone Biology. (Bilezikian, J., Raisz, L., and Rodan, G., Eds.), pp. 647-660. Academic Press, New York.

Bostrom, M.P., Lane, J.M., Berberian, W.S., Missri, A.A., Tomin, E., Weiland, A., Doty, S.B., Glaser, D., and Rosen, V.M. (1995). Immunolocalization and expression of bone morphogenetic proteins 2 and 4 in fracture healing. J. Orthop. Res. 13, 357-367.

Brixen, K., Kassem, M., Nielsen, H.K., Lof, A.G., Flyvbjerg, A., and Mosekilde, L. (1995). Short-term treatment with growth hormone stimulates osteoblastic and osteoclastic activity in osteopenic postmenopausal women: A dose response study. J. Bone Miner. Res. 10, 1865-1874.

Bursch, W., Oberhammer, F., Jirtle, R.L., Askari, M., Sedivy, R., Grasl-Kraupp, B., Purchio, A.F., and Schulte-Hermann, R. (1992). Transforming growth factor β1 as a signal for induction of cell death by apoptosis. Br. J. Cancer 67, 531-536.

Burt, D.W. and Jakowlew, S.B. (1992). Correction: A new interpretation of a chicken transforming growth factor β4 complementary DNA. Mol. Endocrinol. 6, 989-992.

Burt, D.W. and Law, A.S. (1994). Evolution of the transforming growth factor β superfamily. Prog. Growth Factor Res. 5, 99-118.

Canalis, E., Centrella, M., and McCarthy, T. (1988). Effects of basic fibroblast growth factor on bone formation in vitro. J. Clin. Invest. 81, 1572-1577.

Canalis, E., McCarthy, T.L., and Centrella, M. (1991). Growth factors and cytokines in bone cell metabolism. Annu. Rev. Med. 42, 17-24.

Celeste, A.J., Ross, J.L., Yamaji, N., and Wozney, J.M. (1995). The molecular cloning of human bone morphogenetic proteins -10, -11 and -12, three new members of the transforming growth factor β superfamily. J. Bone Miner. Res. 10 (Suppl. 1), S336.

Centrella, M., Horowitz, M.C., Wozney, J.M., and McCarthy, T.L. (1994). Transforming growth factor β gene family members and bone. Endocrine Rev. 15, 27-39.

Chen, D., Feng, J.Q., Feng, M, Harris, M.A., Mahy, P., Mundy, G.R., and Harris, S.E. (1995). Sequence and expression of bone morphogenetic protein 3 mRNA in prolonged cultures of fetal rat calvarial osteoblasts and in rat prostate adenocarcinoma PAIII cells. DNA Cell Biol. 14, 235-239.

Chen, R.-H., Ebner, R., and Derynck, R. (1993). Inactivation of the type-II receptor reveals two receptor pathways for the diverse TGFβ activity. Science 260, 1335-1338.

Cohn, M.J., Izpisua-Belmonte, J.C., Abud, H., Heath, J.K., and Tickle, C. (1995). Fibroblast growth factors induce additional limb development from the flank of chick embryos. Cell 80, 739-746.

Cook, S.D., Baffes, G.C., Wolfe, M.W., Sampath, T.K., and Rueger, D.C. (1994). Recombinant human bone morphogenetic protein-7 induces healing in a canine long-bone segmental defect model. Clin. Orthop. 301, 302-312.

Dallas, S.L., Park-Snyder, S., Miyazono, K., Twardzik, D., Mundy, G.R., and Bonewald, B.L. (1994). Characterization and autoregulation of latent transforming growth factor β (TGFβ) complexes in osteoblastlike cell lines. J. Biol. Chem. 269, 6815-6822.

Dallas, S.L., Miyazono, K., Skerry, T.M., Mundy, G.R., and Bonewald, L.F. (1995). Dual role for the latent transforming growth factor β binding protein (LTBP) in storage of latent TGFβ in the extracellular matrix and as a structural matrix protein. J. Cell Biol. 131, 539-549.

DeChiara, T.M., Efstratiadis, A., and Robertson, E.J. (1990). A growth-deficiency phenotype in heterozygous mice carrying an insulinlike growth factor II gene disrupted by targeting. Nature 345, 78-80.

Dube, J.L. and Celeste, A.J. (1995). Human bone morphogenetic protein-13, a molecule which is highly related to human bone morphogenetic protein-12. J. Bone Miner. Res. 10 (Suppl. 1), S336.

Dunstan, C.R., Boyce, B.F., Izbicka, E., Adams, R., and Mundy, G.R. (1993). Acidic and basic fibroblast growth factors promote bone growth in vivo comparable to that of TGFβ. J. Bone Miner. Res. 8 (Suppl 1), #250.

Dunstan, C.R., Garrett, I.R., Adams, R., Burgess, W., Jaye, M., Youngs, T., Boyce, R., and Mundy, G.R. (1995). Systemic fibroblast growth factor (FGF-1) prevents bone loss, increases new bone formation, and restores trabecular microarchitecture in ovariectomized rats. J Bone Min. Res. 10 (Suppl. 1), #P279.

Eppley, B.L., Connolly, D.T., Winkelmann, T., Sadove, A.M., Heuvelman, D., and Peder, J. (1991). Free bone graft reconstruction of irradiated facial tissue: Experimental effects of basic fibroblast growth factor stimulation. Plast. Reconstr. Surg. 88, 1-11.

Finkelman, R.D., Linkhart, T.A., Mohan, S., Lau, K.-H.L., Baylink, D.J., and Bell, N.H. (1991). Vitamin D deficiency causes a selective reduction in deposition of transforming growth factor β in rat bone: Possible mechanism for impaired osteoinduction. Proc. Natl. Acad. Sci. USA 88, 3657-3660.

Flaumenhaft, R., Abe, M., Mignatti, P., and Rifkin, D.B. (1992). Basic fibroblast growth factor-induced activation of latent transforming growth factor β in endothelial cells: Regulation of plasminogen activator activity. J. Cell Biol. 118, 901-909.

Franzen, P., ten Dijke, P., Ichijo, H., Yamashita, H., Schulz, P., Heldin, C.-H., and Miyazono, K. (1993). Cloning of a TGFβ type-I receptor that forms a heteromeric complex with the TGFβ type-II receptor. Cell 75, 681-692.

Fukagawa, M., Suzuki, N., Hogan, B.L., and Jones, C.M. (1994). Embryonic expression of mouse bone morphogenetic protein-1 (BMP-1), which is related to the *Drosophila* dorsoventral gene tolloid and encodes a putative astacin. Devel. Biol. 163, 175-183.

Gazit, D., Zilberman, Y., Passi-Even, L., Ebner, R., and Kahn, A.J. (1995). In vivo administration of TGFβ 1 increases marrow osteoprogenitor number and selectively reverses the bone loss seen in old, "osteoporotic" mice. J. Bone Miner. Res. 10 (Suppl. 1), #25 (Abstract.)

Ghiron, L.J., Thompson, J.L., Holloway, L., Hintz, R.L., Butterfield, G.E., Hoffman, A.R., and Marcus, R. (1995). Effects of recombinant insulinlike growth factor-I and growth hormone on bone turnover in elderly women. J. Bone Miner. Res. 10, 1844-1852.

Ghosh-Choudhury, N., Harris, M.A., Feng, J.Q., Mundy, G.R. and Harris, S.E. (1994). Expression of the BMP 2 gene during bone cell differentiation. Crit. Rev. Eukaryotic Gene Expression 4, 345-344.

Gimble, J.M., Morgan, C., Kelly, K., Wu, X., Dandapani, V., Wang, C.-S., and Rosen, V. (1995). Bone morphogenetic proteins inhibit adipocyte differentiation by bone marrow stromal cells. J. Cell. Biochem. 58, 393-402.

Givol, D. and Yayon, A. (1992). Complexity of FGF receptors-genetic basis for structural diversity and functional specificity. FASEB J. 6, 3362-3369.

Globus, R.K., Plouet, J., and Gospodarowicz, D. (1989). Cultured bovine cells synthesize basic fibroblast growth factor and transforming growth factor β. Endocrinology 124, 1539-1547.

Gospodarowicz, D. (1974). Localization of a fibroblast growth factor and its effect alone and with hydrocortisone on 3T3 cell growth. Nature 249, 123-127.

Gospodarowicz, D., Ferrara, N., Schweigerer, L., and Neufeld, G. (1987). Structural characterization and biological functions of fibroblast growth factor. Endocr. Rev. 8, 95-114.

Graves, D.T. and Cochran, D.L. (1994). Periodontal regeneration with polypeptide growth factors. Curr. Opin. Periodontol. 178-186.

Hahn, S.A., Schutte, M., Hogue, A.T., Moskaluk, C.A., da Costa, L.T., Rozenblum, E., Weinstein, C.L., Fischer, A., Yeo, C.J. Hruban, R.H., and Kern, S.E. (1996). DPC4, a candidate tumor suppressor gene at human chromosone 18q21.1. Science 271, 350-353.

Hattersly, G., Hewick, R., and Rosen, V. (1995). In situ localization and in vitro activity of BMP-13. J. Bone Miner. Res. 10 (Suppl. 1), S163.

Harris, S.E., Sabatini, M., Harris, M.A., Feng, J.Q., Wozney, J., and Mundy, G.R. (1994). Expression of bone morphogenetic protein messenger RNA in prolonged cultures of fetal rat calvarial cells. J. Bone Miner. Res. 9, 389-394.

Hauschka, P.V. (1990). Growth factor effects in bone. In: Bone-A Treatise, Vol 1, The Osteoblast and Osteocyte. pp. 103-170. Telford Press, Caldwell, NJ.

Hayden, J.M., Mohan, S., and Baylink D.J. (1995). The insulinlike growth factor system and the coupling of formation to resorption. Bone 17 (Suppl. 2), 93S-98S.

Heikinheimo, K. (1994). Stage-specific expression of decapentaplegic-Vg-related genes 2, 4, and 6 (bone morphogenetic proteins 2, 4, and 6) during human tooth morphogenesis. J. Dent. Res. 73, 590-597.

Heldin, C.H., Mayazono, K., and ten Dijke, P. (1997). TGFβ signaling from cell membrane to nucleus through SMAD proteins. Nature, 390, 465-471.

Hock, J.M., Canalis, E., and Centrella, M. (1990). Transforming growth factor β stimulates bone matrix apposition and bone cell replication in cultured fetal rat calvariae. Endocrinology 126, 421-426.

Houston, B., Thorp, B.H., and Burt, D.W. (1994). Molecular cloning and expression of bone morphogenetic protein-7 in the chick epiphyseal growth plate. J. Mol. Endocrinol. 13, 289-301.

Hu, PP., Datto, M.B., and Wang, X. (1998). Molecular mechanisms of Transforming Growth Factor-β Signaling. Endocrine Reviews 19, 349-363.

Hughes, F.J., Collyer, J., Stanfield, M., and Goodman, S.A. (1995). The effects of bone morphogenetic protein-2, -4, and -6 on differentiation of rat osteoblast cells in vitro. Endocrinology 136, 2671-2677.

Hurley, M.M., Abreu, C., Gronowicz, G., Kawaguchi, H., and Lorenzo, J. (1994). Expression and regulation of basic fibroblast growth factor mRNA levels in mouse osteoblastic MC3T3-E1 cells. J. Biol. Chem. 269, 9392-9396.

Isaksson, O.G.P., Ohlsson, C., Nilsson, A., Isgaard, J., and Lindahl A. (1991). Regulation of cartilage growth by growth hormone and insulinlike growth factor I. Pediatr. Nephrol. 5, 451-453.

Jabs, E.W., Li, X., Scott, A.F., Meyers, G., Chen, W., Eccles, M., Mao, J., Charnas, L.R., Jackson, C.E., Jaye, M. (1994). Jackson-Weiss and Crouzon syndromes are allelic with mutations in fibroblast growth factor receptor 2. Nat. Genet. 8, 275-279.

Jackson, A., Friedman, S., Zhan, X., Engleka, K.A., Forough, R., and Maciag, T. (1992). Heat shock induces the release of fibroblast growth factor-1 from NIH 3T3 cells. Proc. Natl. Acad. Sci. USA 89, 10691-10695.

Johansson, A.G., Lindh, E., and Ljunghall, S. (1993). Growth hormone, insulinlike growth factor I, and bone: A clinical review. J. Int. Med. 234, 553-560.

Johnson, E.E., Urist, M.R., and Finerman G.A. (1992). Resistant nonunions and partial or complete segmental defects of long bones. Treatment with implants of a composite of human bone morphogenetic protein (BMP) and autolyzed, antigen-extracted, allogenic (AAA) bone. Clin. Orthop. 277., 229-237.

Jones, J.I. and Clemmons, D.R. (1995). Insulinlike growth factors and their binding proteins: Biological actions. Endocrine Rev. 16, 3-34.

Joyce, M.E., Roberts, A.B., Sporn, M.B., and Bolander, M.E. (1990). Transforming growth factor β and the initiation of chondrogenesis and osteogenesis in the rat femur. J. Cell Biol. 110, 2195-2207.

Kalu, D.N., Salerno, E., Higami Y., Liu, C.C., Ferraro, F., Salih, M.A., and Arjmandi, B.H. (1993). In vivo effects of transforming growth factor β2 in ovariectomized rats. Bone Miner. 22, 209-220.

Kasperk, C., Wergedal, J.E., Mohan, S., Long, D.L., Lau, K.H.W., and Baylink, D.J. (1990). Interactions of growth factors present in bone matrix and bone cells: Effects on DNA synthesis and alkaline phosphatase. Growth Factors 3, 147-158.

Katagiri, T., Yamaguchi, A., Komaki, M., Abe, E., Takahashi, N., Ikeda, T., Rosen, V., Wozney, J.M., Fujisawa-Sehara, A., and Suda, T. (1994). Bone morphogenetic protein-2 converts the differentiation pathway of C2C12 myoblasts into the osteoblast lineage [Published erratum appears in J. Cell Biol. 1995, 128, 713]. J. Cell Biol. 127 (Part 1), 1755-1766.

Kenley, R., Marden, L., Turek, T., Jin, L., Ron, E., and Hollinger, J.O. (1994). Osseous regeneration in the rat calvarium using novel delivery systems for recombinant human bone morphogenetic protein-2 (rhBMP-2). J. Biomed. Mater. Res. 28, 1139-1147.

Kibblewhite, D.J., Bruce, A.G., Strong, D.M., Ott, S.M., Purchio, A.F., and Larrabee, Jr., W.F. (1993). Transforming growth factor β accelerates osteoinduction in a craniofacial onlay model. Growth Factors 9, 185-193.

Kingsley, D.M., Bland, A.E., Grubber, J.M., Marker, P.C., Russell, L.B., Copeland, N.G., and Jenkins, N.A. (1992). The mouse short ear skeletal morphogenesis locus is associated with defects in a bone morphogenetic member of the TGFβ superfamily. Cell 71, 399-410.

Kingsley, D. (1994). The TGFβ superfamily: New members, new receptors, and new genetic tests of function in different organisms. Genes Dev. 8., 133.

Klagsbrun, M. and Baird, A. (1991). A dual receptor system is required for basic fibroblast growth factor activity. Cell 67, 229-231.

Koenig, B.B., Cook, J.S., Wolsing, D.H., Ting, J., Tiesman, J.P., Correa, P.E., Olson, C.A., Pecquet, A.L., Ventura, F., Grant, R.A., Chen, G.X., Wrana, J.L., Massaguè, J., and Rosenbaum, J.S. (1994). Characterization and cloning of a receptor for BMP-2 and BMP-4 from NIH 3T3 cells. Mol. Cell Biol. 14, 5961-5974.

Kulkarni, A.B., Huh, C.-G., Becker, D., Geiser, A. Lyght, M., Flanders, K.C., Roberts, A.B., Sporn, M.B., Ward, J.M., and Karlsson, S. (1993). Transforming growth factor β1 null mutation in mice causes excessive inflammatory response and early death. Proc. Natl. Acad. Sci. USA 90, 770-774.

Lean, J.M., Jagger, C.J., Chambers, T.J., and Chow, J.W.M. (1994). Increased insulinlike growth factor-I mRNA expression in osteocytes precedes the increase in bone formation in response to mechanical stimulation. J. Bone Miner. Res. 9, S142 (Abstract.)

Lee, S.C., Shea, M., Battle, M.A., Kozitza, K., Ron, E., Turek, T., Schaub, R.G., and Hayes, M.C. (1994). Healing of large segmental defects in rat femurs is aided by rhBMP-2 in PLGA matrix. J. Biomed. Mater. Res. 28, 1149-1156.

Letterio, J.L., Geiser, A.G., Kulkarni, A.B., Roche, N.S., Sporn, M.B., and Roberts, A.B. (1994). Maternal rescue of transforming growth factor β1 null mice. Science 264, 1936-1938.

Lin, H.Y., Wang, X.-F., Ng-Eaton, E., Weinberg, R.A., and Lodish, H.F. (1992). Expression cloning of the TGFβ type-II receptor, a functional transmembrane serine/threonine kinase. Cell 68, 775-785.

Liu, J.-P., Baker, J., Perkins, A.S., Robertson, E.J., and Efstratiadis, A. (1993). Mice carrying null mutations of the genes encoding insulinlike growth factor I (IGF-I) and type-1 IGF receptor (IGF1r). Cell 75, 73-82.

Luo, G., Hoffman, C., Bronckers, T., Bradley, A., and Karsenty, G. (1995). BMP-7 (OP-1) -deficient mice fail to develop glomeruli and have skeletal patterning defect. J. Bone Miner. Res. 10 (Suppl. 1), S163.

Lyons, K.M., Hogan, B.L.M., and Roberts, E.J. (1995). Colocalization of BMP 7 and BMP 2 RNAs suggests that these factors cooperatively mediate tissue interactions during murine development. Mech. Dev. 50, 71-83.

Machwate, M., Zerath, E., Holy, X., Hott, M., Godet, D., Lomri, A., Marie, and P.J. (1995). Systemic administration of transforming growth factor β2 prevents the impaired bone formation and osteopenia induced by unloading in rats. J. Clin. Invest. 96, 1245-1253.

Maliakal, J.D., Asahina, I., Hauschka, P.V., and Sampath, T.K. (1994). Osteogenic protein-1 (BMP-7) inhibits cell proliferation and stimulates the expression of markers characteristic of osteoblast phenotype in rat osteosarcoma (17/2.8) cells. Growth Factors 11, 227-234.

Massague, J., Hata, A., and Liu, F. (1997). TGFβ signaling through the Smad pathway. Trends in Cell Biology, 7, 187-192.

Mason, I.J. (1994). The ins and outs of fibroblast growth factors. Cell 78, 547-552.

Mathews, L.S., Hammer R.E., Behringer, R.R., D'Ercole, A.J., Bell, G.I., Brinster, R.L., and Palmiter, R.D. (1988). Growth enhancement of transgenic mice expressing human insulinlike growth factor-I. Endocrinology 123, 2827-2833.

Matzuk, M.M. (1995). Functional analysis of mammalian members of the transforming growth factor β superfamily. Trends Endocrinol. Metab. 6, 120-127.

Mayahara, H., Ito, T., Nagai, H., Miyajima, H., Tsukuda, R., Taketomi, S., Mizoguichi, J., and Kato, K. (1993). In vivo stimulation of endosteal bone formation by basic fibroblast growth factor in rats. Growth Factors 9, 73-80.

McCarthy, T.L., Centrella, M., and Canalis, E. (1989). Effects of fibroblast growth factors on deoxyribonucleic acid and collagen synthesis in rat parietal bone cells. Endocrinology 125, 2118-2126.

McCuaig, K., Racine, M., Nanes, M.S., Titus, L., Catherwood, B.D., Wozney J.M., and Boden, S.D. (1995). Effects of BMPs on osteoblast differentiation in vitro. J. Bone Miner. Res. 10, (Suppl. 1), S309.

Mohan, S. and Baylink, D.J. (1991). Bone growth factors. Clin. Ortho. 263, 30-48.

Muenke, M., Schell, U., Hehr, A., Robin, N.H., Losken, H.W., Schinzel, A., Pulleyn, L.J., Rutland, P., Reardon, W., Malcolms, S., and Winter, R.M. (1994). A common mutation in the fibroblast growth factor receptor 1 gene in Pfeiffer syndrome. Nat. Genet. 8, 269-274.

Nakamura, T., Hanada, K., Tamura, M., Shibanushi, T., Nigi, H., Tagawa, M., Fukumoto, S., and Matsumoto, T. (1995). Stimulation of endosteal bone formation by systemic injections of recombinant basic fibroblast growth factor in rats. Endocrinology 136, 1276-1284.

Nakase, T., Nomura, S., Yoshikawa, H., Hashimoto, J., Hirota, S., Kitamura, Y., Oikawa, S., Ono, K., and Takaoka, K. (1994). Transient and localized expression of bone morphogenetic protein 4 messenger RNA during fracture healing. J. Bone Miner. Res. 9, 651-659.

Narusawa, K., Nakamura, T., Suzuki, K., Matsuoka, Y., Lee, L.-J., Tanaka, H., and Seino, Y. (1995). The effects of recombinant human insulinlike growth factor (rhIGF)-1 and rhIGF-1/IGF binding protein-3 administration on rat osteopenia induced by ovariectomy with concomitant bilateral sciatic neurectomy. J. Bone Miner. Res. 10, 1853-1864.

Nicolas, V., Prewett, A., Bettica, P., Mohan, S., Finkelman, R.D., Baylink, D.J., and Farley, J.R. (1994). Age-related decreases in insulinlike growth factor-I and transforming growth factor β in femoral cortical bone from both men and women: Implications for bone loss with aging. J. Clin. Endocrinol. Metab. 78, 1011-1016.

Niswander, L. and Martin G.R. (1993). FGF-4 and BMP-2 have opposite effects on limb growth.

Ogawa, Y., Schmidt, D.K., Nathan, R.M., Armstrong, R.M., Miller, K.L., Sawamura, S.J., Ziman, J.M., Erickson, K.L., de Leon, E.R., Rosen, D.M., Seyedin, S.M., Glaser, C.B., Chang, R.-J., Corrigan, A.Z., and Vale, W. (1992). Bovine bone activin enhances bone morphogenetic protein-induced ectopic bone formation. J. Biol. Chem. 267, 14233-14237.

Palmiter, R.D., Brinster, R.L., Hammer R.E., Trumbauer, M.E., Rosenfeld, M.G., Birnberg, N.C., and Evans, R.M. (1982). Dramatic growth of mice that develop from eggs microinjected with metallothionein-growth hormone fusion genes. Nature 300, 611-615.

Pfeilschifter, J., Oechsner, M., Naumann, A., Gronwald, R.G.K., Minne, H.W., and Ziegler, R. (1990). Stimulation of bone matrix apposition in vitro by local growth factors: A comparison between insulinlike growth factor I, platelet-derived growth factor, and transforming growth factor β. Endocrinology 127, 69-75.

Pfeilschifter, J., Laukhuf, F., Muller-Beckmann, B., Blum, W.F., Pfister, T., and Ziegler, R. (1995). Parathyroid hormone increases the concentration of insulinlike growth factor-I and transforming growth factor β1 in rat bone. J. Clin. Invest. 96, 767-774.

Postlethwaite, A.E., Raghow, R., Stricklin, G., Ballou, L., and Sampath, T.K. (1994). Osteogenic protein-1, a bone morphogenic protein member of the TGFβ superfamily, shares chemotactic but not fibrogenic properties with TGFβ. J. Cell. Physiol. 161, 562-570.

Postlethwaite, A.E. and Seyer, J.M. (1995). Identification of a chemotactic epitope in human transforming growth factor β1 spanning amino acid residues 368-374. J. Cell. Physiol. 164, 587-592.

Reardon, W., Winter, R.M., Rutland, P., Pulleyn, L.J., Jones, B.M., and Malcolm, S. (1994). Mutations in the fibroblast growth factor receptor 2 gene cause Crouzon syndrome. Nat. Genet. 8, 98-103.

Ripamonti, U, Ma, S.S., van den Heever, B., and Reddi, A.H. (1992). Osteogenin, a bone morphogenetic protein, adsorbed on porous hydroxyapatite substrate, induces rapid bone differentiation in calvarial defects of adult primates. Plast. Reconstr. Surg. 90, 382-393.

Roberts, A.B. and Sporn, M.B. (1990). The transforming growth factor βs. In: Peptide Growth Factors and Their Receptors. (Sporn, M.B. and Roberts, A.B., Eds.). pp. 419-472. Springer-Verlag, Heidelberg.

Rosen, C., Donahue, L.R., Hunter, S., Holick, M., Kavookjian, H., Kirschenbaum, A., Mohan, S., and Baylink, D.J. (1992). The 24/25-kDa serum insulinlike growth factor binding-protein is increased in elderly women with hip and spine fractures. J. Clin. Endocrinol. Metab. 74, 24-27.

Rosen, D., Miller, S.C., DeLeon, E., Thompson, A.Y., Bentz, H., Mathews, M., and Adams, S. (1994). Systemic administration of recombinant transforming growth factor β2 (rTGFβ2) stimulates parameters of cancellous bone formation in juvenile and adult rats. Bone 15, 355-359.

Rosenzweig, B.L., Imamura, T., Okadome, T., Cox, G.N., Yamashita, H., ten Dijke, P., Heldin, C.H., and Miyazono, K. (1995). Cloning and characterization of a human type-II receptor for bone morphogenetic proteins. Proc. Natl. Acad. Sci. USA 92, 7632-7636.

Ross, R., Raines, E.W., and Bowen-Pope, D.F. (1986). The biology of platelet derived growth factor. Cell 46, 155-169.

Rousseau, F., Bonaventure, J., Legeal-Mallet, L., Pelet, A., Rozet, J.M., Maroteaux, P., Le Merrer, M., and Munnich, A. (1994). Mutations in the gene encoding fibroblast growth factor receptor 3 in achondroplasia. Nature 371, 252-254.

Saksela, O., Moscatelli, D., and Rifkin, D. (1987). The opposing effects of basic fibroblast growth factor and transforming growth factor β on the regulation of plasminogen activator activity in capillary endothelial cells. J. Cell. Biol. 105, 957-963.

Schlessinger, J. and Lemmon, M. (1995). Regulation of growth factor activation by proteoglycans: What is the role of the low affinity receptors? Cell 83, 357-360.

Shiang, R., Thompson, L.M., Zhu, Y.Z., Church, D.M., Fielder, T.J., Bician, M., Winokur, S.T., and Wasmuth, J.J. (1994). Mutations in the transmembrane domain of FGFR3 cause the most common genetic form of dwarfism, achondroplasia. Cell 78, 335-342.

Shull, M.M., Ormsby, I., Kier, A.B., Pawlowski, S., Diebold, R.J., Yin, M., Allen, R., Sidman, C., Proetzel, G., Calvin, D., Annunziata, N., and Doetschman, T. (1992). Targeted disruption of the mouse transforming growth factor β1 gene results in multifocal inflammatory disease. Nature 359, 693-699.

Storm, E.E., Huynh, T.V., Copeland, H.G., Jenkins, N.A., Kingsley, D.M., and Lee, S.J. (1994). Limb alterations in brachypodism mice due to mutations in a new member of the TGFβ-superfamily. Nature 368, 639-643.

Takuwa, Y., Ohse, C., Wang, E.A., Wozney, J.M., and Yamashita, K. (1991). Bone morphogenetic protein-2 stimulates alkaline phosphatase activity and collagen synthesis in cultured osteoblastic cells, MC3T3-E1. Biochem. Biophy. Res. Commun. 174, 96-101.

Tashjian, Jr., A.H., Hohman, E.L., Antoniades, H.N., and Levine, L. (1982). Platelet-derived growth factor stimulates bone resorption via a prostaglandin mediated mechanism. Endocrinology 111, 118-124.

ten Dijke, P., Yamashita, H., Sampath, T.K., Reddi, A.H., Estevez, M., Riddle, D.L., Ichijo, H., Heldin, C.H., and Miyazono, K. (1994). Identification of type-I receptors for osteogenic protein-1 and bone morphogenetic protein-4. J. Biol. Chem. 269, 16985-16988.

Urist, M.R. (1965). Bone formation by autoinduction. Science 150, 893-899.

Vukicevic, S., Latin, V., Chen, P., Batorsky, R., Reddi, A.H., and Sampath, T.K. (1994). Localization of osteogenic protein-1 (bone morphogenetic protein-7) during human embryonic development: High affinity binding to basement membranes. Biochem. Biophys. Res. Commun. 198, 693-700.

Wang, X.-F., Lin, H.Y., Ng-Eaton, E., Downward,J., Lodish, H.F., and Weinberg, R.A. (1991). Expression cloning and characterization of the TGFβ type-III receptor. Cell 67, 797-805.

Wolfman, N.M., Celeste, A.J., Cox, K., Hattersly, G., Nelson, R., Yamaji, N., DiBlasio-Smith, E., Nove, J., Song, J.J., Wozney, J.M., and Rosen, V. (1995). Preliminary characterization of the biological activities of rhBMP-12. J. Bone Miner. Res. 10 (Suppl. 1), S148.

Wozney, J.M., Rosen, V., Celeste, A.J., Mitsock, L.M. Whitters, M.J., Kriz, R.W., Hewick, R.M., and Wang, E.A. (1988). Novel regulators of bone formation: Molecular clones and activities. Science 242, 1528-1534.

Wozney, J.M. (1992). The bone morphogenetic protein family and osteogenesis. Mol. Reprod. Devel. 32, 160-167.

Wozney, J.M. (1995). The potential role of bone morphogenetic proteins in periodontal reconstruction. J. Periodontol. 66, 506-510.

Wrana, J.L., Attisano, L., Wieser, R., Ventura, F., and Massagué, J. (1994). Mechanism of activation of the TGFβ receptor. Nature 370, 341-347.

Yamaguchi, A., Katagiri, T., Ikeda, T., Wozney, J.M., Rosen, V., Wang, E.A., Kahn, A.J., Suda, T., and Yoshiki, S. 1991). Recombinant human bone morphogenetic protein-2 stimulates osteoblastic maturation and inhibits myogenic differentiation in vitro. J. Cell Biol. 113, 681-687.

Yamaguchi, K., Shirakabe, K., Shibuya, H., Irie, K., Oishi, I., Ueno, N., Taniguchi, T., Nishida, E., and Matsumoto, K. (1995). Identification of a member of the MAPK family as a potential mediator of TGFβ signal transduction. Science 270, 2008-2011.

Yasko, A.W., Lane, J.M., Fellinger, E.J., Rosen V., Wozney, J.M., and Wang, E.A. (1992). The healing of segmental bone defects, induced by recombinant human bone morphogenetic protein (rhBMP-2). A radiographic, histological, and biomechanical study in rats. [Published erratum appears in J. Bone Joint Surg. Am. 1992, 74, 1111] J. Bone Joint Surg. Am. 74, 659-670.

SYSTEMIC CONTROL OF BONE FORMATION

Toshio Matsumoto and Yasuhiro Takeuchi

I. INTRODUCTION

Bone formation is a complex process involving multiple steps: 1) recruitment and proliferation of osteoblastic cells on the surface of bone, 2) synthesis and accumulation of extracellular matrix by osteoblasts, and 3) mineralization of newly formed extracellular matrix. Using primary cultures of osteoblastic cells, it was demonstrated that the various functions of osteoblastic cells during the bone formation

Advances in Organ Biology
Volume 5B, pages 615-625.

ISBN: 0-7623-0390-5

process are dependent upon the differentiation stages of these cells (Stein and Lian, 1993). Thus, various phenotypes develop in a sequential manner with the differentiation of osteoblastic cells, and factors that regulate the bone formation process exert their effects by affecting the recruitment and/or differentiation of osteoblasts during this process.

Systemic factors that influence bone formation include calciotropic hormones, sex hormones, and aging. These factors may directly regulate the bone formation process, but many of their actions are mediated by affecting the production and/or the actions of local cytokines. Systemic factors and local cytokines can also modulate the synthesis of extracellular matrix, and matrix proteins are shown to affect osteoblast functions by directly modulating osteoblastic differentiation or by modulating the actions of local cytokines. Thus, the effects of systemic factors on bone formation have to be viewed in the context that they may be mediated by alterations in the cell-matrix interactions as well as the production and actions of local cytokines.

II. CALCIUM-REGULATING HORMONES AND BONE FORMATION

A. Parathyroid Hormone

Parathyroid hormone (PTH) stimulates osteoclastic bone resorption via its effect on osteoblasts. However, intermittent injections, but not continuous infusion, of PTH markedly increase bone volume by enhancing bone formation. In the rat, the stimulatory effect of PTH injection on bone formation can be observed in both cortical and trabecular bone (Ejersted et al., 1993; Jerome, 1994), and causes an increase in the volume and strength of femoral neck and vertebral bone (Sogaard et al., 1994). However, in women under medical oophorectomy by gonadotropin-releasing hormone analogue, PTH injections significantly increased vertebral bone mineral density (BMD) but did not increase femoral neck BMD (Finkelstein et al., 1994). The effects of PTH on bone formation are associated with an increase in the number of bone marrow osteoblast progenitor cells as well as an increase in the ratio of alkaline phosphatase-(ALP) positive preosteoblastic cells (Nishida et al., 1994). Histological examinations using autoradiography of thymidine-labeled cells revealed that PTH does not cause proliferation of progenitor cells but appears to stimulate the activation of lining cells to increase the number of osteoblastic cells (Dobnig and Turner, 1995).

PTH has been shown to increase the production of insulinlike growth factor-(IGF) I through a cyclic AMP-dependent mechanism in primary cultures of rat osteoblasts. Although intermittent injections of PTH to rats did not increase serum IGF-I concentration, there was an increase in bone matrix-associated IGF-I and transforming growth factor β (TGFβ) (Pfeilschifter et al., 1995). Therefore, these growth factors may act locally to play a role in the stimulation of bone formation by PTH. However, PTH does not stimulate the production of IGF-I in cultured human bone cells (Okazaki et al., 1995).

It has been reported that subcutaneous administration of human PTH-(1-84) induces a rapid, transient, and sequential expression of the proto-oncogene *c-fos* mRNA in osteoblasts, chondrocytes, and bone marrow stromal cells within 15 to 60 minutes (Lee et al., 1994). An *in vitro* study using an osteoblastlike cell line demonstrated that the effect of PTH on *c-fos* expression is mimicked by protein kinase C activation and is blocked by an inhibitor of protein kinase C, H-7 (Kano et al., 1994), although contradictory results have also been reported (Clohisy et al., 1992). Further studies are needed to clarify the role as well as the mechanism of *c-fos* expression in mediating the anabolic effect of PTH in bone.

B. Vitamin D

Vitamin D is essential for the formation of mineralized bone, and vitamin D deficiency causes an impairment of bone mineralization. Nevertheless, the role of vitamin D and its active metabolite, 1,25-dihydroxyvitamin D [1,25$(OH)_2$D], in bone formation process has not been clear. Using primary cultures of rat osteoblastlike cells, it has been demonstrated that 1,25$(OH)_2$D can both positively and negatively regulate expression of osteoblast phenotypic markers, depending upon the differentiation stage of the osteoblast (Owen et al., 1991). Thus, in premature osteoblastic cells 1,25$(OH)_2$D promotes the differentiation to obtain properties for synthesizing bone matrix. In contrast, in more mature osteoblasts 1,25$(OH)_2$D enhances the expression of osteocalcin and promotes further maturation of osteoblasts to mineralize bone matrix, but does not stimulate matrix collagen synthesis.

Matrix vesicle formation by osteoblasts is important for the mineralization of bone matrix. Matrix vesicle membranes are rich in acidic phospholipids, and these phospholipids, especially phosphatidyl serine, form complexes with calcium and phosphate. Such phospholipid-calcium-phosphate complexes are thought to be an initial step for hydroxyapatite formation (Kirsch et al., 1994). In osteoblastlike UMR 106 cells, 1,25$(OH)_2$D stimulates the synthesis of phosphatidyl serine (Matsumoto et al., 1985). Collectively, the effects of 1,25$(OH)_2$D on osteoblasts appear to promote the terminal differentiation of osteoblasts into mature phenotypes.

Because the mineralization defect in severely vitamin D-deficient rats can be mostly reversed by continuous infusion of calcium and phosphate, it has been postulated that the effect of 1,25$(OH)_2$D on bone mineralization can be explained solely by its effect on intestinal calcium and phosphate absorption (Weinstein et al., 1984). However, in view of the fact that 1,25$(OH)_2$D promotes the differentiation of osteoblasts into more mature cells with mineralizing capability, there is a possibility that 1,25$(OH)_2$D plays a role in the maintenance of bone formation via its effect on osteoblastic differentiation.

III. SEX HORMONES AND BONE FORMATION

A. Estrogen

Estrogen has a potent effect to inhibit bone resorption, and the maintenance of bone mass by estrogen replacement in postmenopausal women is thought to be due to the suppression of bone resorption. There are controversies about the mechanism whereby estrogen loss causes enhanced osteoclastogenesis and bone resorption. The increase in osteoclastogenesis and bone resorption in ovariectomized mice was inhibited by a blocking antibody against interleukin-(IL) 6 (Jilka et al., 1992). In addition, estrogen inhibited the production of IL-6 from some bone marrow stromal cells by inhibiting the transcription of the IL-6 gene via an estrogen-receptor-mediated mechanism (Girasole et al., 1992; Manolagas and Jilka, 1995).

Through these observations, it has been hypothesized that the direct stimulation of IL-6 gene transcription in stromal cells by estrogen deficiency causes an increase in IL-6 production, and IL-6 stimulates osteoclast formation and bone resorption. The effect of IL-6 is mediated by heterodimer formation of IL-6-bound IL-6 receptor and glycoprotein (gp) 130, and gp130 transduces signals of not only IL-6 but also many other cytokines including IL-11 and leukemia inhibitory factor. Although stromal cells and osteoblasts have gp130 but not IL-6 receptors, circulating soluble receptors for IL-6 can transduce the signal via gp130 in these cells (Tamura et al., 1993). The production of IL-1 from peripheral blood mononuclear cells is also reported to be enhanced in postmenopausal women (Pacifici et al., 1991). Administration of IL-1 receptor antagonist prevented the late bone loss in ovariectomized rats (Kimble et al., 1994), while the inhibition of both IL-1 and tumor necrosis factor (TNF) actions was required to completely prevent the early loss of bone after ovariectomy (Kimble et al., 1995).

There are also conflicting observations about the effect of estrogen on bone formation. Westerlind et al. (1993) demonstrated that estrogen decreases histological parameters related to cancellous bone formation in growing rats, and suggested that estrogen does not have an anabolic action on bone formation but has a pronounced inhibitory action on bone turnover. Manolagas and colleagues reported that the numbers of osteoblast progenitors, colony forming unit-(CFU) fibroblast, and of CFU with mineralizing capacity, CFU-osteoblast, increased in ovariectomized mice (Manolagas and Jilka, 1995). Because the expression of gp130 in stromal cells is inhibited by estrogen, and because cytokines that have osteoblastogenic properties such as leukemia inhibitory factor also transduce their signal via gp130, they speculate that the upregulation of these cytokine signals in osteoblastic cells may cause the stimulation of osteoblast formation in estrogen loss (Manolagas and Jilka, 1995). In contrast, administration of estrogen to ovariectomized rats in which bone resorption was suppressed by a bisphosphonate, pamidronate, showed a dose-dependent increase in trabecular bone volume

(Chow et al., 1992). Thus, estrogen appears to maintain bone volume not only through inhibition of bone resorption, but also through stimulation of bone formation.

Receptors for estrogen are present in osteoblasts (Eriksen et al., 1988; Komm et al., 1988). In addition, estrogen stimulates the expression of TGFβ in human osteoblastlike cells (Komm et al., 1988) and IGF-I in rat osteosarcoma cells (Ernst et al., 1989). These effects of estrogen on growth factors may act as anabolic stimulation to bone formation process. However, further studies are required to elucidate the influence of estrogen on bone formation.

B. Androgen

As discussed later, osteoporosis in aged males is usually presented with reduced turnover of bone. Although a reduction in androgen level appears to play a role in the age-related bone loss in men (Wishart et al., 1995), the effect as well as the mechanism of action of androgen on bone is poorly understood. A histomorphometric analysis in hypogonadal men and eugonadal idiopathic male osteoporotic patients revealed that in the hypogonadal men, osteoblast surface, mineralizing surface, and formation rate were modestly increased, and fell after therapy with testosterone and a calcium supplement (Jackson et al., 1987). In contrast, the formation indices were significantly reduced in the idiopathic group. Bellido et al. (1995) reported that orchidectomy in mice caused an increase in the replication of osteoclast progenitors in the bone marrow which could be prevented by androgen replacement. These changes after orchidectomy were inhibited by administration of androgen or an IL-6 neutralizing antibody. In addition, androgen inhibited the activity of human IL-6 gene promoter in HeLa cells cotransfected with androgen receptor. These results suggested that androgen, acting through its receptor, inhibits the expression of the IL-6 gene, and IL-6 mediates the upregulation of osteoclastogenesis and bone loss caused by androgen deficiency as in estrogen deficiency. Goulding et al. examined the effects of medical oophorectomy by a luteinizing hormone-releasing hormone analogue, buserelin, and an anti-androgen, flutamide, on bone turnover of female rats (Goulding and Gold, 1993). The results demonstrated that bone loss caused by buserelin-mediated estrogen deficiency was due to increased bone resorption, but bone thinning caused by flutamide-mediated androgen deficiency was caused principally by reduced bone formation. Thus, it was suggested that adequate androgen action is required for the maintenance of normal bone turnover not only in male rats but also in female rats with normal estrogen status.

The heterogeneity in bone turnover and histomorphometric changes after androgen deficiency among these clinical observations or animal experiments could be due to differences in the age, sex or other endocrinological backgrounds. Therefore, assessment of the effect of androgen on bone appears to require the evaluation of multiple factors interacting with androgen in bone.

IV. AGING AND BONE FORMATION

Major features of age-related bone loss is a reduction in bone formation, and bones from aged rats appear to respond poorly to physical stress (Rubin et al., 1992) or marrow ablation (Liang et al., 1992). Both are strong osteogenic stimuli to young adult skeleton. Note that both the number and function of osteoblasts have been shown to be reduced with aging. The number of adherent colony-forming cells in bone marrow of aged rats is significantly reduced, and the percentage of ALP-positive cell colonies was also lower in aged rats (Quarto et al., 1995).

An ultrastructural analysis of the trabecular bone of rat tibia has demonstrated that there is a 10-fold decrease in the number of osteoblasts per unit length and a 15-fold reduction in the relative bone formation surface per total bone surface, while the number of preosteoblastic cells is not altered (Roholl et al., 1994). Because there was much less decline in resorption surface as well as the number of osteoclasts and their precursors, the authors concluded that the main factor causing age-related bone loss is a diminished maturation of preosteoblasts to osteoblasts. Aging per se also causes a reduction in the functions of osteoblasts. Fedarko et al. (1992) analyzed the ability of human bone cells to synthesize various matrix proteins in culture, and found that the synthesis of collagen and decorin was reduced with age by one-third from the peak level. The age-related reduction in the synthesis of matrix protein was also demonstrated *in vivo.* The expression of type I collagen in periosteal cells from the rat long bone is reported to be reduced with aging, and there is a strong correlation between collagen mRNA level and periosteal bone formation rate (Turner and Spelsberg, 1991). A study using *in situ* hybridization demonstrated that the expression of type I collagen, osteocalcin, and osteopontin mRNA was markedly reduced in the femur of aged rats, and both the activity and the number of osteoblasts expressing type I collagen were reduced by aging (Ikeda et al., 1995). The reduction in collagen synthesis may in turn affect the differentiation and functions of osteoblasts. It has been shown that the increase in ALP activity after long-term cultures of osteoblastic cells requires the presence of ascorbic acid and is dependent upon collagen synthesis (Franceschi et al., 1994).

Our recent observations demonstrate that the effect of matrix collagen on osteoblastic differentiation is mediated by an interaction between the DGEA motif on collagen molecules and $\alpha2\beta1$ integrin on osteoblasts (Takeuchi et al., 1996). The interaction between matrix collagen and osteoblasts also downregulates responsiveness to TGFβ by reducing cell-surface TGFβ receptors (Takeuchi et al., 1996). Because TGFβ has a strong inhibitory effect on osteoblastic differentiation, the accumulation of collagen matrix allows osteoblasts to escape from the inhibitory effect of TGFβ on osteoblastic differentiation, and mineralization of formed matrix can be completed by further differentiated osteoblasts. Thus, the age-related changes in matrix synthesis can also affect the function of osteoblasts (Figure 1).

Growth factors play an important role in the maintenance of the recruitment and functions of osteoblasts, and aging causes a reduction in the synthesis and/or

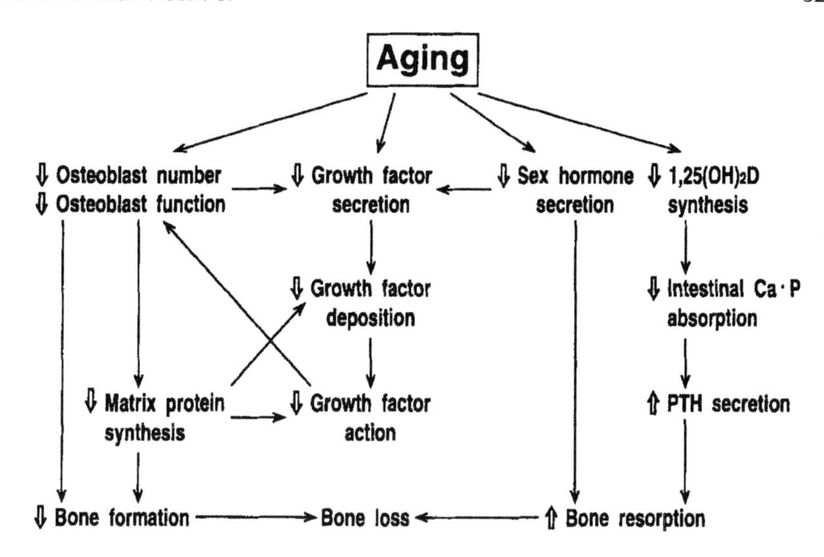

Figure 1. Postulated mechanism of bone loss by aging.

deposition of local growth factors. The content of IGF-I and TGFβ in human bone is reported to decline with aging, and there was a net loss of 60% of IGF-I and 25% of TGFβ from 20 to 60 years of age (Nicolas et al., 1994). No difference was found in the skeletal content of these growth factors between males and females. In addition, there was an age-related decrease in the concentration of IGF binding protein-(IGFBP) 5 in human femoral cortical bone, but no reduction was observed in the skeletal content of IGFBP-3 (Nicolas et al., 1995). In contrast, Johansson et al. (1994) found a strong correlation between BMD at all sites and serum IGFBP-3 concentration in healthy males between 25 and 59 years of age. Because serum IGFBP-3 level is dependent upon growth hormone, and growth hormone secretion is reduced with aging, the authors suggested that the reduction in not only growth hormone but also IGFBP-3 has a role in age-related bone loss. In addition to the reduction in the amount of growth factors in bone, there is a possibility that their actions may also be reduced (Figure 1).

We found that decorin, a proteoglycan rich in bone, binds TGFβ and the binding of decorin to TGFβ enhances the actions of TGFβ in osteoblasts by increasing the binding of TGFβ to its receptors (Takeuchi et al., 1994). Because the synthesis of decorin is enhanced by TGFβ, but is reported to be reduced with aging (Fedarko et al., 1992), a vicious cycle may be created in aging bone: the age-related reduction in the matrix accumulation of TGFβ reduces decorin synthesis which further decreases the actions of TGFβ in osteoblasts.

Although the reduction in bone formation is the principal mechanism for the age-related bone loss, aging is associated with alterations in the secretion and/or the actions of many systemic hormones. Therefore, depending upon the relative impact of these multiple factors on bone turnover, clinical features of each individual may differ (Figure 1). Sex hormone secretion is reported to decrease with age in males as well (Vermeulen, 1991). In parallel with the fall in free androgen index, bone loss in men appears to accelerate from age 50, and is associated with decreased bone formation as indicated by a reduction in bone formation markers (Wishart et al., 1995). These features are different from those of postmenopausal women, and there appears to be a difference in cortical cross-sectional dimensions and trabecular microarchitecture between males and females. Cross-sectional area is greater and vertebral body size is larger in men due to continuous periosteal growth, whereas no age-related compensatory mechanism is observed in women (Mosekilde and Mosekilde, 1990). Such a geometric change may compensate for the age-related loss of bone density in males. A sex difference in the age-related changes in vertebral trabecular architecture has also been observed (Mosekilde, 1989). Thus, females show a higher tendency to perforation especially of the horizontal elements, while age-related reduction in trabecular thickness is observed in both sexes. These architectural differences between females and males may also have a significant influence on the biomechanical properties.

The secretion of $1,25(OH)_2D$ from the kidney in response to PTH is reduced in elderly osteoporotic patients (Slovik et al., 1981). The actions of $1,25(OH)_2D$ on the intestinal calcium and phosphate transport are also reduced by aging due to a decrease in the intestinal vitamin D receptors (Takamoto et al., 1990). Thus, elderly subjects have an inability to adapt to the low-calcium diets, and a reduction in calcium intake which is commonly observed in the elderly can cause secondary hyperparathyroidism. Under these circumstances, an increase in bone resorption superimposes the reduction in bone formation (Figure 1). Decrease in muscle mass and mechanical stress can also influence the remodeling activity of bone. In order to analyze such complex changes in bone metabolism with aging, it is important to evaluate bone resorptive and formation processes separately in each elderly subject. Recent progress in the development of sensitive biochemical markers of bone metabolism should facilitate our understanding of the changes in bone metabolism with aging (Garnero et al., 1994a,b).

REFERENCES

Bellido, T., Jilka, R.L., Boyce, B.F., Girasole, G., Broxmeyer, H., Dalrymple, S.A., Murray, R., and Manolagas, S.C. (1995). Regulation of interleukin-6, osteoclastogenesis, and bone mass by androgens. The role of the androgen receptor. J. Clin. Invest. 95, 2886-2895.

Chow, J., Tobias, J.H., Colston, K.W. and Chambers, T.J. (1992). Estrogen maintains trabecular bone volume in rats not only by suppression of bone resorption but also by stimulation of bone formation. J. Clin. Invest. 89, 74-78.

Clohisy, J.C., Scott, D.K., Brakenhoff, K.D., Quinn, C.O., and Partridge, N.C. (1992). Parathyroid hormone induces c-fos and c-jun messenger RNA in rat osteoblastic cells. Mol. Endocrinol. 6 , 1834-1842.

Dobnig, H. and Turner, R.T. (1995). Evidence that intermittent treatment with parathyroid hormone increases bone formation in adult rats by activation of bone lining cells. Endocrinology 136, 3632-3638.

Ejersted, C., Andreassen, T.T., Oxlund, H., Jorgensen, P.H., Bak, B., Haggblad, J., Torring, O., and Nilsson, M.H. (1993). Human parathyroid hormone (1-34) and (1-84) increase the mechanical strength and thickness of cortical bone in rats. J. Bone Miner. Res. 8, 1097-1101.

Eriksen, E.F., Colvard, D.S., Berg, N.J., Graham, M.L., Mann, K.G., Spelsberg, T.C., and Riggs, B.L. (1988). Evidence of estrogen receptors in normal human osteoblastlike cells. Science (Wash., D.C.) 241, 84-86.

Ernst, M., Heath, J.K., and Rodan, G.A. (1989). Estradiol effects on proliferation, messenger ribonucleic acid for collagen and insulinlike growth factor-I, and parathyroid hormone-stimulated adenylate cyclase activity in osteoblastic cells from calvariae and long bones. Endocrinology 125, 825-833.

Fedarko, N.S., Vetter, U.K., Weinstein, S., and Robey, P.G. (1992). Age-related changes in hyaluronan, proteoglycan, collagen, and osteonectin synthesis by human bone cells. J. Cell. Physiol. 151, 215-227.

Finkelstein, J.S., Klibanski, A., Schaefer, E.H., Hornstein, M.D., Schiff, I., and Neer, R.M. (1994). Parathyroid hormone for the prevention of bone loss induced by estrogen deficiency. N. Eng. J. Med. 331, 1618-1623.

Franceschi, R.T., Iyer, B.S., and Cui, Y. (1994). Effects of ascorbic acid on collagen matrix formation and osteoblast differentiation in murine MC3T3-E1 cells. J. Bone Miner. Res. 9, 843-854.

Garnero, P., Gineyts, E., Riou, J.P., and Delmas, P.D. (1994a). Assessment of bone resorption with a new marker of collagen degradation in patients with metabolic bone disease. J. Clin. Endocrinol. Metab. 79, 780-785.

Garnero, P., Shih, W.J., Gineyts, E., Karpf, D.B., and Delmas, P.D. (1994b). Comparison of new biochemical markers of bone turnover in late postmenopausal osteoporotic women in response to alendronate treatment. J. Clin. Endocrinol. Metab. 79, 1693-1700.

Girasole, G., Jilka, R.L., Passeri, G., Boswell, S., Boder, G., Williams, D.C., and Manolagas, S.C. (1992). 17β-estradiol inhibits interleukin-6 production by bone marrow-derived stromal cells and osteoblasts in vitro: A potential mechanism for the antiosteoporotic effect of estrogens. J. Clin. Invest. 89, 883-891.

Goulding, A. and Gold, E. (1993). Flutamide-mediated androgen blockade evokes osteopenia in the female rat. J. Bone Miner. Res. 8, 763-769.

Ikeda, T., Nagai, Y., Yamaguchi, A., Yokose, S., and Yoshiki, S. (1995). Age-related reduction in bone matrix protein mRNA expression in rat bone tissues: Application of histomorphometry to in situ hybridization. Bone 16, 17-23.

Jackson, J.A., Kleerekoper, M., Parfitt, A.M., Rao, D.S., Villanueva, A.R., and Frame, B. (1987). Bone histomorphometry in hypogonadal and eugonadal men with spinal osteoporosis. J. Clin. Endocrinol. Metab. 65, 53-58.

Jerome, C.P. (1994). Anabolic effect of high doses of human parathyroid hormone (1-38) in mature intact female rats. J. Bone Miner. Res. 9, 933-942.

Jilka, R.L., Hangoc, G., Girasole, G., Passeri, G., Williams, D.C., Abrams, J.S., Boyce, B., Broxmeyer, H., and Manolagas, S.C. (1992). Increased osteoclast development after estrogen loss: Mediation by interleukin-6. Science (Wash. D.C.) 257, 88-91.

Johansson, A.G., Baylink, D.J., af Ekenstam, E., Lindh, E., Mohan, S., and Ljunghall, S. (1994). Circulating levels of insulinlike growth factor-I and -II, and IGF-binding protein 3 in inflammation and after parathyroid hormone infusion. Bone Miner. 24, 25-31.

Kano, J., Sugimoto, T., Kanatani, M., Kuroki, Y., Tsukamoto, T., Fukase, M., and Chihara, K. (1994). Second messenger signaling of *c-fos* gene induction by parathyroid hormone (PTH) and PTH-related peptide in osteoblastic osteosarcoma cells: Its role in osteoblast proliferation and osteoclastlike cell formation. J. Cell. Physiol. 161, 358-366.

Kimble, R.B., Matayoshi, A.B., Vannice, J.L., Kung, V.T., Williams, C., and Pacifici, R. (1995). Simultaneous block of interleukin-1 and tumor necrosis factor is required to completely prevent bone loss in the early postovariectomy period. Endocrinology 136, 3054-3061.

Kimble, R.B., Vannice, J.L., Bloedow, D.C., Thompson, R.C., Hopfer, W., Kung, V.T., Brownfield, C., and Pacifici, R. (1994). Interleukin-1 receptor antagonist decreases bone loss and bone resorption in ovariectomized rats. J. Clin. Invest. 93, 1959-1967.

Kirsch, T., Ishikawa, Y., Mwale, F., and Wuthier, R.E. (1994). Roles of the nucleational core complex and collagens (types-II and -X) in calcification of growth-plate cartilage matrix vesicles. J. Biol. Chem. 269, 20103-20109.

Komm, B.S., Terpening, C.M., Benz, D.J., Graeme, K.A., Gallegos, A., Korc, M., Greene, G. L., O'Malley, B.W., and Haussler, M.R. (1988). Estrogen-binding, receptor mRNA, and biologic response in osteoblastlike osteosarcoma cells. Science (Wash., D.C.) 241, 81-84.

Lee, K., Deeds, J.D., Chiba, S., Un-No, M., Bond, A.T., and Segre, G.V. (1994). Parathyroid hormone induces sequential c-fos expression in bone cells in vivo: In situ localization of its receptor and c-fos messenger ribonucleic acids. Endocrinology 134, 441-450.

Liang, C.T., Barnes, J., Seedor, J.G., Quartuccio, H.A., Bolander, M., Jeffrey, J.J., and Rodan, G.A. (1992). Impaired bone activity in aged rats: Alterations at the cellular and molecular levels. Bone 13, 435-441.

Manolagas, S.C. and Jilka, R.L. (1995). Bone marrow, cytokines, and bone remodeling. Emerging insights into the pathophysiology of osteoporosis. N. Eng. J. Med. 332, 305-311.

Matsumoto, T., Kawanobe, Y., Morita, K., and Ogata, E. (1985). Effect of 1,25-dihydroxyvitamin D, on phospholipid metabolism in a clonal osteoblastlike rat osteogenic sarcoma cell line. J. Biol. Chem. 260, 13704-13709.

Mosekilde, L. (1989). Sex differences in age-related loss of vertebral trabecular bone mass and structure—biomechanical consequences. Bone 10, 425-432.

Mosekilde, L. and Mosekilde, L. (1990). Sex differences in age-related changes in vertebral body size, density, and biomechanical competence in normal individuals. Bone 11, 67-73.

Nicolas, V., Mohan, S., Honda, Y., Prewett, A., Finkelman, R.D., Baylink, D.J., and Farley, J.R. (1995). An age-related decrease in the concentration of insulinlike growth factor binding protein-5 in human cortical bone. Calcif. Tissue Int. 57, 206-212.

Nicolas, V., Prewett, A., Bettica, P., Mohan, S., Finkelman, R.D., Baylink, D.J., and Farley, J.R. (1994). Age-related decreases in insulinlike growth factor-I and transforming growth factor β in femoral cortical bone from both men and women: Implications for bone loss with aging. J. Clin. Endocrinol. Metab. 78, 1011-1016.

Nishida, S., Yamaguchi, A., Tanizawa, T., Endo, N., Mashiba, T., Uchiyama, Y., Suda, T., Yoshiki, S., and Takahashi, H.E. (1994). Increased bone formation by intermittent parathyroid hormone administration is due to the stimulation of proliferation and differentiation of osteoprogenitor cells in bone marrow. Bone 15, 717-723.

Okazaki, R., Durham, S.K., Riggs, B.L., and Conover, C.A. (1995). Transforming growth factor-β and forskolin increase all classes of insulinlike growth factor-I transcripts in normal human osteoblastlike cells. Biochem. Biophys. Res. Commun. 207, 963-970.

Owen, T.A., Aronow, M.S., Barone, L.M., Bettencourt, B., Stein, G.S., and Lian, J.B. (1991). Pleiotropic effects of vitamin D on osteoblast gene expression are related to the proliferative and differentiated state of the bone cell phenotype: Dependency upon basal levels of gene expression, duration of exposure, and bone matrix competency in normal rat osteoblast cultures. Endocrinology 128, 1496-1504.

Pacifici, R., Brown, C., Puscheck, E., Friedrich, E., Slatopolsky, E., Maggio, D., McCracken, R., and Avioli, L.V. (1991). Effect of surgical menopause and estrogen replacement on cytokine release from human blood mononuclear cells. Proc. Natl. Acad. Sci. USA 88, 5134-5138.

Pfeilschifter, J., Laukhuf, F., Muller-Beckmann, B., Blum, W.F., Pfister, T. and Ziegler, R. (1995). Parathyroid hormone increases the concentration of insulinlike growth factor-I and transforming growth factor β1 in rat bone. J. Clin. Invest. 96, 767-774.

Quarto, R., Thomas, D. and Liang, C.T. (1995). Bone progenitor cell deficits and the age-associated decline in bone repair capacity. Calcif. Tissue Int. 56, 123-129.

Roholl, P.J., Blauw, E., Zurcher, C., Dormans, J.A., and Theuns, H.M. (1994). Evidence for a diminished maturation of preosteoblasts into osteoblasts during aging in rats: An ultrastructural analysis. J. Bone Miner. Res. 9, 355-366.

Rubin, C.T., Bain, S.D., and McLeod, K.J. (1992). Suppression of the osteogenic response in the aging skeleton. Calcif. Tissue Int. 50, 306-313.

Slovik, D.M., Adams, J.S., Neer, R.M., Holick, M.F. and Potts, Jr., J. (1981). Deficient production of 1,25-dihydroxyvitamin D in elderly osteoporotic patients. N. Eng. J. Med. 305, 372-374.

Sogaard, C.H., Wronski, T.J., McOsker, J.E. and Mosekilde, L. (1994). The positive effect of parathyroid hormone on femoral neck bone strength in ovariectomized rats is more pronounced than that of estrogen or bisphosphonates. Endocrinology 134, 650-657.

Stein, G.S. and Lian, J.B. (1993). Molecular mechanisms mediating proliferation/differentiation interrelationships during progressive development of the osteoblast phenotype. Endocrine Rev. 14, 424-442.

Takamoto, S., Seino, Y., Sacktor, B., and Liang, C.T. (1990). Effect of age on duodenal 1,25-dihydroxyvitamin D-3 receptors in Wistar rats. Biochim. Biophys. Acta 1034, 22-28.

Takeuchi, Y., Kodama, Y., and Matsumoto, T. (1994). Bone matrix decorin binds transforming growth factor-β and enhances its bioactivity. J. Biol. Chem. 269, 32634-32638.

Takeuchi, Y., Nakayama, K., and Matsumoto, T. (1996). Differentiation and cell surface expression of transforming growth factor-β receptors are regulated by interaction with matrix collagen in murine osteoblastic cells. J. Biol. Chem. 271, 3938-3944.

Tamura, T., Udagawa, N., Takahashi, N., Miyaura, C., Tanaka, S., Yamada, Y., Koishihara, Y., Ohsugi, Y., Kumaki, K., Taga, T., Kishimoto, T., and Suda, T. (1993). Soluble interleukin-6 receptor triggers osteoclast formation by interleukin 6. Proc. Natl. Acad. Sci. USA 90, 11924-11928.

Turner, R.T. and Spelsberg, T.C. (1991). Correlation between mRNA levels for bone cell proteins and bone formation in long bones of maturing rats. Am. J. Physiol. 261, E348-E353.

Vermeulen, A. (1991). Androgens in the aging male. J. Clin. Endocrinol. Metab. 73, 221-224.

Weinstein, R.S., Underwood, J.L., Hutson, M.S., and DeLuca, H.F. (1984). Bone histomorphometry in vitamin D-deficient rats infused with calcium and phosphorus. Am. J. Physiol. 246, E499-E505.

Westerlind, K.C., Wakley, G.K., Evans, G.L., and Turner, R.T. (1993). Estrogen does not increase bone formation in growing rats. Endocrinology 133, 2924-2934.

Wishart, J.M., Need, A.G., Horowitz, M., Morris, H.A., and Nordin, B.E. (1995). Effect of age on bone density and bone turnover in men. Clin. Endocrinol. 42, 141-146.

THE DIRECT AND INDIRECT EFFECTS OF ESTROGEN ON BONE FORMATION

Timothy J. Chambers

I. INTRODUCTION

The central role of estrogen in the regulation of bone mass is well established. Estrogen is crucial to the attainment of normal skeletal mass during development, and to the maintenance of this mass thereafter. Thus, there is a rapid increase in bone density in the vertebrae of females during puberty (Gilsanz et al., 1991), while mutational inactivation of the estrogen receptor results in a low bone mass (Lubahn et al., 1993; Smith et al., 1994). Estrogen deficiency in adulthood, whether natural or pathological, is associated with bone loss, which can be prevented by estrogen administration (Christiansen et al., 1982; Stock et al., 1985; Lindsay, 1987; Wronski et al., 1988b; Stevenson et al., 1989; Kalu, 1991).

Advances in Organ Biology
Volume 5B, pages 627-638.

ISBN: 0-7623-0390-5

It is generally held that estrogen enhances and supports bone mass through suppression of bone resorption. Certainly bone loss in hypoestrogenic states is associated with increased bone resorption, and suppression of resorption by estrogen or other antiresorptive agents prevents osteopenia. It could be argued that the pubertal increase in bone mass by estrogen is also explicable as an effect on bone resorption.

Theoretically, increased bone formation would also increase bone mass, but the prevailing view is that estrogen does not stimulate bone formation. Rather, it is argued that estrogen suppresses bone formation, because estrogen deprivation stimulates, and estrogen administration inhibits this (Wronski et al., 1988a). However, other antiresorptive agents such as bisphosphonates and calcitonin also reverse the increase in bone formation that occurs after estrogen deficiency. For calcitonin at least, the lack of hormone receptors on osteoblastic cells makes it clear that the reduction in bone formation is caused by the reduction in bone resorption. If an antiresorptive action can explain the reduction in bone formation by calcitonin, it can presumably also explain the reduction in bone formation by estrogen. Thus, suppression of bone formation by estrogen is secondary to the strong ability of estrogen to suppress bone resorption. To identify a primary hormonal action of estrogen on bone formation, we need to measure formation under conditions in which such underlying, primary effects on bone formation can be distinguished from the changes in formation that are secondary to changes in bone resorption. These secondary changes are considered to be brought about by the coupling that occurs between resorption and formation.

II. MECHANISMS OF COUPLING BETWEEN THE RESORPTION AND FORMATION OF BONE

There is much evidence that in adult man, under physiological circumstances, the resorption and formation of bone are normally coupled such that an increase in bone resorption leads to an increase in bone formation, and decreased bone resorption causes suppression of bone formation (Christiansen et al., 1982; Parfitt, 1982; Jaworski, 1984; Richelson et al., 1984; Stock et al., 1985; Stepan et al., 1987). A clear microanatomic basis for this coupling has been demonstrated: bone formation follows bone resorption, and occurs at the same site (Hattner et al., 1965).

The mechanisms by which bone resorption and formation are coupled is unknown. There have been several hypotheses which clearly have to take into account the microanatomical basis for the process. The models are not mutually exclusive but include those that invoke increased mechanical signals generated by resorption-weakened bone, the action of growth factors released by bone matrix during resorption, and a series of cellular interactions between osteoclasts and osteoblasts and their precursors, akin to a developmental process (Rasmussen and Bordier, 1974; Parfitt, 1983; Mohan et al., 1988; Rodan, 1991; Turner, 1991).

It has been shown that the trabecular bone of the rat also shows coupling between resorption and formation, such that increased bone formation follows increased bone resorption, and suppression of bone resorption leads to decreased bone formation (Baylink et al., 1969; Wronski et al., 1988a,b, 1991; Hayashi et al., 1989; Kalu, 1991; Seedor et al., 1991; Stein et al., 1991). However the microanatomic basis for this is less clear. Under the special (nonphysiological) circumstances of tooth egression it has been shown that in rat alveolar bone, an episode of bone resorption is followed by an episode of bone formation (Vignery and Baron, 1980; Van Tran et al., 1982). In rat trabecular bone it has been shown that formation can occur at resorptive surfaces (Baron et al., 1984), but it was not determined whether this occurs more commonly, compared to nonresorptive surfaces, than would be expected by chance.

We investigated the microanatomic basis for coupling in rat bone (Chow et al. 1993a) using a strategy similar to that of Hattner et al. (1965). We analyzed the nature of the contour of cement lines in trabecular bone of the secondary spongiosa of the tibia, a bone that shows coupling between resorption and formation, and compared this with the contour of the surface of the same trabeculae. Cement, or reversal lines, are laid down prior to bone formation, and the contour of the cement line thus represents evidence of the type of surface upon which bone formation occurred (Hattner et al., 1965). Thus, if new bone is formed on a site of previous resorption, the cement line shows a crenated or scalloped appearance, similar to that seen on resorbed bone surfaces. If, however, the bone is laid down on a nonresorbed or quiescent surface, the cement line is noncrenated. We found that the majority of cement lines in the trabecular bone of rats of a wide range of maturity, from adulthood to senescence, exhibit a noncrenated contour. This contrasts with the small proportion of noncrenated cement lines in human trabecular bone and suggests that, in the trabecular bone of normal rats, bone formation does not generally occur on the site of previous resorption.

Unlike adult man, the skeleton of the rat, at least at the stage normally used for study (before epiphyseal closure), differs from that of man in that the bones are still growing. The bone formation that occurs as part of this modeling process is not coupled to bone resorption. It is thus likely that at least a component of the bone formation that occurs in the trabecular bone of the rat tibia is part of the modeling process. However, potent inhibitors of resorption suppress bone formation in trabecular bone by an order of magnitude, with an insignificant effect on bone formation in areas of bone modeling activities (Marie et al., 1985; Wronski et al., 1989b; Abe et al., 1992; Chow et al., 1992a). This suggests that the major proportion of bone formation in the tibial metaphysis is coupled to bone resorption. Our results (Chow et al., 1993a) suggest that the microanatomic basis for this bone formation that is coupled to bone resorption in rat trabecular bone is a process whereby most new episodes of bone formation occur on nonresorbed surfaces. Thus, formation is coupled to resorption, but does not necessarily occur at the same site. The lack of site-specificity for the coupling of bone formation to resorption makes models that invoke local mechanisms such as sequential cellular interactions, or localized actions of locally-released growth factors, unlikely to account for coupling in the growing rat.

The microanatomic basis for the coupling of formation to resorption clearly differs in growing rats and in man. This difference might reflect a fundamental biological difference between the species in the coupling mechanism. However, Chow et al. (1993a) noted that in aged rats, in which longitudinal growth had ceased, the proportion of cement lines showing crenated surfaces increased by sixfold compared with growing, 16-week-old animals. The proportion of the most recently formed cement lines that were crenated in such aged rats may have been even higher than this, if the overall figure includes some of the noncrenated cement lines that were laid down when the animals were younger. Even in growing rats, we noted a higher percentage of crenated cement lines (7–11%) than crenated surface (1.1–1.4%), and this suggests that bone formation has a predisposition to occur on previously resorbed surfaces even in young animals. The microanatomic difference between coupling in human and rat bone may thus represent the same mechanism under different circumstances. For example, in the growing rat, modeling might alter strain patterns such that resorption-induced mechanical signals for trabecular bone formation are not necessarily superimposed over the areas of resorption (see Figure 1). If so, the growing rat, with (and because of) its modeling but otherwise

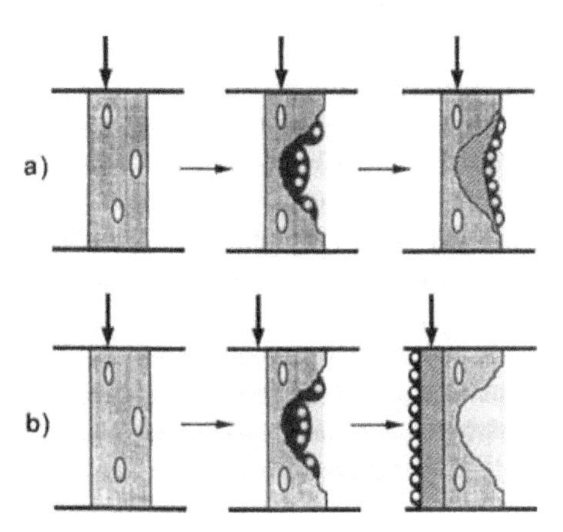

Figure 1. Diagrammatic representation of the response of a bone trabecula (hatched) to mechanical inputs, in direction and position represented by vertical arrow, in a skeleton that is relatively static (man, aged rat) (**a**) and in a skeleton that is still growing (**b**). In (**a**), a systemic stimulus for resorption leads to initiation of resorption on the mechanically least-used area of bone. Resorption increases local strain in the weakened bone, and formation follows at the same site in response to increased mechanical strain. Formation is thus coupled to resorption in a site-specific manner. (**b**) Resorption similarly occurs on the least loaded bone surface, but growth alters the loading pattern in the bone, such that the resorption increases the strain not necessarily at the resorbed surface, but at different site. Thus, bone formation is coupled in a non-site-specific manner.

similar skeleton, might allow an insight into the cell biology of human bone: that site-specificity is not an essential feature of the coupling of bone formation to bone resorption. The implication of this is that the coupling of resorption and formation cannot be ascribed to a local interaction, such as cytokine release during resorption, but must be due to signal such as mechanical signals that can be projected at a distance.

The lack of an essential requirement for site-specificity is not the only evidence which suggests that mechanical inputs mediate the coupling mechanism. Mechanical disuse of the skeleton, whether due to bed rest, immobilization, or spaceflight, results in rapid bone loss, with uncoupling of bone remodeling: bone resorption is increased and bone formation is decreased (Rodan, 1991; Whedon and Heaney, 1993). Similarly, the increased bone formation that occurs after ovariectomy is dependent on mechanical stimuli: it is reduced by underloading in the rat hindlimb (Lin et al., 1994). The ability of mechanical stimulation of bone to induce bone formation without prior bone resorption (Pead et al., 1988; Jagger et al., 1995) is also consistent with a mechanical basis for the coupling mechanism.

III. INTERACTIONS BETWEEN ESTROGEN AND MECHANICAL STIMULI IN REGULATION OF BONE FORMATION

The explanation for the turnover of bone discussed above is unknown. It may function to replace fatigue-damaged bone, or may reflect continuous adaptation of the skeleton to the mechanical environment, or turnover might facilitate plasma calcium homeostasis. The rapid bone loss seen in estrogen deficiency, in women and in rats, is often attributed to increased turnover. However, other conditions in which bone turnover is increased (e.g., hyperparathyroidism) are not necessarily associated with bone loss (Silverberg et al., 1989). Increased turnover does not itself cause bone loss, but facilitates expression of an underlying tendency to lose bone (Parfitt, 1979). The clear deficit between the increased resorption and the coupled increase in formation seen in estrogen deficiency, which accounts for bone loss in estrogen-deficient states, might reflect loss of an anabolic action of estrogen on osteoblasts, such that formation fails to keep pace with resorption, or might reflect an unmasking, by increased turnover, of a preexisting (e.g., age-related) drive towards reduced bone volume. To the extent that increased bone resorption does not of necessity lead to reduced bone volume, the loss of an anabolic action of estrogen might be of greater significance in the pathogenesis of osteopenia caused by estrogen deficiency than the increased resorption.

Analagously, suppression of bone formation by estrogen seems readily explicable through coupling, as secondary to estrogens potent antiresorptive action. Similarly, because coupling need not be site-specific in the rat, suppression of bone formation even on surfaces spatially separated from those on which resorption is suppressed (Turner et al., 1990) is explicable through coupling. Without controlling for the powerful effects of changes in bone resorption on bone formation, the observed responses of bone formation to estrogen administration do not yield insights into the primary action of estrogen on bone formation. We have used three strategies to detect such an action: 1) if bone formation is assessed immediately after a fall in estrogen levels, before increased resorption stimulates formation, a transient reduction in bone formation is observed (Tobias and Chambers, 1993; Lean et al., 1994) ; 2) if bone formation is assessed soon after raising estrogen levels above normal, before turnover is suppressed, an increase is observed (Tobias et al., 1991; Chow et al., 1992b) ; 3) if resorption is inhibited after ovariectomy by bisphosphonates, an anabolic action of estrogen can be detected against a low baseline (Chow et al., 1992a). We found that bone formation was stimulated in the same dose-response range as that which affected recognized targets of estrogen. This stimulation does not represent a hypothetical expedition, by an antiresorptive agent, of the formation phase of a remodeling sequence, because it was not induced by bisphosphonate administration (Chow et al., 1992a). The response of bone to estrogen was therefore quite different from that seen after treatment with other antiresorptives, suggesting that estrogen has an additional action on bone formation, beyond those due to inhibition of bone resorption.

In the above experiments, increased bone formation rates were measured by conventional fluorochrome techniques as an increase in double-fluorochrome-labeled surfaces. It has been suggested that increased fluorochrome label might represent suppression of label-resorption by estrogen, rather than increased formation (Turner et al., 1993). Against this, the bisphosphonate pamidronate reduced rather than increased double fluorochrome perimeters (Chow et al., 1992a). Moreover, it would be surprising if more than a small proportion of surfaces so recently formed that they demonstrate double fluorochrome labels, were resorbed in the few days between the second label and the end of the experiment (Chow et al., 1992b). Nor was increased double fluorochrome label an artifact caused by suppression of longitudinal growth by estrogen: increased fluorochrome labeling was observed even at doses of estrogen too small to affect longitudinal growth (Chow et al., 1992a), and was also seen in six-month-old animals with a very low longitudinal growth rate (Chow et al., 1992b).

A report by Westerlind et al. (1993) argues against an anabolic action of estrogen on bone formation. However, the design of many of the experiments reflects a misunderstanding of the basis for our conclusions. Thus, most involved comparisons of bone formation in ovariectomized rats given estrogen or vehicle. As we would anticipate, bone formation was suppressed by estrogen: any primary anabolic action of estrogen is likely to be masked through suppression of resorption. In experiments using intact animals, the observed reduction by estrogen in proline labeling (Westerlind et al., 1993) would be expected in samples that included the primary spongiosa, due to estrogenic suppression of longitudinal growth, although the robustness of the underlying anabolic action of estrogen might account for the apparent increase in mRNA for type I collagen and osteocalcin shown after estrogen administration. Our finding that osteoblast surfaces are increased in the cancellous bone of rats shortly after estrogen administration (Chow et al., 1992b) was not observed by Westerlind et al. (1993). However, the levels of single-fluorochrome-labeled surfaces and osteoblast surfaces observed by Westerlind et al. (1993) were at least an order of magnitude greater than our experience (Chow et al., 1992a,b) and that of others (Wronski et al., 1988b, 1989a), making interpretation of their significance to our observations difficult. We do not always see an increase in osteoblast surface (Chow et al., 1992a), possibly reflecting the insensitivity of measurements of osteoblast surfaces in adult rats, where osteoblasts are much less plump than in man. Moreover, the anabolic effect is greatest soon after estrogen administration (see below), and osteoblast surface is a later measurement than double-fluorochrome labels, and therefore less reflective of the early increase in bone formation.

We concluded that estrogen has, in addition to its antiresorptive action, an anabolic effect on rat trabecular bone, in the physiological range. This conclusion is consistent with evidence from other species for an anabolic role for estrogen on bone formation. Estrogen has long been known to stimulate medullary bone formation in birds, as part of the egg-laying cycle (Pfeiffer and Gardner, 1938). This mechanism might have been evolutionarily conserved in mammals, where estrogen might play a similar role in providing a skeletal reserve of calcium for reproduction (Chambers and Tobias, 1990). Thus, ovariectomy in beagle dogs has been found to reduce mean wall thickness (Malluche et al., 1986), administration of relatively low doses of estrogen to rabbits causes osteoid accumulation (Whitson, 1972), in mice low doses of estrogen cause an increase in trabecular bone formation (Edwards et al., 1992), and *in vitro,* estrogen stimulates matrix protein gene expression in osteoblastic cells (Ernst et al., 1988, 1989; Komm et al., 1988).

The anabolic action of estrogen in the rat can be detected by four days, reaches a peak after approximately two weeks, declines, and bone formation is then suppressed to the subnormal levels typically observed after prolonged estrogen administration in rats and women (Christiansen et al., 1982; Stock et al., 1985; Lindsay, 1987; Wronski et al., 1988b; Kalu, 1991; Abe et al., 1993). We presume that this reverse is secondary to estrogens ability to suppress resorption. Similarly, other antiresorptives reduce the anabolic action of estrogen (Abe et al., 1992; Chow et al., 1992a). This suggests that the anabolic action of estrogen depends on bone resorption for its expression: resorption entrains formation, which is amplified by estrogen.

It, therefore, seems likely that the antiresorptive action of estrogen suppresses both its own anabolic effect, and the bone formation that occurs in remodeling in general. If so, we should see a clear amplification of the bone formation induced by mechanical stimulation, since if mechanical stimulation is the basis for coupling, mechanical stimulation obviates the need for resorption; and even if not, mechanically-induced osteogenesis appears not to depend on bone resorption. We tested this using an experimental model we have recently developed, in which pins, inserted into the seventh and ninth caudal vertebrae of 13-week-old rats, are used to load the eighth caudal vertebra in compression (Chow et al., 1993b). A single, brief (10 minute) application of external loads, sufficient to cause dynamic strains within the range experienced by bone during physiological activities, increases bone formation on cancellous surfaces. In this model, expression of genes for matrix proteins starts to increase about two days after mechanical stimulation, is maximal by three days, and returns to baseline by five days (Lean et al., 1995). We found (in preparation) that if estrogen is administered to rats three days after mechanical stimulation, when the greatest extent of bone surface shows matrix protein gene expression, estrogen augments the response of bone to mechanical stimulation. A similar synergistic interaction between mechanical stimulation and estrogen on (established) osteogenesis has been observed *in vitro* (Cheng et al., 1996).

However, we were surprised to find, in the same experiments, that if estrogen administration was commenced before, during, or on the day after mechanical stimulation, then the induction of bone formation by mechanical stimulation was completely prevented. We did not find any evidence that this was due to suppression of bone resorption by estrogen: we found no inhibition of mechanical responsiveness by bisphosphonate, and there was no increase in eroded surfaces, which did not exceed 1% of the trabecular surface at any time during the first 48 hours after loading, while approximately 25% of the trabecular surface shows evidence of bone formation by 72 hours. These observations suggest that estrogen suppresses the activation of bone formation in a way that is not shared by other resorption inhibitors, through an unknown mechanism. Thus, activation of bone surfaces not only for bone resorption, but bone formation too, might be directly suppressed by estrogen. The latter characteristic distinguishes estrogen from bisphosphonates.

A substantial body of evidence supports the notion that estrogen maintains bone mass by sensitizing the skeleton to mechanical stimuli (Frost, 1988; Rodan, 1991; Turner et al., 1991). If this were the primary role of estrogen, however, we would not expect the hormone to suppress the induction of bone formation in response to loading, nor should long-term administration of the hormone suppress bone formation to subnormal levels in osteopenic rats and women. Suppression not only of resorption, but also, independently, of responsiveness to osteogenic stimuli suggests that the primary role of estrogen is to cause the skeleton to resist rapid changes in mass—to stabilize rather than to maintain bone mass. Superimposed on this function, estrogen appears to have the ability to sensitize osteoblasts to mechanical stimuli, such that while estrogen deficiency causes rapid bone loss through increased turnover accompanied by relatively deficient bone formation, raised levels of estrogen should gradually increase bone mass, through a net anabolic effect operating on a background of slow bone turnover.

The cell biological basis for these responses remains unknown, but it may be significant that estrogen responsiveness has been documented *in vitro* in bone cells of two distinct phenotypes: committed osteoblastic cells, which respond with increased matrix protein gene expression (Ernst et al., 1988, 1989; Komm et al.,

1988); and bone marrow stromal cells, which have osteogenic potential (Owen, 1985), which respond to estrogen with suppression of synthesis of cytokines (see Manolagas and Jilka, 1995). These cytokines, or related cytokines or responses, might be involved in the activation of responsiveness of bone surfaces to not only bone-resorbing but also bone-forming stimuli.

REFERENCES

Abe, T., Chow, J.W.M., Lean, J.M., and Chambers, T.J. (1992). The anabolic action of 17b-estradiol (E_2) on rat trabecular bone is suppressed by (3-amino-1-hydroxypropylidene)-1-bisphosphonate (AHPrBP). Bone Miner. 19, 21-29.

Abe, T., Chow, J.W.M., Lean, J.M., and Chambers, T.J. (1993). Estrogen does not restore bone lost after ovariectomy in the rat. J. Bone Miner. Res. 8, 831-838.

Baron, R., Tross, R., and Vignery, A. (1984). Evidence of sequential remodeling in rat trabecular bone: Morphology, dynamic histomorphometry, and changes during skeletal maturation. Anat. Rec. 208, 137-145.

Baylink, D., Morey, E., and Rich, C. (1969). Effect of calcitonin on the rates of bone formation and resorption in the rat. Endocrinology 84, 261-269.

Chambers, T.J. and Tobias, J.H. (1990). Role of estrogens in the regulation of bone resorption. In: Osteoporosis: Contributions to Modern Management. (Nordin, B.E.C., Ed.), pp. 21-30. Parthenon Publishing Group, Carnforth, Lancashire, England.

Cheng, M.Z., Zaman, G., Rawlinson, S.C.F., Suswillo, R.F.L., and Lanyon, L.E. (1996). Mechanical loading and sex hormone interactions in organ cultures of rat ulna. J. Bone Miner. Res. 11, 502-511.

Chow, J., Tobias, J.H., Colston, K.W., and Chambers, T.J. (1992a). Estrogen maintains trabecular bone volume in rats not only by suppression of bone resorption but also by stimulation of bone formation. J. Clin. Invest. 89, 74-78.

Chow, J.W.M., Badve, S., and Chambers, T.J. (1993a). Bone formation is not coupled to bone resorption in a site-specific manner in adult rats. Anat. Rec. 236, 366-372.

Chow, J.W.M., Jagger, C.J., and Chambers, T.J. (1993b). Characterization of osteogenic response to mechanical stimulation in cancellous bone of rat caudal vertebrae. Am. J. Physiol. 265, E340-E347.

Chow, J.W.M., Lean, J.M., and Chambers, T.J. (1992b). 17β-estradiol stimulates cancellous bone formation in female rats. Endocrinology 130, 3025-3032.

Christiansen, C., Christiansen, M.S., Larsen, N.-E., and Transbol, I. (1982). Pathophysiological mechanism of estrogen effect on bone metabolism: Dose-response relationship in early postmenopausal women. J. Clin. Endocrinol. Metab. 55, 1124-1130.

Edwards, M.W., Bain, S.D., Bailey, M.C., Lantry, M.M., and Howard, G.A. (1992). 17β estradiol stimulation of endosteal bone formation in the ovariectomized mouse: An animal model for the evaluation of bone-targeted estrogens. Bone 13, 29-34.

Ernst, M., Heath, J.K., and Rodan, G.A. (1989). Estradiol effects on proliferation, messenger ribonucleic acid for collagen and insulinlike growth factor-I, and parathyroid hormoneBstimulated adenylate cyclase activity in osteoblastic cells from calvariae and long bones. Endocrinology 125, 825-833.

Ernst, M., Schmid, C., and Froesch, E.R. (1988). Enhanced osteoblast proliferation and collagen gene expression by estradiol. Proc. Natl. Acad. Sci. USA. 85, 2307-2310.

Frost, H.M. (1988). Vital biomechanics: Proposed general concepts for skeletal adaptations to mechanical usage. Calcif. Tissue Int. 42, 145-156.

Gilsanz, V., Rose, T.F., Mora, S., Costin, G., and Goodman, W.G. (1991). Changes in vertebral bone density in black girls and white girls during childhood and puberty. New Engl. J. Med. 325, 1597-1600.

Hattner, R., Epker, B.N., and Frost, H.M. (1965). Suggested sequential mode of control of changes in cell behaviour in adult bone remodelling. Nature 200, 489-490.

Hayashi, T., Yamamuro, T., Okumura, H., Kasai, R., and Tada, K. (1989). Effect of (Asu^{17}) -eel calcitonin on the prevention of osteoporosis induced by combination of immobilization and ovariectomy in the rat. Bone 10, 25-28.

Jagger, C.J., Chambers, T.J., and Chow, J.W.M. (1995). Stimulation of bone formation by dynamic mechanical loading of rat caudal vertebrae is not suppressed by 3-amino-1-hydroxypropylidene-1-bisphosphonate (AHPrBP). Bone 16, 309-313.

Jaworski, Z.F.G. (1984). Coupling of bone formation to bone resorption: A broader view. Calcif. Tissue Int. 36, 531-535.

Kalu, D.N. (1991). The ovariectomized rat model of postmenopausal bone loss. Bone Miner. 15, 175-192.

Komm, B.S., Terpening, C.M., Benz, D.J., Graeme, K.A., Gallegos, A., Korc, M., Greene, G.L., O'Malley, B.W., and Haussler, M.R. (1988). Estrogenic binding, receptor mRNA, and biologic response in osteoblastlike osteosarcoma cells. Science 241, 81-84.

Lean, J., Jagger, C., Chambers, T., and Chow, J. (1995). Increased insulinlike growth factor I mRNA expression in rat osteocytes in response to mechanical stimulation. Am. J. Physiol. 268, E318-E327.

Lean, J.M., Chow, J.W.M. and Chambers, T.J. (1994). The rate of cancellous bone formation falls immediately after ovariectomy in the rat. J. Endocrinol. 142, 119-125.

Lin, B.Y., Jee, W.S.S., Chen, M.M., Ma, Y.F., Ke, H.Z., and Li, X.J. (1994). Mechanical loading modifies ovariectomy-induced cancellous bone loss. Bone Miner. 25, 199-210.

Lindsay, R. (1987). Estrogen therapy in the prevention and management of osteoporosis. Am. J. Obstet. Gynaecol. 156, 1347-1351.

Lubahn, D.B., Moyer, J.S., Golding, T.S., Couse, J.F., Korach, K.S., and Smithies, O. (1993). Alteration of reproductive function but not prenatal sexual development after insertional disruption of the mouse estrogen receptor gene. Proc. Natl. Acad. Sci. USA. 90, 11162-11166.

Malluche, H.H., Faugere, M.-C., Rush, M., and Friedler, R.M. (1986). Osteoblastic insufficiency is responsible for maintenance of osteopenia after loss of ovarian function in experimental Beagle dogs. Endocrinology 119, 2649-2654.

Manolagas, S.C. and Jilka, R.L. (1995). Bone marrow, cytokines, and bone remodeling: merging insights into the pathophysiology of osteoporosis. New Engl. J. Med. 332, 305-311.

Marie, P.J., Holt, M., and Garba, M.-T. (1985). Inhibition by aminohydroxypropylidene bisphosphonate (AHPrBP) of 1,25(OH) $_2D_3$Binduced stimulated bone turnover in the mouse. Calcif. Tiss. Int. 37, 268-275.

Mohan, S., Jennings, J.C., Linkhart, T.A., and Baylink, D.J. (1988). Primary structure of human skeletal growth factor: Sequence homology with human insulinlike growth factor-II. Biochim. Biophys. Acta 996, 44-55.

Owen, M. (1985). Lineage of osteogenic cells and their relationship to the stromal system. In: Bone and Mineral Research. (Peck, W.A. Ed.), Vol. 3., pp. 1-25. Elsevier Science Publishers BV, Amsterdam.

Parfitt, A.M. (1979). Quantum concept of bone remodeling and turnover: Implications for the pathogenesis of osteoporosis. Calcif. Tiss. Int. 28, 1-5.

Parfitt, A.M. (1982). The coupling of bone formation to bone resorption: A critical analysis of the concept and its relevance to the pathogenesis of osteoporosis. Metab. Bone Dis. Rel. Res. 4, 1-6.

Parfitt, A.M. (1983). The physiologic and clinical significance of bone histomorphometric data. In: Bone Histomorphometry: Techniques and Interpretation. (Recker, R.R., Ed.), pp. 144-223, CRC Press, Boca Raton.

Pead, M.J., Skerry, T.M., and Lanyon, L.E. (1988). Direct transformation from quiescence to bone formation in the adult periosteum following a single brief period of bone loading. J. Bone Miner. Res. 3, 647-656.

Pfeiffer, C.A. and Gardner, W.U. (1938). Skeletal changes and blood serum calcium level in pigeons receiving estrogens. Endocrinology 23, 485-491.

Rasmussen, H. and Bordier, P. (1974). The Physiological and Cellular Basis of Metabolic Bone Disease. Williams and Wilkins, Baltimore.

Richelson, L.S., Heinze, H.W., Melton, III, L.J., and Riggs, B.L. (1984). Relative contributions of aging and estrogen deficiency to postmenopausal bone loss. New Engl. J. Med. 311, 1273-1275.

Rodan, G.A. (1991). Perspectives. Mechanical loading, estrogen deficiency, and the coupling of bone formation to bone resorption. J. Bone Miner. Res. 6, 527-530.

Seedor, J.G., Quartuccio, H.A., and Thompson, D.D. (1991). The bisphosphonate alendronate (MK-217) inhibits bone loss due to ovariectomy in rats. J. Bone Miner. Res. 6, 339-346.

Silverberg, S.J., Shane, E., De La Cruz, L., Dempster, D.W., Feldman, F., Seldin, D., Jacobs, T. P., Siris, E.S., Cafferty, M., Parisien, M.V., Lindsay, R., Clemens, T.L., and Bilezikian, J.P. (1989). Skeletal disease in primary hyperparathyroidism. J. Bone Miner. Res. 4, 283-291.

Smith, E.P., Boyd, J., Frank, G.R., Takahashi, H., Cohen, R.M., Specker, B., Williams, T., Lubahn, D.B., and Korach, K. (1994). Estrogen resistance caused by a mutation in the estrogen-receptor gene in a man. New Eng. J. Med. 331, 1056-1061.

Stein, B., Takizawa, M., Katz, I., Juffe, I., Berlin, J., and Fallon, M. (1991). Salmon calcitonin prevents cyclosporin-ABinduced high turnover bone loss. Endocrinology 129, 92-98.

Stepan, J.J., Pspichal, J., Presl, J., and Pacovsky, V. (1987). Bone loss and biochemical indices of bone remodeling in surgically induced postmenopausal women. Bone 8, 279-284.

Stevenson, J.C., Lees, B., Devenport, M., Cust, M.P., and Gangar, K.F. (1989). Determinants of bone density in normal women: Risk factors for future osteoporosis? Br. Med. J. 298, 924-928.

Stock, J.L., Coderre, J.A., and Mallette, L.E. (1985). Effects of a shot course of estrogen on mineral metabolism in postmenopausal women. J. Clin. Endocrinol. Metab. 61, 595-600.

Tobias, J.H. and Chambers, T.J. (1993). Transient reduction in trabecular bone formation after discontinuation of administration of oestradiol-17β to ovariectomized rats. J. Endocrinol. 137, 497-503.

Tobias, J.H., Chow, J., Colston, K.W., and Chambers, T.J. (1991). High concentrations of 17β-estradiol stimulate trabecular bone formation in adult female rats. Endocrinology 128, 408-412.

Turner, C.H. (1991). Homeostatic control of bone structure: An application of feedback theory. Bone 12, 203-217.

Turner, C.H., Akhter, M.P., Raab, D.M., Kimmel, D.B. and Recker, R.R. (1991). A noninvasive, in vivo model for studying strain adaptive bone modeling. Bone 12, 73-79.

Turner, R.T., Colvard, D.S., and Spelsberg, T.C. (1990). Estrogen inhibition of periosteal bone formation in rat long bones: Downregulation of gene expression for bone matrix proteins. Endocrinology 127, 1346-1351.

Turner, R.T., Evans, G.L., and Wakley, G.K. (1993). Mechanism of action of estrogen on cancellous bone balance in tibiae of ovariectomized growing rats: Inhibition of indices of formation and resorption. J. Bone Miner. Res. 8, 359-366.

Van Tran, P., Vignery, A., and Baron, A. (1982). Cellular kinetics of the bone remodelling sequence in the rat. Anat. Rec. 202, 445-451.

Vignery, A. and Baron, R. (1980). Dynamic histomorphometry of alveolar bone remodeling in the adult rat. Anat. Rec. 196, 191-200.

Westerlind, K.C., Wakley, G.K., Evans, G.L., and Turner, R.T. (1993). Estrogen does not increase bone formation in growing rats. Endocrinology 133, 2924-2934.

Whedon, G.D. and Heaney, R.P. (1993). Effects of physical inactivity, paralysis, and weightlessness. In: Bone. (Hall, B.K., Ed.), Vol. 7, pp. 57-77. CRC Press, Boca Raton, FL.

Whitson, S.W. (1972). Estrogen-induced osteoid formation in the osteon of mature female rabbits. Anat. Rec. 173, 417-436.

Wronski, T.J., Cintron, M., and Dann, L.M. (1988a). Temporal relationship between bone loss and increased bone turnover in ovariectomized rats. Calcif. Tissue Int. 43, 179-183.

Wronski, T.J., Cintron, M., Doherty, A.L., and Dann, L.M. (1988b). Estrogen treatment prevents osteopenia and depresses bone turnover in ovariectomized rats. Endocrinology 123, 681-686.

Wronski, T.J., Dann, L.M., Scott, K.S., and Cintron, M. (1989a). Long-term effects of ovariectomy and aging on the rat skeleton. Calcif. Tissue Int. 45, 360-366.

Wronski, T.J., Dann, L.M., Scott, K.S., and Crooke, L.R. (1989b). Endocrine and pharmacological suppressors of bone turnover protect against osteopenia in ovariectomized rats. Endocrinology 125, 810-816.

Wronski, T.J., Yen, C.-F., and Scott, K.S. (1991). Estrogen and diphosphonate treatment provide long-term protection against osteopenia in ovariectomized rats. J. Bone Miner. Res. 6, 387-394.

Printed and bound by CPI Group (UK) Ltd, Croydon, CR0 4YY

08/05/2025

01865012-0003